EDISON *a biography*

BOOKS BY MATTHEW JOSEPHSON

Galimathias (Poems)
Zola and His Time
Portrait of the Artist As American
Jean-Jacques Rousseau
The Robber Barons
The Politicos
The President Makers
Victor Hugo
Empire of the Air
Stendhal
Sidney Hillman: Statesman of American Labor
Edison

a biography

Matthew Josephson

EDISON

McGraw-Hill
Book Company, Inc.
New York
Toronto
London

EDISON

Contents

Illustrations

Introduction

IN THOMAS ALVA EDISON we have a rough-hewn, old-fashioned American individualist who has been generally recognized as one of the most prolific inventors known to history. The story of his life, filled with human and scientific adventures, and the account of his many and varied works, takes us back to the "heroic age of invention," as the last third of the nineteenth century has been called. It was in its own way a climactic period in the mechanical arts, when America was esteemed as one of the most inventive of nations, and Edison served as a central figure in the technological revolution of his time.

Today we are more than ever concerned with inventors, and with our hopes (or fears) for what they may bring forth. There has been much change, yet it is doubtful that the human element has changed greatly. Edison still has a good deal to tell us. A sort of latter-day Prometheus, he dramatized the life and labor of scientific invention as perhaps no other of his kind had done. One feels that it was with a good instinct that our people made the former trainboy one of the greatest of their folk heroes; to them he was the man who was forever "making things." He was above all creative, hence one of the real builders of America. During a long era when America seemed dominated by men who were great acquisitors, Edison typified the Spirit of Workmanship.

It is curious that, for almost a quarter of a century, very little that is informative has been written about him. While he lived, and during the first years after his death, hundreds of books and thousands of articles about him, in all languages, appeared. The most useful of these works was undoubtedly the authorized biography by Dyer, Martin and Meadowcroft, published originally in 1910 and reissued in a revised edition in 1929; it contained some of Edison's autobiographical notes and his own accounts of his principal inventions. Since this book was published, however, much new knowledge and an immense amount of documentary material has accumulated, making necessary extensive revisions of earlier accounts of the inventor and key episodes in his life. Though one may feel strongly attracted by the rich, human qualities of Edison, one sees no useful purpose in preserving various inaccurate stories and false legends, repeated in one book after another, which obviously demand revaluation in realistic terms. Removing the veils of myth, however, leaves Edison's true stature undiminished. Even his mistakes, it has been observed, were carried out in a big way and make him appear more inspiring to us than if he were regarded, as his hagiographers used to picture him, as an infallible one.

Many misconceptions about Edison, on the other hand, still prevail among contemporary students of science, perhaps reflecting a reaction against the hero worship of yesterday. Some of the "new scientists" tend to depreciate him as one of the breed of pioneering inventors of the last century who had little scientific training and depended largely on cut-and-try methods. But though he liked to pretend that he was only a "practical" inventor, and to make jests at the expense of the men of "pure theory," he worked closely with some of the leading university physicists of America and England, several of whom were constantly employed in his laboratory. The full records of his research show us that he had a wide understanding of the principles of science as known in his time and tremendous faith in the method of scientific experimentation. He was self-taught; but so were some of his most famous predecessors, such as Davy and Faraday.

A just view of Edison's historic role, suggested recently by Norbert Wiener, would be that he was a transitional figure in late-nineteenth-century science, arriving on the scene at the period when the crude mechanical inventors had done their part, and systematic experiment and research was henceforth to be undertaken by skilled and specialized men, on a much larger scale than before. It was, after all, the young Edison who established the world's first industrial research laboratory—in 1876, at Menlo Park, New Jersey—in itself one of his greatest inventions.

Today giant research laboratories, owned by private corporations or by government, stand everywhere as monuments to Edison's innovative spirit. The technicians working in those institutions have been telling us for some time now that the day of the lone inventor, in an attic or small laboratory, has passed forever. Inventions, we have been assured, are henceforth to be produced by scientific teams using more complex instruments than Edison ever knew, and "made to order."

But can we ever dispense with the individual inventor, who is both dreamer and man of action? Good inventions are still not "predictable." Of late, serious doubts have been expressed that the mile-long laboratories and the teams of experts are any guarantee of original achievement. Technical men themselves voice fears that the new conditions of mass research may be less than helpful to independent or nonconformist thought; and that the bureaucratic inventors of today may be losing something of the intuitive skills and the sense of simple things that an Edison possessed. As H. S. Hatfield has written, he had always "his wonderful power of going straight to the practical end, never equaled since Leonardo da Vinci." The very simplicity of his devices is a mark of his greatness. Such art may not be duplicated by the mass production, in our schools, of men of scientific training. In view of the trend of modern scientific work, Edison may remain what he has been often called, "one of the last great heroes of invention."

Acknowledgments

My first acknowledgment should be to the Edison Laboratory National Monument and its staff. This group of ivy-clad buildings in West Orange, New Jersey, which was Edison's habitat for almost fifty years, was presented as a gift to the National Park Service of the Department of Interior in 1956 by the Thomas A. Edison Industries. Here are gathered the memorabilia of his life, the models or replicas of a great many of his inventions, the tools he used, the books he read and annotated. In this collection are included also his original laboratory records containing 3,400 notebooks, his original sketches for inventions, and about 250,000 pieces of correspondence and memoranda. It is mainly from direct study of these materials that this book has been written. That I was able to navigate, in some wise, through this ocean of documents is largely owing to the assistance given me during three years by the devoted staff of the Edison Laboratory, including Superintendent Melvin J. Weig, archivist Kathleen Oliver, and the curators, Harold Anderson and Dr. Norman R. Speiden.

I am also thankful to members of Edison's family for courtesies shown me. The inventor's four surviving children, Mrs. Marion Edison Oser, Mrs. Madeleine Edison Sloane, Charles Edison and Theodore Miller Edison, all gave freely of their knowledge of the subject, and also permitted me use of some unpublished letters of Thomas A. Edison and Mrs. Mina Edison that were in their possession. However, the Edison family are in no way responsible for the opinions expressed in this book.

The Henry Ford Museum Library at Dearborn, Michigan, and its librarian, Kenneth N. Metcalfe, also aided in the work of research by offering its extensive files of Edison material. The same courtesies were shown me also by the staff of the Ford Motor Company Archives. I must also acknowledge the permission granted me by the Houghton Library of

Harvard University to quote from the papers of Henry Villard (one of Edison's principal business associates); and for similar permission granted by the Baker Library of the Harvard School of Business Administration to use their manuscript collection. I am indebted also to Professor John McDonald of Brown University and to Samuel Insull, Jr., for having been given opportunity to study, examine and quote from some of the unpublished papers of Samuel Insull (who was for many years Edison's private secretary).

Finally, I am under large obligations to my friends, Kenneth M. Swezey and Professor Melba Phillips of the Physics Department of Washington University, St. Louis, for their reading of my manuscript and their advice on scientific and technical matters.

Matthew Josephson
Sherman, Connecticut

EDISON *a biography*

chapter I

The Edisons

1   The village of Milan, Ohio, where Thomas A. Edison was born, with its old tree-shaded central square and its quiet lanes of white wooden frame houses, now seems to us a rural backwater, a relic of early nineteenth-century America. The main arteries of Ohio's teeming traffic along the lake shore, like the railway, pass it by. Nonetheless, Milan in its heyday was one of the boom towns of the Middle West and stood for a time almost at the center of America's westward-flowing population. For over twenty years a flood of "internal immigrants" from the East Coast, mainly Yankees, but also recent arrivals from Europe, had been moving up the Erie Canal and along the lake to settle on the virgin soil of the Western Reserve. This region quickly became one of the great granaries of the world. Its little lake ports, which had been villages a decade or two before, such as Toledo and Cleveland, were turned into veritable cities. With loud hammering and sawing the people of "New Connecticut" were putting up homes, shops, and even small factories, especially near the water sites. There was town-booming and canal-digging on every hand—for the canal was the favored medium of transport to the sites of navigable rivers and lakes. The new towns that were laid out by the migrant Yankees, with their white, belfried churches and green commons, and names such as Norwalk and Westport, were of distinctly New England appearance. Such was Milan, a frontier hamlet of a few cottages in the 1820s, but a thriving community of three thousand souls a decade or so later, where a man of energy might well grow up with the country.

Here Edison's father, Samuel Ogden Edison, Jr., a native of Canada, had halted in the course of a season of wandering, in the summer of 1838, and decided to put down his stakes. Like a good many of his fellow immigrants he had left his troubles behind him, at home, and had journeyed to thriving Ohio in the hope of bettering his fortune.

1

Milan was situated on the Huron River, eight miles inland from Lake Erie; but the founding fathers had wisely exerted themselves to build a canal, forty feet in width, extending northward three miles to the navigable part of the river, and so giving access to Lake Erie's broad waterway. At the time when Samuel Edison, Jr., came to Milan, construction work on the Huron Canal and its basin was going on full blast. Looking at the busy scene, Sam Edison, a man of enterprise himself, blessed with the commercial optimism of Mark Twain's Beriah Sellers—hero of *The Gilded Age*—saw in imagination, not mountains of copper or coal, but certainly the shape of the growing local industries of the future, out of which he might turn many an honest penny. He saw Milan transformed almost overnight, once the canal was finished, into a strategic port on America's inland sea. He saw long trains of oxcarts, laden with wheat, pork and wool, converging down all the dirt roads of the neighboring counties upon the new Huron Canal Basin, whence ships and barges of over one thousand tons' capacity would carry freight to the lake, for transhipment via the Erie Canal across the Atlantic. What was more, he saw that warehouses and grain elevators would be needed at the canal terminal. Having had some experience of the lumbering trade, he resolved to take up the business of supplying timber and roofing material for the greater Milan that was to come.

Sam Edison had departed from Canada rather suddenly, about six months earlier, after having been involved in the political disorders of 1837. He was then still in his early thirties. Unhappily, he had been forced to leave his wife and four young children behind him in Ontario Province until he could establish himself in the States. With the help of his family he was able to raise a little money and for seven or eight months he worked zealously to set up his shingle mill. Some help was given him also by an American friend, Captain Alva Bradley (a future merchant prince), whose ships and barges plied Lake Erie between the Ohio lake towns, such as Milan, and Port Burwell, on the Canadian side, a few miles from the former home of Sam Edison. Bradley arranged for the shipment of Canadian lumber of good grade to be cut into shingles at Sam Edison's mill.

By the spring of 1839 Edison had hired a few men and was ready for business. Now at last he was able to bring his wife and children across the lake on one of Bradley's barges. After living in temporary quarters for about a year, the Edisons purchased, for the sum of $220, a town lot of an acre on Choate Lane, situated at the edge of the "hogback," or bluff, about sixty feet in elevation, that overlooked the Huron Canal Basin where the shingle mill stood. Here, in 1841, Sam Edison, with his own hands built a tidy house of wood-frame and brick construction, using lumber from his own mill. It was to be the Edisons' home for many years.

The house, built into the side of the bluff, still stands; it has the simple but pleasing lines of the Greek Revival dwellings of the first half of the nineteenth century. Made up of seven small rooms, its living room is at street level, the bedrooms are in the attic, while the large kitchen and pantry are set below the hill, at cellar level, but open on the sloping orchard behind the house and the view over the canal basin, once the busy scene of sailing ships and barges riding in and out of the roadstead across the flat, green prairie. It is a solidly built and comfortable dwelling, the home of a tradesman who enjoyed some seasons of good fortune.

2 Although Sam Edison was a Canadian by birth, his family had originally settled, more than a hundred years before, in what was then the Crown Colony of New Jersey. After the War of Independence was over, the Edisons, being stanch Loyalists, indeed most obstinate in their devotion to the King, were driven as exiles to Canada.

They were a hardy and restless clan, part Dutch, part English. As a family group they seemed disposed to boldness rather than to prudence; and, far from avoiding trouble or danger, they fairly courted it. Thus they endured periodic misfortunes and were uprooted more often than were most families in those times.

John Edison, the original of the Edison line in this country, according to family tradition was the descendant of peasants and millers living by the Zuyder Zee in the Netherlands. No direct connection, however, with Edisons, or Edesons, in Holland has ever been traced. Such incomplete records as we have show that a Netherlands widow named Edison arrived at Elizabethport, New Jersey, in 1730 with an only child of three named John, and that the widow, "who never married again, deceased and left her valuable estate to her son." These facts come to us from the annals of the well-known Ogden family, of New Jersey, with which John became connected in 1765, upon his marriage with Sarah, daughter of Samuel Ogden.[1] Thereupon he became the proprietor of a farm of seventy-five acres in the township of Caldwell, Essex County, New Jersey, and was possessed of a house, a Negro slave, one black mare worth twenty pounds, fifteen sheep, and other goods and chattels valued by him at a total of 288 pounds sterling.[2] John Edison, almost a patroon in those days, might have prospered in the Passaic Valley; but the insurrection against the British monarch was spreading in America and by 1776 had reached the phase of open warfare.

In New Jersey the division of popular sentiment about the Revolution was extremely close, a large minority remaining loyal to the King. The Loyalists were usually people of property, such as the Ogdens and their relative John Edison, though such families were also often divided in their allegiance. On October 20, 1776, at a period when the Whigs, as the local patriots were called, had taken over in Essex County and were burning the barns and driving off the cattle of their Loyalist neighbors, the Tory John Edison, together with his family and several of the Ogdens, took flight across the Hudson River and found refuge in New York, which was within the King's lines. But when Washington's beaten troops retreated soon afterward southward through New Jersey, the British and Hessian forces followed them, and John Edison came back.

It would have been well if he had remained passive in this civil struggle, for John Edison was of middle age and burdened with a numerous family. But he was stanch in his monarchical principles; he took up arms and joined the forces of General Howe as a scout, to help the British deal with the Continental guerrillas in the wooded ridges of New Jersey. During the confused skirmishing between the two forces in 1777—after Washington's army had rallied at Trenton and moved up again—John Edison, with his brother-in-law Isaac Ogden and three other Loyalist fighters, fell into the hands of the revolutionary militia and was imprisoned in the wretched Morristown jail for more than a year.

John Edison might have won his freedom by taking the oath of allegiance to the Continental Congress. This, however, he and Ogden refused to do. In January, 1778, they were tried by New Jersey's Council of Safety on the charge of high treason, convicted, and sentenced to be hanged, though execution, fortunately, was delayed.[3]

Meanwhile, Sarah Edison in New York moved heaven and earth to win the release of her husband and brother. Through the influence of members of her family who served in the Continental Army—among them her own father—the lives of John Edison and Isaac Ogden were spared. Later in 1778 they were paroled and sent back through the British lines to New York.

The successful outcome of the Revolution was a complete disaster for the Edisons. After having flourished in New Jersey for many years, they saw all their property confiscated and were forced to sail from New York Harbor in 1783 as *émigrés* in the gloomy British fleet that transported 35,000 American Loyalists to the West Indies and to Canada.

Many of the United Empire Loyalists who could prove they had suffered loss of property because of their devotion to His Majesty's Govern-

ment during the late war were given grants of land to aid their resettle-
ment. After several months of painful delay, the Edisons were awarded
some lots in the Hatfield Grant in Nova Scotia, a strip of bleak wilderness
adjoining Digby, on the east coast of the Bay of Fundy. Here for a
quarter of a century they strove against the inhospitable forest and
marshland; marriages, childbirths, and deaths followed in succession. But
they were a virile race. The numerous sons and grandsons of Tory John,
bearing names like Samuel, David, Moses, and Isaac, were very tall and
sinewy men with pale blue eyes and big noses. The eldest son, Samuel
Ogden, was married in 1792 to Nancy Stimpson of Digby; eight children
came to them, the sixth, who was born in August, 1804, being named
Samuel Ogden, Jr. Even the aged patriarch of the clan had more
children; and soon there was a tribe of nineteen Edisons, counting
grandchildren and daughters-in-law, and scarcely enough land to sustain
them. After twenty-eight years, the family decided to give up the struggle
in fogbound Nova Scotia and move on to some new frontier.

Upper Canada now clamored for pioneers; as in the United States, a
stream of immigrants moved westward along the Great Lakes, to clear
the forests of what is now Ontario Province. The Edisons received an
award of a section of six hundred acres of pineland along the Otter River,
about two miles inland from Lake Erie. In the spring of 1811 they were
all on the move, old and young; proceeding by ship to New York, they
continued overland by ox team and wagon for eight hundred miles, the
journey lasting all summer. On arriving at their new homesite, a point in
the wilderness twenty-one miles from their nearest white neighbor, the
sturdy Edison men set to work at once cutting down trees and hewing
logs, throwing up cabins, and uprooting stumps, as they had done in the
Nova Scotia forest a generation before. The Indians here were friendly;
the woods abounded with game; the Edison women preserved and stored
food, and made divers preparations against the coming of winter.

The much-traveled Edisons had just dug themselves in at their "settlement" when, on a January night in 1812, a runner came over the snow
from St. Thomas to give warning that war with the United States was at
hand. Volunteers were called for; and Samuel, the eldest son, responded
by offering to raise a company from among the recent arrivals in the
region. Samuel Edison, Sr., fully shared his father's feelings of loyalty
to the Empire. As captain of a company of the First Middlesex Regiment
under Colonel Thomas Talbot, he acquitted himself well in the strange
war that was waged on the lakes and in the forest—which saw Detroit
quickly surrendered to the British-Canadian forces. After that victory,

in November, 1812, Captain Samuel Edison and his volunteers were released from service so that they might go home to provide for their families in the deep Ontario woods.

With the coming of peace, immigration along the lakes was greatly swelled; in the 1820s the "Edison settlement" became a village, named, oddly enough, Vienna. It boasted a main street, a schoolhouse, a Baptist church, a cemetery—for which the Edison family donated land—and even a tavern, in fact all the amenities of rural civilization.

While the forests were being cut down the region flourished, pine logs in quantity being floated downstream to nearby Port Burwell for shipment to England. The numerous sons of the local war hero worked at the lumbering and carpentry trade as well as on the farm. Captain Edison's own means improved enough to permit him to build a neat clapboard house overlooking the Otter River and having real glass windows and iron nails. In their neighborhood the Edisons were known as hospitable folk whose latchkey was always left hanging out when they were away, so that strangers passing by might enter and take food or shelter.

A simple, hardy, pioneering people, the Edisons labored all their lives in the forests and the fields. Most of them were short on education—John Edison could not sign his name correctly; his grandson Samuel Edison, Jr., could scarcely spell enough to write out a bill. Yet among these people you would find dissidents, an occasional eccentric, even one or two who railed at religion. A marked family trait distinguished them; they were strong individualists, often nay-sayers, and independent-minded to the point of obstinacy—a quality less uncommon in the earlier America than now. The provincial annals of Canada show them to have been long-lived and prolific, "humble tillers of the soil, toiling in contented obscurity." Few traces of them would have remained if the world's attention had not been drawn one day to one illustrious descendant.[4]

3   Even in their frontier village in Upper Canada there was scarcely enough land to provide for all the Edison clan, so numerous were their progeny. John, the old Tory, had finally died in 1814, at almost ninety. Captain Samuel lived on to the ripe old age of ninety-eight, much esteemed by his neighbors—though he was also known at times to have been a very crusty, stubborn, and even rancorous fellow. In 1819, for example, he and his wife Nancy decided to be baptized into the Baptist Church of Port Burwell. But neither proved to be tractable members of their congregation. The church records reveal that "after the steps of labor had been taken with the captain, according to the word of God," the congregation voted that he should be

expelled "for railing and refusing to obey the voice of the church." Soon afterward similar punishment was visited upon Nancy Stimpson Edison for nonattendance.[5]

When he was almost sixty, Captain Samuel, the diverting old sinner, finding himself widowed and the father of eight full-grown children, married again and sired five more. The sight of him, with his big head and snowy beard, made a profound impression upon his American grandchild Thomas A. Edison, who in 1852 accompanied his parents on a visit to the old Ontario homestead.

We crossed [Lake Erie] by a canal boat in a tow of several others to Port Burwell, Canada, and from there we drove to Vienna. I remember my grandfather perfectly as he appeared at 102 years of age when he died. In the middle of the day he sat under a large tree in front of the house facing a well-traveled road. His head was covered completely with a large quantity of white hair, and he chewed tobacco incessantly, nodding to friends as they passed by. He used a very large cane, and walked from the chair to the house, resenting any assistance. I viewed him from a distance and could never get close to him. I remember some large pipes, and especially some molasses jugs that came from Holland.* [6]

When the Captain remarried in 1825, his sixth-born, Samuel Edison, Jr., who was to be the father of Thomas A. Edison, was twenty-one. Born in Nova Scotia, as a small boy he had made the long trek to Upper Canada with his parents; now he stood six feet and one inch tall, was very sinewy and strong, and could outrun and outjump any man in Bayham Township. Through restlessness or ambition, he had been led to try several different trades—carpenter, tailor, and lately tavernkeeper. Moreover, he was much concerned with the troublous politics of the time; indeed, his tavern in Vienna was a gathering place of the local agitators for reform. Samuel junior, something of a hothead, ventured to differ in his opinions with his Tory father and was no less stubborn in upholding them. Like the young men of Boston and Philadelphia fifty years earlier, those of Upper Canada were now loud in talk over their ale about the want of representative government and their resentments against the King's ministers.

* In this passage of a memorandum written for his private secretary, W. H. Meadowcroft, in 1909, Thomas A. Edison speaks of visiting his grandfather (Captain Samuel Edison, then actually eighty-five years old). But his authorized biographers mistakenly name the grandfather as "John" Edison, who, in 1852, had been dead nearly forty years. This error, by which a whole generation of Edisons was omitted, including the mettlesome grandfather, Captain Samuel Edison, who fought against the United States in 1812, has been widely repeated in numerous accounts of Edison's life up to recent times—when a correct genealogical study of the Edison line in Canada was compiled by W. A. Simonds, in his *Edison, His Life, His Work, His Genius* (Bobbs-Merrill, Indianapolis, 1934).

Meanwhile, a small, round-visaged young woman of seventeen, named Nancy Elliott, then serving as a teacher in the recently established, two-room school of Vienna, turned Samuel junior's mind to thoughts of marriage. She was the daughter of the Reverend John Elliott, who had recently come to preach in the Baptist Church, and granddaughter of Captain Ebenezer Elliott of Stonington, Connecticut, a veteran of the Continental Army. There were Rhode Island Quakers also in the Elliott line. After the Revolutionary War the Elliotts had moved from Connecticut to western New York, and thence to Canada. Nancy's two brothers were then studying to become Baptist ministers like their father. The Elliotts were not rich but they were godly and above average in education. Of course, many well-educated young ladies taught in rural schools in the 1820s for a pittance of five dollars a month, or even without pay, so that they might escape the drudgery of a farmhouse kitchen.

The young innkeeper soon went courting Miss Elliott; he was accepted, and in 1828 they were married and he took her home to a new house he had built with his own hands. Four children were born to them in Vienna in the early years of their marriage: a daughter, Marion, in 1829; a son, named William Pitt, in 1832; a second daughter, Harriet Ann, in 1833; and again a son called Carlile, in 1836.

These were again troubled times. An insurrection, aimed at nothing less than the overthrow of the Royal Canadian Government, was being plotted in Upper Canada, none too secretly, by William Lyon Mackenzie. Through the backwoods the followers of Mackenzie journeyed, exhorting the settlers to join them. As early as 1832 the younger Sam Edison was publicly denounced as a leading figure among the would-be revolutionists by his father's former commander, Colonel Talbot, who characterized him as "a tall stripling, son of a United Empire Loyalist, whom they [the insurgents] transformed into a flagstaff." The authorities had evidently learned that the young Edison sometimes drilled armed bands of his Vienna neighbors in the woods after dark.

In December, 1837, Mackenzie, at the head of a few hundred rebels in Toronto, made his desperate bid for power—which was timed with a similar rising in the French-speaking Provinces under Papineau. The insurrectionists, however, were quickly put to rout by regular soldiers, and Mackenzie was driven to flight across the Niagara River.

Sam Edison, Jr., with a file of Vienna and Port Burwell rebels, was marching on Toronto, through the deep woods, when news of the fiasco came to him. What was worse, a large force of six hundred militia was reported to be approaching to give him battle. Edison's small contingent then dispersed through the woods, followed closely by the Government soldiers. Reaching Vienna, Sam paused at his home only long enough to bid good-by to his wife and children, then hid for the night in a barn near his father's house. In after years, stories were told of how militia came to the old man's house to search for the son; while the stern old Loyalist smoked his pipe and said nothing, his wife managed to delude the searchers.

Before dawn the younger Edison was off through the woods, running like a deer toward the United States border more than eighty miles away, hotly pursued by the King's men with Indian guides and dogs. By performing the incredible feat of running for two and a half days, and stopping only for brief intervals of rest or food, this great athlete managed to reach and cross the frozen St. Clair River and find safety on the United States shore at Port Huron, Michigan.

When the father learned that his son had made good his escape, he was said to have remarked dryly, "Well, Sammy's long legs saved him that time." [7]

The disorders in Canada soon subsided when political reforms, especially home rule, were accorded by London. But Sam Edison, Jr., could not return, save under penalty of arrest and deportation to some penal colony. He had no alternative to seeking his livelihood in the United States. Thus the rebellious descendant of British Loyalists exiled from New Jersey in his turn was driven from Canada to the United States. Victims of revolution or civil disorder, the storm-tossed Edison family suffered exile twice in three generations. But Sam Edison, Jr., in effect had come around full circle and repatriated the family in its homeland. It was by this stroke of chance that Thomas A. Edison happened to be born in the heart of the American Republic.

4   After colonial Canada, Ohio seemed immensely "progressive" as well as enterprising. In this corner of the New World, as from the beginning of its settlement by white men, all was changing swiftly; the future meant everything, the past nothing. The country was still predominantly a wood-burning civilization; lake and river steamers and the first small railway locomotives consumed our forests. Farm crops, the stock in trade of eighty per cent of the people, still moved over fearful roads by oxcart. To us the tempo of life may seem languid. It was not so in the eyes of travelers from Europe observing the "restless" Americans under frontier or semi-frontier conditions.

America at the mid-century had become, in the words of Walt Whitman, "a nation of which the steam engine is no bad symbol." Even in the small towns of the Middle West more and more people devoted themselves to the industrial arts, the great end in view being the establishment of home industries, and the manufacture of those articles of common use that formerly had been imported at high cost from Europe. There was too much to be done, and too few hands to do it. To a people in a pioneering society, so engaged, so absorbed, "every new method which leads by a shorter road to material comfort, every machine which spares labor and diminishes the cost of production . . . seems the grandest effort of the human intellect." [8]

It was, after all, a very special kind of world in which the infant Edison would first see the light, one whose ruling idea (in Whitman's words) was to be "all practical, worldly, money-making, materialistic," yet holding the belief that such activity led always toward "amelioration and progress." In such a society, elementary education at last would be free to all, or almost all, but the standards of learning would be low, and the opportunities for higher education few or poor. Seldom would the love of fine arts or purely scientific knowledge be handed down from father to son as in the communities of the Old World. Most young Americans, in fact, would grow up caring little for "the general laws of mechanics," as Tocqueville observed, but thinking mainly of the "purely practical part of science." [9]

On the other hand, a distinctive feature of America's culture a century ago was that it offered a "most inventive environment." [10] Since their earliest days on this continent the American farmers and artisans possessed ingenuity and manual skill said to have been originally fostered by the harsh necessities of food-raising in the stony fields of New England. [11] From subsistence farming the native of Massachusetts and Connecticut turned to a variety of household industries. He made good glassware, pottery, and utensils of copper and iron. The arts of shipbuilding and navigation, as later those of textile manufacture, gunsmithing, and clockmaking, were also advanced. The "whittling boy" on the farm, like John Fitch and Eli Whitney, grew up to be a jack-of-all-trades who thought constantly in terms of tools, machines, and practical appliances, free to ignore tradition and try new ways of doing things. Thus arose the great breed of Yankee inventors, our pioneers in applied science and technology, who, "though their means and knowledge were limited, and troubles beset them at every step . . . founded most of our important industries." [12] Invention, as the economic historian E. L. Bogart has said, became a "national habit." According to a story sometimes attributed to Abraham Lincoln, the typical Yankee baby immediately after being born and placed in the cradle proceeded to examine the cradle to see if some "improvements" might not be worked out for it!

By 1851, the epoch-making inventions of the cotton gin and the sewing machine, the steamboat and the telegraph, among other exhibits of Americans at an industrial exposition held in London, were already winning for Yankee ingenuity the applause of old Europe.

The Yankee inventors, it must be remembered, joined in the movement of internal immigration, through the Erie Canal, toward the interior of the continent. Their skills were quickly disseminated among the pioneers of the Middle Western communities. European visitors were astonished to see how American artisans, such as Cyrus McCormick, inventor of the mechanical reaper, "revolutionized" agricultural tools which in Europe had remained unaltered virtually since the days of the Roman Empire.

Doubtless the English and European literary tourists who, like Dickens or Mrs. Trollope, were hardy enough to voyage to our frontier settlements around that time, counted on finding Americans living much as Rousseau's "noble savages" were supposed to live: free and equal, but nonetheless barbarian. However, they were surprised to see that the American standard of life, even in the new communities, often included the use of many articles of convenience almost unknown to Europe. By the 1850s great numbers of American women had freed themselves from the slavery of needle and thread by acquiring Mr. Howe's sewing machine; many also utilized new-fangled egg-beaters, apple-peelers and clothes-wringers astonishing to English travelers. Even in remote villages the people had a passion for rapid communication and intelligence, reading many newspapers that arrived by post, carrying on commerce by means of the newly invented electromagnetic telegraph.

The 1830s and 1840s marked the height of the Canal Craze in America. Before the coming-of-age of the Iron Horse, states and towns all but ruined themselves to finance the waterways that would give this continental nation a desperately needed transport system. Thus Milan boomed. Six hundred wagons a day arrived from points within a radius of 150 miles, enough to fill twenty big vessels and barges with 35,000 bushels of wheat each day. Soon fourteen grain warehouses lined the basin, most of them roofed with Sam Edison's seasoned pine shingles. Milan was becoming known as the "Odessa" of America and was one of the nation's busiest grain ports. For a period everything and everyone here seemed to prosper, including Sam Edison.

The family reunion in the United States was signalized by the arrival of more children. Nancy had brought four with her in coming to Ohio in 1839; Marion, the eldest daughter, then ten; William Pitt, eight; Harriet Ann, six; and Carlile, four. Now, in 1840, Samuel Ogden II followed, and four years later, Eliza. Unfortunately, the Lake Erie region is subject to severe winter storms; the Edison children suffered sorely from colds and from infantile diseases. Soon after being brought to Milan, in 1841, Carlile died at the age of six. Then soon afterward, to Nancy Edison's great sorrow, the lately born Samuel II and Eliza were taken from her in their infancy. Within a few years half of her children were dead.

Nancy Edison, who was of Scotch-English and Yankee descent, was very different in character from the "Dutch" Edisons; she was small in size, but in her earlier years she showed much steadfast patience and inner strength against adversity. She appeals to us as having been more than usually intelligent and having absorbed from her own estimable family, the Elliotts, a love of learning as well as devotion to religion.

She was already middle-aged when, in the dead of winter, she awaited the birth of her seventh, and last-born, child. It was to be a son with fair hair, large blue eyes, and a round face, who strikingly resembled the mother but seemed unusually frail and perhaps, as the Edisons said, "defective." His head was so abnormally large that the village doctor thought he might have brain fever. Nancy greatly feared for the life of this new child, who arrived in the world during the early hours of February 11, 1847, following a night of heavy snowfall. Later that morning Sam Edison, Jr., ran along the hogback to get medicine at the pharmacist's in the village square, announcing to all his neighbors that a son had been born to him. The parents christened the boy Thomas, after a brother of the father, and added the middle name "Alva" in honor of their friend Captain Bradley. Thus, Thomas Alva Edison.

chapter **II**

Childhood and boyhood

1 Six years passed. The child Alva—or "Al," as he was usually called—though often ailing, grew up to boyhood in Milan. Then, one morning in 1853 a knot of people, some accompanied by their children, gathered together in the cobbled square to look on at a strange rite: Sam Edison, the lumber and feed dealer, was whipping his youngest son, Alva, in the village square. It was like a scene out of the time of the Puritans in New England.

Physical punishment of a child, even in private, has often been found to be of dubious value; Rousseau had condemned it with all his eloquence a hundred years earlier. A public chastisement, carried out in cold anger, in the presence of the neighbors and their children—Sam Edison had advertised the event in advance—was something drastic and could only have been thought condign to some extreme wrongdoing. The boy's mother was said to have been tender-hearted as well as intelligent; neither parent, in fact, could have been considered cruel for those times, when children were regularly whipped at home and in school.

That the youngest Edison boy was something of a problem child, or was at least a difficult one, was generally believed by the neighbors of the Edisons. For he was of a decidedly mischievous bent and was forever falling into scrapes. The latest and most serious of these had occurred a day or two before and had resulted in the burning of his father's barn in the yard below the house. Alva had set a little fire inside the barn "just to see what it would do," as he innocently enough explained it. The flames had spread rapidly; though the boy himself had managed to escape from the barn, the whole town might have gone up in smoke if there had been a strong wind. Hence his father had devised a punishment to fit the crime, one that would be improving (though also no doubt secretly enjoyable) for the neighbors' children.

The public thrashing he received stamped itself on the boy's mind and

13

memory, for Thomas A. Edison described it all, though with wry good humor, sixty years later, when he thought back upon his boyhood.[1]

From his few words we have a picture of the boy, too small not to weep, receiving blow after blow, and early in life learning that there was inexplicable cruelty and pain in this world. (In later times he showed a curious indifference to physical suffering, whether his own or others, and bore pain himself without signs of emotion.)

If some deep and enduring resentment against his father was lodged in his subconscious, he did not show this overtly. While in later years he spoke with marked affection of his mother, there is no record of his having ever said anything complimentary on the subject of his father. To the latter, however, he was always a loyal son. The little that he did say in recollection suggested clearly that there was no understanding between his father and himself. "... My father thought I was stupid, and I almost decided I must be a dunce," was one of Edison's few plain references to this parent.[2]

His father, for his part, said on a number of occasions that he could make nothing of his son and that the boy seemed wanting in ordinary good sense. He was trying, he was vexing, as have been many other unusual children before and after him; he was forever curious, forever asking "foolish questions." The harsh corporal punishment visited so publicly on Alva, meanwhile, did not serve to make him mend his ways, or even to keep him from playing his little tricks on people. While his parents tried their best to rear him properly—his mother especially exerted herself to this end—he seemed in childhood and youth to be one who just grew in his own way.

The tales collected about his childhood in the small Ohio town show him to have been at first a grave infant who seldom cried. For some years he slept in a crib in the tiny windowless attic room under the eaves which he shared with two other Edison children, Harriet Ann and Pitt. They were much older than he—by fourteen and fifteen years—while the children who came in between had died. As for the eldest sister, Marion, she was married when Alva was two years of age. The infant was taken to Marion's wedding and was held in arms during the ceremony.

He seems to have had few childhood companions, and he played alone with his toys a good deal of the time. One of his earliest memories was the sight, in 1850, of long trains of prairie schooners drawn up in the narrow roads of Milan; these people, he was told, were going to the gold fields of California. But where was "California"? and what was "gold"?

As soon as he could talk he began to ask his parents and everyone else his interminable questions. There were so many "whys" and "wheres" and "whats" that Sam Edison said that he often felt himself reduced to exhaustion. Alva's mother, however, was more patient.

"Why does the goose squat on the eggs, Mother?" he would ask.

"To keep them warm," she replied.

"Why does she keep them warm?"

"To hatch them, my dear."

And what was hatching? "That means letting the little geese come out of the shell; they are born that way."

"And does keeping the eggs warm make the little geese come out?" he went on breathlessly.

"Yes."

That afternoon he disappeared for hours. "We missed him and called for him everywhere," it is related in a story that has come down through his sister Marion's family. He had disappeared into their neighbor's barn. At length his father found him "curled up in a nest he had made in the barn, filled with goose eggs and chicken eggs. He was actually sitting on the eggs and trying to hatch them." Such a little goose he was—yet a *logical* goose! [3]

His favorite playground was the terrace outside the kitchen, where he was under the eye of his mother. Here he could gaze down the slope at the canal basin two hundred feet away, with its sailing and steam vessels and barges riding in as if through the fields. Down below were the busy shipyards of Merry & Gay, the Yankees who had laid out the town and later built the canal. There was a large flour mill, a brewery, grain elevators, a tannery, smithies and iron forges, making an animated scene of local industry. The place was loud with the noise of six-horse teams, the crack of bullwhips, the "Gee-haw!" of drivers. Finally, there was his father's wonderful shingle mill and lumber yard, below him, down by the canal. As soon as he could run down the hill, at the age of three or four, he would go there and play with discarded shingles and chips, making plank roads or toy buildings for hours on end.

The shipyard workers and canal boatmen were a rough lot, ready for a brawl on Saturday nights, and uproarious at torchlight processions on election days or holidays. At five Alva knew by heart some of the boatmen's songs, and he could lisp the verses of "Oh, for a life on the raging *canawl!*" [4]

His elder brother Pitt liked to draw and sketch. In imitation of him, Alva, at about five, drew little pictures of all the craft signs hung over the shops in the square of Milan. His mind was decidedly visual; in later years he always made sketches of the mechanical devices that he contrived.

During this period, toward the age of five or six, there was a whole phase of misadventures and scrapes such as the experiment with the barn. Once he fell into the canal and had to be fished out. On another occasion he disappeared into the pit of a grain elevator and was almost smothered before he was rescued. He was nothing if not inquisitive.

With his round face and wide, well-formed chin, Tom Edison was an attractive-looking boy and was characterized by one of his cousins, Nancy Elliott, as being really "a good child most of the time," but very headstrong and willful when he could not have his way. Nancy, at thirteen, acted as his "nanny" and remembered "having spanked him many a time and spanked him hard." [5] His mother also believed that the rod must not be spared. Edison himself later testified that his mother kept a birch switch handy for him behind the old Seth Thomas clock in the living room, and that it had "the bark worn off."

Down at the canal basin there was the big steam-driven flour mill of an eccentric Yankee named Sam Winchester. Often the small Edison boy would be found with his nose pressed against the back window of Winchester's shop watching the strange things being done there. His father scolded him severely for hanging about that place (as the story has come from his sister) and spanked him well for going there again after being warned not to do so. The mysterious Winchester ("The Mad Miller of Milan") was not grinding flour, but was constructing a passenger balloon. The hydrogen he had used for this purpose had burned down his first flour mill. The boy Tom Edison knew about these experiments and about Winchester's first, abortive attempt at flight. Some years later, during a second trial, Mr. Winchester managed to ascend into the air, then was wafted slowly in the direction of Lake Erie, never to be seen again. [6]

Finally there was one tragic accident in which the boy was involved when he was but five, or at most six years old, that deeply troubled Edison's parents and that he himself never forgot. He recalled the whole affair in notes written long afterward, and with remarkable detachment:

When I was a small boy at Milan, and about five years old, I and the son of the proprietor of the largest store in the town, whose age was about the same as mine, went down in a gully in the outskirts of the town to swim in a small creek. After playing in the water a while, the boy with me disappeared in the creek. I waited around for him to come up but as it was getting dark I concluded to wait no longer and went home. Some time in the night I was awakened and asked about the boy. It seems the whole town was out with lanterns and had heard that I was last seen with him. I told them how I had waited and waited, etc. They went to the creek and pulled out his body.[7]

In recalling the strange adventure more than fifty years afterward, Edison showed that it had left an ineradicable impression on him, and "told all the circumstances ... with a sense of being in some way implicated." [8]

Why had he not called for help? Why had he said nothing on returning home? A country boy knew what drowning meant. Was he silent because he felt guilty and feared that, as usual after some mischance, he would be censured and beaten? His parents—his father being usually the more impatient one—could not help showing their distress at his "strange" behavior, so unlike that of other boys. Was this boy "without feelings," as was sometimes said of him? Doubtless they applied the switch again, though whipping did no good. Frequent punishments only reflected poor adjustment between the child and his well-intentioned family, a recurrent failure of understanding and communication between them. The boy could not but sense his father's continuing disappointment and disapproval of him and feel, though dimly, the weight of this.

2 What doom, what blight suddenly fell upon the future metropolis that was Milan, Ohio, the self-proclaimed "Odessa" of our Great Lakes? Today the Huron Canal is a morass, a mud-filled depression, and Milan would be like one of the ghosts towns left by the sudden veering of America's commerce, had she not become a suburban backwater of Norwalk, now a big industrial city, then merely a sister village.

What happened, simply, was that the lake shore railroad came in 1853, but bypassed Milan and ran instead through Norwalk to Toledo. The too canny fathers of Milan who owned its canal had willed that things should be this way. When offered shares of stock and a station on the ʳojected rail line in return for free right of way, they had stoutly refused ⸱⸱ⁿt access to their port. They had no faith in those little stea⸱ ⸱⸱⸱⸱ rattled along iron-plated tracks of wood; they tru⸱ ⸱⸱⸱⸱ʰeir canal and its favored position. Th⸱ followed the line of the railr⸱ ⸱ntually lost 80 per cent of i⸱

ʳent American dream of some strategic ground ⸱st and hopeful specul⸱

who held land, a business establishment, and a house in an area where values were rapidly sinking. Yesterday he was well-to-do; today he was virtually a poor man, though consoling himself with the hope that by moving on he would find new fortune elsewhere.

Sam Edison had initiative, to be sure, and was ever alert for new opportunities all about him. Undoubtedly America was "built up" mainly by pioneers who were as sanguine and as luckless as Sam, and who were forever hunting for that "corner" in a future Chicago, or that mountain of coal, that somehow eluded them.

As Thomas Edison said in later years, there was "a collapse of the family fortunes," caused by the reduction of canal tariffs at Milan in competition with the railroad and a depression in the town that "undermined the social standing of Samuel Edison, forcing him to leave his picturesque home and begin his life anew.... This transpired in the year 1854." [9]

Once more the nomadic Edisons were on the move, by train and carriage to Detroit and thence by a dainty little paddle ship, *The Ruby*, that bore them smoothly up the St. Clair River to Port Huron, Michigan, where they were to settle. It was late spring, the weather sunny and calm, and the scene was full of color, as a young neighbor who accompanied the Edisons described it. To the delight of Alva and his brother and sister, there were not only sailing craft and small lake steamers of every kind all around them, but Indians in feathers and beads, darting their canoes in and out of the ship's course. Along the shore of the blue St. Clair River the Edison children could see smoke rising from the campfires of Indians still remaining in the vicinity. [10]

In the 1840s and 1850s, after the earlier waves of migration to Ohio, Indiana, and Illinois, had come the "Michigan fever" that drew many thousands of settlers to this region. It boasted not only its great forests, but iron, copper, and salt deposits. In this era the population doubled every ten years; land-boomers were confidently selecting future "lake cities" for exploitation. Such was Port Huron, where Sam Edison had first ntered the United States; it was now a small town of four thousand a entrance to Lake Huron from the St. Clair River, and hence c ing the rising traffic to the upper lakes. What was more, Sam E rned that a railroad was being extended northward from miles to Port Huron, while a Canadian line, building ect with the railway in Michigan by a ferry a ame site.

ome was situated at Fort Gratiot, once a ed States Army base, at the norther town. Formerly the home of the purveyor to the military post, it was a large and solidly built house with columned balconies, set in a grove of pine trees adjacent to the parade grounds. From its big windows there were fine views over the river and lake. The rooms were spacious, and there were four huge fireplaces; the grounds, ten acres in extent, included an orchard, a large vegetable garden, and several outbuildings. There was only one thing wrong with this fine dwelling: the Edisons no longer owned their home, but rented it. Sam Edison used his remaining capital to engage in the lumber, grain, and feed trade.

He liked having more than one iron in the fire. One of his ventures at this period, that showed his taste for innovation, was to build a wooden observation tower, one hundred feet high, overlooking the surrounding bodies of water and their curving, wooded shore line. Visitors were to be charged a fee of twenty-five cents for the privilege of climbing its many stairs and peering through an old telescope on the top platform. Here the youngest Edison often played by himself; as time went on, he also acted as gatekeeper, collecting the tolls paid by occasional tourists. At first so few came that no more than three dollars was collected in a whole summer. Later the new railroad brought hundreds of excursioners; but the observation tower (called Sam Edison's Tower of Babel) soon lost its novelty and was neglected, until it finally fell to the ground. The father, by then, had turned his restless mind to other schemes that were scarcely more profitable.

As the Edison fortunes declined, for Sam was no steady provider, his wife emerged as the real head and front of the family. It was she who worked with a compulsive energy, cooking and weaving, sewing and crocheting. It was she who struggled to keep the family afloat, with the aid of her children, and it was she who educated her youngest son.

On their arrival at Port Huron, Thomas Alva fell seriously ill of scarlet fever; the Edison family also suffered greatly from respiratory diseases for which they used to dose themselves with ineffective patent medicines. For this or other reasons the boy's entrance into grammar school had been postponed. In the autumn of 1855, when he was more than eight years old, he was finally enrolled as a pupil in the one-room school of the Reverend G. B. Engle. Mr. Engle, it is related, liked to implant his lessons in his pupils' minds with the help of a leather strap; his wife, who aided in the work of instruction, was said to be even harsher in her methods. It is not surprising, therefore, that the Edison boy, who had been growing up according to his own will, as a sort of child of nature, proved to be somewhat difficult in the classroom, his mind apparently refusing the lessons offered in such form. He said:

I remember I used never to be able to get along at school. I was always at the foot of the class. I used to feel that the teachers did not sympathize with me, and that my father thought I was stupid. . . .[11]

After he had been at the school about three months, he overheard the schoolmaster one day saying of him that his mind was "addled." In an outburst of temper, Tom Edison stormed out of the schoolroom and ran home, refusing to return.

The next morning his mother came with the boy to see the schoolmaster, and an angry discussion followed. Her son backward? She considered him nothing of the sort and believed she ought to know, having taught many children herself in her youth. The upshot was that she removed the boy from school and declared that she would instruct him herself.

The schooling received from the Engles, according to Edison's later recollections, was utterly "repulsive"—everything was forced on him; it was impossible to observe and learn the processes of nature by description, or the English alphabet and arithmetic only by rote. For him it was always necessary to observe with his own eyes, to "do things" or "make things" himself. To see for himself, to test things himself, he said, "for one instant, was better than learning about something he had never seen for two hours. . . ."[12]

The legend has come down to us, through Edison and his family, that it was because of the inadequacy of the teacher, and in the interests of the boy's education, that his mother decided to keep him at home and instruct him privately. Some added facts, however, have come to light recently, showing that his father was either disappointed in his son and therefore reluctant to pay the small school fees, or was unable to pay them. Thirty years afterward, the schoolmaster, having heard a good deal about the later career of his "addled" pupil, wrote him:

Indianapolis, Ind.
August 13, 1885

Dear Sir:

You will remember that some years ago you attended school under my direction (and my wife's) at Port Huron. Your father, not being very flush with money, I did not urge him to pay the school bill. I am now almost seventy-seven years old, and am on the retired clergy list. And as you have now a large income I thought perhaps you would be glad to *render me a little aid.*

Truly yours,
Reverend G. B. Engle [13]

Edison responded with a check for twenty-five dollars, liberal enough in view of his unhappy memories of school and teacher.

Unlike the other boys in the little town, he stayed at home all day, and every morning, after his mother's preliminary housework was done, she called him to his lessons and taught him his reading, writing, and arithmetic. The affection between mother and son was very strong, especially in these years, when Nancy Edison's relations with her husband grew less happy. Her son had the impression she kept him at home by her side, as he said, partly "because she loved his very presence." She taught him not only the three R's, but "the love and purpose of learning... she implanted in his mind the love of learning." [14]

In summer, as in winter, the program of instruction went forward. According to the reminiscence of a Port Huron playmate:

A few of us boys were playing in front of the Edison house one day with Al in our midst, when a lady appeared on the porch, a nice, friendly-looking one, plainly dressed and wearing a lace cap in the style of that period. Looking over the group for a moment, she called out: "Thomas Alva, come in now for your lessons." The boy obeyed without a word; as he went we looked at him in a sort of commiserating way. It seemed rather hard to be called away from the diversions of a beautiful summer day to study dry lessons; and besides it was vacation time! [15]

To a modern educator Alva would surely have been a rewarding subject for study. Among prodigies there are both the precocious ones and those whose minds grow slowly or "unevenly" in boyhood—Isaac Newton's, for instance, as well as Thomas Alva's. In this case, the remarkable mother gave the boy sympathetic understanding that bred confidence. She avoided forcing or prodding and made an effort to engage his interest by reading him works of good literature and history that she had learned to love—and she was said to have been a fine reader. In this sense she was out of the ordinary. In the fifties, most literate women had magazines like *Godey's Lady's Book* as their favorite reading; and if they read to their children it was from the *Rollo Books,* or Peter Parley's *Tales of the Sun, Moon and Stars*. But Nancy Edison had superior taste. Believing that her son, far from being dull-witted, had unusual reasoning powers, she read to him from such books as Gibbon's *Decline and Fall of the Roman Empire,* Hume's *History of England,* or Sears's *History of the World;* also literary classics ranging from Shakespeare to Dickens. Instead of being bored by these works of serious literature, he grew fascinated and at nine was inspired to read such books himself. While immature and ill-disciplined in some respects, he was advanced in others and soon became a very rapid reader.

Nancy Edison, however, could scarcely have been a wholly adequate teacher, her experience having been limited to teaching in a small Canadian village for a year or so before her marriage at the age of eighteen. Her son never learned how to spell; up to the time of his manhood his grammar and syntax were appalling. We see then, as did the Reverend Mr. Engle, that he was indeed hard to teach. Whatever he learned, he learned in his own way. In fact, though his mother inspired him, no one ever *taught* him anything; he taught himself.*

On the other hand, his mother could not but notice that in some of the odd or funny things the boy said he showed a high imagination, sometimes expressing ideas or abstract numbers by some apt visual image. Weighing himself on a scale at the age of ten he is remembered to have exclaimed: "Mother, I'm a bushel of wheat now, I weigh eighty pounds!"

Nancy Edison also sensed, or discovered by chance, the real direction of her son's interests; for one day she brought forth an elementary book of physical science, R. G. Parker's *School of Natural Philosophy,* which described and illustrated various scientific experiments that could be performed at home. Now his mother found that the boy had truly caught fire. This was "the first book in science I read when a boy, nine years old, the first I could understand," he later said. Here, learning became a "game" that he loved. He read and tested out every experiment in Parker; then his mother obtained for him an old *Dictionary of Science,* and he went to work on that. He was now ten and formed a boyish passion for chemistry, gathering together whole collections of chemicals in bottles or jars, which he ranged on shelves in his room. All his pocket money went for chemicals purchased at the pharmacist's and for scraps of metal and wire.

Thus his mother had accomplished that which all truly great teachers do for their pupils: she brought him to the stage of learning things for himself, learning that which most amused and interested him, and she encouraged him to go on in that path. It was the very best thing she could have done for this singular boy.

"My mother was the making of me," he said afterward. "She understood me; she let me follow my bent." [16]

The marriage of experimental activity and knowledge thus came early in life for Thomas Alva. Some of his first "experiments," to be sure, were

* When he was about nineteen, he wrote letters without punctuation or sentences, such as the following (probably in 1866) from Memphis, Tennessee:

DEAR MOTHER—
    Started the Store several weeks I have growed considerably I dont look much like a Boy now—Hows all the folks did you receive a Box of Books from Memphis that he promised to send them—languages [*sic*].

Your son
AL

conceived out of nothing more than sheer mischief. Having read something of Benjamin Franklin's discoveries in static electricity he tried the trick of vigorously rubbing the fur of two big tomcats, whose tails he had attached to wires, the only result being that he was unmercifully clawed. Pondering the problem of balloons that were able to ascend in the air because of the volatile gas in them—like Mr. Winchester's in Milan—he administered a large quantity of Seidlitz powder to his simple-minded playmate, Michael Oates, reasoning that the gas thus generated might set the boy flying through the air. The young Oates, however, became terribly sick to his stomach, and Alva was soundly spanked by his mother. It is noteworthy that when the Edison boy did have companions of his own age he would sometimes play the tyrant with them and they would be his submissive "slaves," as was Oates—though bigger, and older than himself.

The chemicals in his bedroom were a "mess," as his mother complained; the wet-cell batteries sometimes spilled sulphuric acid on furniture and floor. His mother, on such occasions—though ordinarily most affectionate—could show a warm temper, for the son said, "My mother's ideas and mine differed at times, especially when I got experimenting and mussed up things." [17] He was therefore ordered to remove his jars to the cellar of the house. To keep others from using or tasting his chemicals he labeled all his bottles "Poison"—hardly scientific, but good insurance.

A corner of the cellar in the Port Huron house was Thomas A. Edison's first laboratory. There, after the age of ten, he secluded himself, often all day long, absorbed in his study of simple chemicals and gases and in the design of his first homemade telegraph set. Other boys might play in the fields or fish in the river; but Tom Edison buried himself in his cellar laboratory, with his elementary manuals and his chemical and electrical outfits.

His father persisted in his disapproval of the boy's subterranean devotions. He would sometimes offer Alva the bribe of a penny if he would read some book of serious literature. Thus, when he was twelve the boy read Tom Paine's *Age of Reason*, at his father's suggestion. "I can still remember the flash of enlightenment that shone from his pages," he wrote long afterward.[18] The pennies, however, were saved and applied to the purchase of more powders and chemicals.

"Thomas Alva never had any boyhood days; his early amusements were steam engines and mechanical forces," his father commented. On another occasion, in later years, Sam Edison said of his son, "He spent the greater part of his time in the cellar. He did not share to any extent the sports of his neighborhood. He never knew a real boyhood like other boys." [19] Actually, through his own kind of intellectual *play*, the Edison boy was intensely happy—though his father could not understand this. He was also, despite his steam engine models, very much a boy.

Up to the age of fourteen, and even afterward, he continued to show such high spirits that it was hard to hold him down. Though he was older, he could not resist playing practical jokes on people, even at some risk to himself; and so his father continued the whippings from time to time. In 1861, for example, a whole regiment of volunteer troops was stationed at the Fort Gratiot reservation opposite the Edison home, for the Civil War had begun. The place was heavily guarded; there was much drilling and calling of orders between guards and sentinels day and night. Late one night, Tom and another boy had the idea of having some fun with the soldiers. Imitating their tones, Tom called out loudly for "Corporal of the Guard Number One," then ran off in the darkness. The call was duly taken up by the sentinels and repeated, with resultant confusion. When the same trick was tried again the next night, the soldiers pursued the two boys. The other boy was caught, but Tom hid himself in a barrel of apples in the cellar, while the soldiers, having awakened his father, searched for him with lanterns in vain. When all was quiet again, Tom stole back to his room.

"The next morning I was found in bed, and received a good switching on the legs from my father . . ." he related.[20]

Bouts of horseplay alternated with prolonged sessions in elementary physical science and chemistry below stairs. Accidents occurred, the muffled sounds of explosions sometimes reaching the parents from the cellar. "He will blow us all up!" the anxious father would exclaim.[21]

But his mother remained stanch in his defense. "Let him be," she said; "Al knows what he's about."

Above all things, he loved to work over models of the telegraph; as a boy, he was fascinated with the whole idea of electricity, introduced into practical, everyday usage not long before his time by Samuel F. B. Morse. To be sure, it was the earlier experiments and discoveries of the learned American physicist, Joseph Henry, in the field of electromagnetism which had opened the way to Morse's practical invention. That was a fabulous invention in its day. When Morse first exhibited his instrument in New York in 1838, great crowds flocked together in the streets nearby and, in good American fashion, begged to be allowed to look upon it, "declaring they would not say a word or stir, and didn't care whether they understood or not, only they wanted to say they had seen it." [22] By 1848 the telegraph flashed intelligence over a network extending from

New York and Boston, via Albany, as far as Chicago. As a boy, Edison, like many other Americans, followed with intense excitement news of the pioneering bands of young telegraphers who extended their long lines of wire across the great prairies, over the Indian-infested deserts and the mountains, to California—so that the continent was first spanned, not by the railroad, but by the telegraph—in 1861, during the opening months of the Civil War.

What was electricity, Tom Edison kept asking people. A Scotsman, who was a station agent on the new railway that came to Port Huron, finally explained to him in apt words that "it was like a long dog with its tail in Scotland and its head in London. When you pulled its tail in Edinburgh it barked in London."

By the middle of the century hundreds of youths who dreamed of becoming telegraphers could be found almost everywhere around the country, pottering with homemade wet cells and crude sounders of their own device. It was in no way remarkable that Edison, at the age of eleven, had his own homemade telegraph set and had begun to practice the Morse code. Seventy years later he would recall how he had constructed a crude set after having read a popular handbook of experiments in physical science.

I built a telegraph wire between our houses ... separated by woods. The wire was that used for suspending stove pipes, the insulators were small bottles pegged on ten-penny nails driven into the trees. It worked fine.[23]

The Edison of 1857 or 1858, whom we glimpse for a moment running barefoot through the woods to string up his homemade telegraph line, is decidedly an American type—the mechanically ingenious boy, on the farm or in a small town, growing up with a passion for "whittling" or "tinkering." Like the Yankee inventors of the preceding generations, Edison in boyhood cared little for lessons in reading and writing, if he could but "play" with his telegraphs, batteries, and chemicals; and like the others of his kind, he was considered "wayward," or at least "different from other boys," by his parents. At an early age he became absorbed, to the exclusion of everything else, in learning all that he could of the comparatively new electrical science, from his own observation and by making or trying things with his own hands.

In working with his first crude telegraph made of scrap metal, Edison the boy approached, unwittingly, the main stream of electrical experiment since the days of Benjamin Franklin.

More than pocket money, however, was needed for his expanding "laboratory" in the cellar. He was now bent on making a proper Morse sending and receiving set of his own, and these were hard times for the Edisons. At the age of eleven Tom, with the help of the boy Michael Oates, who did chores around the Edison place, embarked upon his first commercial venture. The two boys laid out a large market garden and tried raising vegetables. A horse and cart were hired, and soon Tom was driving about the town trucking onions, lettuce, cabbages, and peas. Evidently the business was sponsored and supervised by his mother; the first summer's harvest, so he claimed, netted all of "two or three hundred dollars." Was he a merchant prince in embryo? However, he quickly tired of this work, as he says, for "hoeing corn in a hot sun is unattractive...." [24]

The big event for Port Huron in 1859 was the coming of the railroad, extending northward from Detroit. Those iron and brass locomotives, those brightly painted carriages held an irresistible attraction for the young Edison. Even before the railway was formally opened at Port Huron, he had learned that there would be a job on the daily train for a newsboy, who, though receiving no regular wage, would have the concession of the "candy butcher" to purvey food and sweets to the passengers. He was only twelve, and small for this work. Though his mother strongly objected to the idea, their situation, as his father admitted, was such that no schooling for the boy could be considered, and Tom was faced by "the early necessity of gaining his own living." [25] In discussing the matter, he promised that during the long layover of the daily train at Detroit, he would use his time to read books; he would also have money with which to continue his scientific self-education.

It has been related that "there was no real need" for him to go to work at twelve, that his parents were not poor, but were, in truth, "well-to-do." He is usually pictured as precocious in his desire to get into the workaday world and make his own way. The evidence, however, points to the impoverishment of Sam Edison, who at this period, it was said, "thought nothing of walking sixty-three miles to Detroit." His business affairs evidently allowed him much idle time, but no horse and carriage. The son remembered that he had to use "great persistence" in bringing his mother around to his idea. It was the father who selected the job for his son and negotiated with the railroad people for his employment. [26] In his autobiographical notes (written in 1885), Edison says that at this time he was "poorly dressed"; and also that, "Being poor, I already knew that money is a valuable thing."

An early photograph, dated 1861, shows him in a worn old cap and roughly clad, yet most attractive and intelligent-looking, with his fine brow, his wide jaws, and his big smile. The local railroad officials, at any rate, gave him the job.

3   The "mixed train" of passengers and freight, with a great huffing and puffing from its tall stack, departed from Port Huron daily at 7 A.M. on a journey of more than three hours to Detroit, waited over there most of the day, and returned again to Port Huron at 9:30 P.M. As the train pulled out of the station, a lively urchin, carrying a basket almost bigger than himself, quickly scrambled on board and then made his way through the carriages, calling out: *"Newspapers, apples, sandwiches, molasses, peanuts!"*

The landscape raced by at the exhilarating speed of thirty miles an hour. There was an ever-varied scene and always the changing crowds of passengers—farmers, workers, immigrants, and sometimes elegant tourists from far-off places. Tom Edison was a cheeky fellow and at first thoroughly enjoyed this new life of movement and active commerce with all sorts of people. In those days he had, as he himself said afterward, a sort of "monumental nerve." Not only did he dispense a stock of candies along with newspapers and magazines, but he also won permission from the conductor to store a quantity of fresh butter, berries, vegetables and fruit in the baggage car of the train, which thus traveled as free freight and which he disposed of at retail along the route.

The report of an old newsdealer of Detroit shows that he was an honest boy who always did business for cash and paid promptly. Also that he would not brook dishonesty in others. Sometimes he would seem so distracted, thinking of one of his elementary experiments, or reading a book, that when a boy who shared his newspaper depot at Port Huron came to give him money, Tom would put it in his pocket without counting it. But when he found that the boy had not been honest with him, he closed up that newspaper depot rather than have anything more to do with the fellow.

Yet there was also the irrepressible spirit of mischief in him. On being questioned, in later years, as to what kind of trainboy he had been, he confessed that he was the sort who "sold figs in boxes with bottoms an inch thick."

He told afterward of his adventures with a zest that showed that inventiveness with him was not confined only to mechanical work. In 1860, just before the Civil War began, he recalled that two elegant young men from the South, accompanied by a colored servant no less elegant, boarded the train at Detroit; on his approaching them, one of them asked, "Papers?" then took all of Tom's newspapers and threw them out the window, saying to his colored man in haughty tones, "Nicodemus, pay this boy." Thomas Alva then brought them quantities of magazines, food, and popcorn; all were disposed of in the same way, until his whole stock was gone, and he had been paid off most liberally. "Finally," he said, "I pulled off my coat, hat and shoes, and laid them out." For this too he was given a high price, while the other passengers roared with laughter at the barefoot boy's antics. Then the Southern swell cried out, "Nicodemus, throw the boy out too!" But Tom Edison ran away.[27]

At a tender age he learned about life from the talk of the railroad workers, farm hands, and immigrants. In the 1850s, child labor was widespread; young boys and girls went to work at an early age in the big towns, as in New York, where one could see girls of ten to twelve sweeping the crossings of main thoroughfares so that ladies and gentlemen might pass dry-shod over the dung-ridden places. So Tom Edison worked on the railroad at twelve, returned home long after dark, and gave his mother a dollar a day from his earnings.

There was so much he could learn merely by using his keen eyes. At the railroad yard in Detroit he could watch the men switching cars, or repairing valves and steam boilers. Waiting in the station he observed closely the operations of telegraphers, then beginning to signal train movements between stations.

In Detroit, already a city of over 25,000, he was left to his own devices all day, wandering about with a little money to spend on equipment or books, talking with men in machine shops, some of them even full-grown inventors, like young George Pullman, then making some of his first railway carriages in a small shop in that city.

"The happiest time of my life was when I was twelve years old," he said afterward. "I was just old enough to have a good time in the world, but not old enough to understand any of its troubles.[28]

He was away from home, he was away from his father, he was on his own. His home, moreover, was hardly a happy place nowadays.

After he had been working on the railroad for a year or so, he thought of a way of occupying his leisure time during the layover in Detroit. Since the baggage and mail car up forward in the train had a good deal of empty space, he had the idea of installing his little cellar laboratory at one end of it. The trainman was won to compliance; soon Tom had transported his stock of bottles, test tubes, and batteries, and ranged them neatly on shelves that were fitted to the back wall of the car. It was said that George Pullman made up the railed shelving that would hold those jars and bottles safely against the bouncing of the train. This was, no doubt, the world's first mobile chemical laboratory, probably installed some time in 1861.

There he was at twelve to thirteen, adrift on the train like Huckleberry Finn on his raft on the broad Mississippi. It makes a lengendary picture that is touched with the insouciant charm of the mid-nineteenth-century era in America. Tom Edison, moreover, restlessly inventive in his pranks as in his elementary mechanical experiments, may also be represented as The Eternal American Boy, with much of the audacity and the imagination of Huck Finn. He too seemed determined to leave home and make his way along the "river of life."

The real picture, however, is not always so pleasing. The hours away from home were long and wearying; leaving at dawn, the young boy would return at ten or eleven at night. There were the sudden blows of life to be borne, even danger to be faced as bravely as he could, at twelve and thirteen. He had courage to spare, though like the old pioneer, Sam Houston, he used to say that he knew no fear save of "the black dark night." In illustration of the rational way in which he tried to conquer his own nerves and imaginary fears, he would tell the story of how Houston, on being accosted in a Texas cypress swamp by a sheeted figure appearing from behind a tree, shouted, "If you are a man you can't hurt me, and if you are a ghost you don't want to hurt me. But if you are the devil come home with me; I married your sister!"

So Tom Edison, driving home from the station toward eleven at night with his horse and cart and leftover newspapers, used to become utterly terrified when he reached a stretch of deep dark woods, containing a soldiers' graveyard. His nerves on edge, he would shut his eyes, whip up the horse, and go thundering past that graveyard, while his heart leaped in his throat. But after many weeks of such anxious flight through the dark thicket, he found that nothing happened; the fear of graveyards finally left his system.

One night, while he was still a trainboy, he was called to the office of a steamship company at Port Huron and given an errand which, as he was told, was of the utmost urgency. A steamboat captain had suddenly died; his big ship, lacking a navigator, was held idle in the port; therefore the Edison boy was to go at once and fetch a certain ship captain who lived in retirement on a timberland property about fourteen miles from the nearest railroad station, at Ridgway, Michigan. There was no other means of reaching the man quickly save on foot; in compensation for that nocturnal journey over the forest trails Edison was offered a generous fee of fifteen dollars.

With another small boy, who had been persuaded to accompany him on the all-night hike—by the offer of sharing his earnings—Tom Edison started off at 8:30 P.M. A heavy rain was falling, and the night was black as ink. They had lanterns with them, but the trail lay through rough, cut-over land and they fell often; every stump looked to them like a Michigan bear. One lantern went out, then the other; after that the imaginative Tom was certain he could hear the movement of wild beasts all about them. When a good many miles had been covered they were utterly spent, and giving themselves up for lost, according to Edison's account, "we leaned up against a tree and cried."

Nonetheless, Tom Edison forced himself to go on and on, with his companion who was even weaker than himself, so that they might reach the old captain in time. Their eyes gradually becoming accustomed to the dark, they managed to stumble along for the rest of the fourteen miles of corduroy road. At the first gleam of dawn they entered the captain's yard and delivered the message. "In my whole life I never spent such a night of horror." [29]

4

Then, as if to make the conditions of life all the harder for him, there came to him the cruel affliction of deafness. Its beginning is placed at the time when he was about twelve, shortly after he began working on the railroad. Long after the event he gave conflicting accounts of how this misfortune came to him "suddenly." The earlier stories of his boyhood, done long ago in the Horatio Alger style, have much pathos but are misleading. From the symptoms of his deafness, as described by himself and others as well, it seems to have been traceable to the aftereffects of scarlatina suffered in childhood, and to have developed through periodic infection of the middle ear that was unattended.

In the earlier tales of how his deafness arose he is described as having been busy one day in his baggage-car laboratory. In those times the iron-plated tracks were so unpredictable that they sometimes curled up and pierced the floors—and seats!—of passing cars. At all events, the train suddenly gave a violent lurch, and a jar holding some sticks of phosphorus in water fell from the shelves to the floor; on being uncovered and exposed to the air the phosphorus soon ignited with a startling white light and burst into flames. The wooden floor of the car took fire, while the boy struggled vainly to smother the flames. The conductor, one Alexander Stevenson, sometimes described as a "dour Scot," came forward in time to douse the little fire. Then, it is related, he lost his head, "cursed Edison roundly and boxed his ears" with such "brutal blows" that the boy soon afterward became deaf. "When a few minutes later the train stopped at Smith's Creek station, the conductor threw the boy overboard, and after him his whole laboratory and printing press." Tom Edison was left weeping beside the railroad track, and permanently injured as well. The details of this story, however—and it has become a legend—are broadly inaccurate. Edison himself tried to recapitulate things toward the end of his life, so as to correct the more romanticized accounts of his boyhood misfortunes. According to these later recollections, he was delayed in getting to the train one morning; it was already leaving the station. "I was trying to climb into the freight car with both arms full of heavy bundles of papers. . . . I ran after it and caught the rear step, hardly able to lift myself. A trainman reached over and grabbed me by the ears and lifted me. . . . I felt something snap inside my head, and the deafness started from that time and has progressed ever since." He remembered that at first he could hear only "a few words now and then," after which he "settled down to a steady deafness." [30]

In retelling the story to his intimates of later years, he also roundly declared that *the ear-boxing incident never happened.* "If it was that man who lifted me by the ears who injured me, he did it to save my life." [31] In truth he considered that Alexander Stevenson, the conductor, had been his benefactor, for they corresponded on friendly terms afterward.*

It seems that Stevenson, far from showing cruelty, took a fancy to the odd but resourceful trainboy; he had even taken some liberties with the rules of the railroad in admitting Alva's laboratory to a corner of the baggage car. "The telegraphers and trainmen were a good-natured lot of men and kind to me," Edison himself said.[32] The accidental fire, which did in fact take place some time in 1862, made Stevenson determined to remove the boy's chemicals and Voltaic jars from the train. The ejection of Tom Edison's laboratory-on-wheels, however, evidently did not take place until more than two years after he first felt himself growing deaf.†

* In 1881, Alexander Stevenson wrote the inventor from Port Huron: "I often think laughingly of the long, and I must say pleasant times we had on the Grand Trunk Railway, thinking over the Printing Press and laboratory . . . and of how near we were of being blown up by the aforesaid laboratory." (Edison Laboratory Collection, 1881 File.)

That Edison may have been cuffed about or pulled by the ears, by Stevenson or others, in view of the rough sort of life he led, is possible. This *may,* conceivably, have "finished it" for him. But medical opinion concerning the "snapping" he heard inside his head, sometimes placed in the autumn of 1859 or the early winter of 1860, attributes this to the breaking of ligaments connecting the external ear to the skull. That would not, of itself, have caused deafness, as the ligaments are not connected with the auditory canal leading to the middle ear. But if his middle ear were already punctured, perhaps as a result of the scarlatina that had kept him out of school a few years earlier, then a sharp blow might have caused congestion or heightened an existing infection in his middle ear.

† The small printing press usually mentioned as having also been thrown out, was not acquired by Edison, as we know definitely, until 1862.

He was, thus, permanently disabled when not yet thirteen. If there had been any thought of returning him to school for higher education or for some formal training, it must now be put aside, for he would not hear his teachers. The prospects for a handicapped boy of a poor family were certainly not good.

Nevertheless, he was able to go on working as a trainboy. When the train was in motion he could hear well enough above the roar of the locomotive and the clanking of the wheels, since everyone in a noisy car tends to raise his voice to a shout. "While the train was roaring at its loudest I would hear women telling secrets to one another. But during the stops, while those nearest to me conversed in ordinary tones, I could hear nothing. . . . Doctors could do nothing for me." [33]

It was his habit to make light of his misfortunes. The loss of one of his most vital senses (as with his lack of schooling) was sometimes represented by him as an advantage or an "asset." He was spared a great deal of the vexation suffered by persons of normal hearing and, so, was the more able to concentrate his thoughts, or think something through without interruption. The uproar in the streets in big towns did not disturb him. But whatever he may have said on this score, his intimates declared that he had never really been glad that he was deaf.

In the privacy of his brief diary he wrote truthfully, years later, these words of infinite sadness: "I *haven't heard a bird sing since I was twelve years old.*"

That the loss of hearing brought him, in effect, to an important turning point in his life was true. He tended to be more solitary and shy; became more serious and reflective; drove himself to more sustained efforts at reading and study, which had been carried on lately, for a year or two, in a rather desultory fashion. He had been only "playing," hitherto, with his books and his "experiments." Now he put forth tremendous efforts at self-education, for he had absolutely to learn everything for himself. And whereas he had earlier, by some boyish traits, appeared immature or lighthearted, he now seemed serious, or rather old for his years, the habit of meditation becoming more fixed.

It was then, one might more readily believe, after his hearing had become impaired, that he carried the paraphernalia of his little cellar laboratory at Port Huron into the baggage car of his train. It was then, while the car stood in the yard, that he came to enjoy spending so many long hours alone, wholly lost in his elementary experiments with wet cells and stovepipe wire and his first crude telegraph instruments.

As he has related, he felt himself "shut off" from "the particular kind of social intercourse that is small talk . . . all the foolish conversation and meaningless sound that normal people hear. . . . Deafness probably drove me to reading."

In Detroit, during the hours of layover, he found his way to the public library. Formerly the reading room of the Young Men's Association, it was reorganized in 1862 as the Detroit Free Library. That year Thomas A. Edison, then fifteen, became one of its earliest members, being given a card numbered 33 and paying the substantial fee of two dollars for it.[34] He relates:

My refuge was the Detroit Public Library. I started with the first book on the bottom shelf and went through the lot, one by one. I didn't read a few books. I read the library. Then I got a collection called *The Penny Library Encyclopedia* and read that through. . . . I read Burton's *Anatomy of Melancholy*—pretty heavy reading for a youngster. It might have been if I hadn't been taught by my deafness that I could enjoy any good literature. . . . Following the *Anatomy* came Newton's *Principles*. . . .[35]

He had formerly read at random books of history, literature, and elementary science, under his mother's tutelage, in an indiscriminate fashion. Now, going it alone, he carried on a sort of frontal attack on books of every sort. Robert Burton, the old seventeenth-century ecclesiastic, on one day, Isaac Newton the next! The title and theme of Burton's whimsical work must have appealed to Tom Edison's mood at this stage; Burton, too, felt himself isolated, sang of his loneliness and poverty, and also of his dreams, of his will to happiness, though in the simplest and humblest circumstances.

In the same way, Edison vastly enjoyed reading Victor Hugo's romantic epic *Les Misérables,* just then translated into English and being very widely read in the United States. The story of the lost children, such as Gavroche, and the figure of the noblehearted ex-convict, Jean Valjean (also an "outsider"), appealed to him strongly. In his youth he spoke of Hugo with such enthusiasm that his companions sometimes called him "Victor Hugo" Edison.

During this period of fermentation, of intellectual excitement and mental growth, he tackled Newton's *Principles,* whose importance he suspected. He was then fifteen—not nine or ten as has been said—but was wholly unprepared to plunge into what he called a "wilderness of mathematics." Without anyone to aid him, he kept at the *Principles* for long hours, until he became baffled and bewildered and threw it aside. He said later, "It gave me a distaste for mathematics from which I have never recovered."

A thousand pities that the young Edison, with his lively imagination and mind voracious for knowledge, was given so little informed counsel and, so, began his studies without the discipline that formal instruction might have imposed. For him there were no devoted teachers who could communicate an interest in the beautiful mathematical constructions of an Isaac Newton—as was done in England for a James Clerk Maxwell or a Lord Kelvin in early youth.

Making his way alone, as well as he could, through masses of books and manuals, he turned to more practical treatises by men of narrower scope. To this period of intensive self-education belongs his reading of books like Karl Fresenius's *Chemical Analysis* and Andrew Ure's treatise on practical mechanics, which he was able to master unaided. In Ure's *Arts, Manufactures, and Mines,* published in 1856, he read passages ridiculing the "academical philosophers" (such as Newton), who were said to be "engrossed with barren syllogisms, or equational theorems ... and disdained to soil their hands with those handicraft operations at which all improvements in the arts must necessarily begin." [36] Ure urged that it was these "men of speculative science" who must be shunned, holding that they neglected for sixty years the steam engine of Newcomen, "until the artisan James Watt transformed it into an automatic prodigy." These were lessons that appealed not only to the public of Victorian England but also to that of "practical, money-making" America.

At an impressionable age, Tom Edison's reading and thinking, pursued alone, without guidance, disposed him to avoid the example of a Newton and follow that of a James Watt or of a factory mechanic like Arkwright.

5   "In my isolation (insulation) I had time to think things out," Edison said of this period.

There was the inconvenience of continued poverty among other disadvantages he suffered. What was to be done about this? He said:

At the beginning of the Civil War I was slaving late and early to sell newspapers; but to tell the truth I was not making a fortune. I worked on so small a margin that I had to be mighty careful not to overload myself with papers.[37]

The years when he rode back and forth on the Grand Trunk Railway coincided with some of the most dramatic events in American history, to him but remote happenings. There was the insurrection and the hanging of John Brown, the election of Abraham Lincoln as President, the firing at Sumter, the opening of the Civil War. He had noticed, however, that when reports of a battle were printed, his newspapers sold faster than on other days, and then he could barely carry enough copies of the papers to the train. He therefore made it his practice to go to the composing room of the Detroit *Free Press* and inquire what the headlines were on the advance galley proofs for that day's edition, so that he might better estimate his needs. Necessity and early experience in hawking produce had given him a shrewd commercial sense.

One day in April, 1862, the first accounts of an immense and sanguinary battle between the armies of Grant and Johnston at Shiloh reached the newspaper office by telegraph. Learning of this before the newspaper was out (and before the evening train left for Port Huron), the trainboy conceived the idea of a splendid little stroke of business for himself. The proofs he had seen showed that the *Free Press* would carry huge display heads announcing a battle in which 60,000 were then believed to have been killed and wounded!

Here was a chance for enormous sales, if only the people along the line could know what had happened. Suddenly an idea occurred to me. I rushed off to the telegraph operator and gravely made a proposition which he received just as gravely.

At Edison's request, a short bulletin was to be wired by the Detroit train dispatcher to the railroad stations along the road to Port Huron, and the telegraphers there would be asked to chalk them up on bulletin boards in the depots, before the train arrived. For this free telegraph service Tom Edison would pay the friendly Detroit telegrapher with gifts of some merchandise, newspapers, and magazine subscriptions. Thus forearmed, the boy went to the office of the newspaper's managing editor, Wilbur F. Storey, to ask for a thousand copies of the paper, an uncommonly large assignment, since usually he took but two hundred, and he asked for this on credit. He had boldly marched up to Mr. Storey himself, because the man in charge of distribution had rebuffed him. The managing editor examined the ragged boy whose expression, however, was resolute enough. "I was a pretty cheeky boy, and felt desperate," he himself recalled later. The authorization was given him and, with the help of another boy, he lugged huge bundles of the newspaper to the train and folded them up.

His device of having advance bulletins telegraphed and posted at the depots worked even better than he had hoped. In his own vivacious way Edison related:

When I got to the first station on the run . . . the platform was crowded with men and women. After one look at the crowd I raised the price to ten cents. I sold thirty-five papers. At Mount Clemens, where I usually sold six papers, the crowd was there too.... I raised the price from ten cents to fifteen.... It had been my practice at Port Huron to jump from the train about one quarter of a mile from the station where the train generally slackened speed. I had drawn several loads of sand to this point and had become quite expert. The little Dutch boy with the horse usually met me there. When the wagon approached the outskirts of town I was met by a large crowd. I then yelled: "Twenty-five cents, gentlemen—I haven't enough to go around!"

The sale of the "extra" was held, as he recalled, in the vicinity of a Port Huron church where a prayer meeting was in progress. He yelled out his news. In a few moments "the prayer meeting was adjourned, the members came rushing out, bidding against each other for copies of the precious paper. If the way coin was produced is any indication, I should say that the deacon hadn't passed the plate before I came along." [38]

"It was then," he added, "it struck me that the telegraph was just about the best thing going, for it was the notices on the bulletin board that had done the trick. I determined at once to become a telegrapher."

It is noteworthy that one of the chief attractions of the telegraph for him was also connected with his bad hearing. *With the telegraph he could hear*. He relates, "... Thus early I had found that my deafness did not prevent me from hearing the clicking of a telegraph instrument. From the start I found that deafness was an advantage to a telegrapher. While I could hear unerringly the loud ticking of the instrument I could not hear other and perhaps distracting sounds...."

After learning all he could by studying the apparatus of the dispatching telegraphers at the railway stations, he had recently made an improved sending and receiving set of his own device. A young neighbor in Port Huron who helped him related:

He [Edison] was always tinkering with telegraphy and once rigged up a line from his home to mine, a block away. I could not receive very well, and sometimes I would come out and climb on the fence and halloo over to know what he said. That always angered him. He seemed to take it as a reflection on his telegraph.[39]

Later, when he was fifteen, he set out a longer line of stovepipe wire running about a half mile, strung out on trees, to the home of James Clancy, who sometimes helped him at news vending. On returning home, though it was 10 P.M., he would stay up and "play" for hours at sending and receiving messages. His father tried to drive him to bed; but Tom Edison, a sly one, knew ways of hoodwinking the old man.

His father, he had noticed, had formed a habit of reading (gratis) one of the unsold newspapers Tom usually brought back with him at night.

The boy therefore took to coming home without any newspapers, reporting that business was humming nowadays. But, he said, there was his friend, Clancy, who had a newspaper that was ordered and paid for by his parents, and he, Tom could obtain the leading news stories by telegraphing him over their little private wire. The elder Edison, much intrigued by this scheme, thereafter permitted him to stay up and practice the Morse code till midnight, or even later. This arrangement was continued until a cow, wandering through the orchard one night, blundered into the low telegraph line and brought it down.

With the thought of possibly getting a job some day as a railroad mechanic, Edison, haunting the roundhouses, learned a good deal about the mechanism of steam locomotives and the working of fireboxes, valves and gears. One day a locomotive engineer, fatigued after a night of revelry, actually allowed him to drive one of the machines pulling a freight train while the older man took a nap. This locomotive, as Edison remembered, "had bright brass bands and beautifully painted woodwork and everything highly polished, as was the custom up to the time old Commodore Vanderbilt stopped it on his roads." Mounted at the levers, Edison cautiously slowed the speed of the train to twelve miles, but before he had gone far a sudden spurt of damp black mud had blown out of the stack and covered part of that beautiful engine, including the boy driver. Stopping at the station where the firemen usually went out to oil the machine, Tom climbed upon the cowcatcher and tried to do likewise, by removing the oil cup on the steam chest. The steam rushed out with a tremendous blast almost hurling him to the ground. He had failed to notice that the driver always shut off steam when oiling was being done.

Somehow he managed to get the oil cup on again, and drove off even more slowly all the way to the junction. But just before he reached it another outpour of mud covered him and the entire engine, so that, as he pulled into the yard, everyone turned out to see him and laughed.

He soon learned that the eruptions of mud were caused by his carrying too much, rather than too little, water, so that it had passed over into the stack and washed out all the soot. "My powers of observation were very much improved after this occurrence," he said.

At the beginning of 1862 he developed a sudden interest in the craft of printing, and thought for a while of becoming a journalist and starting his own newspaper. The important role played by the newspaper he sold every day had not failed to impress itself on him. With some money he managed to save at the time of Shiloh, he purchased a small second-hand press, together with some old type (three hundred pounds of it) found on the bargain counters of Detroit; in a short time he had taught himself to set type and run a hand press. Now, putting aside his chemicals, he undertook the venture of editing, printing, and selling a small local newspaper, which was produced in that same baggage car.

The *Weekly Herald,* issuing from a branch of the Grand Trunk Railroad, at eight cents a copy, with a circulation of about four hundred, was said to have been the first newspaper in the world that was published on a train. It covered local news and gossip, reports of the railroad service, changes of schedule, and occasional bits of war news received over the wires almost before the regular newspapers had them. Certain of its editorial reflections were uncommonly philosophical for an editor-pressman of fifteen, as: "Reason, Justice, and Equity never had weight enough on the face of the earth to govern the councils of men." (It sounded like something his skeptical father might have said.) In the next paragraph there might be the report that some station agent was the parent of a bouncing daughter of seven pounds; or that another had volunteered for war service. The style, grammar, and punctuation were haphazard, the spelling was phonetic, if not worse. An example:

ABOUT TO RETIRE

We were informed that Mr. Eden is about to retire from the Grand Trunk Company's eating-house at Point Edwards . . . We are shure that he retires with the well wishes of the community at large.[40]

We applaud in the fifteen-year-old Edison his manual skill and his spirit of enterprise. But in almost every other line of his text occurred such errors of orthography as "valice," or "villian," or "oppisition," which probably hampered the progress of his gazette.

The entrepreneur-editor of fifteen might have gone on to master the Queen's English and, in time, become a great journalist, if not for an untoward incident that illustrated the serious hazards of this calling. An acquaintance of his, who helped print the Port Huron *Commercial,* after a while persuaded Edison to enlarge his newspaper, turn it into a medium of society news and personal gossip, and rename it *Paul Pry.* But one of its candid little stories, touching a local figure, proved to be so excessively candid that the victim vowed he would have vengeance. One day the embittered man sighted the young editor near the docks of Port Huron, laid violent hands on him, and threw him into the St. Clair River. Fortunately Tom was a strong swimmer; but after this immersion he rapidly lost interest in *Paul Pry,* which ceased publication.

An adventure of another kind, which had more fortunate consequences, also belongs to this period. Late in the summer of 1862, he had descended at the Mt. Clemens station and was waiting on the platform while the mixed train switched back and forth shunting a heavy boxcar out of a siding. The boxcar finally rolled toward the station, when Edison noticed that the little three-year-old son of the stationmaster was playing in the gravel, right on the main track, in the path of the car. In an instant he had thrown aside his bundle of papers, dashed toward the child, and snatched him up in time to avoid the car. The stationmaster, J. U. Mackenzie, was called out, and young Edison handed him his baby, whose life he had saved by his quick action—though, as he recalled it, there was ample time and little risk. Later the story was retold with many embroideries, but it was substantiated by the father.

Everyone present made much of the Edison boy. Mackenzie was filled with gratitude and expressed a desire to repay him in any manner within his power. He had noticed how Tom Edison hung over his telegraph table constantly. On the instant, the happy idea came to Mackenzie of offering to teach the boy to be an operator. Edison fairly leaped at the proposal. He had at that time, in the care of his mother, most of the "immense sum of money," a hundred or a hundred and fifty dollars, that he had won through his business coup after the Battle of Shiloh. Mackenzie invited Edison to come and board with his family at Mt. Clemens for a couple of months, merely paying for the cost of his food, while taking his lessons in telegraphy every night and assisting the train dispatcher. He now gave up half of his newspaper route to another boy so that he could stop off at Mt. Clemens and study. He was only fifteen and a bit undersized. But in a few months he would be a real telegrapher.

At twelve to fifteen Tom Edison seems both boyish and old for his years. His mental growth proceeded at an uneven pace; indeed, in view of his deafness and poverty one would judge that he was under a severe handicap in the matter of continuing his education. If he had been one of your sensitive plants he would undoubtedly have been crushed out as so many unknown talents or geniuses are blighted in the budding.

Talent, or genius, however, shows itself in many different forms, for reasons that are, so far, incalculable. When Tom Edison feels things look desperate, he can become remarkably "cheeky," as he phrases it— in other words, aggressive and competitive. Within him there is a passionate curiosity to learn certain things; he works alone at his elementary experiments for long hours and enjoys himself heartily. But on the other hand he knows a good deal about the world too, and can cope with it, supporting himself as a huckster, helping his family, and even cunningly leading his father by the nose—when he wants to stay up and play with his telegraph set instead of going to bed at a decent hour. Even in boyhood he is a sly one. There is in him both a highly imaginative and contemplative fellow, and a shrewdly calculating one, eager to make his way in the world. Meanwhile the vocation of telegraphy seems just the thing for him—he can hear and communicate with people; he can function.

chapter **III**

The wanderyears

1 To Mackenzie's surprise, Edison arrived at Mt. Clemens for the first day of his lessons with a neat set of telegraph instruments that he himself had made at a gunsmith's shop in Detroit. How patient and skillful with his hands this boy was! He was forever collecting bits of scrap metal or wire at little or no cost; his homemade set for example, had good brass springs. In our material-rich country an inventive fellow could thrive on what others cast off as junk. At about that time, Tom Edison found a pile of corroded and broken-up battery cells in the storeroom of a railroad station; having been given permission to strip the old cells of their electrodes, he discovered that they contained several ounces of sheet platinum!—a perfect treasure-trove, which he used and re-used during many years.

At Mt. Clemens he underwent a rigorous apprenticeship in the autumn of 1862, being trained during five months to send and receive dispatches, and learning also the abbreviated signals used by railroad telegraphers. In the early stages he would be able to serve as a second-class or "plug" telegrapher; to attain the expert's speed of forty-five words a minute would need considerably longer practice at this art. But even as a "plug" the young Edison could find work almost anywhere, because the need for telegraphers was so urgent during the Civil War, many hundreds of them having been recruited by the armies of both sides.

These were his wanderyears that began when he was only sixteen. In those days the strange new tribe of telegraphers were generally young men already noted for their nomadic or Bohemian habits, traveling light, pitching their tents for a brief season at one place, then journeying on to another that seemed to offer greener pastures. There were, to be sure, a few steady young operators who rose from the ranks to become industrial magnates, like Andrew Carnegie and Theodore Vail. But in a nation where opportunity always seemed to beckon elsewhere, and where so

41

many were on the move, none were less stable as a group than the young telegraphers of that era.

"Edison," one of them would suddenly exclaim, "I have only sixty cents in the world, and I am going to San Francisco!" And the next day he would go.

It used to be a tradition, carried over from European folkways, for the journeyman in certain trades to enjoy his *Wanderjahr*, or two, before settling into his career. As a tramp operator Edison traveled thousands of miles in the Middle West, South and East. His apprenticeship was to be uncommonly long, and he was to be well acquainted with hardship, which he bore, however, with the rollicking spirit usually affected by the brethren of the key, many of them hard-drinking, free-living fellows. He was not indisposed to working long hours—as he had done when he was keen to learn his trade under Mackenzie—yet he sometimes showed a marked indifference to routine tasks while his mind wandered to more engrossing subjects. Since his few hours of rest were spent in the reading of scientific or technical books and in "tinkering" for long periods with devices of his own, mechanical or electrical, he was often weary when at the operator's table.

The path of self-instruction and self-improvement was, for him, prolonged by many detours; even after years had passed, without money, without friends, it was not easy to enter upon the true vocation he already dreamed of for himself. His "university" was the rough, workaday world; it was nightwork in shabby telegraph or railroad offices during a turbulent era of war and reconstruction; it was made up also of solitude and study in squalid rooming houses. He might have lost his way or gone downhill, like so many others, in that long-drawn-out, often unhappy interlude of his youth. Yet when these "arduous years" (as he called them) were gone and he had come to manhood, something of iron had entered into him. What he had seen and learned, how he had learned it all, at so much effort, marked him with sharply individual traits of character and mind, setting him apart from those whose youth had been passed in sheltered ease, under wise and solicitous mentors.

By the winter of 1863 he had absorbed all that the excellent Mackenzie could teach him and returned to Port Huron. At the jewelry and book-store of Thomas Walker in that town there was now a little telegraph office, but the operator had gone off to war, and so the place was offered to Tom Edison. Walker himself was a man of some learning and technical skill; he had all sorts of instruments and machines stored in his cellar, and even files of scientific periodicals. In order to practice at his receiving, Tom Edison placed a cot in back of the store and stayed up at night to "cut in" on press copy coming to the local newspaper at high speed. The volume of telegrams handled here was small; but even this Edison would neglect, while he disappeared into Walker's cellar, pored over back numbers of the *Scientific American,* or played with electrical circuits of his own device. What vexed Mr. Walker was that the impetuous boy would make free with his fine watchmaking instruments, using them to cut wires or to work on corrosive primary cells. A more serious issue in dispute, however, may have been the fact that Tom was paid only twenty dollars a month, and his father refused to allow him to sign apprenticeship papers for less than twenty-five dollars.

In the winter of 1864 a heavy ice jam severed the telegraph cable between Port Huron and Sarnia on the Canadian shore of the St. Clair. The ferry service was also cut off, and all communication at this international crossing was out—until the young apprentice telegrapher turned up with one of his bright ideas. All that was required, he said, was that a railroad locomotive and engineer be brought down to the dock. This was done, and Edison stood by while the engineer, under his direction, gave long and short blasts of his whistle in Morse code. Three quarters of a mile distant, on the opposite shore, a curious crowd of Canadians gathered as if wondering what all the whistling was about; then a telegrapher among them caught the idea, a locomotive cab was driven down to the ferry dock on the other side, and tooted back its loud reply.[1]

A little local celebrity as a consequence came to Port Huron's telegrapher; this may have helped to win him his next post, in May, 1864, as a railroad dispatcher at Stratford Junction, on the Ontario side about forty miles from Port Huron. His was the night shift from 7 P.M. to 7 A.M. —for which he always, hereafter, showed preference. The job required only that he await messages and signal trains or stations according to a stated schedule. Traffic, however, was light; Edison found ways of occupying the hours of this night-owl life by reading or by working with his hands over some mechanical device or piece of clockwork.

On the Grand Trunk Railway it was the rule to have the station operators signal to the office of the train dispatcher at certain hours, to guarantee that they were attentive to their duty. Edison's "sixing" signal, so-called, came over quite punctually at the right hours. But after a while it was noticed at the central control office that, when the train dispatcher there tried to reach Stratford Junction following Edison's signal that he was on the alert, no response came. This was queer; the train dispatcher's office soon investigated the new operator.

What was discovered was that the youth had devised a clockwork attachment, having a wheel with a notched rim, which, at a given hour, such as 9 P.M., automatically, like an alarm bell, transmitted by telegraph the dots and dashes making the "sixing" signal. It was a rudimentary version of the district messenger's call box. The device was conceived with a purpose; for in the small hours of the night, with no train movements to signal, he had formed the habit of taking cat naps (which became a lifelong habit); he hardly ever rested in the daytime hours when he was supposed to sleep. The railroad officials, however, were not amused at Edison's ingenuity and administered a severe reprimand.

A short time later he was in trouble again, though, according to his recollection, through no negligence of his own. He had continued to busy himself in the daytime, and to take those cat naps for a half hour between trains or signal times, but had arranged with the friendly night yardman to be awakened at specified hours so that he might take his call at the telegraph table. One night he received a hurried order to hold up a freight train; he rushed out to find the signalman, but before he could do so and have the signal set, the train had run past. He hurried back to the telegraph and reported that he could not hold that train, that the order had not come in time. The telegraphed reply was an oath—the train dispatcher's order had already been given to another train, which was coming up the single track in the opposite direction and by now had left the last station before Stratford. The danger of a serious collision was therefore imminent. Edison recalls that he "ran toward a lower station, near the junction . . . in a forlorn hope of catching the train. The night was dark, and I fell into a culvert and was knocked senseless." [2] Fortunately disaster was avoided, when the engineers approaching on a straight track saw each other's headlights and stopped in time.

Edison was summoned to come by train the next morning to the general manager's office in Toronto. There he was sternly questioned and informed that his presumed negligence could, very possibly, under Canadian law, subject him to a prison sentence. The conference, however, was interrupted by the arrival of an important business delegation, which distracted the attention of the angry manager. Edison was a badly frightened boy; but while the manager was occupied he managed to slip out, jump a fast freight train, and make his way back to Sarnia, feeling relief only when the ferry had landed him again on Michigan soil.

2   In that one year, 1864, when he was seventeen, Tom Edison seems to have held four jobs in as many different towns, as far as we can determine his itinerary. After losing the place at Stratford, he went to a railroad junction at Adrian, Michigan, at seventy-five dollars a month; then, on being fired for some small act of insubordination, he moved on to Fort Wayne, only to be dismissed

soon afterward. Next we find him listed, for three months after November, 1864, as a second-class operator on the payroll of Western Union's office at Indianapolis. By mid-February, 1865, he had shifted to Cincinnati's large Western Union headquarters. He was a rolling stone in those days; his low rating gave him little tenure; and he himself has confessed that, while he could receive messages pretty well, as an apprentice he was very poor at sending. In most cases he was fired, either because he was not amenable to office discipline, or was at times inattentive to his duties. He might cut in on the manager's own wire because he thought the message he was sending was more urgent. Or he would let outgoing messages pile up on the hook above his telegraph table for hours on end while he drew diagrams or read a book.[3] Sometimes while he was receiving messages an idea came to him; Edison would signal the out-of-town operator to stop sending, draw out his notebook, jot down some notes, and then tell the operator to resume.

Another cause of trouble was his continual monkeying with some paraphernalia or "invention" that would make the routine work of telegraphy less irksome or allow it to be done more expeditiously. Finally, his disposition to play practical jokes on people by means of some gadget of his own devising was as marked as ever. All this overflow of playful activity derived, in part, from inward compulsions he himself scarcely understood, or from buried resentments or frustrations; in part, also from an irrepressible, though awkwardly directed, passion to contrive, to discover, to create, and, in short, to invent.

One of his few intimate friends among the roving telegraphers, toward 1865, was Milton F. Adams, whom he met in Cincinnati, a whimsical and carefree young fellow who dressed like a dude and always said the most surprising or devilish things with a perfectly bland expression. Adams later testified that Tom Edison in youth was a queer fish, an eccentric: "The boys did not take to him cheerfully, and he was *lonesome*." At eighteen, the itinerant telegrapher was "decidedly unprepossessing" in dress and uncouth in manner. He is also described as thin-faced, with a prominent nose, an unruly cowlick, and wearing loosely fitted clothes topped off by a soiled paper collar and a straw hat—in short a country "hick."[4]

A photograph taken when he was twenty, however, in Sunday suit and with his face washed, shows no such Lincolnesque homeliness, but reveals a handsome youth with a finely-shaped head, a sensitive mouth, a strong round jaw, and large, expressive eyes. There is an air of pride and a consciousness of strength in this country boy.

Arriving in a strange town to apply for a job, he would usually rent a cheap back room in a boardinghouse, then fill up most of the room with his scientific impedimenta, consisting of tools, balls of wire, batteries, bottles of chemicals, books. His pockets usually bulged with metal scraps, pliers, and other instruments of the electrician's trade. All the money beyond that needed for his wretched meals—which, as he said, amounted to a "system of flesh reduction"—he devoted to the purchase of books and equipment. An hour after receiving his wages, according to another of his co-workers, Walter P. Phillips,

Edison would come strolling in and blandly ask some of us to lend him half a dollar with which to get his supper. When reminded he had received half a month's salary that day he would smile, and taking a brown, paper-covered parcel from under his arm, he would display a Ruhmkorff coil, an expensive set of helices, or something equally useless in the eyes of his comrades; from which we were led to infer that [his] salary had already been exchanged for these apparently useless instruments.[5]

The typical offices in which he worked in the towns of the Middle West were dingy places, usually located in some dilapidated building of the downtown district. From the ceilings a good deal of plaster would have fallen away. On diminutive tables set against the walls were the telegraphic instruments, linked to the small switchboard by worn copper wires. The connections were loose, permitting arcing effects like lightning, or even explosions like cannon shots. In the battery room adjacent would be stands of old Grove cells, whose acid gave forth noxious odors.

Cincinnati was the largest city Edison had ever seen; then the metropolis of the Middle West, it offered many advantages in the way of bookshops, the famous Mechanics Library, theaters, and intelligent company. Yet toward the close of the Civil War, according to the recollections of a telegrapher who served there, the city was rougher and more lawless than it had ever been; fights, street robberies, and murders were daily occurrences.

Discipline among the men of the telegraph fraternity during the war became very lax, as Edison has recalled. A good part of their work was for the war service. Thus in mid-December of 1864, when Edison worked in Indianapolis, it was reported that General Hood's Confederate Army was marching upon Nashville and threatened to surround General Thomas's Union forces there. Urgent cipher messages from Washington passed through Edison's hands, calling for instant communication with General Thomas at Nashville—though the wires leading in that direction seemed to be cut. In efforts to reach Nashville via Louisville Edison tried all kinds of roundabout ways but could get no response from Louisville.

Finally at 1 A.M. . . . . contact was made with a telegraph office along the railroad near Louisville, and a relay of horses was engaged to . . . bring the cipher dispatches to General Thomas. [On December 16 the Union army beat off the Confederate attack.]

A couple of days afterward it was found that there were three night-operators at Louisville. One of them had gone to Jeffersonville, fallen off a horse and broken a leg. Another had been stabbed in a gambling house, while the third had gone . . . to see a man hanged and had got left by train.[6]

The work at the Cincinnati office of Western Union proceeded with a deadening routine; fast press reporters, such as Edison had become by 1865, transmitted long messages at forty-five words a minute; so tensely concentrated were they in their responses, that more often than not they became wholly unaware of the meaning of those messages. One day in April, 1865, the telegraphers noticed from their windows that a great crowd had gathered in the street outside the office of the Cincinnati *Enquirer*. A messenger was sent there and brought back an "extra" sheet of newspaper with great headlines announcing: "LINCOLN SHOT." As Edison used to recall the incident, there had been, all that day, a great rush of press reports; but now the operators in the Western Union office looked at each other wonderingly to see which of them had transmitted this fateful news. Yet no one could recall it, until one man found in his files the copy sheet reporting the assassination. "The man had worked so mechanically," Edison observed, "that he had handled the news without the slightest knowledge of its significance." [7]

As an itinerant telegrapher, Edison often lived in vermin-infested bedrooms. At the Western Union office in Cincinnati, formerly the site of a large restaurant, they were menaced by armies of rats. One of Edison's first essays in invention of a primitive sort was a contrivance made up of two metal plates insulated from each other and connected with a main battery. This machine he laid out in the cellar, designating it as a "rat-paralyzer." When one of the rodents chanced to place its forefeet on one plate and its hindfeet on the other, then, as Milton Adams phrased it, "it would render up its soul and depart this earthly sphere."

It was also while working at Cincinnati that Edison and a companion, to vary the routine, devised another electrical fun-maker. Operated by a Ruhmkorff induction coil, which generated alternating current at high voltage from an ordinary battery, it was secretly connected by one electrode with the washtrough in a railroad roundhouse, the other electrode being connected to the earth. Here the railroad workers, of course, came in to wash up. He relates:

Above the wash-room was a flat roof; we bored a hole in it and could see the men as they came in. The first man dipped his hands in the water. The floor being wet, he formed a circuit, and up went his hands . . . The Ruhmkorff coil, although it would only give a small spark, would twist the arms and clutch the hands of a man so that he could not let go.[8]

Having received their shock treatment and recovered from their pain and surprise, the railway laborers, being good Americans, would then go and wait quietly by the wall to see how others coming after them would react to the same trick. Finally a crowd would have gathered, all excited and puzzled, while Edison, from his peephole, enjoyed the sport hugely.

Most of his associates would hardly have been considered very respectable. They went on jamborees and got gloriously drunk. Two of his acquaintances who were out of work were befriended by Edison, who allowed them to sleep in his room at night while he was working; in return for his hospitality, they appropriated a whole shelfful of books he had recently collected, and pawned them for corn whisky. Then they returned and made a shambles of his place, falling asleep, fully clad, in his bed. Edison "felt that this was running hospitality into the ground, and so I pulled them out and left them on the floor in the hallway to cool off from their alcoholic trance." [9]

One night, when he was working in his office in Cincinnati, there was a loud knock at the door, which an instant afterward was abruptly flung open; a tin box then landed on the floor with a nerve-shattering crash. It was immediately followed by a wandering telegrapher who had left town a few months earlier to manage a frontier post in the Rocky Mountains. He cried out in the carefree manner so often affected by the Lightning Slingers: "Gentlemen, I have just returned from a pleasure trip beyond the Mississippi. All my wealth is contained in my metallic traveling case, and you are welcome to it." The "traveling case" held only a solitary paper collar.[10]

Some time later, in 1866—he was then engaged at his trade in Louisville, and the telegraph management there was as lax as anywhere else— Edison was working at night with one other man, taking reports, when he heard a tremendous commotion at the door.

There appeared one of the most skilled operators of the force, a man whose splendid abilities were crippled by his habitual drunkenness. His eyes were bloodshot and wild and one sleeve had been torn away from the coat. He hesitated, then walked up to the stove without noticing either of us, and kicked it over with all its length of sooty pipe. Then he proceeded to pull down the switchboard and yank all the operating tables away from the wall. Pulling one of them away, he fell with it, cut himself and was covered with blood. After that he went into the battery-room, where he pulled down the shelves, upsetting bottles of nitric acid which ran over . . . eating everything away.

Tom Edison silently signaled to his fellow operator to say nothing. The intoxicated man, when sober, was a great friend of the manager, and few such operators were then to be hired. After he had demolished all he could reach, he disappeared, and the place was left in the condition he had wrought; Edison, however, rigged up the switchboard wires in temporary fashion so that incoming dispatches could be taken.

The manager came at eight o'clock in the morning, looked all about the office and was dumbfounded.
"Edison, who did this?"
"Billy L." I replied unwillingly.
The manager paced the floor for a minute, then said: "If Billy L. ever does that again, I will discharge him." [11]

There was even a trade-union organized at the Cincinnati office in 1865, when Edison worked there. A delegation of five men from Cleveland simply blew into town, recruited a local branch for the National Telegraph Union, then determined to celebrate the event with a spontaneous walkout and many rounds of beer for all hands. Arriving after the workers had gone—and knowing nothing then of labor unions—Edison found only the office boy on hand. Meanwhile the press report wires clamored for attention. All that night he manipulated the wires single-handed, doing a good deal of "guessing" whenever he fell behind; he was still there in the morning when the office manager arrived. The manager was told, though not by Edison, of what had happened. He studied the files of press copy hanging on the hook (which Edison much feared contained figments of his imagination) then turned to the lone operator and said, "Young man, I want you to work the Louisville wire nights. Your salary will be a hundred and five dollars." This was promotion indeed, for he had been receiving much less than that hitherto. It had come in the time of a strike, whose meaning he did not understand or perhaps cared nothing about. But he had long hoped to be advanced to the rank of a first-class operator regularly taking press copy as fast as mind and hand could go.

When he applied for a job nowadays, looking like a young rustic in his old linen duster, and was asked if he could send and receive fast, he would say nothing, but sit down and set the instrument humming. As an operator, it was said, he had few equals; at the age of nineteen he sometimes emerged winner in intercity tournaments between expert operators demonstrating their speed.

3   Out of those vagabond years there would seem to be little of actual accomplishment for Edison to show, save his experience of the world of men, and the knowledge of his trade. The money he earned also melted away quickly, not only in extravagant purchases of electrical equipment, but also in loans or gifts to his itinerant comrades, who, as one of them said, often found themselves "in the last stages of impecuniosity." On the other hand there are many bits of evidence to be seen, during those years of careless youth, of Edison's growing mechanical resourcefulness, of his earnest gropings toward innovations.

In Cincinnati, at any rate, he began to enjoy some of the amenities of civilization. He now had a few good friends and with them sometimes indulged in moderate dissipations, such as visits to the garden cafés "over the Rhine," to hear the music of German bands over draughts of beer with pretzels. There was also the National Theatre, where Edison was entranced at seeing Edwin Forrest and John McCullough in plays of Shakespeare, of which Othello was long his favorite. Milton Adams afterward said that Edison was very fond of tragedy. Yet at any moment the comic spirit might seize him, and he would mimic and clown to the amusement of his companions. After seeing a performance of The Black Crook, Edison, returning to his room with two friends, entertained them by rendering the dances he had just seen in a costume improvised out of his red flannel underwear.

"I cannot avoid a tinge of regret," one of his Cincinnati companions wrote him a dozen years later, "when I think of the loss suffered by dramatic art when you turned your back on tragedy." [12]

At seventeen to eighteen he already attracted the interest of intelligent men. The manager of the Western Union office at Indianapolis, when Edison worked there in 1864, was kind enough to lend him some of his fine electrical instruments so that he might experiment with them. In Cincinnati the following year he made the acquaintance of Ezra T. Gilliland, a telegrapher of some education and with some skill at mechanical invention; their minds seemed to run in the same path, and they became close friends—who were to meet again in later years. Gilliland said then of Edison that he seemed "one of the most wonderful young fellows" he had ever known.[13]

The telegraph (in the days before the telephone was invented) held a place of enormous importance in peace and war ever since it had entered into practical use, about fifteen years before Edison came to it. Technical improvements were being introduced at regular intervals; yet in the sixties installation was still faulty, cables "leaked" current badly under river crossings, and the Morse register which had lately come into use (a receiver with a needle that embossed dots and dashes on a moving paper tape), often faltered under pressure, leaving blanks to be filled in by the operator's imagination.

While he was in Indianapolis in the autumn of 1864, Edison, then seventeen and a half, conceived the idea of using a repeating device which, after recording the impressions received from the Morse register, could be played back at the convenience of the receiving operator, and at a speed he could regulate. The beauty of this scheme was that at periods when a great rush of press copy came at high speed, it could be received at leisure and its messages copied in Roman letters with complete accuracy instead of in haste. Edison described his device as follows:

I got two old Morse registers and arranged them in such a way that by running a strip of paper through them, the dots and dashes were recorded on it by the first instrument as fast as they were delivered ... and transferred to us through the other instrument at any desired rate of speed or slowness. They would come in at the rate of fifty words a minute and we would grind them out at the rate of twenty-five. Then weren't we proud. Our copy used to be so clean and beautiful that we hung it up on exhibition. The manager used to come and gaze at it in silence with a puzzled expression.... He could not understand it; nor could any of the other operators; for we used to drag off my impromptu automatic recorder and hide it when our toil was over.[14]

It was a most ingenious repeating system, but it broke down under pressure.

The crash came [Edison said] when there was a big night's work—a presidential vote, I think—and copy kept pouring in at top speed, until we fell two hours behind. The newspapers sent in frantic complaints, an investigation was made, and our little scheme was discovered.

Use of the duplicating register was forbidden, and Edison was removed from the press copy desk. Yet experiment with innovations in telegraphic apparatus remained an obsession with him.

A year or so later, after the end of the war, he drifted away from Cincinnati, to work at first in Nashville and then in Memphis, Tennessee. These places were then still under military occupation but telegraphers were in high demand and received as much as $125 a month and rations. Here again his passion for technical improvisation undid him.

At that time connections by wire between New Orleans and New York were completed only by long detours, that is, by retelegraphing in roundabout fashion from one city to another, with much loss of time and attendant error. Edison then adapted his old repeater device, made up of two Morse registers, as used by him in Indianapolis. "I was the first person to connect New Orleans and New York directly . . . just after the war. I perfected my repeater, which was put on at Memphis and worked without a hitch."

The manager of the Memphis office, however, had been working on the same problem together with a relative employed in the office. Finding himself bested, he took offense at Edison, accused him of creating disturbances in his office, and summarily discharged him.

The boy was in fact stranded, for no money had been given him for transportation even to the next town along the railroad having a telegraph office.

I managed, with another operator who was in the same boat, to get a railroad pass as far as Decatur, but from there we had to walk to Nashville, one hundred and fifty miles, with only a dollar or two in the world. Then we got a pass to Louisville, where we arrived clad in linen dusters in the midst of a snowstorm. I shall never forget the sensation we made in that city walking through the snow in our airy apparel.[15]

At Memphis a confrere said of him: "He spent his money buying apparatus and books, and wouldn't buy clothing. That winter he went without an overcoat and nearly froze." [16]

It was all very well to make light of his misfortunes; but the streets were icy, his shoes were broken and shapeless, and he was penniless. In truth he was sometimes so utterly spent that he became ill. Fortunately, he found work at the Louisville telegraph office, thanks to his demonstrated skill. Now he conquered his wandering habit for more than a year and, save for a brief excursion or two, remained fixed.

Once more he turned his poor furnished room, which was on Third Street, not far from the Western Union office, into a combined laboratory, machine shop, and library. "I began to frequent secondhand book stores and acquired quite a library," he said. During this stay in Louisville he purchased for two dollars a set of twenty volumes of the *North American Review*, which then published many scientific articles. The bundle of books was left for a while in his office; on deciding to bring them to his room one night after his work was done—and that was 3 A.M.—he slung the heavy bundle over his back and walked along one of the dark streets of the city. Suddenly he became aware of something like bullets flying about his head. He halted and looked back. A policeman was waving at him, had in fact been calling to him loudly and, having no reply (for he had not been heard by Edison), had opened fire. Fortunately, the officer, who thought he had a thief in hand, was a poor shot. When, seizing the young man, he forced him to open the package, he was disgusted to find it contained only volumes of old magazines.

That Edison was now maturing is suggested by the sort of acquaintances he sought out at Louisville. Through handling press copy for the *Courier* he came to know its editor, George D. Prentice, a celebrated journalist of that day who was said to have created the fashion for running columns of humorous paragraphs in American newspapers. Edison thought him highly educated, "poetic," and a fine talker. With him was usually to be found a tall man named Tyler, heading the Associated Press office, who was a Harvard graduate. The two men were much given to lengthy discussions of general, scientific and philosophic ideas. Edison used to invite himself to these nocturnal sessions after the newspaper went to bed and would listen silently through the night, while Prentice dipped crackers in corn whisky as he talked, and Tyler consumed his tumblers of whisky neat and without food.[17]

During this period of more than a year, when his life in Louisville assumed some regularity, his self-education went forward more rapidly. We find that he tried to teach himself foreign languages. At this time he also had the novel idea of training himself in a new style of penmanship, writing characters that were blunt and vertical, while suppressing all curls and flourishes. It was a kind of manuscript-print writing that he found saved considerable time in taking down fast press reports from the sounder. It remained his style of writing for life.

In March, 1867, he won the admiration of all his colleagues by taking down the whole text of one of President Andrew Johnson's long veto messages during a single sitting that lasted from 3:30 P.M. on one afternoon to 4:30 A.M. of the next day—and all in his wonderfully clear manuscript hand.

He was, then, no ordinary "tramp operator" when he was nineteen, during his extended sojourn in Louisville, but a studious fellow with his nose often in books and mechanical treatises. Using the crude instruments of electrical science that were to be found in our provincial cities, he measured and tested things, often trying new combinations of electrical circuits of varying strength and polarity. When there was heavy interference and "clatter" on the telegraph line into Louisville, owing to storms or the leaky cable under the Ohio River, he remembered using several relays, each with a different adjustment, in order to receive the electrical impulses more clearly. A little earlier, in Cincinnati, he had worked with another man on a "self-adjusting telegraph relay," which would have been very valuable if they could have got it.* [18]

The germs of many ideas and stratagems perfected by him in later years were implanted in his mind when he worked at the telegraph key. On returning to Cincinnati for a second stay, after the war, Edison chanced to meet an operator of somewhat secretive character who had formerly been an intelligence agent for the Confederate army, "tapping" Federal telegraph messages, imitating their "style," sending out false reports, and creating much mischief, until he was eventually captured. Now in peacetime the former spy suggested that Edison help him devise a secret method of sending dispatches so that an intermediate operator could not understand them. If they succeeded they might sell such sounders to governments for large sums. Edison went to work and tried to perfect such an instrument, either by varying the strength of the battery current or alternating the direction of its flow—though his recollections on this point are not clear. At any rate, the former spy suddenly disappeared; but the idea of the thing remained with Edison for future development.[19]

As he described this phase of his life afterward, his mind was in a tumult, besieged by all sorts of ideas and schemes. During the fairly stationary year in Louisville the "mysteries of electrical force" as applied to the telegraph, and all the future potentialities of electricity obsessed him night and day. He remembered having written a treatise on electricity at the time. "Even during his sleep, his brain continued to devise the most intricate machines, all traces of which vanished as soon as he awoke. This tendency to dream inventions clung to him all his life." [20]

It was then that he dared to hope that he would become an inventor— like the famous Samuel F. B. Morse, one of America's "kings of fortune." But Morse's telegraph was still only a crude, slow-moving instrument, of limited radius, and operating upon only a single electrical impulse. It fairly cried out for improvements, for refinements of all sorts. In this field Americans were still seemingly backward; in England and Germany systems of duplex transmission over a single wire, though still crude, were already patented by 1865; and the first automatic printing telegraphs, as well, had been under development since 1848.

* A telegraph relay is essentially a receiving instrument containing an electromagnet sensitive to weak line currents. Its armature, in responding to electrical impulses, or signals, coming from a distance, acts as an intermediate transmitter key, alternately closing and opening a separate local circuit in which there is a sounder and powerful battery. As the relay was able to repeat loudly the original signals, it served to extend a telegraph circuit beyond the limits of its own battery power.

There were so many things to be invented! At Memphis, just prior to his coming to Louisville, Edison had seemed so feverish about his schemes, and pottered about so busily with new electrical circuits, over which he hoped to send two messages at the same time, that the manager had called him "the luny one." [21] But during those early, fumbling investigations he observed many things unseen by others and acquired experience that was to serve him well in later years. Even his simple repeating device foreshadowed new developments in automatic telegraphy.

At Louisville's Western Union office he became, once more, passionately absorbed in schemes for reconnecting wiring and batteries for duplex transmission. To the annoyance of the manager he dismantled all the instruments and reconnected them—though without any decisive result. He was thereafter forbidden to monkey with the circuits and batteries, but having no other instruments to work with, and being inflamed with curiosity to learn at last the consequence of his experiments, he defied this order.

I went one night into the battery-room to obtain some sulphuric acid for experimenting. The carboy tipped over, the acid ran out, went through the ceiling to the manager's room below, and ate up his desk and all the carpet. The next morning I was summoned before him and told that the company wanted operators, not experimenters.[22]

He was desperate, feeling himself near a solution, and earnestly appealed to a fellow telegrapher who had some savings. If he could but have the loan of one hundred dollars and a little time he would complete his apparatus, and he offered the other man half of the profits that would flow from this venture. But the other would not entrust his capital to the shabby youth of nineteen. As it happened, a duplex apparatus was not long afterward patented by another inventor, J. B. Stearns, of Boston. Edison always believed that by ill luck and the niggardliness of that man in Louisville he had missed the opportunity to precede Stearns.[23]

Truly, the lot of an aspiring inventor was not an easy one. His insatiable curiosity about new telegraphic techniques, his very spirit of innovation often got the young Edison into hot water. To his employers he seemed indifferent both to his duties and their property rights, and they reacted repeatedly by throwing him out in humiliating fashion—though he occasionally displayed flashing speed as an operator when it pleased him to do so. On being sacked again, he would make his way to another town by means of a railroad pass, or on his legs.

His sense of accumulating frustrations and misfortunes seemed to have reached a climactic phase after he had been working in Louisville about six months. He had ideas for various inventions, though in the matter of the multiplex telegraph they were scarcely clarified for him, and he was burning with anxiety to try them. But to explore, to test his ideas would need much work, equipment, and, above all, money. And who was there who would trust him with some money? He might go without an overcoat to buy some necessary equipment; even then his tests might prove failures. Alternately he was on fire with ambition and reduced to a state of despair. Feeling his situation hopeless—since he would never find time or money while working at his monotonous trade—he did what many other unhappy youths have done: he pinned all his hopes upon some romantic escape from present ills. He resolved that he would go abroad, to some far-off place, and seek his fortune. Thus he committed an act of folly the consequences of which might well have proved tragic for him.

The young ex-Rebels in whose company he was thrown after the war, in the half-ruined cities of the South, were often seized with the fever to set off on adventurous journeys to some distant clime where they might win fortune through acts of courage, or at least escape from their present troubles. Two of these young Southerners, who were working with Edison as telegraph operators, called his attention one day to an advertisement in a local newspaper by the Brazilian government offering appointments at high salaries to experienced telegraphers. They were resolved, they said, to set off for South America and undertake the construction of great telegraph networks in the Brazilian jungle, where many Confederate Army veterans were already transplanting themselves to new slave plantations. Desperate and reckless as he was, Edison vowed that he would go with them.

In the early months of 1867 he began to prepare for his departure on this hazardous voyage, writing his mother that he was studying Spanish industriously with the aid of a Spanish-English dictionary and hoped to be able soon "to speak Spanish and read and write it as fast as any Spaniard." He had already taught himself French in this way.[24]

He then made a flying visit to Port Huron, by means of a railroad pass, in order to see his mother again before leaving the country. Both his parents, on hearing of his plans, tried earnestly to dissuade him. Nancy Edison had already lost four of her seven children, and she feared she would never see Alva again. Despite her pleadings he remained stubborn in his determination, and theirs was a sorrowful parting.

After rejoining his Southern friends in Louisville, he threw up his job and left with them for New Orleans, in great haste. They were to take passage on a specially chartered steamer that was to transport a whole contingent of former Confederate Army officers to Brazil. Arrived in New

Orleans, however, they found the city under martial law following recent mob riots against the "carpetbag" regime, so that their ship's departure was delayed. Edison, meanwhile, learned from a friendly Spaniard who had long lived in South America that the prospects of meeting with anything but danger and hardship in the forbidding Amazon region were dismal indeed. His informant painted the picture so darkly that Edison was persuaded to abandon the whole adventure. Returning to Louisville, he was lucky enough to regain his former job—only to lose it again several months later. His two friends, he learned afterward, had proceeded by boat only as far as Vera Cruz, Mexico, where they both died during a great plague of yellow fever.

In the late autumn of 1867, Edison made his way back to the parental home in Port Huron, Michigan, as penniless and hungry as ever. It was, indeed, a sad home-coming. His mother, whom he greatly loved, was transformed by grief. They were about to lose the big "house in the grove" which had been the Edisons' spacious home these thirteen years; a change in authority at the Fort Gratiot reservation had resulted in the place being taken over by the military. To Nancy Edison's bitter disappointment, they were forced to move to temporary quarters until a new dwelling could be found. The whole family now made great moan at the evil days that had befallen the "Ea-di-sons," as they were called in Port Huron.[25]

Hard work, anxiety, numerous deaths in the family, had all combined to make Edison's mother old and ill before her time. Her illness was attended with symptoms of mental derangement, as her neighbors noticed.[26] The father, never a reliable party, now betook himself elsewhere as often as possible.

It was not a home to remain in any longer than Tom Edison could manage. On the other hand, there was nothing for the mother to be happy over in the erratic career of her Alva. Though he earned a good living for those days, in spells, and aided his parents whenever he could, he usually had nothing to show for his labors. In contrast, his elder brother Pitt, who was employed for long years in the office of the local horsecar line, seemed a solid citizen.

"After stopping for some time at home, I got restless, and thought I would like to work in the East," Edison recalled. A friend of Cincinnati days had written him from Boston, and Edison wrote inquiring if there was an operator's job there. The reply advised him to come at once to the Western Union office in that city.[27]

Boston was then not only America's "hub of culture," but also her foremost center of scientific learning. At this stage of his development

Edison sensed that he needed contact with better sources of technical knowledge and equipment. In the 1850s there were not many more than two hundred graduate engineers in the United States, and few of them were knowledgeable in electricity. But Boston, the home of the Yankee inventors, had already well-established manufacturers of fine electrical equipment. There he would likely find the answers to many problems in electrical science that now tormented him.

How would he travel almost a thousand miles in winter without money? He had lately rendered some assistance to the Grand Trunk Railway, while in Port Huron, by helping to repair their broken electric cable across the St. Clair; and the railway repaid him with a free pass over their line to Boston.

Well might Mr. Horace Greeley of the *Tribune* urge our young men to "go West" to seek their fortune; Edison, a son of the West, determinedly rode eastward. Unlike his early wanderings, this remove was purposeful.

chapter IV

The free-lance inventor

1
The long journey to Boston was made even more protracted and hazardous than might have been expected, when a fierce March blizzard set in, halting the train in big snowdrifts midway between Toronto and Montreal. After waiting for twenty-four hours, the trainmen set off across the drifts on snowshoes to organize the rescue and feeding of the passengers. Fortunately an inn, almost buried in the snow, was found a half mile off, and the frozen travelers were brought there. Eventually they reached Montreal four days late. There, as often before, Tom Edison received the generous hospitality of one of the brethren of the keys, one Stanton, whom he had known in other days. That is, he was invited to sleep on the floor of Stanton's hall bedroom, with the airy assurance that the recent cold wave had caused the usual "animal kingdom" infesting such lodgings to go into hibernation. The next day he arrived in Boston, walked to the office of the Western Union, and applied for a job.

I had been four days and nights on the road, and having had very little sleep, did not present a very fresh or stylish appearance, especially as compared to the operators of the East, who were far more dressy than their brethren of the West. ... My peculiar appearance caused much mirth.[1]

The truth was that Edison had a will to be "different." With his wide-brimmed hat, his long hair and baggy clothes, and a plug of tobacco in his jaw, he looked in fact like a real jay from the woolly West. George Milliken, the manager, however, was a man of discernment. On being shown a letter in Edison's neat manuscript writing and receiving assurance that the youth could take press copy in this fashion for hours on end, and at breakneck speed, he was impressed. After a five-minute interview he hired Edison, proposing that he begin work at 5:30 P.M. that same day. His fellow workers thought to initiate the ragged stranger from Michi-

59

gan by "putting up a job" on him. It was arranged that he should be assigned to the Number 1 New York wire to receive voluminous press copy for the Boston *Herald,* without making it known to him that he would be expected to keep pace with one of the fastest senders in New York. Edison suspected that something was afoot from the way the Boston operators stood by, looking over his shoulder, "their faces shining with fun and excitement." The New York operator began slowly, then ran on at top speed, while Edison not only kept pace with him, but sometimes, with studied deliberation, paused to sharpen a pencil, letting the sounder run on. Then the sender began to slur his words, sticking the signals, but Edison was used to this style. Finally, as he told the story, he felt the fun had gone far enough and, opening the key, he made a telegraphic aside: "Say, young man, change off and send with the other foot." That "broke up" the New York man, and, in disgust, he turned his instrument over to someone else.[2]

After that it was the sly rustic Edison, who, turning the tables, played tricks on the Boston operators. To prevent his colleagues from carelessly forgetting to return their common water dipper to the pail that stood near him, he wired it to a battery and electrified it, so that they received a mild shock and replaced it in haste. At midnight an old "cake man" used to bring lunch; but this office being infested with insects, an army of cockroaches would appear regularly at the hour of refreshment. Edison then brought forth a variant of his "rat-paralyzer," setting out along the tables little tin-foil strips that were connected with the positive and negative poles of telegraph batteries. By this mechanism the cockroaches were swiftly "oxidized." These exploits of the queer country boy from the West were signalized in humorous paragraphs of a Boston newspaper, giving him some eminence in his office.

Soon he was established in a hall bedroom on Exeter Street, together with an old acquaintance, Milton Adams, following his singular routine of working at his telegraph sounder at night and staying wide awake most of the day. In the old bookshops on Cornhill he could browse over books, which together with the chemical and metallic junk he accumulated soon filled up his room. Back in the telegraph office by evening, "he would sit and draw and dream, until reminded by the chief operator that he must attend to his work," one of his companions recalled. He would be thinking of new combination of rheostats, polarized magnets, and currents of varying potentials, spending a great deal of his time, when on duty, "making diagrams to show how wires could be operated in a multiplex way...." Though he tended to be reserved at this period, he could become excited and vehement in expounding his novel schemes. He would hold forth with undeniable eloquence "on every conceivable subject excepting that relating to the prompt dispatch of messages the company had on file for transmission." [3] It was around this time that the quip was made by another Boston co-worker that "Alva is shy both mainspring and escapement."

His job was to him a distasteful bondage—a long slavery of Jacob for Laban—but offering no avenue for his real desires and ambitions. It was at this time that he bought a secondhand copy of Michael Faraday's two-volume work, *Experimental Researches in Electricity*. It was a red-letter day for him. Edison said in after years that this first encounter with the great English scientist's journals of his experiments was one of the decisive events of his life. His own education, up to now, had proceeded slowly, perforce, through direct observation reinforced by haphazard or miscellaneous readings. But here at last, for his guidance, was the lucid exposition of Faraday's long searches in the field which most fascinated the young Edison; and it was steeped in that spirit of truthfulness and humility before nature that was always in Faraday's character. Best of all for Edison, the account of the experiments was wholly free of complicated mathematical formulas. ("His explanations were simple," Edison said.) For Faraday, the natural laws were revealed above all through experiment. Was it not well said of him that he *"lived with his facts* like a naturalist with his animals"? To Edison he appeared as "the Master Experimenter," whose laboratory notes communicated the highest intellectual excitement—and hope as well.

Michael Faraday (who had died the year before, acclaimed by all the world) had been poor, like Edison, and though virtually without schooling had taught himself everything. Since the 1820s his had been a career of marvelous discoveries in many different fields, that of induced electricity appearing as the most remarkable of all. With his large horseshoe magnet and disc rotating between its poles, he had demonstrated before the Royal Society in 1831 how mechanical energy could be converted into electrical energy—presenting the world with the first, or miniature, model of the modern dynamo. In later years Edison often spoke with veneration of Faraday's "selflessness" and of his indifference to money or titles, as exemplifying the ethics of the true man of science, which he too would have liked to emulate.

Though he had worked up to an early hour of the morning at the telegraph office, Edison began reading the *Experimental Researches* when he returned to his room at 4 A.M. and continued throughout the day that followed, so that he went back to his telegraph sounder without having slept. He was filled with determination to learn all he could by repeating the experiments recorded in the master's journals. To his roommate, Adams, he exclaimed, "I am now twenty-one. I may live to be fifty. Can I get as much done as he did? I have got so much to do and life is so short, I am going to hustle." With that he snatched some pie and coffee, and went off to work on the run.[4]

From repeated allusions he made in later times, we know that Edison's mind was haunted for ten years by the problems and possibilities of multiplex telegraphy—ever since the idea of it first came to him, when he was eighteen. He now proceeded to carry out a continuous series of cut-and-try experiments with telegraphic apparatuses. According to Adams, he was "immersed" in these activities in Boston; "his brain was on fire." If he could but send two messages over one wire telegraphy would be made cheap!

Invention is not so much bred by old Mother Necessity as by a growing reservoir of knowledge of various arts and industries, by the rising level of culture and technics. Boston fairly teemed in those days with "addicts" of science and with skilled artisans who were would-be inventors. It was probably the country's leading center of "light engineering." Here were fine clockmakers and opticians who had turned to making, by hand, telegraphic and other fine electrical instruments, or parts of them, as in the shop of Charles Williams, Jr., in Court Street. (A year or two later another unknown experimenter, Alexander Graham Bell, would take up quarters in the shop of Williams to work on a "speaking telegraph.") There was, for instance, Thomas Hall, of Palmer & Hall, who had produced and exhibited a miniature electric-motored train (though not the first) as long ago as 1852. And there was J. B. Stearns, who introduced numerous innovations in the telegraphic art. Visiting these shops, studying their products, talking with their makers, Edison was received in friendly fashion by the other enthusiasts, and was greatly stimulated by observation of their work. At the Williams shop he found to his delight that some of America's first fire-alarm telegraphs were being turned out, a most useful adaptation of the original instrument. And though Edison might be unkempt, or look like a peasant from the West, the Bostonians noticed the stamp of his mind. At least two of them advanced him small sums of capital to be risked in his enterprises.

Soon he had planted himself in a corner of the Williams shop, during the daylight hours when he was not occupied at Western Union. Experiment and study in all sorts of fields was carried on, as Edison said later, in "helter-skelter fashion," and led to some surprises and contretemps. For example he had read that the Swedish scientist, Alfred Nobel, the year before, had perfected a method of combining nitroglycerin with inert matter so as to permit relatively safe handling of the explosive compound. Edison relates that he was "fired up" by the wonderful properties of Nobel's dynamite and resolved, with the help of Adams, to investigate them, though such reports as he had read in a scientific paper were incomplete, and (in this case) dangerous to rely on:

We tested what we considered a very small quantity [of Nobel's compound] but this produced such terrific and unexpected results that we became alarmed, the fact dawning upon us that we had a very large white elephant in our possession. At 6 A.M. I put the explosive into a sarsaparilla bottle, tied a string to it, wrapped it in paper and let it gently down into the sewer, corner of State and Washington Streets.

Soon afterward he was working with a large induction coil to be used in circuit with his proposed multiplex telegraph. He tells us:

One day I took hold of both electrodes of the coil, and it clinched my hands so that I couldn't let go. The battery was on a shelf. The only way I could get free was to back off and pull the coil, so that the battery wires would pull the cells off the shelf and break the circuit. I shut my eyes and pulled, but the nitric acid [of the batteries] splashed all over my face and ran down my back. I rushed to a sink . . . and got in as well as I could.[5]

His face and back were yellowed by the burns, and for two weeks his appearance was such that he did not venture into the streets by daylight.

In June, 1868, a journal devoted to his trade carried an article on the subject of "Mr. Thomas A. Edison, of the Western Union Office, Boston," and his invention of a "mode of transmission both ways on a single wire . . . which is interesting, simple and ingenious." [6]

This overoptimistic article, though signed by his friend Milton F. Adams, had been written under Edison's supervision.

Jubilant over the first public notice of what may be considered his initial attempt at invention, he gathered together numerous copies of the article in *The Journal of the Telegraph* and posted them off to persons in Memphis and Louisville who, a year before, had ridiculed his "luny" experiments. This youthful experiment was hardly of prime importance, consisting of little more than refinements of, or innovations in, existing mechanisms. But the interest it aroused, the fact that a little capital was now offered him to pursue his experiments, made him determined to throw up his job at Western Union and devote his full time to his own projects.

Whenever Edison grew bored with his job as telegrapher, the spirit of mischief seized him. He treated the receiving of stock market quotations as a joke, copying the lengthy reports in a hand so fine that two thousand words covered only a single sheet. When the night manager protested that this was unreadable, Edison wrote down the stock market report in characters so huge that a few words used a whole page and it needed a bundle several feet thick to contain the entire message. At this the newspaper's press foreman flew into a rage and threatened to assassinate the whimsical operator with his own hands. Edison was then demoted from the press copy table to "sending" and thereupon resigned.[7]

2   In January, 1869, a brief notice in a small telegraph trade journal reported that Thomas A. Edison, formerly an operator, "would hereafter devote his full time to bringing out his inventions."[8] Other notices or advertisements indicated that he had models of his "double transmitter" for sale at four hundred dollars; that he was associated with George Anders in the manufacture of alphabetical dial instruments for private telegraph lines; and that he was located at Charles Williams's shop on Court Street.

From now on he was, at his own risk, an inventor of the free-lance type. He had cast off from the safe shores of steady employment. There was, perhaps, no vocation in life he might have chosen that would have been more perilous. It was proverbial that inventors for the most part starved, or even went mad, like poor John Fitch, or at least were ruined in the end, like that talented Yankee Charles Goodyear, creator of vulcanized rubber, who had recently published the story of his life (which included a sojourn in a debtors' prison) under the revealing title, *The Trials of an Inventor*. Nonetheless Edison must take the same stony road.

To make an invention, even to possess the talent to do this, was, however, not enough. Capital and plant and the commercial ability to win acceptance for one's product from the public were needed. Now, the "business talent" for promoting an invention and bringing it to market, as Jeremy Bentham, the philosopher of utilitarianism, had written long ago, seemed to occur in men "in inverse proportion to the talent for creating inventions." As Bentham defines the problem, your typical "poor inventor" must somehow "penetrate the antechamber of the rich or the noble whom it may be necessary to persuade.... Admitted to their presence, how will the necessitous man of genius behave when he has arrived there? Often he will lose his presence of mind, forget, stammer ... and retire, indignant that his merits should be misappraised." Obsessed with his overruling idea, he remains unaware of related problems and practical conditions which must be dealt with before his novel product can be brought into general use. Novelty itself is a disadvantage, inasmuch as most men are wont to cling to antique equipment still useful to them, while fearing to "waste" money on some device of uncertain value and future. The inventor, meanwhile, thinks only of what is in his own mind and not of the calculations and anxieties of his prospective patrons. "Thus," Bentham concludes sagely, "in every career of invention . . . minds should be attended by an *accoucheur*," one who has, primarily, the gift of persuasion, one who "knows the world, half-enthusiast, half-rogue." [9] On such matters wiser words were never uttered. Edison had never read them. Many years were to pass before he was to find a good midwife-promoter to help him body forth his creations.

He had, from the days of his youth to his old age, an immitigable enthusiasm for every project he undertook. Shy or retiring he might be, but once he began to talk about an idea or an invention that was close to his heart, he could become surprisingly lucid and even eloquent. Shortly after he had arrived in Boston, he was invited to give a demonstration and lecture on the telegraph before a young ladies' finishing school. Though, to begin with, he had been fairly frightened out of his wits, he ended by giving an excellent talk.

At this period numerous "addicts" of electrical experiment in Boston were devising all sorts of machines that were merely adaptations of the Morse telegraph, such as fire alarms, district messenger call boxes, dial telegraphs, printing telegraphs, and stock quotation tickers. Having observed these things carefully Edison had by now got several irons of his own in the fire. What was all-important was that he had been able to persuade one Boston capitalist to advance him five hundred dollars, in return for a half share in future profits from his duplex telegraph. At about the same time he had persuaded another man to invest one hundred dollars as his partner in the development of a telegraphic vote-recording machine, his first invention that was operative, for which patent was filed in October, 1868, at the United States Patent Office in Washington. This was granted, June 1, 1869, as No. 90,646, his first recorded patent.

His development had been gradual, no doubt for want of formal training or education. Now he was moving forward with impetuous strides; in all, three Boston "capitalists" invested money in his schemes, small sums to be sure, yet more than he had ever seen in all his life.

The electric vote recorder was a simple, though bulky, telegraphic contraption designed to take down votes from an audience or legislative gathering by electromagnetic impulse. Having observed the great loss of time attending roll calls for voice votes in Congress (when he used to report them on the press wire), he had conceived of this machine, a wooden model of which he "whittled" with his own hands. It was the crude forerunner of present-day voting machines. As Edison described it, wires were connected from the seats of persons voting to the central receiving instrument:

In front of each member of the House [would be] two buttons, one for aye and one for no. By the side of the Speaker's desk was erected a square frame, in the upper part of which were two dials, corresponding to the two classes of votes. Below the dials were spaces in which numbers appeared. When the vote was called for, each member pressed one or another of the buttons before him and ... the number of votes appeared automatically on the record. All the speaker had to do was to glance at the dial and announce the results.

Having obtained his patent, Edison and his associates hastened to approach a committee chairman of the Massachusetts legislature. They counted on selling their machine to numerous state legislatures, as well as to both houses of Congress, and were sure they would reap a fortune of fifty thousand dollars. To their disappointment the spokesman for the Massachusetts legislature promptly rejected Edison's apparatus, on the ground that it would infringe upon the sacred right of the minority to "filibuster."

Nothing daunted, Edison and his partner went on to Washington and demonstrated the machine before a committee of Congress that was authorized to purchase such equipment. Their patent attorney, Carroll D. Wright, a learned and upright man who had served with much distinction as a Federal official, tried to help their case, but found Edison all too young and ignorant of the ways of the world. To the aspiring inventor the committee chairman said chillingly:

Young man, that is just what we do *not* want. Your invention would destroy the only hope that the minority would have of influencing legislation. . . . And as the ruling majority knows that at some day they may become a minority, they will be as much averse to change as their opponents.

The idea of hastening the process of vote-taking was abhorrent. "I saw the force of his remarks," Edison recalled, "and was as much crushed as it was possible to be at my age. The electric vote recorder got no further than the Patent Office." [10]

On the sad return journey to Boston, as he remembered afterward, he did a heap of thinking and came to the conclusion that hereafter he must confine his efforts to inventing products that were certain to be in "commercial demand." Otherwise his chances of survival as a free-lance inventor were nil.

The form of restriction he now placed on his own inventive activities brings into sharp contrast the attitude of the empirical inventor and the spirit of the scientific discoverer. Nobody "needed" Newton's or Galileo's discoveries; and there was no market demand for Faraday's history-making induction coil, which he, regarding himself as a man devoted only to advancing scientific knowledge, refused to have patented. However, the young Edison was not as coldly utilitarian as he pretended to be; he had, in truth, no disposition for money accumulation and could be gloriously "impractical." The desire to impose himself, to make his mark in the world, by creating things never known before, in short, the passion for glory, rather than money, and the desire to gain preeminence through *acts of skill* are usually the ruling motives of the inventor type. What great new devices may not be discovered or perfected, the inventor seems to ask himself, if only I can exert my mental faculties to the uttermost?

But Edison's blundering efforts to launch his first invention taught him a sobering lesson, since it resulted in loss. His next undertaking would be on safer ground. Meanwhile, his interest, like that of many other contemporary investigators, remained focused on variants of the telegraph instrument that would serve to speed up commercial intelligence.

During the "Gilded Age" after the Civil War, a period of prolonged money inflation, speculators in gold and securities on all the financial exchanges were of course greatly dependent on Morse's telegraph. Market quotations were rushed by wire to distant cities and to newspapers. The transatlantic cable of Cyrus Field was at last repaired, and also technically improved in 1868, thanks to the assistance furnished him by William Thomson (afterward Lord Kelvin). But the mounting pace of speculation required still more complex and swift mechanisms; thus, in 1866, the very active Gold Exchange of those days in New York had seen the appearance of a telegraphic gold indicator, invented by Dr. S. S. Laws, a former gold broker who had studied electrical science at Princeton University under Joseph Henry. The Laws gold indicator transmitted numerals as electrical impulses—denoting the momentary fluctuations in the market price—to dial instruments placed in subscribing brokerage houses in the financial district, which were connected by wire to the indicator, or central transmitter, in the Gold Exchange.

A year later, one of the Boston inventors, E. A. Callahan, developed a superior telegraphic apparatus called the "stock ticker"—because of the sound it made—which was, in reality, a variation upon earlier printing telegraphs invented in Europe. Having a little electrically propelled typewheel, this machine was able to print full stock market quotations as well as gold prices by wire on a moving paper tape. A large firm, the Gold &

Stock Telegraph Company, was soon formed to exploit the Callahan instrument and began to compete successfully with Laws's service in New York.

After giving much study to the Callahan apparatus, Edison in 1868 conceived of a number of refinements which he incorporated in his own model of the stock ticker. This was his second invention in order of patents applied for (January 25, 1869); for its development he acquired another backer who lent him some money.

Constant experimenting with telegraph circuits had already given him a wide-ranging knowledge of the behavior of electric currents. One of his important improvements on the Callahan machine was to do away with the need of a special attendant at the receiving end in the office of each subscriber, and also to reduce the large amount of wiring used. Placing two electromagnets in the same circuit, he had one controlling the rotation of a little type-wheel, while the other activated the little printing hammer. Connected with these two last instruments was a polarized relay which automatically reversed the current, now passing it through the magnet of the type-wheel, and now through that of the printing hammer.[11]

The new Boston company exploiting Edison's stock ticker succeeded in persuading about thirty subscribers to rent it from them. For a while Edison himself helped to erect a number of private lines leading from the subscribers' offices to the Gold Exchange in Boston, as had already been done on a larger scale in New York. At this time he also engaged in the business of making and selling private telegraph systems using an alphabetical dial (not his own invention) on which words could be transmitted to another dial apparatus.

At the room he had hired for his workship, he already had several men working with him to make up the dial telegraph apparatus; he would go out with them to apply to the proprietors of stores or residences for permission to string wires over their roofs. "I used the roofs of houses just as the Western Union did," he said. Since permission was readily granted, the sky of downtown Boston, like that of New York and other large cities, was then a crazy network of crisscrossing telegraph or ticker wires, strung up on anything that could hold them, and often breaking and creating fire hazards.

In his early twenties Edison was already a "whirlwind of activity," sleeping little and working at all hours of the night and day. His relations with those who backed his first stock ticker, however, became unpleasant; after differences arose between the partners, his patent rights were sold off to a large telegraph company and, thanks to his want of business experience, Edison received almost nothing for his work on this venture.

But he had other irons in the fire and soon turned all his attention to them.

3   The embryonic inventions of this period were essentially imitative, constituting refinements on others' handiwork, though by such activity Edison's grasp of existing technology was rapidly being enlarged. The stillborn vote-recording machine alone showed some sparks of originality. More ideas, however, were in the back of his mind; they related to the development of an efficient multiplex telegraph, one that would be capable of carrying two messages over a single wire. Such a duplex telegraph instrument had been invented in earlier years by Europeans, but its impulses remained slow, since they were affected by electrostatic charges retained in the telegraph wire. Joseph B. Stearns, of Boston, in 1867 had greatly improved the operation of the duplex. With Stearns's apparatus two messages could be sent over the same wire at the same time, but only in opposite directions.*

At this period, Western Union had already purchased Stearns's patents and was trying out his device on an experimental basis. Duplexing, if successful, would halve the cost of all telegraph traffic. Edison hoped that his own apparatus would have features that would make it more effective than the older inventor's.

No patents were taken out by him on his first duplex instrument of 1869, and no records of it have been found, so that it is impossible to judge what progress he had made up to that date. From the brief description contained in an agreement of that time with E. B. Welch, of Boston, we gather that his 1869 version of the duplex was not as yet markedly different from Stearns's and was also designed to send two messages in opposite directions. The contract, dated April 7, 1869, assigned to Welch a half interest "in any instrument or method which I may invent to be held for the transmission of messages on telegraph lines both ways simultaneously." [12]

When he felt himself ready for a large-scale trial of his duplex system, he asked for permission to dismantle the whole Boston office of Western Union and reconnect its wiring and batteries according to his new system. This privilege being refused him, he applied to the managers of a rival

* Described in general terms, the duplexing of telegraphic messages through the Stearns apparatus was accomplished by varying the strength of the current, the main feature of this system being the use of the differential, or neutral, relay. Eventually, after several years, Edison was to add a new and wholly different method of duplexing. We shall return to his work in this field below, in Chapter VI, where he will be seen to have reached an advanced phase in the investigation of this problem.

concern, the Atlantic & Pacific Telegraph Company. The head of their Rochester office showed interest and invited him to try his apparatus over the company's wires between that city and New York.

He raised eight hundred dollars on a loan from Welch, to whom he gave both his personal note, promising repayment in any case, and an agreement allowing the other man a big share of the profits in the event of success. A model of the new duplexing apparatus was completed and, with the highest hopes, he set off for Rochester, bearing with him all his baggage of rheostats, condensers, magnets, and relays. Before leaving, he had sent minute written instructions to an expert telegrapher who was to work at the other end of the line, in New York. Edison arrived in Rochester on Saturday, April 10, 1869, and waited at the Atlantic & Pacific Telegraph Company's office in the Reynolds Arcade until the early hours of Sunday morning, when the wires would be entirely clear, so that his experimental trial could begin. A prematurely optimistic report (perhaps inspired by his backer) announced that:

On Tuesday evening last (April 13) a new double transmitter, on an improved plan, invented by Mr. Thomas A. Edison, was tried between New York and Rochester, a distance of over 400 miles by wire, and proved to be a complete success...." [13]

What actually happened was nothing so good; when he began to signal, Edison could get no response whatsoever from New York. As he later described the trial, the assisting telegrapher at the New York end failed to do his part, or could not be made to understand what should have been done. The trouble may have been due to electrostatic charges set up in the telegraph wire; or there may have been other "bugs" he had not yet eliminated. After repeated trials carried out in the early morning hours of several successive days, Edison gave up and made his way back to Boston.

The moment was a desperate one. His more or less profitable undertakings had been disposed of for a bagatelle. He was deep in debt thanks to the fiasco of his duplex transmitter, and was finished so far as Boston was concerned. Of late, he seemed to have gone from failure to failure, step by step, and now his credit in Boston was virtually ruined. There was no chance of finding new patrons for his experiments. His only hope was to get out of Boston, reach the bigger center of New York, and try his luck there. An expert electrical engineer of that city, in recent correspondence, had given him some encouragement along those lines.

As a young inventor, grief at the failure of an experiment may have penetrated him deeply. Later he would become very familiar with experimental failures, but as milestones along the road to successful inventions. His was a sanguine temperament and gloomy thoughts were usually not with him very long.

In Boston, being already heavily in debt all around, he blandly borrowed a few dollars from one of his telegrapher friends and set off for New York by boat, arriving the next morning with not a coin left in his pocket. All his possessions, tools and equipment had been left behind in Boston; he would have to send for them when he could afford to do so. In effect he had burned his bridges behind him, and not for the first time.

New York, already a vast metropolis extending up to the woods of Fifty-ninth Street, and full of uproar, thanks to its endless traffic of horsecars, drays, and carriages, exhibited what must have seemed a loudly proclaimed indifference to the hungry nomad who, on a fine day of spring, landed at a West Side dock to seek his fortune here.

He might look like a country boy still, but he was already one who had functioned in his craft; he had experimented, contrived some patentable devices, and had even, at his own cost, learned something of the ways of business. There were probably not ten persons in America who knew the tricks he had with currents, relays, and resistances. His mind buzzed with ideas for improvements in electrical communication. He was, so to speak, full of his genius; or, to paraphrase the words of Stendhal, "he was a fire with no chimney through which his smoke might escape." That morning he was also as hungry as he had ever been in his life and had nothing with which to buy his breakfast. The exercise of walking, moreover, made the pangs of hunger exquisite.

On a side street near the Washington Market he passed a wholesale tea house where, through the window, he saw a man tasting tea. The irrepressible Tom Sawyer in him prompted him to go in and ask the man if he might have a sample package of the tea in order to judge of it himself. The taster obliged. Edison then managed to exchange his package of fine tea for a very skimpy meal at Smith & McNeill's, a well-known downtown restaurant hard by the market. All that he had to eat was some freshly baked apple dumplings and coffee, which he afterward avowed to have been "the finest repast" he had ever enjoyed. Apple dumplings thereafter were his favorite dish.

Much heartened, he set off on a long walk uptown to find one of his friends of former days who might lend him a small sum of money or even allow him to sleep on the floor of his room, until he found a job. According to one of the varying accounts he gave of his experiences, the only friend he had in New York was not at home, and on his first day there he had to "walk the streets a whole night long."

chapter V

Pluck-and-luck Edison

1 The adventures that befell the emaciated young inventor shortly after he arrived in the great city unnoticed and unwelcome were so singular and led to such an extreme change in his fortunes that one might well believe they were imagined by Horatio Alger. However, Edison turned up many years before Alger began his famous series of moral fictions about poor country boys winning sudden wealth in the streets of New York; one might say that Edison's success story, on being made known some years later through the popular press, served to inspire Alger's romances.

Tom Edison did not save a beautiful lady from a runaway horse; nor did he use his muscles to rescue a banker from footpads; nor did he perform any of the extraordinary or fantastic deeds described by some of the early and ingenuous chroniclers of his life.* But he was certainly quick-witted and energetic when opportunity knocked at the door.

By now he was not unused to being down on his luck. After much walking about New York he found a telegrapher he knew, but was able to borrow only one dollar of him, since the other was also out of work. One of his next calls was on Franklin L. Pope, electrical engineer and telegraph expert, who was then employed at the headquarters of the Gold Indicator Company, on Broad Street. Pope already knew of Edison, as one who had made a stock ticker of his own in Boston.

Talking with the young stranger, Pope could not but appreciate his mettle. There was great need for such a resourceful fellow at this time, the very dawn of technology in America. As William James has phrased

---

* J. B. McClure's *Edison and His Inventions* (1879) was the first of several biographies giving wholly misleading accounts of how the "ragged" Edison sud-denly appeared in the financial district of New York at the time of the Black Friday Panic, on September 24, 1869, and, miraculously enough, "saved" Wall Street!

72

it, both "the hour" and "the man" were being chosen by the changing environment.

At this first meeting Pope invited Edison to make his headquarters temporarily in his, Pope's, office and to work at his experiments in the machine shop of the Gold Indicator Company. While he was waiting for a job to turn up, Edison could study the mechanism of the Laws central transmitter, which was under Pope's superintendence and was still undergoing improvements.

By giving close attention to his diet, Edison hoped to make his one dollar last several days. But where would he sleep in the meantime? Franklin Pope kindly suggested that for the moment he could use a cot in the Gold Indicator's battery room, down in the cellar of the Mills Building on Broad Street. There he slept two or three nights, as he recalled, living frugally on five-cent meals of apple dumplings and coffee, while above his head millions on millions of dollars were swiftly changing hands. Meanwhile, during the daytime, Edison inspected and familiarized himself with the Laws indicator, a rather ponderous piece of telegraphic clockworks.

The gold indicator itself symbolized the union of technology and capital in the postwar era of expansion and inflation. During the war when the country's credit had suffered so greatly, and during the postwar boom, speculation in gold reached its highest pitch; all commodity and security values were naturally affected by changes in the dollar price of gold. Thus any machine that facilitated or sped up communication concerning this key market assumed immense importance. By the 1850s the telegraph was the prime instrument for rapid financial intelligence. Changing quotations in the prices of stocks or of gold at the New York Stock Exchange were, however, still distributed momently among the brokers of the Street by hundreds of messenger boys, or "runners." In 1865 Laws had conceived his idea of using the key of the telegraph to operate an electrical indicator, showing price changes on a centrally located board in the Gold Room of the Exchange. A year later he converted this simple apparatus into a central transmitting instrument connected by wires to the offices of outside brokers, where the telegraphed impulses were received on a dial instrument resembling the fare registers in trolley cars. So successful was this venture that three hundred brokers subscribed, at high fees, for Gold Indicator's wire service, and imitations, in the form of Callahan's more elaborate gold and stock quotation ticker, and even Edison's first model, followed.

Edison had reached New York in May, or at least by the beginning of June, 1869, as indicated by a letter dated there.[1] At this season speculation in gold was rising again to the fever pitch of wartime, the price of gold fluctuating between 133 and 150 per cent of parity with the dollar's official value. The country seemed launched upon a new inflationary boom, and virtually everyone engaged in commerce, or even farming, followed the changing price of gold.

The famished Edison, though knowing nothing about the complexities of the money market, was examining the mechanism of Laws's cumbersome indicator one day, when suddenly the whole central transmitting instrument came to a stop with a great crash. Pope was on hand and quickly went to work to find out the cause of stoppage and to repair the machinery. But in the excitement of the moment, he could detect nothing visibly wrong. Dr. Laws himself came rushing forth from his office to learn what the trouble was, and, being even more agitated than Pope, was no more helpful.

Meanwhile, it was as if half the heart of Wall Street had suddenly stopped beating (the other half was now served by a rival wire service). Laws's clients, finding the dial instruments in their offices had halted, grew frantic and sent messenger boys, as of old, to bring back reports of current market quotations. Within a few minutes a mob of three hundred messengers, shouting and jostling one another, were trying to fight their way into the Gold Indicator Company's office, which could not contain them. Franklin Pope, in the melee, seemed to forget everything he knew. Dr. Laws, whom Edison remembered as "the most excited person I have ever seen," completely lost his head and began to yell at everyone. A business Laws had struggled to build up during four years, and now having attained an annual revenue of $300,000, was threatened with immediate and total ruin.

Edison, in the meantime, had been quietly engaged in looking over the machine himself and soon ascertained the cause of the trouble. He had a pair of sharp eyes and often saw what others could not see. He then went up to Pope and Laws, who seemed to be holding a shouting match, and with some difficulty managed to convey to them his belief that he had spotted the trouble. One of the many contact springs had broken off and dropped between two gear wheels, stopping the whole mechanism, he said.

"Fix it! Fix it! Be quick, for God's sake!" Laws yelled at him.

The young stranger then carefully removed the broken contact spring, and proceeded to reset the dial back to zero; linemen, in the meantime, were sent out to the different subscribers to set their receiving instruments in unison with the central transmitter. Within two hours the Gold Indicator's quotations were being ground out again and communicated all over the Street.

The tempestuous Dr. Laws, having become much more composed, inquired about Edison, learned that he was out of work, and invited the youth to see him at his private office the next morning.

Laws was a man of wealth and some learning and had traveled much; his office was filled, as his visitor noticed, "with stacks of books on philosophy and science." Edison, by contrast, seemed like a country boy lost in the big city. In response to Laws's questions, he overcame his shyness and owned up that he himself had invented and patented a stock quotation instrument, and he had no hesitation in making suggestions as to how the Laws gold indicator might be simplified and improved. The upshot was that Dr. Laws offered to engage him at once, at a larger salary than he had ever received, to work over the mechanical maintenance of his plant, assisting Pope. Though feeling "rather paralyzed," as he put it, Edison maintained his aplomb. Yesterday he was starving; today he was comparatively rich. The important thing for an itinerant telegrapher, however, was to seem carefree and debonair.

Throughout that summer he worked on mechanical changes in the gold indicator. In July, Pope, who had befriended him, resigned from his position with the temperamental Dr. Laws, to go into business on his own as a consulting electrical engineer, and Edison took his place.[2] His earnings were now advanced, as he remembered it, to three hundred dollars a month, a princely salary! In August, 1869, he applied for a patent on a second version of his own printing telegraph and in the next month made patent application for a third and improved model of the same instrument, this time in partnership with Pope.

A fruitful and prosperous period had opened for Edison, who now perfected a whole series of new devices, mainly refinements upon the telegraph. This instrument was Thomas A. Edison's first great love, the primary invention out of which many other innovations were to flow, some of them eventually destined to supplant Morse's instrument itself.

2   In the midst of the fantastic world that centered at Broad and Wall Streets, a world of "bulls" and "bears," of "Robber Barons" and "Gold Bugs," the free-lance inventor of twenty-two now sat entrenched, yet knowing nothing of the meaning or folly of all the money-changing transactions going on around his head. Edison's sole concern was that of the good technician: to see to it that the electromagnetic machinery recording all those operations functioned properly, or to improve it. By such service, he was to be linked to the financial community during many years of his life.

In late September there was more and more excited trading in gold, which rose gradually to a peak of 150. Now came the climactic phase of one of the boldest and most scandalous financial conspiracies in American history, the attempt to "corner" the gold market in America by the "Erie Railroad Ring," headed by Jay Gould and James Fisk. In those long-gone days when the manipulation of capital was entirely free and unregulated by the government, the Machiavellian Jay Gould had conceived the scheme of bidding up the price of gold by heavy buying operations of his syndicate, and thus seizing control of the nation's currency from the United States Treasury. The objective was supposed to have been the engineering of a great inflationary boom that would force gold up to a price of 200 (in terms of greenback dollars), prevent resumption of specie payment against the dollar at the official par value, and raise the prices of all commodities and foodstuffs. To this end the Gould syndicate (as revealed later in a Congressional investigation) had actually made corrupt bargains with high officials of President Grant's Administration.

In Wall Street wild rumors ran about that the available American supply of gold had been "cornered" by the Erie Ring, that the Federal government faced a crisis, and that the Treasury would be forced to take some action, though none knew what it would be. On Friday morning, September 24, 1869—about four months after he had come to New York—Edison was at his post in the balcony of the Gold Room, presiding over the keyboard controlling the Gold Indicator's transmitting mechanism. That morning the leaders of the "gold conspiracy" openly showed their hands by suddenly bidding up the price of gold from 150 to 165. Now the wheels of the indicator operated by Edison moved quite slowly, requiring a certain number of revolutions to register each fractional change in price by eighths, which was enough for normal markets. But on this day of madness the machine fell far behind, and it was afternoon before Edison was able to bring it up to the soaring 165 quotation. By then the whole financial structure of the country was imperiled by the wild gyrations of gold prices.

Edison, while pursuing his own tasks in a spirit of calm detachment, observed how the gold brokers around him seemed suddenly to turn into a mob of madmen. He was an eye-witness to one of the most stirring events in America's financial history. In the memorandum he dictated long afterward he said:

On Black Friday we had a very exciting time with the indicators. . . . New Street as well as Broad Street was jammed with people. I sat on top of the Western Union telegraph booth to watch the surging, crazy crowd. One man came to the booth, grabbed a pencil, and attempted to write a message to Boston. The first stroke went clear off the blank. . . . Amid great excitement Speyer, the banker, went crazy and it took five men to hold him; and everybody lost their head. The Western Union operator came to me and said "Shake, Edison, we are O.K. We haven't a cent." I felt happy because we were poor. These occasions are very enjoyable to a poor man; but they occur rarely.[3]

He saw Jim Fisk, clad in his velvet corduroy coat and cream-colored vest, arrive in a carriage, accompanied by ladies of the theater who were to celebrate his expected triumph, and bearing with him baskets of champagne. Edison also heard telegraph reports indicating that the Gould-Fisk pool seemed to have the entire nation's economy at its mercy on that day. People whose businesses or securities were affected by the rise of gold and the corresponding fall of the dollar were losing millions hour by hour; others were fighting to buy gold, though in fear and trembling. Some of the brokers, he recalled, "had the complexion of cadavers," for all but a few gave themselves up for lost.

Then toward the end of that day the United States Treasury finally took action, selling some of its reserves of gold through its New York bankers. The corner was broken, but now the brokerage firms controlled by Gould suddenly repudiated all contract payments due from them for gold they had purchased that day, and pandemonium broke loose as gold fell back again to 132. "It took us all night to get the indicator back to that quotation," Edison said. Throughout the financial district, offices remained lighted that night, and all hands were at work.[4]

The aspiring inventor, however, took it all easily enough. He, for his part, could never understand or become infected with the mania for speculation, or for "mere money-getting," that seemed to possess the crowds around him. All that he earned he spent as quickly as it was received; that is, he "invested" it in new apparatus to serve his inventive undertakings. He could be only a maker, an inventor—*Homo faber*.

Earlier that summer he had improved the Laws indicator by incorporating in its mechanism an alphabetical wheel and printing device. The Gold Indicator could now offer a full stock quotation service, such as its competitor, Gold & Stock Telegraph—thanks to Callahan's invention—already boasted. That rival company had been bought out by the all-powerful Western Union, the "octopus" of telegraph companies. Rather than contend against Western Union, Laws wisely chose to accept an offer of consolidation and retired from the field with an independent fortune.

At all events, Edison had fulfilled his assignment so well that he seemed to have worked himself out of a job. The head of the new consolidated company, General Marshall Lefferts, to be sure, offered Edison a similar position. But this really meant working again as an employee of a Western

Union subsidiary, and Edison had had his fill of that. He had other ideas; indeed he was on the verge of realizing his long-cherished dream of establishing himself on his own as a free-lance inventor.

On October 1, 1869, *The Telegrapher* displayed a half-page advertisement announcing the formation of the new firm of Pope, Edison & Company as "electrical engineers" and "constructors of various types of electrical devices and apparatus" useful in the telegraphic art. The firm would also undertake to build private telegraph lines, make a variety of scientific or experimental apparatus on order, or carry out tests of instruments and furnish written reports. This is believed to have been the first announcement of a professional electrical engineering service in the United States, the idea of it being credited jointly to Pope and Edison. The two men also had a sleeping partner, J. L. Ashley, publisher of *The Telegrapher,* who donated his journal's advertising space to the enterprise, in lieu of cash. Pope, though under thirty, was a recognized telegraphic expert; he had had some scientific education and was to become, in 1886, one of the founders and a president of the American Institute of Electrical Engineers.

In the interests of economy, it was arranged that Edison should board with his friend Pope at the latter's home in Elizabeth, New Jersey. This, as it happened, was the very region where Edison's ancestor, Tory John, had first settled in America, almost a hundred and fifty years earlier. Part of an old shop was rented in Jersey City, near the yards of the Pennsylvania Railroad, in which Edison was to carry on experimental work. He would rise at 6 A.M. to breakfast with Pope and his wife, catch the seven o'clock train to nearby Jersey City, and stay at his work until nearly one in the morning, when he would take a train back to Elizabeth. This iron regime was maintained all that winter; there was the half-mile walk from the station at Elizabeth to Pope's house, and Edison remembers being nearly frozen on his early morning and nightly walks in the winter of 1870.

What did it matter if it was cold? He was truly busy; he was doing the things he had longed to do. And nowadays he attracted the serious interest of men of skill in his field, such as Pope, and men of affairs, such as General Marshall Lefferts, a former Army telegrapher who had become the Eastern Superintendent of Western Union.

Absorbed in his work though he was, Edison after a time became aware of a horrible stench pervading the Jersey City shop. In a room adjacent to his quarters there was an eccentric old fellow, one Dr. Bradley, an experimenter of the old school, who was one of the first to make precise electrical measurements with the galvanometer. In those early days of electrical science it was believed that one could perform almost any miracle by passing an electric current through an object. During Dr. Bradley's absence, Edison, forever curious, explored his quarters to determine the source of the terrible stink. This original man's hobby, he found, was the artificial aging of raw whisky. The old scientist had filled numerous jars with whisky, connected them with platinum electrodes held in place by hard rubber, and given them a strong charge of current with a powerful battery. The hard rubber, being full of sulphur, had produced hydrogen sulphide, which smelled like rotten eggs.[5]

Despite those hideous perfumes, Edison managed to complete his work on a new type of printing telegraph, or "gold printer" as it was called, which would report by wire current gold quotations and sterling rates. The machine was to be made by Pope, Edison & Company and rented to subscribers, who were mostly importers and currency dealers, through a new commercial wire service company established for the purpose. The rental price, set at twenty-five dollars a week, moreover, was low and offered competition to similar services of Western Union. As so often happened in such cases, Western Union took steps to eliminate this competitive threat by buying up the new service within six months. Pope, Edison & Company realized fifteen thousand dollars by this transaction, Edison receiving five thousand as his share.

He was already flourishing in his trade of inventor, but this was the first large sum he had realized from such work. He had recently begun to send money to provide for his parents, and he now wrote one of his terse, telegraphic letters describing his recent progress:

DEAR FATHER AND MOTHER:
   I sent you another express package saturday, enclosed you will find the receipt for the same.
   I C Edison writes me that mother is not very well and that you have to work very hard. I guess you had better take it easy after this. Don't do any hard work and get mother anything she desires. You can draw on me for money. Write me and say how much money you will need till June and I will send the amount on the first of that month. Give love to all folks. and write me the town news. what is Pitt doing? ...

Your affec. son
THOMAS A.[6]

In the year that followed his arrival in New York, a period of incessant labor, he took out seven patents covering a whole series of innovations in the telegraphic art. Although these early inventions were of minor importance they showed distinctive skills and insights. From these patents one may discern how systematically he attacked the problems set for him. The practical end in view was to improve upon the printing-telegraph apparatus, for which there was now a strong demand. But the electrical impulses to drive the printing wheel and ink it were still weak and left faint impressions in existing machines. Edison solved these problems with much mechanical ingenuity, using small relay magnets and an electrically driven escape wheel (as in a clock mechanism) to obtain more effective transmission.

He could see that even his invention of small parts or refinements for so "commercial" a machine as the printing telegraph possessed a considerable value. At the mere threat by Dr. Laws to introduce Edison's stock quotation printer in competition with Western Union's service, the bigger company had been persuaded to buy out Laws. So also with the "gold printer" service initiated by Pope, Edison & Company. What galled Edison, however, was that while he did almost all the work, his two partners got the lion's share, the silent partner Ashley receiving as much as Edison, though he had not even invested any capital. In the summer of 1870, Edison therefore terminated his partnership agreement with the other two men, in friendly fashion. "I got tired," he said dryly, "of doing all the work with compensation narrowed down to the point of extinguishment by the superior business abilities of my partners."

His real "market," after all, was the giant Western Union Telegraph Company, or its subsidiary, Gold & Stock Telegraph, headed by Lefferts. Organized out of a combination of smaller systems, Western Union had spanned the continent with its wires during the Civil War, then moved on ruthlessly to absorb one regional concern after another, becoming an industrial corporation of 41 million dollars capital, the "mother of trusts." Its capital investment in equipment and plant was low, its tariff rates as high as possible. Among its earlier directors were Ezra Cornell and Amos J. Kendall, who had quickly become multimillionaires, while Samuel Morse, on whose patents the company was originally based, had been given a modest competence.

By 1870 it was noteworthy that Western Union employed a whole team of inventors, among them Elisha Gray, J. B. Stearns, D. E. Hughes, and G. W. Phelps, who was its technical director. Sometimes more than one of these inventors were assigned to work independently, without knowledge of each other, on the same problem, the company, in the end, taking over "the best features" of each one's effort, as the phrase goes. Now General Lefferts, one of the best informed of the Western Union executives, and an engineer himself, decided to bring Edison, the youngest of the new crop of inventors, into the company's stable. For Lefferts had observed that it was Edison who did all the work for Pope, Edison & Company and that he showed a real expertise. This smooth-shaven youth with the big head and broad brows might look like a rustic or a plain mechanic, but one could feel the cunning behind his devices, and it was amusing to hear him talk about his craft.

Though usually reserved or retiring in manner, he had nowadays, despite his deafness, a marked air of self-confidence. His speech, studded with homely illustrations of his ideas, was most pungent. He would address himself to his associates in the following fashion: "The troubles with telegraphic appliances can only be gotten out in the same way as the Irish pilot found the rocks in the harbor—with the bottom of his ship." [7]

One of the worst "bugs" that troubled the early wire-printing services was their way of suddenly "running wild," and printing crazy figures. While employed by Laws in 1869, one of Edison's best tricks had been the contriving of a central mechanism by which the receiving instruments in outside offices could be quickly regulated down to zero, and thus put in unison, from the home office. That was one of the reasons why Western Union had bought out Laws. General Lefferts is reported to have said at the time, "Edison is a genius and a very fiend for work."

After Edison had turned in a whole series of minor inventions to Western Union, he was asked to work on the improvement of the Callahan stock printer, owned by them, which also occasionally ran wild. In order to reserve Edison's entire services for his company, Lefferts now offered to advance him money at regular intervals for the cost of all research and experimentation on their assignments. Edison, who was then at the point of terminating his partnership with Pope and Ashley, accepted the proposal. It was typical of him, at this period, that he did so without specifying what he was to be paid by these gentlemen for his inventive services, or who was to control his patents.

Within about three weeks he was back in Lefferts's office at Gold & Stock Telegraph, and before a group of the company's directors he gave a demonstration of a successful unison-stop device that promptly brought stock tickers in outside brokers' offices into alignment with the central station transmitter, whenever they began to print wild figures. The device made a profound impression on the capitalists; it seemed to them positively uncanny and would, it was obvious, save much labor and trouble. Even Callahan, whose machine Edison thus completed, said afterward in somewhat grudging praise that a ticker without the unison device would be impracticable and unsalable. At this stage of affairs money had been owing to Edison for some time on account of a whole series of telegraphic inventions.

"Well, young man"—thus Lefferts addressed him—"the committee would like to settle up the account.... How much do you think they are worth?"

Edison remembered afterward that he was uncertain about what to demand. He had thought he ought to ask five thousand dollars, but might manage with three thousand. "General, suppose you make me an offer," he said at length.

Lefferts then said, "How would forty thousand dollars strike you?"

Feeling "as near fainting as I ever got," Edison managed to stammer that the offer seemed fair enough. He was asked to come back in three days, when a contract would be ready, and the money also.*

The contract, when it was presented to him, seemed as "obscure as Choctaw," and he signed without reading it. That it conveyed to him a sum of money rivaling in his mind the fortune of the Count of Monte Cristo was, however, plain enough. A check with his name and those immense figures on it was handed him, and he was instructed to proceed to a bank on Wall Street. He had never in his life held a check in his hand. It seemed to him, as he liked to tell the story in later years, that he was dreaming and that his vision would soon vanish. Surely "there was some Wall Street trick about this thing. I had been reading about Wall Street tricks for years.... I had never been in a bank before."

When he arrived at the window of the teller, that official "yelled out a large amount of jargon, which I failed to understand on account of my deafness." Edison retreated, bewildered, heartbroken. For the piece of paper in his hand he had signed away all his rights, and now who would give him fifty dollars for it?

On his return, empty-handed, to the Western Union office, the long face he made provoked much laughter. Obviously the executives there had played a typical New York joke on their "hayseed." What the bank teller had demanded of him before making payment, was some identification, which could have been provided in the first place. Now Lefferts' private secretary accompanied Edison to the bank wicket, and the money was duly paid over to him, in a big pile of ten- and twenty-dollar bills about a foot in thickness, which he could hardly stow away in his clothing.[8]

---

* Although other records of transactions with Western Union are in evidence, none of this one has been found. From the contents of a (preliminary) letter-agreement in Leffert's hand, as if to be signed by Edison, dated October 19, 1870, it seems that the inventor was to receive $30,000, not $40,000, so that he may have remembered incorrectly; part of the consideration was his binding himself to reserve all future inventions in the form of stock tickers for the Gold & Stock company. (Edison to M. Lefferts, ms., October 19, 1870, Edison Laboratory Archives.)

One feels that Edison, in later years, tended to draw the long bow in his stories picturing himself as an innocent lamb among the cunning wolves of Wall Street—a lamb who nonetheless made off with more than one bag of gold or greenbacks. According to his own account, on receiving his precious bundle, he drew his big loose overcoat about him and departed by ferry and train for Newark, where he was living at the time in a dingy boardinghouse. There he sat up all night, unable to sleep for fear of being murdered for the sake of his great hoard. In the morning he returned to New York to inquire about some way of disposing of his heavy encumbrance of legal tender. Friends advised him to go back to the same bank and open a deposit account; which he did, receiving a little book testifying to the transaction. Oh, the paradoxes and mysteries of money! Thomas Edison, who could carry in his head all the intricate design of a complex printing-telegraph apparatus and all its ramifications of wires, relays, and compound-wound magnets, could never understand the mysterious ways in which the money-men worked, with cash or mere credit, their wonders to perform. And yet he possessed his own sort of shrewdness and, one might even say, "a nose for money." Was not his inventive work directed nowadays toward products which could pass the test of a real "commercial" demand? One large money prize appeared, as if by magic, to bring more in its train, until the greenbacks seemed to be raining down from heaven upon the same unwitting head that, only yesterday, as it were, did not know where to lay itself—if not in some cellar beneath one of the great casinos of the money-changers.

chapter VI

"A bloated eastern manufacturer"

1

In the winter of 1871 Edison wandered about the industrial
section of Newark, New Jersey, looking for a good location for
the manufacturing shop he was to establish. He had received
from Western Union orders for 1,200 stock tickers, to be manufactured over a period of several years—orders totaling nearly half a
million dollars. After climbing many stairs in dingy buildings having
"To Rent" signs, he found what he wanted, at his price, on the top floor
of a three-story structure at 4–6 Ward Street. Soon horse-drawn vans
arrived with a quantity of machinery he had purchased. Many mechanics
came to Ward Street in response to advertisements he had inserted in the
newspapers. To their surprise they found the proprietor to be a young
man of only twenty-four. In this district of small shops, it was rumored
that he had struck it rich.

"Mine was too sanguine a temperament," he said, "to keep money in
solitary confinement." Within thirty days he had spent virtually all the
fortune he had just won for equipment of all sorts.[1]

His mood, nevertheless, was of the highest optimism, as shown by his
one letter to his parents in that winter of 1871. He would help them with
much money. But his great anxiety was on the score of his mother, reports
of whose ill health reached him. He had not been home in three years
and did not know when he would find time for a visit. Perhaps he even
dreaded to return and find his mother so sadly changed, her mind clouded
in the last years of her illness.* Her life, in fact, had been filled with
tragedy; three of her seven children had died in infancy, and a fourth,
Harriet Ann, in youth. Tom Edison's letter ended on an apologetic note:

---

* A Port Huron neighbor wrote to Edison May 18, 1878: "I remember when I
used to come about your father's house,
when you were a boy, when your mother
was well, before disease had destroyed her
mind." (Edison Laboratory Archives,
1878, "Personal.")

84

... I have a large amount of business to attend to. I have one shop which employs 18 men and am fitting up another which will employ over 150 men. I am now what 'you' Democrats call a "Bloated Eastern Manufacturer." [2]

Indeed, he longed to see his mother once more—the woman who "was the making of me," as he said—and show her that her hopes in him had not been misplaced. But that winter he was unable to go to her, and after that it was too late. On April 11, 1871, amid the rush of his new affairs, a telegram arrived from Port Huron telling him that Nancy Edison had died two days before. He hurried home to Michigan to attend her funeral, and kneel beside her grave in the cemetery by Lake St. Clair; then he returned quickly to Newark to lose himself in work.

Later he learned that a few weeks after his mother's death, his father, aged sixty-seven, had formed a connection with one of the local dairymaids, aged seventeen, whom he was eventually to marry.

The allusion to himself as a "Bloated Eastern Manufacturer" in his last letter to his parents reminds us that the inventor's background was that of Western agrarian radicals, who still reflected the Jacksonian Democrats' mistrust of the "Eastern money power," and of the monopolistic organizations rising in the railroad and telegraph industries. Sam Edison, the old nonconformist and political firebrand, was always "agin"; he was opposed to Lincoln and the War for the Union, as he had been opposed to the King's rule in Canada. On one occasion, early in the Civil War, Sam had had to run away on his long legs from an infuriated mob of Unionists in Detroit. He hated the Grand Trunk Railway, one of whose locomotives had killed a cow belonging to him (owing, as it was said, to his fences being neglected); for ten years he stubbornly conducted a lawsuit against the railroad for the price of that cow, though in vain. After the war, rural Michigan became a hotbed of the radical Greenback and Antimonopoly movement; but the son of that old rebel Sam Edison was the protégé of the Western Union Telegraph Company, the nation's prime example of an all-powerful monopoly. It then owned 25,000 miles of telegraph wires; its reputed control of newspaper wire services, through affiliation with the Associated Press, also gave it a sinister fame that was augmented by public scandals over its finances and high rates. Small businessmen feared its secret power, and leading members of Congress urged that Western Union should be nationalized and made part of the Post Office system, as had been done in Europe. The company's directors, however, argued that their industry functioned "naturally" as a unified organization, and they reinforced their political arguments, during the General Grant era, by the widespread distribution of complimentary books of telegraph franks to members of state legislatures and of Congress.[3]

Young Edison knew something of the repute of Western Union, to whose services some of his most skillful inventive labors in the telegraphic medium, during the next seven years, were mainly consecrated. It used to be said in those days that the telegraph "trust" held Edison in thrall, just as it did various other inventors and technicians; and there is a core of truth in the assertions of Lancelot Hogben and other popular historians of applied science that Edison gave some of the best years of his life to advancing the efficiency of the monopolized communications system.

At this period after the Civil War, as Henry Adams has said, the whole nation seemed to turn its energies to the task of building up its industrial plant, its factories, its transport net, and its modern communications system. The railroad "barons," driving for monopoly, not only violated the law by engaging in wholesale political bribery, but sometimes were even involved in violent combat with each other's organizations. The telegraph "barons" (often linked with the railroad people) were apparently no less unscrupulous in their drive for complete control of their field. But ruthless and unconscionable though they might be in the pursuit of gain, these new industrial potentates presided over a brilliant period of economic and technological expansion. America needed heavy industries and a continental railway system and got it in its own reckless fashion. Likewise the country needed a unified telegraph network, then most vital to all commerce—the "nerve" of industry. The job of Thomas A. Edison and other technicans was to see to it that the system was the most efficient that current scientific skill could produce.

In its heyday, the great Western Union was an outstanding example of the marriage of capital and applied science; it was indeed the principal initiator of the new electrical industry in this country. Among all the capitalist groups who participated in America's industrial progress, Western Union's directors distinguished themselves in the two decades following the Civil War by subsidizing scientific invention more actively than did any other industrialists. The company encouraged inventors by paying them; but while thus sponsoring their work it also took control of their products. The valuable patents of Edison, as of other men, became the property of the giant telegraph company. And when Edison went off to Newark to manufacture stock printers for the company, Marshall Lefferts saw to it that the young man took as his partner one William Unger, said to have been Lefferts' business associate.[4] The firm name was, therefore, Edison & Unger.

If Edison's condition was that of a "captive scientist"—as was sometimes said of him in the early stages of his career—it is doubtful that he worried about it very much at that time. His ruling idea was to function in his own field, to do his own part as well as he could, and to be paid for it. Toward the political and social issues involved he felt then a detachment amounting almost to indifference. In any case, the relations of inventor and patron, like those of the artist and patron, are often ambivalent. While it was good to have financial supporters (and Edison had had his troubles finding such) they could also be irksome; the motives of the scientific worker and the money man were by no means the same, as Bentham had said. In the privacy of his notebooks and correspondence Edison sometimes gave vent to his dislike of "small-brained capitalists"—as if in revenge for their repeated complaints against "harebrained" inventors.

Now began an intensely active and happy period of Edison's life, lasting about five years, when he worked at the Newark shops as both inventor and manufacturer of diverse electrical products. This period also, especially in its earlier phases, was one of schooling for him. Although his main task was to perfect—often by simplifying—the crude inventions of others, his telegraphic work required scientific and technical knowledge and a growing experience of the industrial arts. Beginning thus as a sort of master technician who refined upon, or added minor innovations to, others' devices, he developed within two or three years, with increasing resourcefulness, into an original creator with a flair for truly strategic inventions.

At Edison & Unger's he began assembling about fifty men to turn out stock tickers, serving as their foreman himself. In hiring workers Edison looked above all for skilled craftsmen, advertising especially for men who had "light fingers." The electrical industry was then in its infancy; the workers he first drew into it had had previous training mostly as clockmakers and machinists.

There was, for example, John Ott, a youth of twenty-one whom he had engaged as an assistant at the smaller shop he had in Jersey City. When Ott was first interviewed, Edison pointed to a heap of the disassembled machine parts of a stock printer on the floor and said, "Can you make this machine go?"

Ott said, "You needn't pay me if I don't." When he was done, Edison exclaimed impulsively, "Here, you come and take charge of this place." Ott thereafter served as a sort of assistant foreman at Newark.

A more remarkable mechanic still was Charles Batchelor, a genial, black-bearded young Englishman who had been sent to the United States to install special machinery at the Clark Sewing Thread Mills in Newark. Having decided to remain in this country, and hearing of the new electrical equipment factory, he came to Edison and was promptly engaged when he demonstrated his manual skill and his ability as a mechanical draftsman. Another hard-working and well-loved member of the original force was John Kruesi, a Swiss clockmaker, who had undergone a long and varied apprenticeship in Europe and who could construct almost any form of instrument or machine he might be called upon to make. Then there was Sigmund Bergmann, a German, who could then scarcely speak English, but who showed himself a keen and diligent mechanic. When Bergmann's want of language was remarked upon, the young master exclaimed, "What difference does it make? His work speaks for him!" Bergmann and another of Edison's Newark mechanics, Johann Schuckert, in later years returned to their native Germany to found two of the largest electrical manufacturing concerns in Europe, the latter becoming head of Siemens-Schuckert of Berlin.

These skilled men made up the core of the original crew who learned to manufacture electrical equipment under Edison, who was himself still in the learning stage; they, in their turn, trained great numbers of other workers. They made products that were both solid and finely contrived, which Edison always delighted in. Though their wages were average, usually on the piecework basis, they worked hard and long hours under the command of him whom they already called the "Old Man."

He was only twenty-four, younger than most of his men, yet he directed his people and assumed complete responsibility for everything going on in his plant. For Edison by now had a sense of the uniqueness of his inventive talent. Like his obstinate ancestors, he was decidedly the egoist, but one who knew how to ally the devotion of other men with his ego drives, for he had an intuitive knowledge of human psychology.

John Ott's first impression of young Edison was of one "who was as dirty as any of the other workmen, and not much better dressed than a tramp. But I immediately felt that there was a great deal to him." [5] He was like no other shop boss those men had ever known; approachable to all and sundry, full of humorous sallies, and yet, in the next breath, very severe with them, and bent on keeping them "hustling" perpetually.

His intuitive powers and his endless curiosity about everything also amused and intrigued his co-workers. If he were suddenly struck by some new idea, all hands would be shifted from making stock printers to something entirely different. Or a visitor might come in with a new product —the first crude model of a typewriter by Sholes was brought to Edison's shop in Newark—and he would forget everything else to experiment with this mechanical novelty and determine whether it could not be improved. His shop was a theater where the scenery was constantly changing.

It was said of Pasteur that he showed "great impetuosity" in his experimental work. Now that he was his own master in his own shop, Edison likewise showed an irrepressible enthusiasm, like that of a child. He kept his men at work long hours, but he himself would work even longer, and his own dedicated efforts were a challenge to them.

When undertakings of unusual moment absorbed him, he would go about his shop "laying bets with the men" or offering them prizes to continue their work without stopping until some experimental model he had assigned them was brought to working order. When the results of some prolonged period of labor were unusually pleasing to him, he might declare a holiday and invite all hands to go fishing with him. At moments of good fortune he showed a "joyous nature," or indeed "a boyish hilarity," as one of his assistants said. One day, after having completed successful negotiations with some New York capitalists, he "entered the workshop with a whoop, fired his silk hat into an oil pan, and was preparing to send his fine coat after it, when someone laughingly pinned him down...." [6]

He could be the complete eccentric, too, in his business methods. "I kept only payroll accounts. I kept no books. I preserved a record of my own expenditures on one hook, and the bills on another hook, and generally gave notes in payment. The first intimation that a note was due was the protest; after that I had to hustle around and raise the money. This saved the humbuggery of bookkeeping, which I never understood, and besides, the protest fees were only one dollar and fifty cents."

His allergy to bookkeepers had begun, he claimed, after he had employed one to balance his accounts at the end of a twelve-month period and learned that he had a clear surplus of $7,500. Thereupon, in the highest spirits, he announced that there was to be a "jamboree" for all hands in his shop. But the day before the party he looked over the bookkeepers' report himself and had some afterthoughts, recalling certain large recent expenditures that were not included. After a whole night of figuring, he arrived at the chilling certainty that there was no surplus, but a *deficit* of several thousand dollars. The order for the festivities was therefore abruptly canceled. Henceforth Edison held to the belief that life was simpler without the torment of double-entry bookkeeping. [7]

From time to time, after neglecting his contractual work for Western Union in favor of some other experimental undertaking, Edison would find himself out of funds. Then he would turn back to making more of his Universal Stock Printer, working his people on two shifts, night and day, to realize cash.

On one occasion in the early 1870s he had a rush order for thirty thousand dollars' worth of the Edison stock printers, then being manufactured according to a much improved model he had lately developed. But these new instruments turned out to have "bugs" in them and refused to perform as they were supposed to. Edison then called together a half dozen of his devoted helpmates, among them Batchelor, "Honest John" Kruesi, Bergmann, and Ott, and informed them that he intended to shut himself up with them in the laboratory on the top floor of the Ward Street factory until they had located the cause of the trouble. "Now, you fellows," he declared determinedly, "I've locked the door, and you'll have to stay here until this job is completed. Well, let's find the 'bugs'." [8]

During sixty hours of physical and mental exertion, without sleep and with little food, they all stayed at their task until the new machine ran smoothly again. His laboratory, as Edison later told the story, was thus "turned into a prison." Several of the workers had affectionate wives who pined for their husbands and came to the barred doors, wailing, or tried to convey parcels of food for them through the windows. It was of no avail; the young master was relentless.

In recalling those busy years of his youth in Newark Edison used to say, "At least I did not have *ennui*."

2   Only a few months after he had organized the shop at Ward Street, he set aside a room on the top floor as his laboratory, which he provided with far more elaborate equipment than he had ever possessed before. During much of the day he might be busied with his duties as a manufacturer, yet he could not give up the habit of experimental investigation in many different departments of applied science. As before, curiosity drove him to research; to this he now brought a more extended knowledge, based on much reading as well as practice. At Newark, where he was his own master, experimentation became a daily and systematic, rather than intermittent, pursuit. From the summer of 1871 he began to keep a laboratory notebook, setting down almost day by day the exact record of his investigations, ideas, and even random reflections. On its first page he wrote:

Newark, N.J., July 28, 1871
This will be a daily record containing ideas previously formed, some of which have been tried, some that have been sketched and described, and some that have never been sketched, tried or described.

Among the first entries are the subjects of work on hand, all relating to printing telegraphs and type-wheels, such work being done under contract with Gold & Stock Telegraph over the next five years. But he stipulates:

Reserving to myself any ideas contained in this book which I do not see fit to give to said Gold & Stock Telegraph.[9]

There follow descriptive notes accompanied by rough sketches for a variety of proposed electrical inventions, including one of a multiplex telegraph, which he designates as being for his own exclusive use and profit:

Invented by & for myself and not for any small-brained capitalist.[10]

In the years at his Newark shops, Edison the electrical instrument manufacturer gradually developed into one of the most accomplished technicians of the telegraph industry. The Edison Universal Stock Printer, improved by him over a period of several years and incorporating the earlier patents of Callahan and others, came into use in almost all financial offices and security exchanges in Europe as well as in America, speeding up the process of all speculation and investment.

When the devices of others were brought before him for inspection, it was seldom that he could not contribute his own technical refinements or ideas for improved mechanical construction. As he worked constantly over such machines, certain original insights came to him; by dint of many trials, materials long known to others, constructions long accepted, were "put together in a different way"—and there you had an *invention*. This would follow usually upon an extended period of patient observation and testing that ended with some act of insight coming "suddenly," as it appeared, but in reality derived from an accumulated body of technical knowledge stored in the searcher's subconscious mind.

Because he worked at this time in a highly competitive field, often adding his innovations to the patented devices of other men, the idea got about that Edison did a good deal of borrowing and "using" of other inventors' work. But those inventors also built upon the many investigations that others had done before them. It may be said that all invention is wrought within the continuum of man's total technical and scientific knowledge; hence the regular recurrence of parallel or simultaneous invention and of identical scientific discoveries by different men in different lands.

Edison's first sponsors noticed how, given some other inventor's product to analyze and improve, he usually completed his task promptly, or within a surprisingly short time. In the year 1872 alone he brought out thirty-eight patents for new models or new parts of the stock ticker; there were telegraph type-wheels, various printing devices, relay magnets, unison-stops, and automatic telegraphs. In 1873 he brought forth twenty-five more of such patents, several of them being of marked originality. Our patent laws, to be sure, allowed the issuance of letters of patent for minor improvements or refinements upon existing mechanisms, which was an encouragement to continued technological advance.

Even those early variants of the telegraph, as one of his old-time laboratory assistants wrote long afterward, sufficed to place him above the average inventor of his time.

One can see with what refinement he devises compound-wound electro-magnets, and how he creates local circuits through relays; and with what nicety he solves the problem of shifting mechanisms and unison regulation. He displays cunning also in the way he neutralizes or intensifies electro-magnets, applying strong or weak currents, and commands either negative or positive directional currents to do his bidding.[11]

At twenty-four his growing professional repute brought many an important entrepreneur to his Newark shop. Not long after he had begun making stock printers, a young railroad and telegraph engineer named Edward H. Johnson turned up at the Ward Street shop bringing with him a model of a new type of automatic telegraph—invented by one George D. Little—which Edison was asked to examine and test. Johnson explained that he was acting as consultant for a group of financiers who were trying to promote Little's apparatus. But he himself had found the machine disappointing in its performance and had been unable to improve upon it; after inquiring in Washington and New York for some expert who might help him, he had been directed to Edison as a "young wonder," who was doing great work for Western Union.

Little's device used a moving paper tape having perforations corresponding to Morse dots and dashes; when fed into a transmitting instrument the tape's perforations controlled signals that were sent by wire to a receiving instrument at the far end of a circuit, where the same impulses were printed automatically as dot and dash signals on another ribbon of paper tape. Over a short line, the apparatus apparently worked rapidly and well. In tests over a circuit of two hundred miles distance, however, Little's automatic instrument had been found to print its messages all too slowly and indistinctly.

Edison tested the machine thoroughly over wires of the Pennsylvania Railroad and diagnosed its weaknesses. Instead of being faster than the ordinary Morse manual sender, it proved to be extremely "sluggish" because it was highly subject to electrostatic interference. The needle at the receiving end dragged or smeared things. Then Edison, after some study, made a number of proposals for correcting the defects of this apparatus. The mechanism, in his judgment, had splendid possibilities, and he appeared to Johnson both confident and very clearheaded in his plans for its "cure."

The group that had purchased Little's patent had already formed a stock corporation named the Automatic Telegraph Company to promote the invention. They had been on the verge of abandoning the whole project, in which no one would take any stock, when Johnson delivered an ecstatic report on the young inventor in Newark and on his plans for making their machine practical. Whereupon the company's directors hurried over to see Edison and, on April 24, 1871, entered into a contractual agreement with him. One of the directors of Automatic Telegraph was George Harrington, formerly Assistant Secretary of the United States Treasury, but reputed to be the financial agent of the sinister Jay Gould. Another director was Josiah C. Reiff, former newspaper correspondent and currently Washington lobbyist for the Gould-controlled Kansas & Pacific Railroad.

These people were all for technical progress, on the one hand; but on the other (though Edison knew nothing about this) they were in hopes that a smart piece of invention could help them carry out a good stock-jobbing venture. Even at this early stage of Edison's career, it was being said of him, as by Patrick Delany, a well-known expert in telegraphic instruments: "Edison's ingenuity inspired confidence, and wavering financiers stiffened up when it became known that he was to develop the automatic." [12]

Harrington, on behalf of the Automatic Telegraph Company, advanced Edison the sum of forty thousand dollars on account for an assignment of all rights connected with the invention of automatic telegraphs. Edison, thereupon, in partnership with Joseph T. Murray, of Newark (as Edison & Murray), opened another shop in Newark, spent most of the money he received on special equipment and material, hired more men, and set to work. One of his stipulations was that Edward H. Johnson, the railroad engineer, a vivacious and well-spoken character, be assigned to assist him for the duration of this project. Johnson said afterward that he had come only to consult Thomas A. Edison but remained to work with him as an associate for more than twenty years.

Edison's methods of investigation at this period were described by Johnson as requiring complete knowledge of everything that had ever been done before in such a field as automatic telegraphy. Books and periodicals describing all that Wheatstone and Bain and others had attempted in England, and indeed all records of previous experiments in automatic mechanisms were acquired. Johnson relates:

There were numerous theoretical solutions in French books, but none of them enabled him to exceed the rate of 200 words a minute. . . . I came in one night and there sat Edison with a pile of chemical books that were five feet high when laid one upon another. He had ordered them from New York, London and Paris. He studied them night and day. He ate at his desk and slept in a chair. In six weeks he had gone through the books, written a volume of abstracts, made two thousand experiments . . . and produced a solution, the only one that could do the thing he wanted.[13]

A great part of the autumn of 1871 was devoted by Edison to the automatic, or high-speed, telegraph. The transmitting end worked rapidly; it was at the receiving end that trouble was encountered, mainly in the form of electrostatic interference. After much searching, Edison discovered that by using as a shunt around the receiving instrument a coil of wire enclosing a soft-iron core, he brought about a striking improvement in speed over long circuits. Self-induction would produce a momentary reversal of the current at the end of each impulse, so that each signal was very sharply defined; thus the stylus of the receiving instrument was made to function accurately and no longer smeared.

In tests carried out during the winter of 1872 over lines between New York and Philadelphia, he was eventually able to transmit one thousand words a minute in Morse code signals. Meanwhile, he had also greatly improved the chemical preparation of the paper tape at the receiving end —with a cheap ferric solution. Then he perfected the mechanism of the punching machine that made the perforations in the tape. On August 16, 1872, a patent application covering these improvements and additions to the Little apparatus was filed and assigned to George Harrington.

It was Edison's most skillful performance so far and clearly foreshadowed the early development of a commercially practicable automatic of tremendous speed. Yet he was not satisfied. A big idea had come to him, one that had great logic, and he must work it out at all costs. Since no human being could receive the messages coming at such speed over the lines, why not make the receiving instrument entirely automatic? Why send messages by an automatic transmitter in dots and dashes when they might be printed out at the other end of the line in Roman characters by combining the automatic telegraph with the mechanism of his own Universal Stock Printer? In his laboratory notes for the day he writes:

At the present writing I am inclined to believe that I am the originator of the idea of using perforated paper for controlling or working printing telegraphs. September 15, 1871. Witnessed by Joseph T. Murray [Edison's partner].

He pushed on with his work, commenting on each experimental trial with expletives such as "N.G." or "not good enough," or "damn," or some plain Anglo-Saxon term in four letters. A happier occasion would be commemorated by notes such as:

A bully experiment by T. A. Edison, assisted by Charles Batchelor.... The sentence we took: 'Now is the winter of our discontent'—with this we got 250 words per minute, but counting five letters to a word we got 228.[14]

By the winter of 1873 the Automatic Telegraph people had reason to be jubilant over Edison's progress. He had performed more than he had promised at the beginning of his investigations. The completed models of the new automatic instrument incorporated at the dispatching end of the system a machine for punching holes in a strip of paper, somewhat resembling a typewriter. A perforated tape, running over a metal cylinder, permitted electrical contacts to be made or broken through the perforations, thus flashing telegraphic signals to the far end of the circuit and guiding the receiving instrument there so that it printed letters on tape.

Before he finished this task he gave the new system tests over long lines for 120 successive nights. The messages were usually dispatched by Edison from a loft on lower Broadway, in New York, to Washington, where Johnson and Patrick Delany handled the receiving instrument. "Each succeeding night Edison opened another bag of tricks," as Delany recalled. On a number of occasions Edison delivered his automatically transmitted messages as far as Charleston, South Carolina, something of a feat in those days of poor wire. In these experiments he also used some of the newly improved telegraph wire, known as "compound wire," which was made of a steel core with copper ribbon wound spirally around it. Harrington later testified (in a suit of equity) that in every instance Edison had made good his promises of specified improvements.[15] His development of Little's imperfect apparatus "made high-speed telegraphy over long lines possible, and gave great impetus to the telegraph industry...."[16]

The automatic had certain limitations: the punching of characters in perforations at the transmitting end had to be done manually, at the speed of manual telegraphy, or twenty to fifty words a minute; it was only then that lengths of combined tape could be fed into the automatic transmitter at two hundred or more words a minute, or about six times as fast as manual dispatching and receiving by Morse code.

An economic revolution of some sort, seemingly, impended for the telegraph industry toward 1873—when Edison & Murray were given orders to start manufacturing the automatic apparatus. It was at this point that the mysterious Jay Gould, already known as "the destroying angel of Wall Street," stepped from behind the scenes and took over from Harrington *et al.* the ownership of the Automatic Telegraph Company, together with its new telegraphic patents that now promised such remarkable achievements.

After his conspiracy to corner gold, and his profitable manipulations of the stock of the Erie Railroad—then at war with the Vanderbilt-owned New York Central—the calculating and ambitious Gould had determined to force his way into the rich telegraphic field. Along the several railroads he controlled, almost spanning the continent, Gould began to build up an independent and competing telegraph system whose real object was to harass and raid Western Union's empire, until they would be forced to pay him his price—nothing less than a controlling share of that company's stock. By reimbursing Harrington, the principal stockholder of Automatic Telegraph, for his payments to Edison, and by ultimately acquiring that company, Jay Gould also acquired the services of Thomas A. Edison and rights to his automatic patents, a powerful pawn in the campaign for the conquest of the nation's telegraph system. Of all this grand design the exalted young inventor probably knew less than anyone else.

3   By 1871, when Edison had first appeared in Newark as a young man of capital, with a top hat and a Prince Albert coat, the former tramp telegrapher certainly seemed to have come up in the world. But there were arrears in terms of private life of which he was increasingly aware. Prosperous though he was, he still lived in a furnished room in Newark; he had neither a home of his own nor a wife. It was said of him in those days that he was "timid in the presence of women"—but clearly not through lack of interest in them. The Edisons were a virile clan. That scapegrace father of his—whose shennanigans, oddly enough, always amused his son—when nearly seventy sired three more children by his young wife.

Three years earlier, when he was in Boston, the Western Union manager had given Edison, one day, the assignment of demonstrating the Morse telegraph before the students of a female academy. Edison went off with Milton Adams and set up a circuit in the auditorium of the school, which was a most elegant establishment. As he had come directly from work he was roughly clad; being also unaccustomed to female company, he was consternated when the door opened and the head mistress led in, no mere children, but twenty full-blown members of the senior class, the handsomest and most beautifully gowned young ladies he had ever seen. According to Adams, Edison "looked as if he were going to faint." At any rate his tongue clove to the roof of his mouth, and his sight swam. The two young men were accustomed to the uninhibited social manners of railroad and telegraph workers, but here they were lost. Edison, at the moment, was on the little stage adjusting his sounder, while Adams was engaged with wiring at the back of the room.

As Edison related in his own version of this affair, he managed to signal to Adams in code, beseeching him to take the floor. But Adams was so weak-kneed that, as he came forward, he fell over an ottoman, which set all the young ladies tittering. Thereupon, seeing how helpless Adams had become, Edison, after an interval of vocal paralysis, proceeded with his discourse and, once plunged into his subject, "talked and explained better than I ever did before or since." [17]

From that day on, whenever those proper young ladies happened to meet Edison on the street, they made a point of greeting him with polite nods or smiles—for he was quite a personable lad—but as this led to his being much ridiculed by his fellow workers, he dared not pursue the acquaintance.

Science, alone, had been his chosen mistress for long years. But now he found himself exposed to the electromagnetic effects of daily contact with a beautiful young girl who worked in his shop. Her name was Mary Stilwell and she was then barely sixteen years of age; but she was tall and full of figure, and had a great pompadour of lovely golden hair. More and more often, his mind was distracted from his experiments by the person of Miss Stilwell.

The accounts of how it all began are conflicting; according to that of his laboratory assistant, W. K. L. Dickson, she was "a member of the inventor's working force when they first met, and retained always an affectionate interest in her co-workers." [18] Another account holds that the inventor first chanced to meet Mary and her older sister Alice when, in the midst of a summer downpour, they took refuge in the doorway of his shop just as he was going out. He addressed them politely, invited them to come in out of the rain, noticed the beauty of the younger sister, and made it his business to find out her name and seek her acquaintance. The two sisters, he learned, were of a poor but respectable family, and taught Sunday school. Then, after he met her again, the story goes on to say, the young inventor offered her employment in his shop for the pleasure of having her near him. She was one of a number of women engaged in punching perforations into telegraph tape; but Edison did not look at the others. She was then

...a rather demure young person who attended to her work and never raised her eyes to the incipient genius. Edison would stand nearby observing her as she drove down one key after another with her plump fingers, until, growing nervous under his prolonged stares, she dropped her hands idly in her lap, and looked up helplessly into his face.[19]

On one such occasion (as the people in the shop noticed) when Edison had been standing beside her saying nothing and just breathing down her neck, Miss Stilwell, after having borne this as long as she could, exclaimed timidly, "Oh, Mr. Edison, I can always tell when you are near me!" [20]

His courtship was conducted in his own odd way and with his own "characteristic humor," as his laboratory assistants later testified.

Though he had scarcely ever spoken to her before, one day he gave her a sudden smile and inquired rather abruptly, "What do you think of me, little girl, do you like me?"

"Why, Mr. Edison, you frighten me. That is—I—"

"Don't be in a hurry about telling me. It doesn't matter much, unless you would like to marry me."

He had a strange pallor; his eyes were steel-blue, very large, very unusual eyes that some called "fisheyes."

The young girl was both frightened and disposed to laugh at the same time. But Edison went on impetuously, "Oh I mean it.... Think it over, talk to your mother about it and let me know as soon as convenient; Tuesday, say. Next week, Tuesday, I mean." [21]

To Edison's advancing deafness may be attributed some of his odd mannerisms, such as investing the young woman as closely as possible in order to hear her better. But in recalling these amorous maneuvers of his first romance, he has told how he slyly turned his very weakness into a tactical advantage:

Even in my courtship my deafness was a help. In the first place it excused me for getting quite a little closer to her than I would have dared ... to hear what she said. If something had not overcome my natural bashfulness I might have been too faint of heart to win. And after things were going nicely, I found hearing unnecessary.[22]

Soon Edison and his partner Murray were seen riding out in a carriage to call on Mary and Alice Stilwell at the Sunday school—though the inventor cared nothing for church—and at her home as well. When he duly asked her father for her hand, he was told that he would have to wait about a year because of her extreme youth. There is a tradition that

Thomas Alva taught his Mary the Morse code, so that while sitting with her in the parlor, in the presence of her parents, he conversed with her in complete privacy by tapping on her palm with a silver coin. They did not wait out the year, but within a few weeks, on Christmas Day of 1871, they were married at a small family ceremony in Newark. After the wedding lunch Edison brought his bride to the solid private house at Wright Street, Newark, which he had bought a few days before their marriage.

They were no sooner alone together than some defective stock tickers began preying on his mind. He related that "Just about an hour after the marriage ceremony had been performed," he could think only of those stock tickers! "I told my wife about them and said I would like to go down to the factory. She agreed at once. . . ." His claim was that he returned around dinnertime.

The more famous legends about Edison's wedding day, that have been handed down in the family, touch upon his habit of complete forgetfulness when working on some technical problem. It seems that he stayed on at his laboratory throughout the afternoon of his wedding and even far into the night, while his young bride, frightened and tearful, repined alone in her strange new quarters. It was twelve when his friend Murray rushed into the laboratory, exclaiming, "Tom, what are you doing at this late hour?"

"What time is it?" Edison asked vaguely.

"Midnight!"

"Midnight," he said in a dreamy way. "Is that so? I must go home then, I was married today." And with that he seized his hat and went off.

The next day the couple left for Niagara Falls. According to Mary Edison's daughter, who was born a year later, it seemed her mother "felt so young and inexperienced that she insisted that her elder sister Alice company her on her honeymoon trip." [23]

Their first child, fair and blue-eyed, was named Marion, after Edison's eldest sister. In honor of the telegraph which occupied all his days, he also nicknamed her Dot. The next child, Thomas junior, arriving in 1876, was nicknamed Dash—though on the ground of vivacity the titles should have been reversed. In 1878 a second son was born who was given the name of William Leslie.

For a brief season, reform and a measure of order and comfort were brought into the inventor's private life. On Sundays he usually managed to rest and often played with the children, becoming wildly gay with them and mischievous almost to the point of cruelty. But as often as not he would be entirely distrait, his mind completely separated from his charming wife and his infant children.

He often returned from work late for supper and, when his work was very pressing, not until the early hours of the morning. Sometimes he would be away from home for two or three days, sleeping on a bench or cot in his laboratory—as in the famous sixty-hour "prison-shift," which occurred during the first year of his marriage. Then he would appear at last, disheveled, pale, and too utterly worn to speak to anyone; and soiled as he was, he would throw himself upon a couch or bed in all his clothes.

Like one possessed, he would carry in his head the entire plan of some new and elaborate invention, in all its complex details, for days on end. He had the gift of total recall. His memory was so extensive that he would work out many aspects of a difficult problem in his mind, oblivious to his surroundings, forgetting the time, the place, and even his own identity. In short his ability to divert all his faculties from less important considerations and concentrate them on the mental work at hand, his very power of memory made him at times distracted or absent-minded in the extreme.

In the early years of his marriage he was much obsessed by the problems of multiplex telegraphy, which required that he imagine many different mechanisms operating simultaneously in different ways. While thus engrossed, he was once notified that unless he paid his real estate taxes in Newark the next day (the end of a term of grace), he would be compelled to pay a large additional fine. To comply with this stern order he hastened to the city hall the next afternoon and took his place at the end of a long line, with a hundred persons ahead of him. During the wait that followed, the brain-twisting problem that haunted him soon diverted his mind, and he became lost to everything around him. The last hour had struck; he finally found himself before the officer collecting taxes who, having addressed him without receiving any response, said to him gruffly, "Now young man, look sharp. What is your name?"

Edison said later:

I had lost my composure completely, and all recollection of my name as well, for I stared at the official behind the counter in blank perplexity and answered, "I don't know sir." Jumping to the conclusion, I suppose, that he had an idiot to deal with, he waved me impatiently aside. Others poured into my place, the fatal hour struck, and I found myself saddled with an extra charge of twelve and a half per cent.[24]

Mary Edison was scarcely the woman who might have "improved" him. A gentle and affectionate being, whose head was as yet only slightly furnished, she found herself wedded to a "great man" such as she had never imagined. He was older than she by almost ten years, strong-willed, and often transported or exalted by ideas she could never understand. She could be only his "sweet companion." Indeed she never required that he alter his way of life to conform with the needs of domesticity, but submitted entirely to his wishes. One of Edison's early associates wrote of Mary Edison, "She was a helpful spouse; she revered her husband and thought him almost a god." [25]

For Thomas Alva there were no hobbies nor recreations. On the many nights when her husband was away Mary Edison had only the company of a few women friends and her sister Alice, who lived with the Edisons for several years prior to her own marriage. In this family the rule was that "father's work always came first." [26]

Mary Edison, then, could not "manage" her husband; she could never make a stand against him. He would show a fierce temper when aroused by opposition. Several years after he was married, he happened to write his father inviting him to come and stay at the Edison home in New Jersey. One feels that since his childhood Edison's attitude toward his father was ambivalent—he resented him, and he also retained a lingering affection for the wayward old fellow who was much bigger physically than he was, and who used to thrash him. It seems, however, that his young wife did not enjoy the company of her father-in-law and had indicated as much. Edison thereupon assured his father, "My wife does not nor never can control me, and you can have anything that I have." [27]

In the early months of marriage the enraptured young inventor no doubt made some attempts to explain to his young wife something of what he was doing and what he aspired to do. But he must have found soon enough that he could not draw her attention to his problems. In his laboratory, in the early days, he often thought of her in moments of revery, and as he worked with his notebooks before him, sometimes scribbled affectionate nicknames for her. Thus on one page, among notes of various electrochemical tests, there is the spontaneous and revealing phrase, "My wife Popsy-Wopsy can't invent." [28]

4   The Automatic Telegraph Company had been negotiating for some time with the British postal authorities for the sale of English rights to Edison's inventions. The prospects of sale seemed promising; but it was thought that only Edison could see to it that the apparatus was properly demonstrated. Thus, on April 29, 1873, in response to the pleas of his associates, the inventor gathered together a trunkful of equipment, kissed his young wife and infant daughter good-by, and hurriedly sailed for London on a journey lasting about six weeks.

On this, his first transatlantic journey, he thought nothing of the beauty of England in the spring; but only of means of overcoming the difficulties in the way of a full demonstration of his automatic instrument. Invited to make a test of a London-to-Liverpool circuit, and to record messages at the rate of one thousand words a minute, he was allowed only some old wire and "sand batteries" (obsolete even then) with which to connect and operate his instruments, although the apparatus needed good conduction and high power.

Edison was discouraged. At his small hotel near Covent Garden, the dull fare of roast beef and fried flounders further depressed him. "My imagination was getting into a coma." After finding a French pastry shop in High Holborn Street, however, his imagination improved, and he went off to see the London representative of the Automatic Telegraph Company, a Colonel George Gouraud, to whom he applied for help. Somehow Gouraud, a breezy man with influential connections, managed to obtain for his use a powerful battery of a hundred cells that had been used by John Tyndall at the Royal Institution. Thus the unknown American inventor was able to pass a strong current from the sending machine in Liverpool to the receiving end he had set up at the Telegraph Street headquarters in London—where, as he claims, the Automatic recorded swift-running messages in an imprint "as clear as a copper plate."

He was then asked by the British authorities if he could adapt his automatic for use on submarine cables and obtain a greater speed than by regular methods. There were some twenty-two hundred miles of cable (intended for a line to Brazil) then stored under water near the docks at Greenwich; and there he went to set up his apparatus and make the test. He first sent a dot, and it was recorded, not as a character a thirty-second of an inch high, but a smear "twenty-seven feet long! If I ever had any conceit it vanished from my boots up," he said in recollecting the experiment. Try as he would, during two weeks of nighttime labor, he could send no more than two words a minute. Edison's admitted mystification at such phenomena has suggested that he did not fully understand the theory of electrical self-induction. What he did not know at the time was that a coiled cable was far more subject to self-induction than one laid out straight, since the cable wire tended to retain, as an electrical field, a portion of each electrical impulse passing through it.*

* Three years later, Lord Kelvin, visiting America for the Centennial Exposition at Philadelphia in 1876, first inspected a model of Edison's automatic telegraph and gave it exhaustive tests. In his reports to the Exposition's jury, he recommended Edison's instrument for an award "as a very important step in land telegraphy. The electromagnetic shunt with soft iron core, invented by Mr. Edison, using Professor [Joseph] Henry's discovery of electromagnetic induction in a single circuit to produce a momentary reversal of the live current when the battery is thrown off . . .

Gouraud came down one night to visit Edison at the lonely cable works by the Greenwich docks and at daybreak breakfasted with him at a dreary inn nearby that was frequented by longshoremen. The place was infested with roaches; the coffee and cake were all too evil-tasting for Gouraud. "He fainted. I gave him a big dose of gin and this revived him. . . . He lost all interest in the experiment after that, and I was ordered back to America." [29]

Things moved slowly in the British Isles. Though the English were among the first to give recognition to young Edison's accomplishments, their Post Office eventually adopted another automatic telegraph apparatus that was a variant of the Little-Edison instrument. Edison always felt that, in this instance, they had taken whatever was needed from his system without paying him or his sponsors; but there were a dozen or more British inventors having prior patents in this very field, so that, as often happens in such cases of concurrent technical research, absolute priority is hard to determine.

He returned to the United States in the last days of June with little to show for his long hard voyage. Western Union had also refused to adopt his automatic system, its managers at this period having no faith in it. But Jay Gould soon had it put to the test on the long lines of the Atlantic & Pacific Telegraph Company. Some of the major cities on its lines, after installing the new apparatus, attained tremendous speeds. A long message of the President of the United States running nine thousand words, for example, was dispatched in one hour and two minutes. According to J. D. Reid, an early historian of the telegraph industry:

The Atlantic & Pacific Telegraph Company had twenty-two automatic telegraph stations in major cities, such as Buffalo, Chicago and Omaha. . . . The through business during nearly two years was transmitted in this way. The perforated paper was prepared at the rate of twenty words a minute. Whatever its demerits, this system enabled the company to handle a much larger business than it could otherwise have done with its limited wires.[30]

Nevertheless the automatic was to be abandoned several years later, or used thereafter only as an auxiliary. To Edison it was a grievous disappointment as well as a mystery that telegraphy in America should be allowed, at the will of a Jay Gould, to go back to hand transmission and remain that way for some thirty or forty years thereafter. Nor did Edison

---

is the electrical secret of the great speed he has achieved." He also remarked on the originality of Edison's solutions in handling other problems in automatic telegraphy, by means of his improved perforator, contact maker, and special chemical solution of ferrocyanide of iron.

(after 1871) receive any further money rewards or royalties for "three years of very hard labor" devoted to Gould's organization in connection with his invention.

What Edison did not then realize, except dimly, was that the decision as to the commercial acceptance or refusal of inventions, and much of the control of industrial technology, turned not upon the question of merit or usefulness, but upon the outcome of intermittent wars or peace negotiations between the rival "barons" in the railroad and telegraph fields, such as the Goulds and the Vanderbilts. (After 1871 the Morgan-Vanderbilt group played a leading role in Western Union.) In this strange financial warfare, this ceaseless maneuvering and intriguing of the money lords of the Gilded Age, Edison, the technician, would be unwittingly and hopelessly involved—nay pulled hither and thither, as he tried to serve first one side and then the other, until he himself would scarcely know where he was, or if he owned his own soul.

On his return to Newark, from England, he had found his young wife in a state of great mental distress. Thanks to his peculiar ways of doing business—spending what should be reserve funds, and securing credit by postponing payments—the sheriff had threatened to close down Edison's Ward Street shops and put up a red flag; he promised also to dispose of Edison's machinery, and even his laboratory equipment, unless creditors were paid.

Once more, in 1873, the country was in the throes of one of its periodic Great Depressions. Hundreds of thousands were being dismissed from their jobs. The New York Stock Exchange itself suspended business for ten days. The Wall Street magnates who had found it profitable to employ Edison were now stony to him. During his absence Mary Edison, by appealing for help to one of Edison's partners, Joseph Murray, had managed to raise enough money to delay the sheriff and stave off disaster —for the moment.

In this perilous time Edison exerted himself with unheard-of energy to develop new products and raise funds. From day to day, for many months, he faced a desperate situation in which all might be lost. The crisis prolonged itself into 1874. "I was paying a sheriff five dollars a day to withhold judgment entered against me in a case which I had paid no attention to . . ." he recalled.[31]

In such times the lot of the free-lance inventor was not an easy one. He had some new ideas for what might prove to be brilliant inventions in the electrical communications field. If he could only come through with one of them he would be afloat again. And in truth, it was by his wits that he was saved.

chapter VII

The inventor and the "Barons"

1   In 1873 America was on the threshold of an era of unrivaled material progress. The resourcefulness of her engineers and inventors, among whom Edison was one of the younger figures, was to be bravely displayed only three years later at the great Centennial Exposition in Philadelphia. Here were giant Corliss steam engines of 1,400 horsepower, two crude examples of the electric arc light, George Westinghouse's air brake for railway trains, automatic printing and multiplex telegraphs by Edison, new gas stoves, fifty-ton locomotives, and new devices for cultivating the land, mining the earth, and making shoes, cloth and buttonholes. Surely if men of past times could have foreseen the coming of such engines, mechanisms, and workshops, as Henry George wrote, they would have imagined that "all these muscles of iron and sinews of steel" would abolish drudgery, make the laborer's life a holiday, and realize "the golden age of which mankind have always dreamed."

The country, nevertheless, was in dire straits in 1873; neither its natural riches nor the industry of its population spared it the "hard times" which periodically recurred. At such periods capital seemed to disappear, and not even the strongest of financiers seemed to have money to invest in the wealth-creating discoveries of inventors.

Edison at this time was on the verge of disaster. It was always difficult for him to manage the exact costing of new products, partly because of his peculiar ideas of bookkeeping. Nor did he easily brook criticism or interference by partners and associates. After having opened a third small factory in Newark in 1871, in partnership with George Harrington, for the manufacture of automatic printing telegraphs, Edison had designed an improved model of this instrument, hired men, and arranged for its production. But within a few months he discovered that Harrington had placed a man above him as superintendent of the Edison & Harrington shop. This was more than Edison could take and he walked out. As he told Joseph Murray, he had terminated his partnership with Harrington because "he felt like an old coat that had been used until it was worthless and then hung up." [1]

At about the same period, in the summer of 1872, he also wound up the firm of Edison & Unger, which had been producing stock tickers, giving Unger $5,000 in cash, and notes for $7,100, in satisfaction of all claims. His partners in business ventures had their troubles owing to Edison's unorthodox and whimsical procedure; he, on the other hand, often became suspicious and angry, believing (with good reason) that he was being gulled by some of them. Meanwhile heavy debts weighed upon him again in those lean years.

Fortunately, Harrington offered to patch things up with him, promising that he would not interfere with Edison in the future, or allow him to be interrupted while engaged in experimental work. On these terms Edison resumed working for Harrington in 1873 and permitted himself to draw small sums of money on Harrington whenever he was hard pressed. His only regular business partnership now was with Joseph Murray, whom he trusted.

One of a series of minor inventions with which he strove to keep himself afloat was that of a district messenger call box system. Along with E. A. Callahan, Edison was one of the early arrivals in this field. In the early seventies he formed a concern called the Domestic Telegraph Company, which made and rented signal boxes of his design. Such devices gained a passing popularity because they seemed to offer protection against fire, burglary, and sudden illness. Sensational newspaper stories of how a deranged maid servant in a mansion on Brooklyn Heights had held a woman and child prisoner, without food or water, for twenty-four hours, threatening to kill them if any one entered, caused much excitement at this period. This brought in several hundred subscribers to the signal-box service Edison had launched, with the result that he and his backers were soon able to sell it out for a few thousand dollars to the Automatic Telegraph Company.

Another minor, though diverting, invention of 1874 was the Edison Electric Pen, a by-product of the perforating machine he had devised a year or two earlier. The pen was operated by a small electric motor of the impulse type, powered by a battery of two wet cells. As he described it in his patent application it "rapidly punctured a sheet of paper with numerous small holes, filling such holes with a semi-fluid ink, and pressing this same upon the surface to be printed . . . against a steel plate." [2] The advantage was that after letters or records had been thus written in perforations, many copies could be made. The pen cost eight dollars, the battery five, and a roller three dollars. In the days before the Remington-Sholes typewriter came into commercial use, the electric pen (later modified into a stencil pen) achieved a considerable popularity, some sixty thousand being used in government and business offices for rapid copying. On a few occasions Edison even ventured to appear in public and give demonstrations of the working of the pen. It had a tiny motor, about the size of an egg, fastened to its top, and a long wire running to the battery. This is said to have been the first small electric motor ever manufactured and sold in large quantities. In using this device many Americans received their first lessons in applied electrical science.

While experimenting with chemically prepared telegraph tape Edison tried paraffin paper; he may well have been the first to perceive how effectively such material could serve for stenciling purposes. He also relates that "toward the latter part of 1875 I invented a device for multiplying copies of letters called the Mimeograph." It was a simple machine for automatic stencil duplication using paraffin paper, and he later sold it for a modest sum to A. B. Dick, of Chicago. Out of it grew one of America's largest office equipment industries.

He was, above all, prolific. A score of new devices or refinements upon others' inventions would be under way in his shops at the same time. In the seven years after he came to New York he applied for and was awarded two hundred patents, and in consequence he was described as "the young man who kept the path to the Patent Office hot with his footsteps." But while small sums ventured by small businessmen helped him in the initial development stages, there were usually long delays before profits were realized from these minor inventions.

Though the times were bad, his heart was set on a truly big invention, that of a multiplex telegraph that would surpass all that had been done before in this field. After about a quarter century of exploitation the Morse circuit was still essentially a single-track affair operating at the speed of manual sending; but traffic swelled year by year, and it was difficult to keep pace with it.

The importance of the telegraph in 1850 to 1880 (before the day of the telephone), the mighty impetus it gave to the American economy, can scarcely be measured nowadays. It was commonly said that it "broke down isolation" and "conquered distance"; it hastened the rescue of many who met with some disaster, or who needed prompt medical attention, while performing a vital service to all trade, including agriculture. Yet control of this key industry fell into the wrong hands at a period when inventors,

Edison in particular, could have extended its usefulness and cheapness. "It was not in itself wholly a force for good," a scholarly historian, Dr. R. L. Thompson, concludes in his modern survey of its early history. Its control was to be linked to the control of great railway systems, newspapers and press wire services; and thus it became in time, as Thompson relates, a means to "foster selfish monopolies detrimental to the interests of the American people." [3]

Edison, however, thought only of realizing the full technical potential of the telegraph. He had great visions of adding "phantom wires" to the single Morse sending and receiving circuit, doubling and quadrupling volume, while increasing the speed of the operation with the aid of his automatic. The idea, as we have noticed, had haunted him since he was eighteen, working as a second-class operator in Cincinnati and Louisville. Among the noted inventor-scientists who had worked on this problem since the 1850s were Sir Charles Wheatstone and Lord Kelvin in England, Werner Siemens in Germany, and, lately, J. B. Stearns in America. Edison thoroughly familiarized himself with all their devices by reading abstracts of their patents. Yet he felt that discontent with the current solutions which is one of the strongest stimuli to further inventive struggle. None of the others had really mastered the problems involved, not even Stearns, who had finally made the duplex telegraph operable. The opportunity was still open to invent a truly multiplex telegraph. What was galling to the impetuous Edison was the technical conservatism of those who controlled the telegraph industry. Though the telegraph was his first love, the financial and scientific "reactionaries" in the end would drive him from the field.

Shortly after the Civil War, William Orton had become the president of Western Union; though not a telegrapher, he was a man of considerable education and of an impressive, even puritanical, mien. In former times he had shown business ability in publishing religious books on a large scale. His methods in running the Western Union monopoly, however, were not noticeably more refined than those of his predecessors, who had brought disrepute upon the company. The distribution of complimentary books of telegraph franks to politicians was continued—as before, in thirty-seven states and nine territories—and it paid off politically, as Orton frankly avowed. [4]

Armed with a letter of introduction, Edison had first approached Orton in 1871, to suggest that Western Union back his experiments in duplex telegraphy. But nothing was done for him; a year later the company, instead, bought the patent rights to Stearns's duplex service. Then, early in 1873, Edison again wrote to Orton, saying he had had some further thoughts on the duplex that might be of value. Orton thereupon invited

Edison to come to his office in the granite headquarters of Western Union at 145 Broadway and asked him what he could do to improve on Stearns's devices. The truth was that the Stearns apparatus, though workable, had not turned out to be a profitable affair. In discussing it with Orton, Edison referred to it with scorn.

He had a good deal of the Gascon in him; on occasion he would expand, he would swagger even in the presence of the imposing President Orton. "He treated the business of making the duplex as a very trifling affair," Orton testified. "He said he could make a dozen, he could make a bushel. 'Very well,' I said, 'I'll take all you can make—a dozen or a bushel.'" A few days later, Edison was back with a little book of drawings, containing twenty-one sketches of a duplex machine, saying of one of them, "Here is one you can have for five dollars. I made it the night before." Was he expected to avoid infringement on Stearns's patent, he asked? Orton replied that if he could make improvements without infringing on Stearns that would be just as well. Edison then asked that he be allowed to use the facilities of the Western Union headquarters so that he could make experiments over their wires, and that several telegraphers be assigned to help him. This was agreed upon, with the understanding that such inventions in the duplex as developed from these experiments be sold by Edison to Western Union.[5]

All the winter and spring of 1873 the young inventor worked like a madman at a whole series of duplex devices in a large room in the basement of the Western Union building in New York, often staying there all through the night when the wires were free. Edison recalled afterward: "I brought my apparatus over and was given a separate room with a marble tiled floor, which, by the way, was a very hard kind of floor to sleep on. . . ."[6]

The Western Union building was heavily guarded. Edison wanted to feel free to come in at all hours to test his contrivances. "Want order go in W.U. night to feel the pulse of my patients," ran one of his telegrams to Orton's secretary.

It was in his character to report the difficulties he encountered with frankness. He ran down a whole series of duplex telegraphic "combinations" and, though admitting they were not workable yet, suggested they might have a "negative value" for Western Union, as protection against others who might turn up with a new duplex. He wrote:

If I run across another duplex I will take steps to confine it to the patent office immediately, so that duplex shall be a patent intricacy and the intricacy owned by the W.U.

EDISON [7]

Although Tom Edison invariably posed as a country boy who knew little about the ways of the wicked world, he already appreciated how inventors' patents, even if without operational value, could be used as pawns or bargaining counters by the moneyed patrons of applied science.

He was working, however, on a new principle, entirely distinct from those of his predecessors, such as Stearns, whose device could send two messages on one wire only *in opposite directions*. Early in 1873, Edison (according to Orton) asked "if a duplex that sent *two messages in the same direction* would or would not be as desirable as one that sent two messages only in opposite directions." Orton replied that "it would be more useful, as with streetcars, everybody wants to go downtown in the morning and uptown at night." To be able to double traffic flowing in one direction when needed would be a vast improvement over Stearns's device.

After two months Edison announced promising results:

I experimented 22 nights—tried 23 duplex systems, 9 failures, 4 partial success, 10 all right, 1 or 2 bad—Several experiments made on Washington wire after heavy rain—played-out wire. . . .

Or again he reported tersely that he had tried seven "makeshift instruments," and that "six of them work charmingly," while the seventh was "a satisfactory failure." But there were fifteen more combinations to be tested.

Another of his communications runs:

I have been sick in bed—have had the most interesting features of 4,000 nightmares in the daytime. Cause: root beer and duplex. EDISON [8]

At some time during the series of 1873 experiments, when Edison discussed them with Orton and George B. Prescott, who was the chief engineer of Western Union, the idea came up that "if he succeeded in sending two messages in one direction, why would it not be as easy to duplex both [transmitters] as it would be to duplex one? And then there would be four?" Edison expressed the opinion that if one could be done the other could be done. Thus the real possibility of the *quadruplex*, which he had vaguely outlined for himself in earlier years, arose again and fairly haunted him.

His laboratory notebooks of 1873 contain a rough drawing (undated) labeled by him "Fourplex, No. 14," and bearing the interrogatory remark in his handwriting, "Why not?"

As he told an associate some years later, none of the difficulties he encountered in later experiments, even with incandescent light, proved as tantalizing as the quadruplex telegraph. He said:

This problem was of the most difficult and complicated kind, and I bent all my energies to its solution. It required a peculiar effort of the mind, such as the imagining of eight different things moving simultaneously on a mental plane, without anything to demonstrate their efficiency.[9]

In the spring of 1873 Edison interrupted his experiments on the "diplex" to make his first trip to England, in an attempt to demonstrate and help sell patent rights to the British for his other important invention, the automatic printing telegraph. Just before leaving for England on April 23, 1873, he filed application for a patent (issued later as No. 162,633), which covered all the work he had done thus far toward perfecting a new type of duplex—or rather, "diplex," as he called it.

On his return from England, he found the panic on, the sheriff after him, and no money in sight. When he tried to resume his series of experiments on the "diplex" telegraph at the Western Union headquarters, he could not find Orton in—for he was away on a long business trip—and could not interest any of the other executives in his experiments. Western Union itself was having financial troubles; its stock was being pounded down in the market. For the time being the Western Union people seemed to have lost all interest in putting money into new inventions.

Finding himself in extremities Edison had gone back to work on other projects for the smaller, rival telegraph company headed by George Harrington and Josiah Reiff. Either he did not know or did not care that these men were really the agents of Jay Gould, the unconscionable war lord of American finance; nor did Edison in the least suspect that he himself was soon to become a central and controversial figure in the "great telegraphic war" that now began between Gould and the Western Union interests.

2   Wall Street fully believed that Jay Gould might go under during the Panic of 1873. But out of the depths of depression he rose again, with the mastery of a whole group of big Western railways in his hands: the Kansas & Pacific, the Missouri Pacific, the Wabash, and even the Union Pacific, which made up a sprawling system almost spanning the continent. Along its tracks he gave right of way for telegraph lines to his own Atlantic & Pacific Telegraph Company, of which the Automatic was a subsidiary. The Gould-controlled telegraph venture was smaller and less profitable than Western Union but gave it increasing competition and cut down its rival's revenues.

Gould's motives in making war upon the Western Union from 1874 on were explained by him as deriving from his conviction that

The telegraph and the railroad system go hand in hand, as it were, integral parts of a great civilization. I naturally became acquainted with the telegraph business and gradually ... kept increasing my investment.[10]

Indeed, his carefully laid schemes involved capturing not only railroads, but also large newspapers and press wire services, thus to control the very intelligence going to all financial markets—and manipulate them at his will—thanks to the telegraph, the "nerve" of industry. With superb effrontery Gould publicized his war against the giant telegraph company as "a crusade against monopoly." His strategy was directed at weakening the financial position of Western Union in order to capture it more easily. He attacked it in the stock market, offered intense competition through his rival company, and then sought ownership of Edison's valuable new patents as a means of defeating the enemy. While engineers and inventors like Edison strove to play a creative, constructive role in society by applying their scientific skills, a Jay Gould functioned mainly as a "disturber of the economy," as Thorstein Veblen defined his role.

Early in 1874 there took place in the "telegraphic war" an early skirmish in which an Edison invention was used by one side as a weapon against the other. The Automatic Telegraph Company boasted of Edison's new automatic printing telegraph as giving it a vast superiority over Western Union, which then had only manual sending and receiving. The Gould-controlled company thereupon challenged the other to a speed contest in which the merits of their different systems would be demonstrated to the public. On one day Automatic Telegraph arranged to transmit an 11,130-word message by President Grant over a single wire, by means of Edison's new printing telegraph. To send this elephantine dispatch from Washington to New York, Ed Johnson, who was in charge of the operation, employed ten clerks to punch paper tape perforations recording the Presidential message; only two Morse operators were needed to run the automatic sender; but thirteen copyists were required to take down the copy as it came to New York. This whole job was performed in the record time of sixty-nine minutes.

On the same day, Western Union raced against the Automatic Company, using old-fashioned manual methods, but opening up eight lines between Washington and New York and employing eight expert operators in each city. They managed to record the whole Presidential harangue in seventy minutes flat. Edison's automatic telegraph had won out—but only by an eyelash—and at the higher cost of twenty-five personnel against

Western Union's sixteen. The speed contest, as reported in the country's newspapers, revealed the bottlenecks that handicapped the rapid automatic printer at the tape-perforating and copying end, where the work had to be done manually.

Edison's interest in the outcome of the conflict between the "barons" was somewhat academic. (Alternately he did work for both sides.) What disturbed him was that Western Union, during the depression and its struggles with Gould, showed a loss of interest in backing his new work—and this at a time when he was desperate for money. Thus, a good many months passed during which the tantalizing problems of the "diplex" and quadruplex were allowed to rest.

But they continued to obsess him. He wrote to Orton complaining that he had received no further cooperation from Western Union officials when he tried to renew these experiments in multiplex telegraphy; but he got no reply. Then, in May, 1874, he wrote to George B. Prescott, chief engineer of Western Union:

You will probably think it strange that I have done nothing with the duplex. The fact is Mr. Orton's sudden disappearance took the bottom out of my boat, as I can do nothing without his or your cooperation.[11]

After having studied the problem a good deal in his mind and having worked out some new combinations, he was ready to go to work again, but he wanted to use the facilities of Western Union and eight expert telegraphers as assistants. He now begged that Prescott intervene in his behalf. To gain his support, Edison offered Prescott a partnership in his invention-in-progress. Though the other man was no inventor, his personal influence with the Western Union might be decisive in arrangements to complete the necessary experiments.

Now Orton reappeared and gave orders that the fullest cooperation be extended Edison; he also approved of the idea of Prescott becoming Edison's partner—though Prescott was his subordinate and a high officer of Western Union, to whom the invention, if successful, would be sold. The conduct of both Prescott and President Orton, who advised him, reflected low standards of business ethics even for those Robber Baron times. It was later stigmatized in court by enemies of Western Union as showing the intention of fraud against Edison.

His own recollection of the affair in some measure bore out this view of the case:

I wanted to interest the W.U. Telegraph Company [in the quadruplex] with a view to selling it, but was unsuccessful, until I made an arrangement with the Chief Electrician of the Company, so he could be known as joint inventor and receive a portion of the money. At that time I was very short of funds and needed it more than glory. This electrician appeared to want glory more than money, so it was an easy trade. . . . [12]

In a suit brought in 1877 by the Jay Gould interests against Prescott, Western Union, *et al.* (for control of the Edison quadruplex patents), evidence was introduced to show that Prescott had no real part in the inventive work, aside from giving Edison some advice and aid in conducting his experiments.

After the deal with Prescott, experiment on the new type of duplex conceived by Edison went forward more rapidly, at Western Union's expense and with the help of its personnel. To the basement room at 145 Broadway were brought Edison's "queer collection of instruments" that were such a puzzle to ordinary telegraphers at the time. As one of his assistants described the proceedings, these would be set out on the floor in everybody's way until

Along about midnight Edison came in and, gathering up his paraphernalia, began to arrange it by connecting up the various parts with a fine copper wire which he unwound from a small spool that he produced from his pocket. He was our companion by day and by night for nearly a week, during which time he never went to bed or had any regular meals. When he was hungry he visited a coffee and cake establishment in the neighborhood for what he pleased to call the Bohemian Diet; and, returning with an unlighted cigar between his lips, he would begin his experiments anew. After a while he would throw himself into a chair and doze, sometimes for an hour. . . . He used to say that while thus napping, he dreamed out many things that had puzzled him while awake.[13]

By early summer Edison was sending jubilant notes to Orton.

After an almost infinite amount of experiment I have a duplex working in shop —two messages in the same direction. . . . I want the loan of three duplex sounders and one Phelps relay for a week. I am ready to put the new duplex in operation between New York and Philadelphia.

And again, at about the same period, he announced joyously:

I have struck a new vein in duplex telegraphy. Think 't will be a success.
Two messages can be sent in same direction.
In opposite directions.
Way stations can work on it. . . .

The design of the "quadruplex" was now complete—indeed the very word which was destined to become famous was coined by Edison in

1874. At one of his tests Edison rigged up two rooms constituting the near and far ends of the circuit, with a wire running out to Albany, a hundred and fifty miles distant, connecting them; it worked fairly well. "Boys, she's a go!" he exclaimed to his assistants. But it was still a new thing and full of "baffling perversity" at times. "It must be used," he said, "to find out where the bugs are." He paused significantly. The eight telegraphers working under him enjoyed hearing the exuberant young man, and waited politely for him to continue. Then he said suddenly, in his tone of badinage, or playful cruelty—or of both, mixed— "You don't seem to *tumble*. Every man jack of you is fired after today." [14]

The turning point had come when Edison got the two messages going in one direction on one wire in that summer of 1874 and combined his own apparatus with the Stearns system, thus transmitting, all together, four messages on one wire (two in each opposite direction) at a rate of 142 telegrams per hour.

Edison, Prescott, and Orton then met at Orton's office, and Edison signed a preliminary agreement dated July 9, 1874, in which he designated Prescott as his "coinventor," the patent for the quadruplex to be filed jointly in their names. If the invention proved on further testing to be successful, Western Union was to buy the patent rights from Edison and Prescott on conditions it usually accorded to inventors, with any disputes that might arise over actual terms to be settled by an outside arbitrator.

Now, Edison, for want of money, had had some involved business transactions at about the same period with the rival group in which, it was charged later, he had *also* promised to sell rights to the multiplex telegraph invention to George Harrington. Before closing the affair with Orton he therefore sent his friend Joseph Murray to tell Harrington that he intended to dispose of his new invention to Western Union. Harrington, extremely angry, made vehement objections to Murray, saying that "other parties" associated with him in backing Edison would never consent to such an arrangement. The "other parties," not named, were obviously the enemy of Western Union, Jay Gould, and the independent telegraph system he controlled.

A small news item reporting the successful trial of Edison and Prescott's quadruplex instrument and its assignment to Western Union appeared in the New York *Times* for July 10, 1874. The night before, Harrington, having received this news in advance, sent a frantic letter by messenger to Edison:

Midnight. I returned this afternoon. Having learned what was going on, have been all evening investigating ... *Beg* of you to see me before you sign any more papers. Come to 80 Broadway. I am in hopes I can relieve you. At the moment, adverse action will cause a loss of $100,000.[15]

Up to this time Edison had received no money from Western Union for the all-important quadruplex experiments. He had, however, been drawing advances from Harrington and the Automatic Telegraph Company for performing other tasks for them. Josiah Reiff, another of Jay Gould's henchmen, and president of the Automatic, testified later that he had strongly urged Edison not to take any money from the Western Union people, because his past agreements with the Harrington-Reiff group would be in conflict with these new commitments. From the tangled state of his business as a free-lance inventor, dealing as he did with the two rival interests in the telegraph industry, arose the oft-repeated charge that Tom Edison had "harum-scarum notions about his contractual obligations." [16]

Meanwhile, he was desperately in need of ten thousand dollars to save his home in Newark. His friend Murray is reported to have said to him on such an occasion, pointing to his golden-haired child Marion, "If you don't watch out that daughter of yours will be a *pauper!*"

Beside himself with anxiety, the impecunious inventor rushed to President Orton's office to beg him for a loan of ten thousand dollars, against a mortgage on his house and shops or against an assignment of remaining rights to his other concurrent invention, the automatic printing telegraph. But Orton coldly turned him down, holding that there was not sufficient security for such a loan and advising him to go to the Automatic people. Moreover, Orton considered the automatic telegraph a "cripple." Edison then obtained the needed ten thousand dollars from Reiff, according to Reiff's later testimony, the money being provided from the treasury of Automatic Telegraph and its parent company, Atlantic & Pacific.[17]

But the money received from Reiff had really been paid at the order of Jay Gould, who at the time remained behind the scene. Several years later, Gould, in reply to a letter of Edison's complaining of the treatment he had received from Gould's organization, reminded the inventor of the money he had given him in the summer of 1874, saying:

I have always felt kindly toward you and assisted you financially, when but for it you would have been ruined, at least so you told me.[18]

These payments, however, were not made out of charity; as far as can be judged from Edison's highly raveled accounts, they were for a new type of telegraph relay. For, just as he was completing work on the quadruplex, the versatile inventor was called upon by the Gould forces to help them in a new battle of the "telegraphic war."

In the summer of 1874 Western Union struck a powerful blow at Gould's company, instituting an injunction suit in which it was charged that the Gould telegraph network was infringing upon patents for the Page relay owned by Western Union. The Page relay was then the only known means for stepping up, that is *relaying*, messages going over long-distance telegraph lines. Essentially it was a standard telegraph receiver, the armature of which acted as a switch in another electrical circuit. Its function was to repeat signals, however weak, over the next section of line by means of renewed battery power.

After long litigation, the assignment of Page's patent to Western Union had been upheld in the Federal courts. Now, by obtaining an injunction for infringement of patent rights against Gould's lines, Western Union hoped to close down all of its adversary's sounders, throughout the country, for seventeen years.

Gould's henchmen fought a delaying action in the courts. But time was running out; in the crisis that faced them, they soon called in Edison. Rescue us, they cried. He must now drop everything else and improvise or invent some substitute, "some means of evading Page's patent," in case Western Union's claims were finally sustained in the higher courts. Edison promised that he would attack the problem that very night.[19]

At the time, something long forgotten flashed back into Edison's mind. In his Newark laboratory, two years earlier, he had discovered, while experimenting with chemically prepared paper for his automatic telegraph, a most curious phenomenon: that a strip of moistened paper, passing under the telegraphic stylus while transcribing a message, ran more quickly and with less friction whenever the current passed from the stylus to the paper, than when no current passed. If he rubbed the stylus with a piece of moistened chalk, he noticed how the friction diminished even more abruptly when the current passed through, and how it increased sharply when the current was cut off.

In the summer of 1874, being confronted with the new problem of evading Page's patent, he recalled the action of that moistened chalk on the telegraphic stylus. "I substituted a [drum-shaped] piece of chalk rotated by a small electric motor for the magnet [of the Page relay], and connected a sounder to a metallic finger resting on the chalk. It made the claim of Page worthless." [20]

What he had here was a differential friction device, in place of the spring and electromagnet relay. At the make or break of the current, the yielding surface of the chalk disk slipped back and forth, making and breaking contact in the secondary circuit and communicating signals precisely as fast as electric impulses passed through the primary circuit. Patent for this device, which Edison called the "electromotograph" or sometimes simply "motograph," was applied for on August 13, 1874, and was granted as No. 158,787, in January, 1875. His chalk-drum apparatus, Edison claimed, could work any form of relay, sounder or telegraph printer.

We have the recollection of an eye-witness, a telegrapher in a downtown office of the Atlantic & Pacific system, of the day in the autumn of 1874, when by court injunction all the keys were silenced—and Edison suddenly appeared, grinning confidently, to introduce his new device:

First he detached the Page sounder from the instrument, an intensely interested crowd watching his every movement. From one of his pockets he took a pair of pliers and fitted [his own motograph relay] precisely where the Page sounder had been previously connected, and tapped the key. The clicking—and it was a joyful sound—could be heard all over the room. There was a general chorus of surprise. "He's got it! He's got it!" [21]

The industrial monopolists at that stage had not settled the boundaries of their "empires" (as they have today) and were carrying on the fight to the bitter end, for the stakes were high. But then into the breach would come Edison, a sort of scientific soldier of fortune, who in an emergency could improvise any engine needed—in order to "evade" some patent. His services to Gould were of inestimable value; in fact he had saved the Gould telegraph empire from extinction.*

* In recalling the incidents of the "telegraph war" for his authorized biographers, F. L. Dyer, T. C. Martin, and W. H. Meadowcroft, authors of *Edison, His Life and Inventions* (1910, and 1929), Edison gave them an entirely misleading account of the affair; in that account the roles of the different actors were completely reversed. In reminiscences of events dating back forty years he indicated that the "respectable" Western Union people were "saved" from the unscrupulous Jay Gould by means of the invention of the chalk-drum relay. As we have noticed he was actually estranged from the Western Union management for almost a full year in 1874–1875 and derived his income principally from Gould and his subordinates.

Nevertheless the incorrect account of this affair has been widely repeated in almost all existing books on Edison, as in G. S. Bryan's *Edison, The Man and His Work,* where it is remarked in conclusion: "The spectacle of unscrupulous force confounded by applied science is not displeasing" (p. 78).

In the 1880's, when these incidents were still fresh in his mind, Edison told W. K. L. Dickson: "during the controversy between competitive telegraph companies *one of the companies* owned the Page patent," which controlled the use of a magnet acting as a relay. The other company "was therefore compelled to purchase the 'motograph' patent from me." He did not say then which of the companies it was. Evidently he did not care to admit at the time that it was the piratical Jay Gould whom he had rescued—Gould, who later swindled Edison. The narration of his transactions with Western Union and Jay Gould has been completely reconstituted here in accordance with the sworn testimony recorded in *Atlantic & Pacific v. Prescott, Western Union, and Serrell,* New York Superior Court, 1877.

Nevertheless, in the lean summer of 1874 Edison had only ten thousand dollars from Gould on account of the chalk relay—just enough to save the mortgage on his home in Newark—and nothing at all from Western Union. Though he felt bitter at the Western Union officers because of their delay in compensating him, he continued to work hard over the installation and testing of the new quadruplex apparatus on their long lines.

Later that summer preparations were made for a full-dress exhibition before a board meeting of Western Union. Edison relates:

Under certain conditions of weather one side of the quadruplex would work very shaky, and I had not succeeded in ascertaining the cause of the trouble.... The day arrived, I had picked the best operators in N.Y. and they were familiar with the apparatus. I arranged that if a storm should occur and the bad side got shaky to do the best they could and draw freely on their imagination. They were sending old messages—about twelve o'clock everything was working fine, but there was a storm somewhere near Albany—Mr. Orton, the president, and Wm. Vanderbilt and other directors came in. I had my heart trying to climb up around my oesophagus.... But the operators were stars, they pulled me through. The N.Y. Times (get it) came out next day with a full account....[22]

On October 14, 1874, a published report of Western Union showed that the quadruplex device had been in operation over part of its lines, and that results thus far indicated it would double the company's existing facilities and yield economies amounting to many millions. Yet Orton still delayed closing the transaction and paying Edison, who was desperate.

In a letter of early November to his father which bespeaks his mental torment, Edison writes, "Everything is approaching a climax, and I expect to get my money from Automatic and Western Union within six weeks." [23]

As the big deal with Western Union seemed about to be closed, Edison, who was still harassed by creditors, asked Orton for a temporary advance, sending him what Orton termed one of his characteristic and peculiar requests, "I would like 10, 9, 8, 7, 6, 5, 4, 3, 2, one thousand, as you may choose to advance." Orton put his finger in the middle, suggesting five thousand, and Edison agreed.[24]

But it was not enough; the money had to be paid out immediately to creditors. A week later Edison was back urging Orton to draw up their final contract for himself and Prescott. The company, he pointed out, had over 25,000 miles of wire which could be profitably quadrupled. "It will create 50,000 miles of wire in addition." There would be large savings, of which a fraction, some 5 per cent, should adhere to the inventor as royalty. Edison therefore asked for $25,000 down, in behalf of Prescott and himself, certain royalties to be paid him on each quadruplex instrument in use, and $25,000 in final payment after six months. The letter was signed by both Edison and Prescott. Orton afterward claimed that he informed Edison he had to leave for a trip of several weeks through the Middle West and would close up the whole matter upon his return. Edison's version is that he was told nothing and, in fact, was led to suspect that he would never get anything at all for his pains. As he recalled these memorable events in his life:

At that time General Superintendent of the W.U. was General T. T. Eckert. It seems that there was great friction between Eckert and President Orton. . . . (Eckert was getting ready to resign.) One day Eckert called me into his office and made inquiries about money affairs. I told him Mr. Orton had gone off and left me without means and I was in straits. He told me I would never get another cent, but that he knew a man who would buy it. . . ." [25]

A few days later the news circulated in Wall Street that the able young inventor, Thomas A. Edison, had suddenly gone over with his entire bag of inventions to the camp of Jay Gould and the rival Atlantic & Pacific Telegraph Company. When Orton returned from Chicago in January, 1875, Edison could not be found. There was panic at Western Union, whose stock was falling in the market to its lowest levels in recent times.

3   Edison had a quadruplex set at his Ward Street shops in Newark. On the evening after he had made his strange proposal to the inventor, General Eckert suddenly appeared in Newark accompanied by a small black-bearded man wearing an overcoat whose collar was turned up to conceal all but his dark eyes. It was Jay Gould himself; this was the first time Edison had ever laid eyes on the Machiaevelli of American finance. The quadruplex apparatus was exhibited and explained, and the two men departed without making any promises.

The next day Eckert sent for me [actually a week later] and I was taken up to Gould's house. . . . In the basement, he had an office; it was in the evening and we went by the servant's entrance as Eckert probably feared that he was watched. Gould started in at once and asked me how much I wanted. I said— make me an offer—then he said, I will give you $30,000. I said I will sell any interest I have for that money, which was somewhat more than I thought I could get. The next morning I went with Gould to Shearman & Sterling's office and received a check for $30,000, with a remark by Gould that I had got the steam yacht *Plymouth Rock,* as he had sold her for $30,000 and had just received the check.[26]

A few days after this secret transaction with Edison, the conniving General Eckert, an old-time telegrapher who had risen from the ranks, resigned from Western Union to become the president of Gould's Atlantic & Pacific Telegraph Company.

Gould had been informed by the inventor about his contract with George B. Prescott as "coinventor" of the quadruplex. The great financier expressed the opinion that this was an outrageous fraud upon the real inventor; he had been "entrapped," but Jay Gould would "save him." In any case Edison conveyed to Gould (actually to one of Gould's agents) only his own half interest in the quadruplex invention in return for those thirty thousand "pieces of silver." Gould, the head of the rival Atlantic & Pacific, now held equal ownership rights with Western Union's Prescott in the patent on the vital quadruplex invention.

Gould was the last man in the world that Orton, Vanderbilt, and company would have cared to have as their partner, for obviously he would seek only to obstruct and torture them in the courts. No one was more adept than he when it came to harassing one's financial opponents by litigation; he retained for such work the nation's strongest lawyers and continued until the enemy cried for mercy and paid up. It was said at the time that, when the news came out that Edison and the disgruntled Eckert had been lured into Gould's camp, the stock of Western Union fell sharply under heavy "short-selling" by Jay Gould, so that he was believed to have gained in a single day twenty or thirty times that which he paid the young inventor.

The famous "telegraphic war" between the Gould and Vanderbilt interests raged for about seven years, between 1874 and 1880, and was fought mainly in the courts. Western Union opened hostilities by entering suit to have the assignment to Gould of Edison's patent rights in the quadruplex canceled by court order. Before the suit was begun, however, President Orton had made belated offers to Edison to pay him what had been agreed upon, to which the inventor had answered in a letter of January 26, 1875, that he "rescinded" his agreement with Prescott and Western Union.

As a free lance Edison had a right to sell his professional services or patent rights to anyone he pleased, even to a financier who was notorious for his betrayals of associates in Wall Street deals. Edison owed no special loyalties to Western Union, which despite his great efforts on their behalf had thoughtlessly delayed compensating him for his quadruplex work—the small installment of five thousand dollars given him in December, 1874, had not even covered his expenses in manufacturing twenty quadruplex instruments for their recent tests. In short, as he remarked on one occasion, he had found that the Western Union people, who were regarded as America's financial aristocracy, could be as pitiless, in business dealings with him, as any other money men.

One cause of his bitterness toward them was undoubtedly the mean bargain, forced upon him under duress, by which their man Prescott was to share both credit and money rewards as "coinventor." Edison had labored like one of the Titans of old, for three years, to develop his quadruplex invention, while Prescott had contributed virtually nothing. Moreover, Tom Edison was never one to share the glory he believed rightfully due himself with another and lesser man—and this was a work he considered the crowning achievement of his career. He foresaw the commercial importance of this device. But was it not also for love of such labor and for his own sense of glory that he had struggled so long with its difficulties?

The court battles of 1875 were exceedingly unpleasant for Edison, caught in the crossfire of the warring factions. The eminent attorneys of Western Union, United States Senator Roscoe Conkling and Grosvenor P. Lowrey, called him to the witness stand and publicly stigmatized him as one who had "basely betrayed" his kind patrons, the officers of the Western Union, and who made agreements to sell the same patent rights to others twice or even three times over. He was assailed as a "rogue inventor" and a "professor of duplicity and quadruplicity." [27] On the other side, Gould's counsel charged that Western Union had attempted a conspiracy to defraud Edison, so that, like a simpleton, he assigned away half of his rights "without valuable consideration."

Why did these great "barons" contend with each other in the courts for possession of Edison's quadruplex? It was not only a superb invention, but one of the most important contributions to the telegraphic art and would bring dominance of the great industry to whoever owned it.

4   During the modern era of electronics the applied science of old telegraph days has been all but forgotten; the industry itself, in unsuccessful competition with the telephone, suffered a decline. Nonetheless, Edison's quadruplex was a strategic invention in its time and impresses us, as it did his contemporaries, as the masterwork of his youth. Up to now the twenty-seven-year-old inventor had been known to a small circle of clients who appreciated his budding talents. But when reports of his perfected quadruplex were published in newspapers and technical journals in the autumn of 1874, his reputation

began to spread among a mainly scientific public abroad as well as at home. An English technical journal now commented:

Americans are maintaining their ground for ingenuity and enterprise in the electrical world, especially in telegraphic inventions. . . . The benefits arising from the Duplex System bid fair to be multiplied still further by the joint labors of G. B. Prescott and T. A. Edison.[28]

(How galling it was for Edison to have Prescott's name coupled with his own, hereafter, in all references to this invention!)

The quadruplex was surely the most complicated job Edison had done thus far, and therefore marks a decided jump in his own intellectual capacity. He had never been one of your precocious young men of science. Growing slowly in his own undisciplined way, Edison by now had reached a stage where his mind functioned in scientific work with great clarity as well as ingenuity. He had to use the most elaborate circuits—combining strong and weak currents with rapid changes in the direction of their flow—that were ever attempted in the era before electronic science was born. The quadruplex apparatus, though now obsolete, appears to us as a wonderfully ingenious system of sounders, circuits, condensers, batteries, relays of both differential and polarized type, and "pole changers" effecting reversals of polarity. Yet all this array of telegraphic paraphernalia—which may be described with accuracy of detail only in terms of a detailed diagram—worked precisely, harmoniously and unfailingly.

Edison himself had the utmost difficulty in working out in his head a scheme of electrical currents moving in two directions at the same time and using four different modulations of the current. "He had really very little power of abstraction and had to be able, above all, to *visualize* things," one of his sons said of him in later years.[29] In order to see the thing with his own eyes he constructed an "analogue," or a model hydraulic apparatus using a pump to force fluid backward and forward through pipes and valves, in the pattern of the wires and controls planned for his quadruplex system.

The duplex telegraph of J. B. Stearns employed a "differential" or "neutral" relay, as a means for registering variations in the strength of the current. The differential relay connected at each station would not respond to signals sent out of the home station, but would respond to signals sent from the other end of the line. The receiving instrument at each end of the circuit would respond to incoming signals, but not to outgoing signals, and so messages could be sent in opposite directions at the same time.

Now Edison proceeded to add his own distinctive "diplexing" device to Stearns's apparatus, so that he could send two messages simultaneously *in the same direction* over one wire, as well as two in opposite directions. To accomplish this he introduced as a wholly new feature a second relay, called a *polarized* relay, which responded only to changes in the direction of the current passing through it. With this he combined a mechanism called a "pole-changer," which was used in place of the regular sending key and brought about instantaneous reversals in the direction of flow of the current as it was pressed and released.

The Edison quadruplex system thus doubled the capacity of the Stearns duplex; it consisted of two sets of sending instruments at each terminal, one of the pair working to vary the strength of the current, the other to vary the direction of its flow; at the far terminal a pair of operators did likewise in sending messages over their two instruments. The two pairs of receiving instruments at each terminal were so constructed and adjusted that one set would respond only to a change in the strength of the current, while the other set responded only to a change in direction of current.[30]

The beauty of Edison's device was that it not only solved a difficult problem in applied electricity (which famous inventors of Europe had wrestled with for many years) but also doubled traffic facilities for peak loads of dispatches moving in one direction. After the apparatus was fully installed on Western Union lines, in 1876, the company's report showed economies of over $500,000 annually—which amounted, with growing volume, to over $20,000,000 in the next thirty years. It is therefore no surprise that the Gould and Western Union interests were locked in conflict for control of Tom Edison's patent.

Experts of telegraphy in the decade that followed this invention regarded it as "the chief product of Mr. Edison's genius"; so James G. Reid characterized it in his authoritative manual on the subject. Other specialists in this field paid tribute to his "beautiful combinations of currents . . . so quickly made, broken up and reformed." *

---

* William Maver, Jr., in the *Encyclopedia of the Telegraph* (1884) elaborates: "For example . . . during the making of a simple dash of the Morse alphabet by the neutral relay at the home station, the distant pole-changer may reverse its battery several times; the home pole-changer may do likewise; and the home transmitter may increase and decrease the electromotive force of the home battery repeatedly. Simultaneously . . . and as a consequence of the foregoing actions, the home neutral relay itself may have its magnetism reversed several times, and the signal, that is the dash, will have been made partly by the home battery, partly by the main-line current, partly by the distant and home battery combined, partly by the

To revert to the great telegraphic war. Soon enough Edison discovered what almost everyone who ever worked for Jay Gould learned to his sorrow: that the inscrutable financier swindled his associates no less than his opponents, whenever it suited him to do so. Early in 1875, Gould took steps to merge the several telegraph companies under his control. By a written agreement drawn earlier, the Automatic Telegraph Company's stockholders were to exchange their shares, at a valuation of about four million dollars, for Atlantic & Pacific securities. Edison had been promised payment for his services in stock, which was supposed to total about $250,000 in new securities. That is, he had been assured in writing that he would receive "one-tenth" of the benefits accruing to the Automatic Telegraph from his inventions, and also that he would be named the chief electrician of the new amalgamated company at a good salary. To his chagrin, none of these promises was kept. Jay Gould made cash payments only to Harrington, his lieutenant, and then by some legal chicane discriminated against the other stockholders of the Automatic Telegraph, allowing them nothing for their stocks. Among those who were thus defrauded was Edison. His former patron, Harrington, simply took the money he got from Gould and fled to England to avoid the expected lawsuits.

By the summer of 1875 Edison was through with Jay Gould forever. About two years later, having received none of the compensation promised him, Edison addressed Gould an impassioned letter protesting at the treatment he had received, "which, acting cumulatively, was a long unbroken disappointment to me." He "had to live somehow," he said, and so had gone back to work for Western Union, which was "only too glad to get him back." [31]

No reply came from the great man. Edison, with other creditors and stockholders of the Automatic company, all of whose assets had been taken away, joined in a suit against Jay Gould which dragged itself along for nearly thirty years and ended only in 1906! At that time a Federal court judge finally ruled for the plaintiffs and ordered an accounting—Gould was then long dead—but the referee who made the accounting set the damages to the plaintiffs at only one dollar!

For a quarter of a century, Gould was one of America's most brilliant money lords and best hated individuals. A fanatic for power, driving for

---

'artificial line' current, partly by the main-line static current, partly by the condenser static current; and yet, on a well-adjusted circuit, the dash will have been produced on the quadruplex sounder as clearly as any dash on an ordinary single-wire sounder."

supreme monopoly over rail transport and wire communications, Gould epitomized the Robber Baron era; he was the prototype of Theodore Roosevelt's "malefactors of great wealth." The telegraph workers who waged a series of fierce strikes against his company used to march to such songs as "We'll Hang Jay Gould to a Sour-apple Tree!"

In January, 1881, after William Orton had died, the younger Vanderbilt, who lacked the starch of his father the "Commodore," sent Gould a brief note inviting him to a conference at the Vanderbilt residence. It was really the white flag of surrender. An exchange of stock between Atlantic & Pacific and Western Union assured Gould's dominant control of the entire amalgamated telegraph system. Soon afterward Gould moved his office to the massive headquarters of Western Union and ruled over his industrial empire from that place of vantage up to the time of his death in 1892.

In his reminiscences Edison left some discerning observations on this singular man. "Gould had a peculiar eye," Edison noticed, and he was convinced "there was a strain of insanity somewhere" in him.

He had no sense of humor. I tried several times to get off what seemed to me a funny story, but he failed to see any humor in them. I was very fond of stories and had a choice lot . . . with which I could usually throw a man into convulsions.

What was impressive about Gould, however, was the way in which he seemed to collect every kind of information and statistics possible, before entering upon some venture. He was without doubt a hard worker. "His connection with men prominent in official life, of which I was aware, was surprising to me," Edison remarks; and he adds, in objective spirit:

Gould took no pride in building up an enterprise. He was after money, and money only. Whether the company was a success or failure mattered little to him.

His conscience seemed to be atrophied, but that may have been due to the fact that *he was contending with men* [of Western Union] *who never had any to be atrophied.* . . . [32]

Though Gould had swindled him of large sums, Edison, surprisingly enough, concluded his remarks by saying, "I held no grudge against him, because he was so able in his line." All that Edison cared about was that his own part, as engineer, should be "successful," while "the money with me was a secondary consideration." The last statement, made in his late years, seems sincere; no one to whom money was the primary consideration would have spent his whole life tormenting himself to invent things.

"When Gould got the Western Union," he said in conclusion, "I knew no further progress in telegraphy was possible, and I went into other lines."

5   At the impromptu research laboratory on the top floor of the shop at Newark, Edison often permitted himself to be drawn by his curiosity into new and unexplored fields of science—despite his resolution to limit himself to "commercial" inventions. Whenever something unexpected was observed in the course of his experiments, he would stop and examine such phenomena from every aspect. He was still learning, and his assistants were also learning, sometimes at considerable cost—there were explosions through careless handling of chemicals, and one fire that needed the Newark Fire Department to extinguish it.

One novel experiment undertaken by Edison on his own account was devoted to making an electric arc light of exposed carbon points that he connected with a battery of thirty-one cells. The light did not last more than a minute or two.[33] Edison had been prompted to attempt this experiment by reports, in the autumn of 1875, of successful arc lights being introduced commercially in Russia, Germany, and France. But he soon dropped the arc lights to pursue investigations of a quite different and highly mystifying character.

He had been working several months over the problems of "acoustical telegraphy," under renewed subsidies from his old friends at Western Union, to whose service he had returned following his sad experience with Gould. The objective, also pursued then by Elisha Gray and Alexander Bell, was to transmit sound vibrations of tuning forks over a telegraph wire, in order to achieve new methods of multiple transmission. It was in fact one of the last stages before the invention of the telephone. On November 22, 1875, his laboratory notebooks relate, "In experimenting with a vibrator magnet consisting of a bar of Stubbs steel fastened at one end and made to vibrate by means of a magnet, I was astonished to see peculiarly bright, scintillating sparks issuing from the core of the magnet." He had previously noticed such sparks in telegraph relays, and in the loose iron filings between the armature and magnetic core of a stock printer, but, as he relates in his laboratory notebooks, he had always assumed such sparks were caused by strong induction. Yet they seemed so strong that it struck him that they might be caused by "something more than induction." His notes read:

We now found that if we touched any part of the vibrator or magnet we got the spark. The larger the body of iron touched to the vibrator the larger the spark. We now connected a wire to the end of the vibrating rod, and we found we could get a spark from it by touching a piece of iron to it. . . . By connecting to the gaspipe we drew sparks from the gaspipes in any part of the room. . . .

Applying a gold-leaf electroscope he found it unaffected, indicating no electric charge present. Testing of the insulation and connections showed they were perfect. This spark, then, was like no other he had seen, and the whole manifestation seemed to Edison *nonelectric*. (He had evidently obtained electromagnetic oscillations so rapid that the slow-moving leaves of the electroscope could not follow them.) His notes, dictated to Charles Batchelor, who was always by his side, ended in a tone of high excitement, "This is simply wonderful, and a good proof that the cause of the spark is *a true unknown force.*" [34]

He made many tests of the spark by forming different circuits with the vibrator. In one of them he had his laboratory workers holding hands, while one man touched a wire to the magnet, drawing bright sparks which they saw but did not feel. Edison even brought in our old long-suffering friend the frog, object of Galvani's classical experiments and, after placing it in circuit, could detect no movement. He then had a black box constructed in which there were two carbon points which could be adjusted by micrometer screws. When the dark box was placed in circuit with the vibrating device, the thick sparks jumping across the gap between the carbon points could be observed clearly through an eyepiece at the top of the box.

The name Edison gave to the new type of spark he had observed moving across that gap or space was "etheric force" or "etheric current." Such sparks, however, had been detected years before by Joseph Henry; earlier, Faraday had speculated upon the possibility of such phenomena. James Clark Maxwell in 1863 had "predicted" theoretically and had outlined exactly the production of electromagnetic waves passing through space and between the atoms of all known substances—which Heinrich Hertz, in 1887, was to demonstrate by producing and detecting them, thus opening the road to radiotelegraphy.

It is a pity that Edison, groping with his "sparks," or his perhaps not cleanly produced electromagnetic waves—which were generated without sufficient energy for detection—did not study the matter further. He was, of course, entirely wrong in concluding that those sparks or oscillations were "nonelectric." While his capacity for experimentation was remarkable indeed, he was neither a mathematician nor a theoretical scientist, and he did not attempt to extend his knowledge of the "new force." He saw, however, that if such results, the big sparks, were genuine, they denoted the passage of energy through space and implied the possibility of communication without wires. He then rushed to the newspapers with an announcement of his discovery. The New York press had recently begun to see in Mr. Edison an intriguing and picturesque character. Under the heading, "Edison's Discovery of a Supposed New Force," a friendly reporter, evidently quoting the inventor's statements, made prophecies concerning communication systems of future times that seemed no less remarkable at the time than the laboratory notebooks covering his experiments:

The cumbersome appliances transmitting ordinary electricity, such as telegraph poles, insulating knobs, cable-sheathings may be left out . . . and a great saving of time and labor accomplished. Ocean cables [may be] operated by "etheric force." . . . Wires may be laid in the earth or water. The existing methods or mechanisms may be completely revolutionized.[35]

The *Scientific American* also opened its pages to discussion pro and con of Edison's discovery in its issue of December 25, 1875. Edison himself appeared before a scientific association of New York, the Polyclinic Club of the American Institute, reporting his observations and giving a demonstration of the black box and the new force. Many challenged him on this occasion, though a Dr. George Beard, a well-known physicist, stoutly defended Edison as one who had observed "new principles until now buried in the depths of human ignorance."

In Philadelphia two professors of physical science, Elihu Thomson and E. J. Houston, vigorously assailed Edison's claims. Elihu Thomson reported experiments of his own showing that such "excess waves," generated by a Ruhmkorff induction coil, with sparks jumping across gaps, could be opposed or neutralized by the interference of "resonators," devices which sent out opposing waves tuned to the same frequency.[36]

A minor scientific tempest blew up. Even in London, Professor Silvanus Thompson, the physicist, gave a demonstration before a scientific body in June, 1876, which was intended to prove that Edison's "etheric" sparks could be accounted for by known principles and were therefore "spurious." Sir Oliver Lodge, who was later to play a notable role in the development of radio, recalled the brief agitation in scientific circles over Edison's curious claims to have drawn sparks from insulated objects in the neighborhood of an electrical discharge. They were, after all, by no means spurious, but could not be understood by anyone at the time, Lodge explained. "Edison did not pursue the matter, for the time was not ripe; but he called it 'etheric force' which rather set our teeth on edge."[37] None of Edison's learned critics, either in Philadelphia or London, any more than Edison himself, realized that the possibility of producing and detecting high-frequency electric waves was "around the corner." Meanwhile, the experience of being ridiculed by university professors implanted in Edison's mind a long-lasting prejudice against all theoretical scientists.

He had been on the threshold of the hidden world of electronic science and turned back, abandoning those perplexing experiments to devote himself to other, more immediately rewarding investigations. At this stage of his career he was prosperous again, but he was also determined to bring about a sweeping change in his way of life. He must, at all costs, quit "business" and manufacturing and live in some quiet retreat where he could give himself entirely to the vocation he loved—inventive research. When, not long afterward, he left Newark and woke up deep in the country, "etheric force" was clean forgotten; there were other things on his mind.

chapter **VIII**

Menlo Park

1
In the early spring of 1876 residents of the tiny hamlet of Menlo Park, New Jersey, saw with some surprise a new and rather oddly shaped building going up in an open pasture on a hill overlooking the main railway line between New York and Philadelphia. It was a plain wooden structure of two stories, rectangular in form and looking for all the world like a country meeting house or tabernacle. A tall white-haired man with a long nose directed the building operation; he was our old friend Sam Edison, the father of the inventor, who had been called from Michigan for this purpose. On weekends a slightly built, negligently dressed young man with a silk hat perched on his big head used to drive out to see how the work was going here.

By the end of May the barnlike structure was all but completed; it was 100 feet in length by 30 in width, with white-painted clapboard sides, tall windows, and a porch in front. Surrounding it was a stout picket fence to keep out cows and pigs. Before the paint was dry, great horse-drawn trucks thundered up from Newark, bringing equipment and more equipment—boxes of chemicals, rolls of wire, loads of books, a Brown steam engine, and a gasoline converter that would supply gaslight. Finally the younger Edison came to install himself in the new building; and with him came his faithful retinue, Batchelor, Kruesi, John Ott, and a dozen other manly looking, full-bearded workers.

Edison had obeyed a sudden whim: to move to an isolated spot in the country about twenty-five miles southwest of New York, and set up a new laboratory there. He would give up his manufacturing interests in Newark and devote himself entirely to the "invention business," in the seclusion of a rural village of only a half-dozen dwellings.

The remove of Edison to the country, his establishment of a lavishly equipped research laboratory as a sort of "factory" for inventions of all

131

sorts, aroused passing notice in the press. Some predicted that the venture would turn out badly; Edison, it was pointed out, lacked the formal education that would enable him to conduct exhaustive investigations of scientific problems. A few newspaper reporters made pilgrimages to Menlo Park to look at the place and describe it. In that homely barnlike structure, the inventor had partitioned off a small office, a little library, and a drafting room on the lower floor; the floor above was a single long room furnished with many tables covered with instruments, machines, and batteries, the walls being lined with shelves holding a great variety of materials and chemical jars of every color. On this floor there was a force of thirteen skilled mechanics working with fine steel and brass instruments, or electrical products of every kind. At their head, bent over a table, was Edison. A journalist wrote:

There is a general appearance of youth about his face, but it is knit into anxious wrinkles, and seems old. The hair beginning to be touched with grey, falls over his forehead in a mop. The hands are stained with acid, his clothing is "ready-made." He has the air of a mechanic, or, with his peculiar pallor, of a night-printer.... When he looks up his attention comes back slowly as if it had been a long way off. But it comes back fully and cordially. A cheerful smile chases away the grave and somewhat weary look. He seems ... almost a big, careless schoolboy.[1]

Why had he removed himself to this isolated place? There is no doubt that he strongly wished for a certain measure of solitude and an atmosphere of repose and security. He had the habit of meditation; the distractions of life in a large industrial suburb of New York were enemies to sustained analysis and reflection. In Newark, moreover, he was vexed by disputes with thievish landlords; his business ventures were unrewarding; he was often hounded by creditors. As soon as he had found himself with a modest accumulation of capital, his debts paid off and some twenty thousand dollars to the good, he had thought at once of building a place of his own from which he could never be dispossessed.

As he told a friend of this period, "the combined work of manufacturing and inventing" taxed even his superhuman energies. Indeed, the two occupations proved irreconcilable; if a new idea struck him he felt it had to be tested at once, with the help of every man and instrument within call. But this could hardly be done in a factory run on a regular time schedule, where his impromptu laboratory work held a minor place.[2] At the end of 1875 he therefore decided to wind up the last of his manufacturing shops, Edison & Murray, commissioned his father to look for a good site in the country, and there set up his own laboratory. There, in effect, he hung out his shingle, announcing to the world that he, Thomas A. Edison, would undertake research and development work on *any and all inventions*.

It will be recalled that on first coming to New York he had formed the partnership of Pope, Edison & Company, as consultants in electrical engineering, perhaps the first concern of its kind. The establishment at Menlo Park was more ambitious; it was, in fact, the first industrial research laboratory in America, or in the world, and in itself one of the most remarkable of Edison's many inventions.

As a scientific worker with accumulating experience, Edison appreciated the need for fine and precise instruments of measurement; he acquired a costly reflecting galvanometer, several powerful induction coils, photometrical equipment, a finely contrived electrometer, and, all in all, a stock of tools and materials the cost of which, a year or two after his arrival at Menlo Park, amounted to forty thousand dollars. With equipment he was always extravagant. At that time, though interest in scientific education in the United States compared well with that in Europe, there were only a few poorly equipped laboratories at some of the leading universities or at the new engineering colleges such as Massachusetts Institute of Technology, devoted mainly to teaching purposes. There was, to be sure, the all-important laboratory Joseph Henry had created at the Smithsonian Institution, but this was for purely scientific research. Like Faraday, Henry considered it infra dig for a true scientist to pursue primarily practical inventions, or even to bother with patenting them; he labored only to "increase the sum of human knowledge." We were still then at the period of (misguided) controversy between the amateurs of "pure" science and the practical or "empirical" inventors—such as Edison believed himself to be—who applied their scientific skills to making things of use and convenience to mankind.

In any case, no one had ever heard of a man setting up a center of research, a sort of "scientific factory" in which investigation by a whole group or team would be organized and directed solely toward *practical* inventions. Such things came, in those days, by chance, and by the happy "accidents" of feats of technical skill performed in attics or cellars or at some bench in the corner of a factory. Moreover, the hazards of the inventor's profession were all too well known. Alexander Graham Bell, struggling with his "speaking telegraph," found himself in 1875 on the verge of ruin. "The cares and anxieties of being an inventor," he wrote to a friend, "seemed more than flesh and blood can stand." Yet, before he departed from Newark, Edison told the physicist Dr. George Beard, in all seriousness, that he proposed to turn out at Menlo Park "a minor invention every ten days and a big thing every six months or so." At the time, Beard had been dumfounded and had thought Edison was blowing. But the recent list of his patents ran to about forty a year. A number of them, covering the stock printer and quadruplex, had already brought him sizable revenues; there were others under way; and he knew in addition that leading capitalists of New York would seek him out deep in the country, if only for the purpose of hiring him to test, improve, or perfect someone else's invention.

In the spring of 1876, for example, he was being paid a monthly retainer of five hundred dollars by Western Union, to do research on the "speaking telegraph" with which Elisha Gray and Alexander Graham Bell had been occupied for several years.

He was truly happy, he felt a blessed sense of freedom, when at last he found himself in the country in the month of June; and in this spirit he wrote to a patent lawyer of his acquaintance about his "brand-new laboratory . . . at Menlo Park, Western Div., Globe, Planet Earth, Middlesex Country, four miles from Rahway, the prettiest spot in New Jersey, on the Penna. Railway, on a High Hill. Will show you around, go strawberrying." [3]

What of his devoted assistants, his acolytes, whom he led as a little flock of the faithful to that isolated hilltop at Menlo Park? The best of his workers in Newark had come with him to the country. But there were no amusements here; only a few farmhouses and one saloon with a billiard room down by the railway depot, where old Sam Edison used to hold forth when he visited the place. It was going to be a fairly monastic existence for the staff. But Edison, after all, was a dedicated being; he had his "religion" of scientific work. His own wife, Mary, born and bred in a large town, hated the silence and the dark nights of the tiny village. Nonetheless Edison purchased a plain farmhouse of six rooms, hard by the site of his laboratory, and installed her there with his two children, "Dot" and "Dash." Two of the other six houses in the village were occupied by his principal assistants, Batchelor and Kruesi, who also had families. The one boardinghouse in the place, a Mrs. Jordan's, adjacent to the laboratory, was soon filled with those laboratory mechanics who were bachelors. In effect, the whole small hamlet became a community devoted to experimental science. Edison Village, it was humorously called. Here Edison was, in fact, monarch of all he surveyed.

What would these people do to entertain themselves here in this quiet backwater? They would invent things, he told them. And after that? They would invent more things.

Despite Edison's materialistic principles, the older inhabitants of Menlo

Park persisted in believing that the spirit of magic hung over his laboratory. Even one of his assistants, describing the picturesque appearance of the array of scientific and chemical apparatus, the many-colored jars and phials of chemicals and cases of semi-precious metals, wrote, "When evening came, and the last rays of the setting sun penetrated through the side windows, this hall looked like a veritable Faust laboratory."[4]

As at Newark, his men worked long hours, starting with the whistle at seven in the morning, sometimes sleeping on benches or on the floor during all-night vigils. On occasions when they worked late at night, a pleasant midnight collation was sent in by Mrs. Edison, and they would make merry, the place resounding with laughter.

Not only did the young man with the old head exert a powerful attraction upon his co-workers, but the ambitious program of the next five years gripped their imagination more and more. Young workers, hearing of the remarkable deeds performed in this place under the leadership of the young inventor, eagerly came to apply for jobs. Thus William J. Hammer, then a boy in years, later to be a power in the electrical industry, appeared one day before Edison.

The inventor remarked to him gruffly that almost everyone who applied for a job "wanted to know only two things: how much we pay and how long we work. Well, we don't pay anything, and we work all the time." Young Hammer promptly agreed to work on these terms.

John Ott, the machinist and draftsman, who served under Edison for half a century, at the end of his life described the "sacrifices" some of Edison's old co-workers had made, and he commented on their reasons for so doing.

"My children grew up without knowing their father," he said. "When I did get home at night, which was seldom, they were in bed."

"Why did you do it?" he was asked.

"*Because Edison made your work interesting. He made me feel that I was making something with him. I wasn't just a workman.* And then in those days, we all hoped to get rich with him," he added wryly.[5]

At this period of the onrushing industrial revolution, the engineers and the practical inventors were truly coming into their own. The theoretical scientists, to be sure, still assumed an attitude of condescension toward the practical men of science, as did James Clerk Maxwell toward Alexander Graham Bell, when he characterized him as a mere elocutionist who, "to gain his private ends [money], has become an electrician." A distinguished American physicist such as Professor Henry A. Rowland of Johns Hopkins would say (as he did in 1879) that "he who makes two blades of grass grow where one grew before" might be a benefactor of society, but "he who labors in obscurity to find the laws of such growth is the intellectual superior as well as the greater benefactor of the two." Nevertheless, Edison, the epitome of the practical engineer and inventor, proudly declared himself such and also insisted that his standards were frankly "commercial," that is, aimed at that which was useful. He even nursed some grievances against the "theo-retical scientists," who were to him dilettantes, playing out their little games in their ivory-tower studies. He saw them much as Charles Dickens satirized them in *Little Dorrit*, as the futile members of the Mudfog Association (the Association for the Advancement of Science of the 1840s), while identifying himself, no doubt, with the character of Daniel Doyce, "the common-sense craftsman, with the broad thumb of the hand worker ... the honest man who is always facing facts or, at any rate, dealing with the facts of science rather than its principles or theories." [6] Since his contretemps before a scientific club, suffered when he gave his lecture on "etheric force," Edison overtly showed contempt for all mathematicians and physicists and represented himself, almost defiantly, as "an industrial scientist."

He showed, however, an increasing devotion to the spirit and discipline of scientific research, especially after the age of twenty-seven. One perceives the sternness and self-critical attitude with which he continues lengthy tests in order to assure himself of the results he sought, and his reluctance, even after years of vain struggle, to leave any work uncompleted. In short, as he comes to the prime of his life, he gives what R. C. McLaurin afterward defined as "a brilliant exhibition of the method of science, the method of experimentation—which embraced all his faith." [7]

As he conducts experimental studies on a larger scale, we note that he works not only with skilled mechanics, but also with expert chemists and even (heaven preserve us!) a mathematical physicist. The latter was Francis R. Upton, who came to Menlo Park in 1878 as a specialist in mathematical physics, after studying at Princeton and for a year in Germany, under the great Helmholtz. Edison enjoyed teasing Upton about the time he "wasted" in precise mathematical calculations; but he fully recognized the aid Upton rendered him in complex investigations. After all, as he said on a later occasion, it was just as well "to have one mathematical fellow around, in case we have to calculate something out."

Edison has been called the last of the "independent" inventors of the nineteenth-century school. But as Norbert Wiener has written so perceptively, this is not true; he was in no sense a lone-wolf inventor of the old school, but was a "transitional figure" who pointed the way toward the systematic research of the technological age. The significance of the

Menlo Park laboratory was that its master worked with a whole team, comprising not only machinists and technical men but also several persons with formal scientific training. He still adhered to his cut-and-try methods. But at Menlo Park, industrial invention depended not on the insights of the shopworker alone, but on a careful comprehensive search by a whole team under him. It is for this reason that his little organization in the country served as a pilot model for the huge industrial research laboratories organized in later years, such as that of the Bell system and General Electric.

The young Edison thus takes his place as a central figure in the new age of technology that began in the United States after the Civil War. It moved, to be sure, in step with industrial and scientific progress in Europe, especially in Germany. In steel, mining, transport, and electrical communication, big industry increasingly adopted scientific standards and depended on scientific methods as showing the way to profits. Science and engineering, on the other hand, had reached a stage where new practical experimenters like Edison were *required* and could find their fullest play. In more than one field of industry he had already offered his services, and those of his organization—his "invention factory"—in *making inventions to order*. The Germans, who were the world's most advanced technologists in the latter part of the nineteenth century, recognized that, as the eminent German economist Werner Sombart said, "Mr. Edison was, perhaps, the outstanding example of a man who *made a business of invention itself*." [8]

The traditional view of invention assumed that it was something like an act of God, a "divine accident"; like "the poet, in a fine frenzy rolling," the inventor and the scientist were supposed to discover things by a stroke of genius. Had not Alexander Pope written, "God said, Let Newton be! and all was light."

Edison was a genius who held that there was no such thing as genius. With his bustling organization at Menlo Park he worked to dispel all the old myths about the accomplishments of inventors. Like a good Darwinian, he believed that inventions arose out of man's developing culture, his environment, his social and industrial relations. His busy workshop was turned by him into something far removed from those elegant laboratories of the earlier epoch, often placed in a pavilion in some formal French garden, where aristocratic amateurs of science demonstrated their superior intellectual capacities or their "superhuman cleverness," without regard to the needs of industry or human welfare.

Edison's decision not to undertake inventions unless there was a definite market demand for them was of great historical importance, as a modern commentator, James G. Crowther, has written in a very discerning paper: "He was the first great scientific inventor who clearly conceived of invention as subordinate to commerce." In thus dedicating his work to the advancement of industry, in making his inventions conform to "commercial demands" as to the necessities of human use and convenience, he also established a social and democratic criterion for applied science. Under a capitalist system, market acceptance was the standard of the useful and the practical in the industrial arts; therefore, he would make his inventions meet that standard. He was par excellence the inventor of the democratic era.

To be sure his social thinking was far less original than his workmanship and seemed to stop at the belief, so prevalent in those times, that the acquisitive were the fittest to survive and that the rights of capital, as well as those of the individual, were sacred. Nevertheless his work tended to destroy the old traditions about undirected inventive or scientific inspiration. As Crowther observes, "He began the advance toward a democratic theory in which invention would be cultivated in order to increase human happiness. . . ." [9]

In choosing the "commercial" standard, Edison was by no means being "vulgar" or mercenary, as compared with such fastidious characters as Faraday, or Joseph Henry, or Willard Gibbs. He was determined to make his formerly hazardous profession, so to say, both businesslike and respectable. But no one was more satisfied than he, once there was a little money on hand, to make that little do for his needs and to live for the joy of inventive work.

How little Edison, the new type of inventor, depended on Providence or "luck," and how much on systematic search, may be judged by the way in which he attacked new problems posed for him after coming to Menlo Park.

2   Since July of the preceding year a wholly new field of investigation had absorbed him: that of the acoustic, or harmonic telegraph, which by use of tuning forks or reeds transmitted musical notes over a wire. For years the experts of the telegraph, such as Stearns, Elisha Gray, G. W. Phelps (scientific director of Western Union), and Edison, had been "educating" the Morse instrument so that it performed all sorts of tricks, printing, repeating, sending double and, finally, quadrupled messages. By 1875, Elisha Gray, after years of experimenting, had contrived an acoustical, or "harmonic" telegraph that carried nine to sixteen different musical notes, taken from tuning forks and transmitted to a receiving instrument the parts of which

were attuned to respond separately and distinctly to each of those different notes. But if different musical notes could be sent over an electric wire, why not the sound of the human voice—why not a *speaking* telegraph? the restless searchers asked themselves. Gray, in fact, called his harmonic telegraph a "telephone" in lectures delivered early in 1875. Alexander Graham Bell, a well-educated Scottish immigrant residing in Salem, Massachusetts, who was Edison's age, had had the same idea, and for several years struggled with the problem of making a metal disk, vibrating in response to sounds, convey and reproduce those sounds over an electrified wire.

The convergence of numerous "parallel" inventors at about the same time upon a common problem is now an old story. Bell was an elocutionist who, like his father, had studied phonetics and was skilled in teaching deaf-mutes to talk; at the outset of his experiments he had no knowledge of electrical science. "If I can make a deaf-mute talk, I can make metal talk," he is reported to have said; and so for five years he experimented with harps, disks, and mechanical ears.

At the beginning of 1875, Bell came to New York to test his harmonic telegraph on the wires of Western Union, with Orton's permission—but learned, to his disappointment, that Elisha Gray had already patented such a device and assigned it to Western Union. After that Bell concentrated his efforts, more or less secretly, on the speaking telephone.

In the meantime, Orton seems to have been alerted about the promise of the experiments pursued by Gray, Phelps, Bell, and others. He now called in the young "professional," Edison—of whom it was already said, "He never fails in anything he seriously undertakes"—and invited him to investigate acoustic telegraphy for Western Union. By the summer of 1875 Edison had severed his relations with Jay Gould and his enterprises. Whatever hard feelings the Western Union millionaires may have entertained toward him during their recent legal battle, they swallowed them; Orton and his associates were gratified to have this talented fellow detached from the dangerous Gould and back in their own stable of inventors.

The records indicate that Western Union had assigned Professor Phelps and Gray to the same investigation. By 1875 it seems to have been assumed by technical men that a speaking telegraph was a distinct possibility within "the next ten or twenty years." In fact, as long ago as 1861, a German, Johann Philipp Reis, had demonstrated at Frankfurt-am-Main a primitive telephone made of wood and a vibrant membrane of pig's bladder.

It is interesting that in July, 1875, Orton had put in Edison's hands a detailed abstract of a report on Reis's machine, made for the Royal Prussian Telegraph Department at Berlin.

Edison now went to work on his own and constructed a similar machine for himself, in order to study the principle of the thing. It incorporated the initial discovery that the sound-induced vibrations of a diaphragm could open and close an electrical circuit (by the make-and-break contact principle) thus acting on an electromagnet at the receiving station and causing it to give forth a corresponding sound or pitch. Edison found that while words were only indistinctly perceptible on the Reis apparatus, "the inflections of the voice, the modulations of interrogation, wonder, command, etc., attained distinct expression." [10]

He attempted at first some electrochemical experiments with the simple Reis transmitter, then realized that one of the main problems would be to control and balance the variations of current. Shortly before moving to Menlo Park he had devised an apparatus for analyzing the various waves produced by different sounds. It consisted of two hollow metallic cylinders, one inside the other with a metallic base acting as a diaphragm. To this diaphragm was attached a magnet, which ran through the center of a coil and acted as a miniature generator, inducing slight currents in the coil according to the sound vibrations of the diaphragm. On January 14, 1876, he filed a caveat and drawings with the United States Patent Office, giving warning of this invention-in-progress and describing it. Then he put the matter aside, busying himself with other aspects of acoustical telegraphy and with several completely unrelated projects.

Edison was late in entering the race for the telephone; half a dozen expert men had begun the investigation of the speaking telegraph years before him. He also showed great courage in undertaking such work in view of the fact that he was very hard of hearing. His sister-in-law related that at this period the inventor "was a great sufferer from earaches, and I have seen him sit on the edge of the bed and fairly grind holes in the carpet. . . ." [11] Evidently his hearing loss was that of the eardrum and middle ear, for he seemed to be able to hear sound instruments by biting his teeth into them, thus allowing vibrations to be conducted through the bones of his head to the inner hearing nerve. On the other hand, his severe handicap led him to produce an instrument which could transmit sound and speech clearly.

Alexander Graham Bell, meanwhile, had been making rapid progress. One afternoon in June, 1875, his ear had caught a first feeble sound of words, the voice of his assistant speaking to him over the wire of a crude magnetotelephone. This clue was followed up closely until the morning of February 14, 1876, when Bell thought he had invented enough to apply for a patent covering his telephone. A few hours later (or earlier?) that same day, his nearest rival, Elisha Gray, walked into the Patent Office to file a caveat outlining exactly the same invention transmitting human speech. It is probably the most famous example of "parallel" invention made by two men working independently. Both men had proceeded along almost exactly the same path in their investigation; both had taken as their point of departure the harmonic telegraph, and both had been inspired with the same purpose of transmitting human speech. The patent law, however, accorded the entire authorship and rights over one of the most valuable patents ever granted to Bell alone.*

After the details of Bell's telephone were made public, in March, 1876, Edison went back and examined his own device for measuring sound waves, described in his caveat of January. He now found that it was capable of transmitting sound, though crudely. (A model of it has been used to transmit speech in modern times.) If only he had had the ears to hear faint sounds of speech, he might have won the race for the telephone. He acknowledged himself fairly beaten. But his physical handicap of deafness did not prevent him from making his own distinctive contribution to telephony, and a decisive one at that.

3   The telephone was the direct offspring of the telegraph. Bell, in need of money after the long years of experiment, approached Western Union with an offer to sell his patents at a reasonable price. President Orton, however, turned him down, saying, "What can we do with such an electrical toy?" Bell's prophecies that telephone lines would one day replace the telegraph system seemed to Orton merely the usual ravings of an exalted inventor. Western Union acted with customary bureaucratic caution in rejecting the letters of patent that might have given them, for $100,000, a monopoly of the world's future telephone industry. The idea of making heavy investments in developing the new telephone and scrapping their present equipment was also repugnant to Western Union. "Pioneering don't pay," was a familiar saying, attributed at the time to Andrew Carnegie, who waited over ten years, until Bessemer furnaces were in profitable operation in this country, before investing in steel mills.†

* Bell's patent won priority by a hairsbreadth, though under peculiar circumstances that somewhat clouded his glory as an inventor. In the famous patent suit over the rival claims of Bell and Elisha Gray, it was alleged that the order of priority as between Gray's caveat and Bell's patent application may have been rearranged by a dishonest Patent Office employee.

† In discussing inventions and the working of the patent laws, Mr. Justice Louis D. Brandeis remarked in 1912: "The great organizations are constantly *unprogressive*.

A powerful corporation, moreover, had many ways of getting control of a new product, once it was proved practicable: by imitation; by evasion; or by the erosive effect on the original inventor of lawsuits conducted at enormous costs, at the end of which, as Edison himself observed, the invention might fall to the corporation.

In July, 1876, at the Centennial Exposition in Philadelphia, Bell's demonstrations of his telephone made a tremendous public sensation. The price of Bell's patent went higher. Now the Western Union magnates began to feel uneasy, and decided to call in Mr. Edison, of Menlo Park, New Jersey. As he put it simply, they had found the 1876 telephone a crude affair, and "wanted me to make it commercial." [12] It consisted only of two receivers used alternately for hearing and speaking; while one person talked the other listened. Bell thought it was pretty wonderful when the human voice was carried for a distance of two miles. The man "sending" had to shout each sentence three or four times in order to be heard above a good deal of "interference."

Bell's financial backers, nevertheless, were already preparing to establish the first local telephone exchanges. Western Union, therefore, was considering doing the same, if another and better telephone could be invented.

The first model of the magnetotelephone was different in principle from Reis's contact-breaking device in that it communicated speech by varying the current in the line in accordance with the vibrations of a diaphragm. The diaphragm was a thin soft-iron disk, placed near the pole of a permanent bar magnet. Around one pole of the magnet was wound a coil of fine copper wire, one end of the wire being grounded. There was, however, no battery or other electric generator in the circuit. The mere sound waves of the voice striking upon the diaphragm made it vibrate, inducing increased and decreased impulses in the magnetic coil, which caused the current in the line to vary. At the end of the line the receiving diaphragm was set to vibrating in sympathy with the electrical variations in the line. Thus the receiving diaphragm vibrated exactly as the sending diaphragm vibrated, when words were spoken into it, producing sound waves detected by the ear of the listener. In other words, as Bell put it, he first changed sound into an undulating electrical current and then changed the current

---

They will not take on the big thing. Take the gas companies, they would not touch the electric light. Take the telegraph companies, the Western Union, they would not touch the telephone." And neither telegraph nor telephone companies later cared for radio.

back into sound. The range of the instrument, however, was extremely limited, as there was not enough electric power in Bell's first circuit.

During the autumn of 1876 and the winter following, Edison puzzled over the problem of raising the telephone's volume while still controlling and balancing the current variations it induced. No one then knew more about varying and balancing electrical forces than he. This was one of the main problems he had mastered in his quadruplex telegraph. At that time he had devised a carbon rheostat (made up of fine silken disks saturated with particles of graphite and a screw to adjust their pressure) which he used to control the currents of his multiplex telegraph apparatus with great precision. Why could he not vary the resistance of this rheostat by means of voice waves impinging on an attached diaphragm, connect it in series with a battery and a telephone receiver, and so produce a telephone more powerful and different from Bell's? All through that winter he kept trying an assortment of materials to be incorporated into the "sending" disk.

On January 20, 1877, as his notebooks show, he first "succeeded in conveying over wires many articulated sentences." This new apparatus had a diaphragm with a small spring attached to the center of it, connected in turn to three platinum points immersed in a dish of loose carbon granules. Current from a battery, passing through the platinum points and the carbon granules, varied in strength as the granules were compressed and released by the vibrations of the diaphragm. "The talking, though poor in quality, was of a sufficient volume, so that it could be heard through the teeth [by Edison] when an ordinary Morse relay magnet, the magnet of which was included in a circuit, was held against the teeth." His notebooks indicate, however, that the instrument was constantly getting out of adjustment.

At this period he wrote to his father in Port Huron:

I am having pretty hard luck with my speaking telegraph, but I think it is O.K. now ... I am at present very hard up for cash. Tom and Dot are well and thriving.[13]

It was at the end of the first winter at Menlo Park that Edison had his big idea. The Bell magnetotelephone, for many reasons, could not be used effectively as both receiver and transmitter. What was needed was a separate transmitter in circuit with Bell's receiver, which when used alone for both purposes was too weak.

He came down to New York in March and made his report of progress before the Western Union executives. A memorandum agreement was then drawn up allowing him more money for his experiments. He had already, in crude form, the essentials of a new transmitter, with a metallic diaphragm and disk of graphite, the surfaces of the disk and diaphragm being in contact. He also introduced two electrodes of (partly) carbon composition connected with the diaphragm, against which the sound waves struck, thereby effecting a variation in the current. The idea of using a closed circuit in which a battery current was constantly flowing, thus raising the power, was also a feature of Edison's first "speaking telegraph transmitter" patent (filed April 27, 1877, but granted, as No. 474,230, only after a delay of fifteen years). He said of this machine to an officer of Western Union:

As yet it is not sufficiently perfect for introduction. It is, however, more perfect than Bell's. You need have no alarm about Bell's monopoly as there are several things that he must discover before [the telephone] will be practicable. When my apparatus is perfect you will be informed.[14]

He exuded confidence, an amazing confidence, though the road ahead of him was actually to be quite hard and long.

He had conceived of a second strategic innovation (besides using a separate transmitter)—introducing an induction coil. Thus the battery current flowed not over the line, but through the primary circuit of an induction coil, so that electrical impulses of an enormously higher potential could be sent out on the main line, by the secondary circuit of the coil, to the receiving end. Instead of being limited to a few miles' range, the sound vibrations of Edison's transmitter could be heard over long distances.

What was needed at this stage was an improvement in the variable-conducting substance for his transmitter, one that was sensitive and at the same time stable.

Now he began a broad search for the right substance, in line with (what became henceforth) his characteristic method of the drag hunt. He divided this labor among his staff at Menlo Park and, as he related, beginning at the end of his stock of chemicals tried every one of them—some two thousand.[15] If it took all summer, he would continue the search by such methods. And after that, through the autumn and the winter. The signed testimony of the old laboratory notebooks record in thousands of yellowed pages the laborious and patient testing of one model after another—sometimes in vigils of fifty-two hours—while he watched sharply for clues. Once he found a clue, he would "hang on" and work in that area.

Improvement in the articulation of consonants, especially the hissing of sibilant ones, engrosses him. He writes down sibilant phrases which are to be tried on the transmitter under development: "The vibrations of the oscillations"; and "Physicists and Sphynxes in majestical Mists." (Some of these nonsense words show the persistence in his unconscious mind of the current controversy with the devotees of pure science.) Then occasionally there is a happy exclamation:

TELEPHONE—July 17, 1877: Telephone perfected this morning at 5 A.M. Articulation perfect—got ¼ column newspaper every word. Had rickety transmitter at that. We are making it solid. . . . Hemi-Semi-demi-Quaver. . . .[16]

But despite this note of triumph there was more work to be done.

He was thoroughly familiar with the wonderful properties of carbon by which variations in the pressure applied to it cause similar variations in its electrical resistance. The rheostat he had used in connection with his quadruplex-telegraph work had particles of graphite, or carbon granules, along with a device for raising or lowering pressure and thus varying its conductivity. His first rude transmitter with its graphite-impregnated electrodes had operated fairly well; but it was no accident that he searched for a purer carbon.

As he related at the time, "At last, on one day, an assistant brought him a piece of broken glass incrustated with carbon black from a smoking lamp chimney. He scraped off the carbon, pressed it into a little cake and tried that." [17] It was molded in the form of a button and introduced into the transmitter, placed between two metal plates (adjacent to the diaphragm), the metal plates being connected with the battery circuit. That carbon button proved to be wonderfully elastic in varying its resistance in accordance with the pressure exerted upon it by the diaphragm. In progressive experiments he also discarded the tiny spring he had originally placed between the carbon button and the two outer metal disks, and he found that this eliminated some interference. He now made the surprising discovery that this transmitter did not require the vibration of a thin sensitive diaphragm, like Bell's original goldbeater's skin. Instead, a metal outer disk, one sixteenth of an inch thick, was fastened tightly to his carbon button. Now the volume of sound was several times larger than in Bell's telephone and the articulation was sharp. After all, it was all a matter of pressure to be communicated to the carbon granules, he realized. Edison's carbon transmitter was, in fact, a *microphone*.

His work had been made all the harder because of the handicap of poor hearing. As he wrote to an acquaintance in the early winter of 1878, when near the end of the trail, Bell had had "a very soft job" compared to his own. Since Bell had first exhibited his telephone at the Centennial Exposition in July, 1876,

There ain't much improvements or change been made [in] over a year; whereas I had to create new things and [overcome] many obscure defects in applying my principle. Besides I am so deaf that I am debarred from hearing all the finer articulations & have to depend on the judgment of others. I had scarcely gotten the principle working before [there] was pressure in New York to introduce it immediately. I made 2 or 3 pair but found they were unhandy.... That delayed. I have finished a new pair and they have been working two days without no change or adjustment.... I have my man making a model for the Patent Office, which is essential I should get in, and I and Batchelor must go to New York to show it there so you can see I have my hands full— Even working 22 hours per day.[18]

In February, 1878, he applied for a patent covering the use of a carbon lampblack button in the telephone transmitter (Patent No. 203,015). In his carbon transmitter Edison introduced two radical changes in the magnetotelephone of Bell. Instead of having the sound waves of the human voice generating the electrical impulses as they struck against the diaphragm, as in Bell's instrument, in Edison's transmitter the sound waves actuated an "electrical valve," so to speak, by varying the resistance between two electrodes. His second important innovation was the use of the induction coil which extended the useful length of the telephone line by hundreds of miles.

In a test that was a sort of full-dress rehearsal over a line 107 miles long, between New York and Philadelphia, in March, 1878, in the presence of Western Union's directors, among them Orton, William H. Vanderbilt, and H. M. Twombly, Edison's transmitter delivered speech "loudly." The effect was sensational. Everyone present felt that he had liberated the art of telephony—so hoarse and stuttering in its first two years—and won a tremendous advantage over the limited Bell apparatus. One spoke into the separate Edison transmitter and at the same time also listened, through the receiver held at one's ear, to the speaker at the other end of the line. On that day in 1878 the telephone received its fundamental form, which was to remain virtually unaltered for almost half a century. In the opinion of present-day Bell System engineers, Edison's "great and lasting contribution to telephony was his introduction of the variable resistance factor by use of carbon." [19]

4   After the summer of 1877 Western Union finally saw in the telephone a future threat. The Bell group had raised some new capital in Boston and began advertising and selling private telephone lines to business concerns (which found them marvelously convenient); soon afterward they began to install the first central telephone exchanges. In November, Western Union, on the strength of enthusiastic reports from Menlo Park and Edison's first graphite-type transmitter, formally entered the field by organizing a subsidiary called the American Speaking Telephone Company, with a capitalization of $300,000. The new company picked up various telephone patents, besides Edison's all-important carbon transmitter—the receiver invented by Elisha Gray, and a "condenser" telephone of Professor A. E. Dolbear, of Tufts College. They proceeded at once to install a switchboard at Western Union headquarters in New York and tried to provide a telephone service of some sort over their heavy telegraph wire. Later, they strung up more suitable telephone wire on their telegraph poles, and the "war" for the telephone monopoly was on.

At the beginning of 1878 Western Union enjoyed a tremendous advantage in owning rights to Edison's loud transmitter. The Bell Telephone Company had nothing to compare with it; unless they obtained a transmitter equally as good they were faced with ruin. But from the depths of despair they were suddenly rescued, several months later, when they came upon an obscure, self-taught inventor residing in Washington, one Emile Berliner, a German-Jewish immigrant who had made a good, clearly articulating telephone transmitter of his own, on the "loose-contact" principle.

Berliner, moreover, had filed a caveat with the Patent Office, giving fair warning of his invention-in-progress (according to the rules in force prior to 1910). Edison, who had been completely ignorant of Berliner's current work, had filed a full application for a patent on his first carbon-type transmitter, as a complete invention on April 27, 1877, two weeks after Berliner's caveat was recorded. The Bell people, who quickly bought rights to Berliner's transmitter, were now in a position to start an "interference" procedure at the Patent Office against Edison and Western Union. This legal action became a deadlock that was protracted for fifteen years, which was why the final grant of patent for Edison's telephone transmitter was delayed until 1892. At that time a Federal Court decision finally awarded him his full rights, ruling that

... Edison preceded Berliner in the transmission of speech. ... The use of carbon in a transmitter is, beyond controversy, the invention of Edison. The carbon transmitter displaced Bell's magnetic transmitter. . . . The advance in the art of telephony was due to the carbon electrode of Edison.

The Berliner claims to a prior patent were then declared invalid.[20]

Meanwhile, in the late seventies, long before that tardy court decision on the transmitter arrived, there was another of those "wars" for a patent monopoly that so often centered about Thomas A. Edison. As Edison put it bluntly, "the Western Union [was] pirating the Bell receiver, and the Boston company was pirating the Western Union transmitter"—which Edison had invented.

In the historic suit in Federal court in Boston, beginning in September, 1878, of Bell Telephone Company against Peter A. Dowd, American Speaking Telephone, et al., the issue that created the bitterest dispute was that of the conflicting claims to "priority" in the invention of the receiver by Alexander G. Bell and Elisha Gray (whose invention rights had been assigned to Western Union). After extended trial hearings running more than a year, Western Union's counsel advised that Bell's rights would most likely be sustained in the end and that a compromise settlement out of court would be the wisest course.

Under an agreement negotiated between both parties in October, 1879, the validity of the Bell receiver patent was acknowledged by the defendant; and Western Union also agreed to withdraw from the new and already growing telephone field, selling its existing lines to the Bell Company and receiving from it, for seventeen years thereafter, a royalty of 20 per cent of all telephone rentals on the lines it had sold out.*

By this shrewd bargain—or so it seemed at the time—Western Union gained 3.5 million dollars in royalties; yet Theodore Vail, the rising managerial head of the Bell System, won for it undisputed control of the future American telephone industry. But whatever strength there was in Western Union's position in these final negotiations was mainly due to its confident claims on behalf of Edison's transmitter patent, which, with the settlement, passed to Bell Telephone.

Several months before this happy consummation was brought about, in the spring of 1878, Edison had begun to feel that "he ought to be taken care of . . . and threw out hints of this desire." Orton sent for him and

---

* Elisha Gray, who had sincerely believed in the priority of his own inventive work, withdrew his charges and signed an admission of Bell's priority, receiving a consolation payment of $100,000. Testimony given in court had accused a U.S. Patent Office examiner of irregularity of conduct, through having possibly placed Bell's application ahead of Gray's caveat —though there was no indication that Bell had knowledge of such action, if it occurred. A sensitive man, Bell was made so unhappy by these events that the whole subject of the telephone became distasteful to him, and he did little inventive work thereafter.

asked him how much he wanted. Edison says he thought $25,000 might have been enough, but blandly invited Orton to make him an offer. Orton thereupon said his company would pay the inventor $100,000, a tiny percentage of the wealth his invention was to bring in. Very well satisfied, Edison agreed on condition that the money be paid him in installments, during the life of the patent, of $6,000 a year for seventeen years. Orton had no trouble agreeing to this, since the interest alone on $100,000, at 6 per cent, would have netted the inventor a similar sum. But to Edison it was just as well, since he believed that "inventors didn't do business by the regular process." He feared that if he received all the money at once, it might disappear quickly.[21]

5   Nothing reveals better the technical resourcefulness of Edison than the incidents connected with the sale of his telephone transmitter patent in England—at the time of the great trial in Boston. Early in 1877, Bell had gone to England and demonstrated his telephone, with such effect that a Bell Company was promptly organized to exploit his invention in Great Britain. Very soon after that Edison had his first carbon-type transmitter patented in England, in July, 1877; and a little later filed for patent on his improved, carbon-button transmitter. When this was demonstrated over a line between London and Norwich, a distance of 115 miles, it made a great sensation. In England it was seen at once that Edison had a grip on half of the telephone invention and that his own apparatus held a great advantage over Bell's in volume and articulation. Venture capital quickly appeared to back a rival telephone company, based on his devices, which was named the Edison Telephone Company of Great Britain, Ltd.[22]

The irrepressible Colonel George Gouraud—who was last seen in 1873 at a dreary waterside tavern in Greenwich, together with his young American friend, downing great drafts of gin to dispel the effects of bad weather, worse food and a disappointing trial of the automatic telegraph —now reappears as Edison's promoter, publicist, and benefactor. A genial transplanted American of the "boomer" type, Gouraud acted as the London representative of American banking groups; and he had useful acquaintances among British capitalists. Thanks to Gouraud's recent publicity work in connection with another invention by Edison, the master of Menlo Park was known throughout England.* It was also Gouraud

---

* In January 1878, on receiving the first news of Edison's invention of the phonograph, Gouraud undertook to promote it in England and raised capital for its commercial exploitation. An account of this concurrent invention will be found in the following chapter.

who raised up the British syndicate that backed Edison's telephone inventions with an initial capital of £100,000.

Both the Bell and the Edison companies in England therefore began negotiations with the British Post Office at about the same time for a franchise authorizing the establishment of telephone lines and exchanges. Again industrial warfare impended; the Bell interests threatened suit against the British Edison Company, and the latter accepted the challenge. Fortunately Gouraud retained as counsel the brilliant Sir Richard Webster, who was later to become Chief Justice of Britain; Webster urged that *pourparlers* be initiated with a view to a peace of compromise between the conflicting parties.

Edison later described the imbroglio over the British telephone venture as a sort of lark:

In England we had fun. Neither the Bell people nor we could work satisfactorily without injuring each other. They infringed on my transmitter and we infringed on their receiver, and there we were cutting each other's throats. Well, of course, this could not go on forever, and consolidation had to come, although a second fight over the terms of consolidation was also bound to come. In a measure, they had the whip hand over us; so I was not surprised to receive one day from our representative in England a telegram the gist of which was that the Bell people wanted more than their fair share of the receipts in case of consolidation, and that our agent was at his wits' end what to do. I cabled back at once something to this effect: 'Do not accept terms of consolidation. I will invent new receiver and send it over.' Then I set to work.[23]

At the time Gouraud had as his secretary a young man named Samuel Insull, who soon afterward came to the United States to work in the same capacity for Edison. He later confirmed Edison's recollections, saying, "I remember writing the cable [from London] and receiving a reply to it the next day, stating that if we would wait sixty or ninety days he would supply a new form of receiver!"[24]

Was there ever such self-confidence, one might say cheek, displayed by an inventor before? Bell had worked for at least six years over his receiver. Edison, putting all other work aside, in December, 1878, opened up his whole bag of tricks, and tried "a thousand chemicals," as his notebooks for Christmas Day, 1878, indicate. Within about three months— not three weeks, as some exaggerated accounts have held—he turned out an entirely new type of telephone receiver which, he claimed, "worked better than Bell's."

The condition to be met in this case was that the receiver must absolutely avoid use of a magnet in order not to infringe on Bell's patent. The idea of his nonmagnetic, or "electromotograph," relay, which in a similar emergency he had contrived for Gould's telegraph lines in 1874, now came back to him, and was put to use in a new form. He wrote:

I had recourse again to the phenomenon discovered by me years previously, that the friction of a rubbing electrode passing over a moist chalk surface was varied by electricity. I devised a telephone receiver ... afterward known as the "chalk receiver." There was no magnet, simply a diaphragm and a cylinder of compressed chalk about the size of a thimble. A thin spring connected to the center of the diaphragm extended outwardly and rested on the chalk cylinder, and was pressed against it.... The chalk was rotated by hand. The volume of sound was very great. The voice, instead of furnishing the power, merely controlled the power, as an engineer working a valve could control a powerful engine.* [25]

Headquarters of the Edison Telephone Company of Great Britain had been opened in Cannon Street, London. Gouraud wrote in mid-January:

MY DEAR EDISON:
Receiver. Pray do not lose a moment unnecessarily in forwarding this, as until it is received I can make no substantial progress with my negotiations.... The moment I have this new receiver I shall move with vigor and have no doubt of satisfactory results.[26]

The first chalk receivers made brave sounds, but they had bugs in them. The moment a prospective customer for Edison Telephone stock arrived in Gouraud's office, the new receiver would behave in silly or balky fashion. Gouraud arranged to have John Tyndall lecture on the chalk receiver and demonstrate it before the Royal Society; but the lecture had to be postponed twice, until an improved instrument arrived in the spring.

At Menlo Park Edison had his nineteen-year-old nephew, Charles P. Edison, son of his brother Pitt, and a very clever electrician, working under him on the telephone. "Charley" was therefore dispatched to London in charge of six of the improved receivers. The first tests were most promising. The affair began to seem so important that Edison also sent Edward H. Johnson, his principal business lieutenant, and Charles

---

* In its later form, in 1879, a chalk drum, mounted on an axis and covered by a band of paper soaked in a solution of caustic potash, was turned under a spring, the end of which was in contact. The spring was attached to the center of the diaphragm, so that when the drum was turned, friction between the spring and the paper deflected the diaphragm. The current from the line passed through the spring and paper to the cylinder, causing friction to be diminished. As the undulating telephonic current passed through the apparatus, the constant variation of the friction of the spring caused the deflection of the receiving diaphragm to vary in unison with variations of the electrical current, so that sounds were given out corresponding in pitch and quality with the sounds produced at the distant transmitting station. (U.S. Patent No. 221,-957; executed March 24, 1879)

Batchelor, his cleverest mechanic, to help manage things at the London end.

Skilled men were urgently needed in England to install and operate private telephone lines, already in demand for business use; but there were none there, and almost none in America who knew the telephone. Edison, thereupon, hastened to New York, engaged about sixty likely subjects, brought them back with him to Menlo Park, and began to train them. A telephone exchange was set up in the laboratory; its wires ran from wall to wall, upstairs and downstairs, connecting ten of his "loud-speaking" telephones. After having trained the men to become expert with the instruments, Edison says:

I would then go out and get each one out of order in every conceivable way, cutting the wires of one, short-circuiting another, destroying the adjustment of a third, putting dirt between the electrodes of a fourth, and so on. A man would be sent to each to find out the trouble. When he could find the trouble ten consecutive times, using five minutes each he was sent to London. About sixty men were sifted to get twenty.[27]

The laboratory thus was turned into a perfect bedlam, with all this strenuous testing and shouting into the instruments going on at the same time. "Being hard of hearing, Edison went on with his work undisturbed," one eye-witness has related, "while the rest of us were nearly deafened as 'Hello—Hello—Hello' echoed from corner to corner."[28]

The demonstration of the new chalk receiver in London on March 15, 1879, was a complete success, as cabled reports showed. Charley Edison had kept the still moody apparatus running in fine form while the Prince of Wales and the Prime Minister, Mr. Gladstone—thanks to Gouraud's able public relations work—appeared to speak over it before an admiring public. The London press was enthusiastic; even the conservative *Times* described Edison's chalk receiver as far superior to Bell's and as being, in fact, a "microphone-receiver" or "loud-speaking telephone." "Every inflection of the voice is audible," reported the London *Standard*. To be sure, there were some amusing contretemps. When Mrs. Gladstone, who accompanied her husband, was invited to speak over the line, "she asked the man at the other end whether it was a woman or a man at our end, and the reply came in loud tones that it was a *man!*"[29]

Among the contingent of employees engaged in promoting the Edison telephone at the London office, there happened to be a decidedly odd, lanky and red-haired Irishman of twenty-three, named George Bernard Shaw. This was one of Shaw's first commercial jobs in London. In those days, when he was a poor young man, and before he had begun to write his comedies of manners and morals, Shaw was as much engrossed in physical science as in music or literature. He had read Tyndall and Helmholtz and in consequence believed that he was "the only person in the entire establishment who knew the current scientific explanation of telephony."

From the memory of his transient connection with the Edison enterprise in England, Shaw some years afterward drew the material for one of his early novels, *The Irrational Knot,* whose hero was an English inventor of working-class stock. In a characteristic preface to the novel the future playwright indicates that Edison was the inspiration of this work, and then with some humorous exaggeration, describes the Edison telephone apparatus as "a much too ingenious invention . . . of such stentorian efficiency that it bellowed your most private communications all over the house, instead of whispering them with some sort of discretion."

Having been hired to demonstrate the operation of the telephone before curious visitors, Shaw, as he declares unblushingly, proceeded to do so in a manner which

laid the foundation for Mr. Edison's London reputation; my sole reward being my boyish delight in the half-concealed incredulity of our visitors (who were convinced by the hoarsely startling utterances of the telephone that the speaker, alleged by me to be twenty miles away, was really using a speaking trumpet in the next room).

Shaw's recollections also yield us some graphic impressions of the high-spirited band of American technical workers dispatched by Edison to install some of London's first telephone lines and exchanges:

Whilst the Edison Telephone Company lasted [Shaw relates] it crowded the basement of a high pile of offices in the Queen Victoria Street with American artificers. These deluded and romantic men gave me a glimpse of the skilled proletariat of the United States; and their language was frightful even to an Irishman. They worked with a ferocious energy which was out of all proportion to the result achieved. Indomitably resolved to assert their republican manhood by taking no orders from a tall-hatted Englishman, whose stiff politeness covered his conviction that they were, relative to himself, inferior and common persons, they insisted on being slave-driven with genuine American oaths by a genuine free and equal American foreman. They utterly despised the artfully slow British workman who did as little for his wages as he possibly could; never hurried himself; and had a deep reverence for anyone whose pocket could be tapped by respectful behavior. Need I add that they were contemptuously wondered at by this same British workman as a parcel of outlandish boys, who sweated themselves for their employer's benefit instead of looking after their own interests. They adored Mr. Edison as the greatest man of all time in every possible department of science, art and philosophy, and execrated Mr. Graham Bell, the inventor of a rival telephone, as his Satanic adversary; but each of them had (or pretended to have) on the brink of completion an improvement on the telephone, usually a new transmitter. They were free-souled creatures, excellent company; sensitive, cheerful, and profane; liars, braggarts, and hustlers; with an air of making slow old England hum which never left them even when, as often happened, they were wrestling with difficulties of their own making; or struggling in no-thoroughfares from which they had to be retrieved like strayed sheep by Englishmen without imagination to go wrong.[30]

The moistened chalk-drum receiver was a somewhat awkward affair; it had to be cranked by hand as you listened to the apparatus. In England, at any rate, it literally drowned out the Bell telephone. On receiving a later shipment of the "water-chalk" telephones, Ed Johnson reported, "They belched right out. Very loud speaking. I am inclined to think we have reached the goal." [31]

Though it was soon to be abandoned, following the settlement with the Bell interests (as being more difficult to manufacture, or less economical than the magnetic receiver) it served its purpose by inducing the opposition group to yield satisfactory terms for a merger of the Bell and Edison companies into the United Telephone Company of Great Britain. Another important factor making for this result was the successful legal defense of the British patent for Edison's carbon transmitter, conducted by Sir Richard Webster; the British courts promptly and completely sustained Edison's claim to priority in that invention.

The closing stages of this commercial war were not reached without some serious casualties. James Adams, one of the skilled craftsmen sent from Menlo Park, turned out to be an alcoholic of the extreme type, and amid the strain of the work in London suddenly died. The promising young Charles Edison, one of the inventor's favorites, after going off to Paris to demonstrate the new telephone apparatus, was also suddenly stricken with illness and, to Edison's great sorrow, died. The inventor arranged for the transport of his nephew's remains to Port Huron, Michigan. Then the indispensable Batchelor was taken ill in London, but managed to get back to Menlo Park and recover his health. Finally, Johnson reported from London that he too was sick.

Edison, busy shipping out telephone sets in large wooden cases, took note of all these untoward events, and in the autumn of 1879 wrote to Johnson with a touch of his sometimes macabre humor:

... Hope you are not going to "kick the bucket." If you men are going to die off so, it would be better to have the large boxes made a little longer, so that you can send back the corpses in them....[32]

For the inventor himself the final outcome of the British venture was most fortunate. To the end he always maintained the pose of being a simple country boy who was repeatedly dumfounded at finding money come to him like manna from heaven after one of his "commercial" inventions had been completed. To be sure, he was often distracted by his work and in the midst of it gave no thought to finances, but only to the technical objective. Still, at thirty, the telegrapher who had come to New York threadbare ten years earlier was already the highest-paid professional among American inventors, one who, in a pinch, could be trusted to deliver almost any contrivance needed. Such had been the somewhat ephemeral device of the chalk-drum receiver. He has related:

One day I received a cable from Gouraud offering "30,000" for my interest. I cabled back I would accept. When the draft came I was astonished to find it was for 30,000 pounds sterling. I had thought it was dollars.[33]

It was indeed a golden age for inventors. For "curing" Bell's "magneto-telephone" (as Edison phrased it) he gathered in altogether a quarter of a million dollars. After the coup in London and the merger, Edison was to abandon this field. Other able inventors, such as Emile Berliner and Francis Blake, around that time contributed very handsome improvements in the telephone transmitter. Then technological progress in the telephone virtually stopped, for some inscrutable bureaucratic reason, for about a quarter of a century, while the industry grew to be one of the world's greatest privately owned monopolies. Edison himself made public his own explanation in 1890, declaring that

many extremely useful improvements on the telephone are in the possession of those controlling the invention, and are safely locked up from the world because of the great extra expense which would attend their application to existing instruments.[34]

chapter IX

The legend of the "Wizard"

1

The years at Menlo Park, from 1876 to 1881, were by all odds the happiest and most fruitful of Edison's life. He was in his early thirties and at the very height of his creative power. The business of inventing was humming along; the combats of great Wall Street money men, as he called them, for possession of his patents testified to their importance to the industrial system.

The alarums of those "wars" over his telegraph and telephone patents, however, did not seriously affect the atmosphere of peace and freedom he enjoyed at Menlo Park. Here, at any rate, he was in the happy condition of being at liberty to absorb himself in a whole variety of inviting studies. Here he could allow himself to meditate, permit his mind to wander, even to "play," without fear of interruption. In other words he could work, if he wished, at the leisurely pace of a man of the study and the laboratory; he could be now reflective and dreamy, now energetic and rapid in his pursuit of an objective. Despite his professions of being only an empirical and practical inventor, he had a disposition that drove him repeatedly to be more than that; he was immitigably curious about the secrets of nature, and his mind often turned toward untrodden paths, as fresh insights into experimental science came to him. He possessed naturally a great power of concentration and at the same time was highly conscious of all the movements of his imagination. Thus he encountered sometimes unexpected and unwonted illustrations of natural law hitherto unknown.

There was a charm about life in this village of applied science that many who came here noticed. "Edison is always absolutely himself," one visitor writes, and "possessed by the *joie de vivre*." When he wished to inform himself on some special subject of moment to him, he would sometimes gather together a great mass of books, lay them out on the floor of his library and, flinging himself down among them, "pore over

156

them for hours on end ... after which he would go back, refreshed, to the manual part of his task." [1]

His life here was simple and yet complete. His co-workers were sworn friends and disciples who shared in his devotions. Moreover, his wife and children were close by. The eldest, Marion, as a girl of five or six, often played in the laboratory at her father's feet. On evenings when he returned home early for supper—rarely enough—he sometimes entertained his family and the young children with mechanical or musical toys. One day, to amuse his wife, he connected one of his loud-speaking chalk telephones through his private wire to Western Union headquarters in New York, which in turn was connected to a concert hall. The music was clearly heard in Menlo Park, twenty-three miles away.

In his laboratory it was noticed that his mind was so compartmented that he was capable of pursuing a large variety of inventions under development at the same time. At one period, early in the eighties, there were said to have been forty-four of these under way, with different assistants in charge of each. Edison would study their progress daily, passing rapidly from one to another as he inspected them. In 1877, for example, we know that technical improvements were being worked out for various telegraph and submarine-cable devices; different types of telephone were being improvised; electric pens and mimeograph machines were being made; various sound-measuring instruments were being developed; many chemicals and drugs were under investigation; even a crude incandescent lamp was contrived, with much trouble, in the autumn of that same year, but was abandoned when it burned out in a few moments.

Did his mind wander off, losing itself, when so many different paths were offered? The lack of direction may seem to us to reflect uncertainty in the mind of the searcher; yet while *seeming* to stray, and failing to reach what he is seeking directly, he may come upon something else that is, after all, even more valuable.

This indirect or apparently wandering course of investigation, often leading to important scientific discoveries, has been defined as "serendipity"—the gift of "finding valuable or agreeable things not originally sought for." The term was originally coined out of an expression used in a letter of Horace Walpole, alluding to the old Persian fairy tale of "The Three Princes of Serendip" (Ceylon), which described those mythical travelers as "making discoveries, by accident or sagacity, of things they were not in quest of." In his paper "The Role of Chance in Discovery," the American physiologist Dr. Walter B. Cannon has recently revived interest in this suggestive idea, which illuminates the importance of that reservoir of the subconscious mind where facts, experiences, fancies, and memories, seemingly confused and long-submerged, generate some novel, unexpected thought or combination of ideas.

As Edison himself was quoted as saying:

"Look, I start here with the intention of going there"—drawing an imaginary line—"in an experiment, say, to increase the speed of the Atlantic cable; but when I have arrived part way in my straight line, I meet with a phenomenon, and it leads me off in another direction—to something totally unexpected." [2]

Edison, at any rate, appears to have sensed clearly the dependence of invention and discovery upon the total accumulation of knowledge, including that which seems forgotten. He was forever collecting curious and miscellaneous facts, and squirreling them away in his memory, in the many folds of his large brain, which was, in truth, a capacious storehouse of such miscellanies. In his work, to be sure, we find all the evidence of methodical and close-gripping deduction. But he himself also stressed the important role played by chance or accident in discovery. Pasteur, who had a deep understanding of the mental processes of discovery, said, "Chance favors the mind that is *prepared*."

On a number of occasions Edison discussed the role of "accident" in his life of systematic inventive research. It seemed as if accidental discoveries were recurrent, though he sometimes tried to differentiate between these and inventive work, saying:

Discovery is not invention, and I dislike to see the two words confounded. A discovery is more or less in the nature of an accident. A man walks along the road intending to catch the train. On the way his foot kicks against something and ... he sees a gold bracelet imbedded in the dust. He has discovered that—certainly not invented it. He did not set out to find a bracelet, yet the value is just as great. [3]

In the achievements of men of science, chance discovery seemingly played a big part. Edison believed, for instance, that Bell had discovered the principle of the telephone when he was looking for "something else." But had not Bell's mind also been alerted to greater opportunities? And chance, as Edison thought, played an important role even in discoveries concerning natural law, as in the case of Newton. But, he concluded, "Newton had been at work on the problem [of gravitation] for many years"—which shows that Edison was in accord with Pasteur's view that such magnificent "accidents" come often to those who are prepared.

Edison, too, was all attention when, as he confessed, he met, "by the merest accident," with the opportunity for a fundamental invention, the most original he had ever conceived, one that would open up a wholly new and marvelous art to mankind.

2
Bell's telephone invention had drawn many minds to the problems of the reproduction of speech. The study of sound fascinated Edison—all the more in that he was partially deaf.

At this time, he has said, "my mind was filled with theories of sound vibrations and their transmission by diaphragms." He knew something of the important studies by Helmholtz analyzing the nature of sound and hearing, which inspired the various experiments in acoustical and "speaking" telegraphs, from Reis to Bell; he knew of Leon Scott's "phonautograph" (1857) which used a diaphragm and hog bristle to trace a record of sound vibrations on lampblacked paper, though without reproducing any sound itself. In working with Bell's telephone he also became familiar with the elastic properties of disks, which enabled them to vibrate in tune with the vibrations of the voice.

Some time before he had worked out the carbon transmitter, on February 3, 1877, he applied for a patent covering "An Improvement in the Automatic Telegraph" (No. 213,554), which was designed to permit the recording and repeating of telegraph messages at speeds of two hundred words a minute. This device consisted of a disk of paper laid on a revolving platen and rotating around a vertical axis, quite like the modern phonograph disk. The armature of a telegraph receiver was connected to an embossing point at the end of an arm, which traveled over the disk, embossing on it the dots and dashes of incoming telegraph messages in a volute spiral. When this disk was removed and put on a similar machine provided with a contact point, the indentations of the embossed record caused the signals to be repeated on another telegraph wire. Here we have several features of the disk-record contrivance, used to reproduce telegraphic impulses.

Later on, in the summer of 1877, while working with Bell's telephone receiver, he noticed how its diaphragm vibrated in tune with the voice. On studying the amplitude of these vibrations, he observed that they were of considerable size and could be made to do mechanical work. Since he could not judge amplitude with his own faulty hearing he used to test a diaphragm by attaching a short needle to it, resting his finger on the needle, and speaking into the diaphragm, with the result that the needle pricked his finger. To illustrate the power of the diaphragm, he made a little toy at this time which was vastly amusing to his daughter Marion. It was a mechanical doll with a funnel on top of its head.

When you recited loudly into the funnel, [it] would work a pawl connected to the diaphragm; and this, engaging a ratchet wheel, served to give continuous rotation to a pulley . . . connected by a cord to a little paper toy representing a man sawing wood. Hence, if one shouted, the paper man would start sawing wood.[4]

Returning later to the embossing telegraph repeater, he tried to improve its performance by changing over to an arrangement using continuous rolls of paper tape, paper that was coated or paraffined; he also introduced a steel spring in proximity to the paper tape, to keep it moving in better adjustment. When this instrument raced along at high speed, the indentations of dots and dashes striking against the end of the spring sometimes gave off a noise, "a light musical, rhythmical sound, resembling human talk heard indistinctly." Working over his new telegraphic repeater in June, 1877, he made a memorandum in his notebook to use "thin copper or other metallic foil," and to make the grooves wider.

These indistinct mutterings in the machine haunted him; the sounds it made were *almost human,* like those issuing from the first weak Bell telephones. He was working on other jobs at the time, trying ways of strengthening the hissing sounds over his own telephone transmitter; he was also trying to discover what made a vowel sound, or trying to reproduce one mechanically. E. H. Johnson, who was then in Philadelphia demonstrating the Edison transmitter to Professor George Barker, of the University of Pennsylvania, wrote his chief that summer:

Now as to the latest idea of mechanically speaking the letters of the Alphabet, Professor Barker is delighted and says: "It looks as if you might reach the end sought by scientists for ages. . . ."[5]

He had been thinking of repeating devices since he was a boy telegraph operator in Canada. In following the development of his sound-reproducing studies we note that he pursued *both* the method of deduction and that of chance trials. His reasoning was that if he could record the movements of a diaphragm and attached point on some sort of disk or strip and then use the indentations thus made to set another diaphragm in motion, the second diaphragm should reproduce the sounds which had struck upon the first. And yet his direction was set neither by deductive reasoning nor by chance alone, he recalled later, but by something of both. When, in the course of an experiment, "You come across anything you don't thoroughly understand," he would say, "don't rest until you run it down."

It *may* have been accident (such as some additional current thrown into the motor he was using for his telegraph recorder) that caused the musical sound in the telegraph repeating instrument. But diaphragms were very much in his mind now. One day he took a tape of paraffined paper and placed it underneath a diaphragm having a small blunt pin attached to its center. As he related:

I rigged up an instrument hastily and pulled a strip of paper through it, at the same time shouting "Halloo!" Then the paper was pulled through again so that its marks actuated the point of another diaphragm, my friend Batchelor and I listened breathlessly. We heard a distinct sound, which a strong imagination might have translated into the original "Halloo." That was enough to lead me to further experiment.[6]

They had heard the first strangled cries of the infant talking machine struggling to be born.

The laboratory notebooks (under the heading "Telephone") contain a note scribbled by Edison on July 18, 1877, saying:

Just tried experiment with a diaphragm having an embossing point and held against paraffin paper moving rapidly. The speaking vibrations are indented nicely and there is no doubt that I shall be able to store up and reproduce automatically at any future time the human voice perfectly.[7]

There have been legendary accounts of this historic experiment that describe him as having worked in his most impetuous fashion for several days and nights on end to make a talking machine; and that in one all-night session he managed to perfect the thing, then, on the next morning, rushed forth to make a public demonstration of the machine before newspaper reporters.

The little-known true facts are more fully in accord with Edison's own later definition of genius as being "99 per cent perspiration and one per cent inspiration." He was in truth engaged for about four months in a patient search for a device quite different in purpose from that of the magnificent invention he would actually contrive. Edison's friend Johnson has indicated that the inventor at the time was working on a commercial project: to record and reproduce sound coming over Bell's telephone. As he described it to Johnson, "it would be a *telephone repeater*—it would transmit, repeat, be of great practical value, like the telegraphic repeater." It did not dawn upon him that what he was contriving was a talking machine.[8]

This work was kept somewhat secret. There is evidence that he approached the Western Union people with his idea of reproducing and recording the human voice, but they saw no conceivable use for it! By late October, rumors of his experiments on a strange new machine, nevertheless, were reaching some of Edison's business friends. From Washington, General Ben Butler wrote him:

MY DEAR EDISON:

Tell me something about your wonderful invention in recording the human voice. I need not say that you had better keep it perfectly secret. It is so remarkable that I do not understand it at all. . . ."

Butler still referred to some means of recording the voice "on paper" that might be of great commercial value in preserving telephone conversations.[9] There was gossip of the forthcoming "phonograph" in a New York newspaper on November 5, 1877.

Then, Edison's exuberant "advance agent" Johnson sent out the first public announcement of a "talking machine," which was published as an article in the *Scientific American* for November 17, 1877. It told of Mr. Edison's original idea of recording the human voice on *a strip of paper*, without an electromagnet or current, and by mechanical means solely. The object was to record telephone messages and transmit them again by telephone. Edison was still said to be meeting with difficulties in reproducing the finer articulations, but the first crude results indicated "that he will have the apparatus in practical operation within a year." The accompanying drawing showed a strip of paper tape traveling under a needle extending from one diaphragm that embossed it, and then passing on to a second diaphragm that reproduced sound—it was a replica of the telegraph repeater, with diaphragms instead of electromagnets.

It is under the date of August 12 that we first come upon an entry in Edison's notebooks using the word "phonograph" (from the Greek for "sound" and "writing"); and by early November there were indications of the pristine form of the talking machine:

I propose having a cylinder . . . 10 threads or embossing grooves to the inch . . . cylinder 1 foot long.
I have tried wax, chalk, etc.[10]

A fortnight later, November 29, 1877, the first accurate sketch in his own hand of the original talking machine is entered into the notebooks. While he was mistaken or confused about the actual time of the invention, he recalled incidents accompanying it with great vividness:

Instead of using a disc I designed a little machine using a cylinder provided with grooves around the surface. Over this was to be placed tinfoil, which easily received and recorded the movements of the diaphragm. A sketch was made and the piecework price $18 was marked on the sketch.

Here it should be noted that Edison paid his mechanics according to a minimum-wage and piece-work system. If the job cost more than was estimated for it, the mechanic received the minimum wage; if it took less time and cost less, he received in addition to his wage the difference saved.

The workman who got the sketch was John Kruesi. I didn't have much faith that it would work, expecting that I might possibly hear a word or so that would give hope of a future for the idea. Kruesi when he was nearly finished, asked what it was for. I told him. . . . He thought it absurd.[11]

In explanation Edison had said laconically: "the machine must talk." Kruesi scratched his head to indicate disbelief. Others present bet the "Old Man" cigars that the contraption would not work. When the thing was completed, it was a solid job of brass and iron, with a three-and-a-half-inch cylinder on a foot-long shaft and a hand crank to turn it; two diaphragms, each with stylus, were mounted in adjustable tubes at opposite sides of the cylinder. Edison deliberately fixed a sheet of tin foil around the cylinder, began turning the handle of the shaft, and shouted into one of the little diaphragms:

> Mary had a little lamb,
> Its fleece was white as snow,
> And everywhere that Mary went
> The lamb was sure to go.

Then he turned the shaft backward to the starting point, drew away the first diaphragm tube, adjusted the other in position to reproduce sound, and once more turned the shaft handle forward. Out of the machine came forth what everyone recognized as the high-pitched voice of Thomas A. Edison himself, perfectly, or "almost perfectly," reproduced, reciting the little Mother Goose rhyme. Kruesi turned pale and made some pious exclamation in German. All the onlookers were dumfounded.

Edison declared afterward, "I was never so taken aback in all my life. Everybody was astonished. I was always afraid of things that worked the first time." After that, he tells us, they sat up all night fixing and adjusting it so as to get better and better results—talking into it and singing, testing different voices, then listening with unending amazement to the words coming back.

From instructions issued for the use of the first phonograph, it is plain that it needed skill to keep its tin foil under control and its styluses in adjustment. "No one but an expert could get anything intelligible from it," the inventor admitted.

On December 6, 1877, Edison was satisfied with the performance of the "talking machine." Batchelor's diary of that date states: "Finished the phonograph. Made model for the P.O. [U.S. Patent Office]" and relates also that Edison and he took the machine to New York the next day and brought it to the office of the editor of the *Scientific American*. Unwrapping his package, Edison declared that he had "a machine that would record and reproduce the human voice." Numerous persons gathered around to watch him, while he set up the phonograph, recited into it, and then played it back. "They kept me at it until the crowd got so great that Mr. Beach was afraid the floor would collapse. The next morning the papers contained columns."

The first demonstrations of Edison's phonograph made most people believe either that they had taken leave of their senses or that it was all a ventriloquist's trick. What made it uncanny was that the apparatus was so utterly simple.

The first informed account of the phonograph, published in the *Scientific American* for December 22, 1877, indicates that Edison's visit occurred on December 7, the day when he executed his patent application. The editor states that "the machine began by politely inquiring as to our health, asked how *we* liked the phonograph, informed us that *it* was very well, and bid us a cordial good night." *

There had been various prophecies of such an invention, principally by men and women of a literary imagination. Ralph Waldo Emerson, for one, had said long years ago, "We shall organize the echoes!" Within a brief period, Bell had overcome distance in transmitting the human voice; and now Edison recorded and reproduced the fleeting sounds of the human voice so that even the dead might speak or sing to us from the grave.

When, on December 15, 1877, he filed a patent for the phonograph, his most original invention, nothing remotely resembling such a machine was ever found to have been mentioned in all the voluminous records of

* Much confusion as to the date of this invention is reflected in all earlier accounts of it. A reproduction of a drawing by Edison of an alleged first model of his phonograph, with instructions written in his hand, saying: *"Kruesi, Make this—Edison. Aug. 12. '77,"* has often been wrongly cited as representing the date of invention and original model. However, this sketch is now known to have been drawn by the inventor some time after the event, from memory—and without date or any written instructions. This sketch, with its wrong date ("Aug. 12, '77"), does not furnish the mechanical information from which Kruesi could have made the first working model of the tin-foil phonograph. The recent discovery of Charles Batchelor's diary for 1877 and 1878, which has been deposited by his daughter in the Edison Laboratory National Monument, establishes the exact date of the phonograph invention beyond all doubt as December 6, 1877.

the United States Patent Office, and the grant of patent was made in the unusually brief time of fifty-seven days.

It is noteworthy that Edison, like the Three Princes of Serendip, was willing to go "off course" when he came upon some unusual phenomena; and that he often encountered such unusual phenomena because "he conducted his laboratory work so that he was directly in touch with all the details of work done, and did not leave this to be carried on, unobserved, by his assistants." [12]

The 1877 phonograph, it has been often said, is a superb example of the practical and original applications of scientific knowledge of sound dynamics up to that date. As one commentator wrote in a popular family journal at the time, the invention was not a tenth as intricate as the sewing machine, and in truth was "so simple in its construction, so easily understood, that one wonders why it was never before discovered." [13] But as R. C. McLaurin, considering the phonograph invention nearly forty years later, observed, "one of the most impressive things about Edison, beside "the enormous range of his activities [is] the wonderful simplicity of many of his devices. After all, simplicity of device is always the sign of the master, whether in science or art." [14]

When he had finished with his invention, Edison hardly knew what to do with it. These days his inventions were usually made *on order;* but no one had ordered this. Was it only a scientific toy, a curiosity? Yet the first raucous croaks of "Professor" (so he was honorifically titled in the press nowadays) Edison's phonograph were heard round the world. People did not yet understand the 1876 telephone; and the next year they were confronted with the phonograph, which seemed even more astounding.

3   Fame entered the door of the Menlo Park laboratory at the end of 1877; thenceforth Edison was never to escape the attentions, flattering or irksome, which the great public pays to an accepted national hero. He had gradually become known among men of science and businessmen interested in the new electrical industry, in Europe as well as in America, as one of the most ingenious of practical inventors. But the acclaim suddenly given the phonograph and its author was almost unprecedented. A "phonograph craze" flared up. Newspaper reporters and writers and artists of popular illustrated magazines flocked out to Menlo Park in large numbers, and described the "nineteenth-century miracle" of the phonograph, as *Leslie's Weekly* termed it, and its maker. The new talking machine, it was promised, would "turn all the old grooves of the world topsy-turvy and establish an order of things

never dreamed of even in the vivid imaginings of the Queen Scheherazade in the 1001 Nights' Entertainments." [15] This was a fair sample of the mingled expressions of wonder, amusement, and excitement that greeted the invention which "bottled" sound and music and gave it forth again.

A leading journal in America saluted Edison in an editorial as being conceivably "the greatest inventor of the age," adding:

We are inclined to regard him as one of the wonders of the world. While Huxley, Tyndall, Spencer and other theorists talk and speculate, he produces accomplished facts, and with his marvellous inventions is pushing the whole world ahead in its march to the highest civilization. [16]

In England, both the new telephone transmitter and the Edison phonograph were introduced to the public at about the same time, with sensational effect. In January, 1878, Sir William H. Preece, the electrical consultant of the British Post Office, demonstrated the first model of the phonograph and lectured on it before the Royal Institution in London. It is significant that not only common men but persons of learning and cultivation were immensely impressed with the potentialities of this new and undreamed-of machine. In Edinburgh, the brilliant Fleeming Jenkin, one of England's greatest engineers, having failed to obtain a model of the phonograph, quickly made one of his own after published descriptions of the invention, and demonstrated it before a scientific society in Edinburgh. In Paris, the directors of the International Exposition of 1878 received and accepted an Edison phonograph for display, though the scientific jury could not determine what category of the industrial arts the machine belonged to.

Du Moncel, the electrical scientist, read a paper on the lessons it taught, in which he declared that *"cet étonnant Edison"* possessed in himself alone "more genius than a whole scientific senate."

The high estimate of Edison's achievement among scientists of his day is reflected in an article that appeared in a leading British scientific periodical, *Nature*:

Mere ingenuity in contriving machines does not add to the sum of human knowledge, and if Mr. Edison were merely a clever inventor and nothing more, I should feel less interest in the man. It is, however, a noticeable feature of his inventions that they, in general, contain some new principles, some original observation in experimental science, which entitles him to the rank of a discoverer. [17]

After the press had done its duty in telling the millions about the "speaking phonograph," the crowds who had read about "the New Jersey Columbus" came to Menlo Park to see him and his works. They came from cities and farms, by carriage or wagon and by train; indeed the Pennsylvania Railroad organized excursions bringing hundreds of persons at a time to flood the tiny hamlet that had grown famous overnight as "the village of science." It became the Mecca of a continuous pilgrimage of scientists and curiosity hunters. Foreigners arriving in New York by transatlantic steamer would ask their way to Menlo Park. Once arrived there they were astonished to find that it was neither a "park" nor a town, but a flag station on the railroad, having only six houses and a laboratory. But some would ask, "What is manufactured here?" And the reply invariably given was, "Nothing." It was, in truth, a place for *spending* money rather than making it.[18]

People crowded into the lower floor of the "tabernacle" and saw nothing but books, blueprints and mechanical designs; then they tramped upstairs and gaped openmouthed at all the array of chemical jars, metals, batteries, and electrical or sonic machines, such as the "aërophone"—a megaphone with great horned ears that would permit a conversation to be held over a distance of two miles. And overhead were tangled lines of telephone and telegraph wires that made strange cobwebs everywhere. Were not all these books, papers, and instruments private matters, some among those who were admitted so freely inquired. "Oh, no," one of Edison's staff answered. "Nothing here is private. Everyone is at liberty to see all he can, and the boss will tell him all the rest."[19]

The "professor" himself stood among the crowd, a surprisingly young man, smooth-shaven, with a thick mop of hair falling forward over his face, who good-humoredly submitted to being stared at without reserve. He himself demonstrated his invention in an affable, modest, and informal manner. He answered all questions promptly, speaking with a Western twang but making his explanations remarkably clear. It was said that "Mr. Edison's explanations pleased people greatly. His quaint and homely manner, his unpolished but clear language, his odd, but pithy expressions charmed and attracted."[20]

At first Edison, who had been so much alone, enjoyed the "bath in the multitude," as hundreds and even thousands came to Menlo Park. The exaggerated compliments of rustics, who told him he was the most sought-after man in America, he brusquely waved aside. There were also fools and bores; one man to whom he had explained everything with great patience, said at last, "Yes, I comprehend perfectly," but then added to Edison's dismay, "I understand it all, except how the sound gets out again!"

Among the visitors there were a good number who came to see with their own eyes if there were not some prestidigitator's trick about this new invention. One of these was Bishop John Vincent, cofounder of the Chautauqua Association. After looking sharply all about the laboratory for a hidden ventriliquist, the bishop began to shout into the phonograph's recording tube a long string of jaw-breaking Old Testament names, doing this with such rapidity that none could follow him. When the tin-foil record was played back to him, he announced emphatically that he was now satisfied there was no fraud by Edison, since not another man in the whole country could recite those Biblical names with such speed as he had used.[21]

Like a comedian, Edison entertained the crowds by showing them all sorts of tricks the talking machine could perform, if it willed: he whistled popular airs, such as "La Grande Duchesse," a song hit of the day; or he rang bells, coughed, and sneezed before the recording tube, then reproduced these assorted sounds. Also, by superimposing upon a vocal duet the strident interruptions of angry listeners, he made the phonograph simulate a first-class street brawl, with shouts of "Oh, shut up!" or "Go away, if you can't sing any better!" and "Help! Police! Murder!" Betimes, the "professor" would pat his machine affectionately and say, "Well, old phonograph, how are we getting on down there?" The apparatus would then growl back at him in harshly metallic tones, uttering scraps of Spanish, German or Latin. Such were the pleasantries that cast a spell over the uninitiated audiences of 1878 during the first "phonograph craze." As one commentator wrote, Edison might be no man of letters or stage performer, but he was a kind of artist who knew how to "dramatize" his inventions, and his machines were his "characters."[22]

Nothing would suffice, after all the newspaper stories of his miracle, but that he must come to the nation's capital to exhibit his *speaking* phonograph before the notables of government and science. Since he was not minded to discourage the universal interest in his invention, he accepted urgent invitations to Washington, arriving there April 18, 1878. First he paid a call on Joseph Henry at the Smithsonian Institution and demonstrated the phonograph in the venerable scientist's parlor. Later in the evening he made his appearance before a large scientific gathering and allowed his machine to introduce itself. As Edison turned the crank its voice was plainly heard saying, "The speaking phonograph has the honor of presenting itself before the American Academy of Sciences."

Earlier, on that same afternoon, he had exhibited the talking machine before members of both houses of Congress who had gathered at the home of Miss Gail Hamilton, a well-known Washington hostess, who was the niece of Speaker James G. Blaine. Senator Roscoe Conkling, the

Republican boss of New York State, a famous dandy, noted for his handsome figure and curly auburn hair and beard, was among those present when Edison slyly played the ditty:

> There was a little girl, who had a little curl
> Right in the middle of her forehead;
> And when she was good, she was very, very good
> But when she was bad, she was horrid.

The curls of Conkling, one of the most controversial figures of those stormy Reconstruction days, were regularly featured in newspaper cartoons; and everyone present laughed at the allusion by the phonograph to a point on which the powerful Senator was sensitive. Edison may well have played this record deliberately, for three years earlier, Conkling, as lawyer for Western Union, in a telegraph patent suit, had assailed him as a "rogue inventor."

The last stop during that triumphal tour of the Capital was at the White House. As Edison relates:

About 11 o'clock word was received from the President that he would be very pleased if I would come. . . . I was taken there and found Mr. Hayes and several others waiting, among them I remember Carl Schurz who was playing the piano. . . . The exhibition continued till about 12:30 A.M., when Mrs. Hayes and several other ladies who had been induced to get up and dress, appeared. I left at 3:30 A.M.[23]

Little known hitherto, except in the circle of his profession, and given to a secluded life, the youthful-looking inventor of the phonograph, during this first season of world fame, was now seen or talked about everywhere. At first sight the great public took him to its heart as it had never done with other men of science or any other inventor—at least since Ben Franklin's day. Morse had known fame for a time, but he was a Boston Brahmin. The brilliant Henry, a shy and fastidious man, never knew popular acclaim. But Edison was a plain, rough-hewn, democratic type— and yet, as all acknowledged, he certainly "knew his stuff." Above all, the people of America admired a worker, a "doer," who had mechanical ingenuity; and he seemed to possess such skill in a measure far above all other men. Henceforth, as he was written about constantly in the press, exaggerated tales were spread of the extreme poverty of his boyhood and youth. Without schooling, without the help of friends or family, the former trainboy was said to have come to New York in his rags and conquered. He was the very type of the Self-made Man, whom so many Americans fervently believed in and sought to emulate. So was that picturesque old Commodore Vanderbilt, who had recently died as America's richest individual, leaving a fortune of 90 million dollars. But while some envied and others feared Vanderbilt, few loved him.

Edison's was a success story of another and more admirable sort. His exploits and the fortune he had won were much magnified by popular rumor. Fables were made for children about his habits of work, his indifference to rest or sleep, and his repeated triumphs through his sheer wits. Did not men who had worked closely with him say that "Edison can evolve from his own brain any invention required"? Was it not said of him in the press that he cared only for his work and would not stop even to attend a banquet in his honor if he were paid $100,000! As one of his old acquaintances of telegraph days wrote him from Michigan, in the terms of that universal homage that was accorded him by millions of Americans, "At least you have made your success by virtue of your hard work and brains, and not by exploiting other men's work."

Popular traditions about great inventors and discoverers have long been woven around old race memories or legends of "magicians" or beings endowed with superhuman powers, from the ancient Titans in their caves to the witches and sorcerers of medieval times. In effect, a kindred folklore grew up in the popular imagination around Edison, though he exhibited no transports, had a plain, down-to-earth manner, and embraced principles that tended to confound all such mystifications.

"Aren't you a good deal of a *wizard*, Mr. Edison?" a metropolitan newspaper reporter asked him one day.

"Oh no!" he answered with a laugh. "I don't believe much in that sort of thing." [24]

Nevertheless the rustic neighbors of Menlo Park and nearby Metuchen gossiped about his having machines that could overhear farmers talking or even cows munching the grass in the fields a mile away. It was said that he had another machine which was supposed to measure the heat of the stars; and that illuminations of meteoric brilliance were seen blazing up through the windows of his laboratory and were extinguished as suddenly and mysteriously as they had appeared. Catching glimpses of figures gliding about the fields near his laboratory at midnight with lights and equipment, bent on missions none of them could understand, the simpler inhabitants of the region, according to one reporter recording "A Night with Edison," "were minded of the doings of the powers of darkness." [25]

The appelation of "Wizard" was affixed to him; henceforth he was to be the Wizard of Menlo Park. As one of his laboratory assistants said of this period, "A species of glorified mist soon enveloped [Edison]," thanks to the grotesque and exaggerated reports of his powers. He was

regarded with a kind of uncanny fascination, similar to that inspired by Dr. Faustus of old; no feat would have been considered too great for his occult attainment. Had the skies been suddenly darkened by a flotilla of airships [bringing] a deputation of Martians, the phenomenon would have been accepted as a proper achievement of the scientist's genius.[26]

That Edison had some of the spirit of the actor and the showman was plain enough; like Barnum he was not afraid to advertise his wares. But on the other hand he also relished his solitude and was sometimes irked by the inconveniences of glory. For a time he considered wiring his picket fence with a strong battery, saying, "I shall blow somebody up yet!"

For a season the "professor" seemed never to tire of experimenting with his phonograph, it was observed. "You have made so many inventions," a newspaper man remarked to him in 1878. "Yes," he replied, "but this is my baby, and I expect it to grow up and be a big feller and support me in my old age." [27]

Meanwhile there was widespread speculation, both fanciful and serious, on the future usages of his invention. Actors, statesmen and orators rejoiced at the thought that their mortal voices could now be preserved after they had been turned to dust. Other more diverting suggestions were set forth: that life-sized statues of great personages, such as Henry Ward Beecher, might have phonographs stuffed inside them with which to address crowds in public squares. Clergymen, as some wag proposed, might take their rest while their sermons were repeated for them automatically. Illustrated periodicals showed organ-grinders bearing phonographs instead of barrel organs; or dying men recording their last wills and testaments before lawyers and family; and the Statue of Liberty with a phonograph established in her torchbearing arm, giving "salutes to the world" at the entrance to New York's harbor. At this period the illustrated periodicals often represented the typical American family seated in an overstuffed parlor, with bewhiskered men and women in bustles all gathered in a devout circle to attend Mr. Edison's phonograph.

Edison himself caused an article to be prepared and published in his name in a leading magazine, in which he outlined his own ideas of the future usefulness of his machine:

1) Letter writing, and all kinds of dictation without the aid of a stenographer.
2) Phonographic books, which will speak to blind people without effort on their part.

3) The teaching of elocution.

4) Music.—The phonograph will undoubtedly be liberally devoted to music.

5) The family record; preserving the savings, the voices, and the last words of the dying members of the family, as of great men.

6) Music boxes, toys, etc.—A doll which may speak, sing, cry or laugh may be promised our children for the Christmas holidays ensuing.

7) Clocks, that should announce in speech the hour of the day, call you to lunch, send your lover home at ten, etc.

8) The preservation of language by reproduction of our Washingtons, our Lincolns, our Gladstones.

9) Educational purposes; such as preserving the instructions of a teacher so that the pupil can refer to them at any moment; or learn spelling lessons.

10) The perfection or advancement of the telephone's art by the phonograph, making that instrument an auxiliary in the transmission of permanent records.[28]

A surprising number of the inventor's predictions have been borne out by modern developments. But his own immediate procedure at that period, his first steps in commercializing the new machines, show us that he failed to comprehend or foresee what was to constitute the magnificent destiny of his invention: the introduction of great and serious music, formerly the luxury of a small privileged class, into the everyday life of the common man.

Promptly in January, 1878, a group of venture capitalists had come to Edison and, after a conference, reached an agreement to form the Edison Speaking Phonograph Company, to which sole rights were given by him to exploit, manufacture and sell his apparatus as a *music box primarily*. C. A. Cheever, Uriah C. Painter, a former newspaper man of Washington, and Gardner G. Hubbard, the father-in-law of Alexander Bell, were among the principal stockholders of the company, which agreed to pay Edison ten thousand dollars to begin with and a royalty of 20 per cent on each phonograph sold. James Redpath, founder of the celebrated Redpath Lyceum Bureau of Boston, a popular lecture service, was chosen to give exhibitions of the machine in various parts of the country, whose territory was to be parceled out to branch managers. Offices were established at 203 Broadway, New York; the instruments themselves were manufactured at the electrical equipment shop of Edison's old friend, Sigmund Bergmann, on Wooster Street.

In its first advertisement the company announced that since the adaptation of its product to "the practical uses of commerce" had not yet been completed, it would offer the apparatus to the public only in the form of a "novelty."

The phonograph was, of course, still a crude affair; words were hard to distinguish, and the tin-foil record could be played only a few times. Although the inventor had the belief that the principal future use of the phonograph was to be as a machine for business correspondence, it was at first to be presented mainly as a scientific curiosity. Under Redpath a show business was launched and five hundred phonographs were exhibited around the country at popular entertainment halls or at special amusement centers, where a small admission fee was charged. While the "phonograph craze" lasted, for a year or so, crowds came, and a swarm of old telegraph operators and "barkers" exhibited the machine, which played a few popular airs, or recorded the jokes of music-hall comedians on tin-foil records running a minute and a half. In the first days so much excitement was caused by the exhibitions that Edison's royalties for one week's gate receipts in Boston alone ran to $1,800. Then the "bubble" burst; after a large part of the country's population had seen and heard the tin-foil phonograph, its curiosity seemed satiated and audiences at the music-box parlors dwindled away.

Edison was quite aware that the tin foil was not a satisfactory material for recording, being awkward to install and remove; that some of the consonants came out "soft" or were wanting altogether; and that rotating the cylinder by hand at an even rate of speed was difficult, the sound of speech and notes of a song being wholly altered if the hand turned too rapidly or slowly. The thing gave, in fact, a "burlesque or parody of the human voice," as William H. Preece reported from London after the first wave of enthusiasm had died down.

As one student of the talking machine's history has concluded: "The phonograph, in truth, had been launched prematurely. It was all very well to talk about dictating letters ... or using it to read *Nicholas Nickleby* to the blind, but not when a tin-foil cylinder would play for scarcely more than a minute." As for providing whole concerts of inspiring music, as Edison himself had predicted, the grating quality of the machine forbade that.[29]

For a while, the inventor at Menlo Park investigated new substances for recording, such as wax, and a new form of record, shaped like a plate, or disk (described in the 1878 patent in England). For the disk records, however, there was a mechanical problem of the stylus being altered in performance as it approached, in diminishing orbits, the center of the circle. A scheme for reproducing cheap copies of records en masse from a master matrix also came to his mind, as well as one for improving the suspension of the needle in relation to the record. In short, the new machine greatly needed development before it could be brought to realize its immense possibilities—which no one foresaw, least of all Edison, who had poor hearing and was unmusical. What troubled him was that the first phonograph could not be quickly adapted for use in business dictation.

Though he loved the phonograph and called it his favorite invention, Edison laid it aside, *abandoned* it for ten long years, leaving all the potential of this highly original device undeveloped. Here, it has been remarked with hindsight, was one of the inventor's greatest blunders, one that in the end was to cost him dearly.

Within a year or two the phonograph was regarded only as a scientific curiosity. Edison himself in a newspaper interview described it as "a mere toy, which has no commercial value." Rival inventors, in his judgment, would not even bother to pirate his invention.[30]

He had other ideas that engaged his full attention in the summer of 1878. One undertaking particularly gripped his imagination; it was described by one of his intimates as being that of dispelling "night with its darkness . . . from the arena of civilization."

chapter X

Toward the light of the world

1. In the late spring of 1878, to his own and everyone else's surprise, Edison felt very tired and ill. He had been working at the development of a dozen or more inventions at the same time. In seven years, since his brief honeymoon journey to Niagara Falls, he had had no real vacation. When an invitation came to him, through Professor George F. Barker, to join an expedition of scientists going to the Rockies to observe the total eclipse of the sun that summer, he gladly agreed to make the trip.

In his intervals of "serious play" he had invented a "tasimeter," an instrument for measuring minute changes in temperature. It was so sensitive that Edison believed he could measure changes in heat down to one-millionth of a degree Fahrenheit. It was his intention to test an improved model of his tasimeter by trying to record changes of heat from the sun occurring during the eclipse.

To go west to a point near the Great Divide, at Rawlins, Wyoming, the rendezvous of the group of scientists and government officials, was then still a diverting adventure. With a crowd of astronomers from many nations, the inventor journeyed in a special car of the Union Pacific, his traveling companion being Marshall Fox, the well-known correspondent of the New York *Herald,* who often wrote about Edison's doings. Arriving late in the night at Rawlins, a mere flag stop on the railway, they found the place flooded with scientific visitors and were quartered together in a small room at the only hotel there. Edison relates:

After we retired and were asleep a thundering knock on the door awakened us. Upon opening the door, a tall, handsome man with flowing hair, dressed in western style, entered the room. His eyes were bloodshot and he was somewhat inebriated. He introduced himself as "Texas Jack" ... and said he wanted to see Edison as he had read about me in the newspapers.[1]

The two visitors were fairly disconcerted when Texas Jack, in his efforts to entertain them, drew a Colt revolver and, aiming through the window at a weather vane above the depot down the street, neatly shot it off. Only by pleading that he needed sleep and agreeing to see the man in the morning could the inventor be rid of his unruly admirer. In this frontier country with its high-spirited, free-shooting citizenry, fame could have its drawbacks.

The next day, July 29, 1878, the day of totality, Edison, though heavy-headed for lack of sleep, went out with the astronomers—who did not forget to take their rifles along—to observe the eclipse. The astronomers busied themselves for hours determining their exact location on the earth and filling big sheets of paper with mathematical calculations which, as the incredulous practical inventor said, "looked like the timetable of a Chinese railroad." Edison meanwhile made himself ready for his own test, installing his tasimeter apparatus in a hen house. The hens, he noticed, "all went to roost just before totality." A storm arose, and the shelter began to disintegrate, while Edison struggled to level a telescope at the sun and hold on to his other instruments. The heat from the sun's corona proved to be many times beyond the index capacity of his tasimeter, and so the trial was marked "no results."

After the eclipse, he joined a hunting party for a day or two, then decided to visit the California coast together with Barker. Thanks to his celebrity, Edison was given permission by the local officials of the Union Pacific to ride the cowcatcher of the locomotive as it bumped along over the mountain grades. As he later described this boyish adventure:

The engineers gave me a small cushion, and every day I rode in this manner, from Omaha to Sacramento Valley, except through the snowshed on the summit of the Sierras, without dust or anything else to obstruct the view.[2]

The much-needed vacation lasted two months, after which he returned to Menlo Park greatly refreshed and ready to undertake a new enterprise that was more difficult by far than anything he had ever tried.

During their journey together, Professor Barker, who had become passionately interested in the possibilities of electric lighting, talked long and earnestly with Edison of recent developments in this field and urged him to investigate it for himself. "Just at that time I wanted to take up something new . . ." Edison recalled.[3]

The problem of turning electric current into illumination had haunted men of science throughout the nineteenth century, ever since Sir Humphry Davy's famous demonstration before the Royal Society of London in 1808, when he ran a strong electric current (furnished by a battery of two thousand cells) through a small gap between two carbon rods. As the carbon was oxidized by the current, it created a brilliant blue-white light in that gap, in the form of an arc. But for many years after, progress in this field was slow for want of an abundant source of electrical current. The dynamo of Faraday, invented in 1831, and based on his discovery of the principle of induced electricity, converted mechanical power into electrical energy as a conductor was passed or rotated through the field of a magnet. It seemed to provide the answer to the problem of a cheap supply of electric current. Then the dynamo, driven by steam engines, gradually underwent improvements by many hands so that by the 1860s lighthouses began to be illuminated by big arc lights along England's coast. In America, somewhat later, at the Philadelphia Exposition of 1876, Moses G. Farmer exhibited three glaring arc lights, burning in the open air, powered by his own rude dynamo. A year later Edison received a report on the new "electric candles" which Paul Jablochkoff, a former Russian officer of engineers, had successfully introduced as street lamps in Paris. This report Edison pasted in his scrapbook.[4]

At the time (1877), he himself set to work experimenting with open arc lights having carbon strips as burners. He also investigated incandescent lights, such as many early investigators had tried to perfect. The incandescent light was entirely different from the arc light in principle: it used a slender rod or pencil enclosed in a glass globe from which the oxygen had been exhausted, more or less. Electric current heated the pencil to incandescence, the absence of oxygen preventing the metallic or carbon rod from burning out or melting. As yet, only poor results had been obtained with incandescent lamps, after fifty years of desultory experimenting. Baffled by the perplexities of the task, Edison had dropped his electric-light experiments to devote himself to the phonograph.

On his return from the West, at the end of August, 1878, he found a file of papers sent him by Grosvenor P. Lowrey, general counsel to Western Union, reporting on the Paris Exposition that summer and especially on the new electric candles of Jablochkoff. By now a half mile of the Avenue de l'Opéra had been illuminated by those big arc lights; they were said to have provided the finest artificial light ever seen. According to the testimony of the American physicist, Professor Benjamin Silliman, Jr., new dynamos invented by Z. T. Gramme were used to provide a source of constant current; as the carbon rods, or "candles" burned out they were replaced by hand or by an automatic feeding mechanism.[5]

Now Lowrey joined Barker in urging Edison to undertake an investigation of the electric light. In the United States, Charles F. Brush of Cleveland, as well as Moses Farmer, had already begun to introduce arc lights on a commercial scale for street lamps and for illuminating factories and shops—Wanamaker's store in Philadelphia already used arc lights. Early in September Edison agreed to go with Professor Barker to Ansonia, Connecticut, to visit the brass-manufacturing shops of William Wallace, partner of Moses Farmer and coinventor of the first American electric dynamo. After receiving the inventor and his party warmly, Wallace exhibited eight brilliant arc lights of 500 candlepower each as well as the Wallace-Farmer dynamo of 8 horsepower that supplied them.

As an eye-witness related:

Edison was enraptured. . . . He fairly gloated. . . . He ran from the instruments to the lights and then again from the lights back to the electric instruments. He sprawled over a table and made all sorts of calculations. He calculated the power of the instruments and the lights, the probable loss of power in transmission, the amount of coal the instrument would use in a day, a week, a month, a year. . . .

He then turned to Mr. Wallace and said challengingly, "I believe I can beat you making the electric light. I do not think you are working in the right direction." They shook hands in friendly fashion and, with a diamond-pointed stylus, Edison signed his name and the date (September 8, 1878) on a wine goblet served by his host at dinner.[6]

2   He had merely played with the idea of making electric lights before this time—at Newark in 1876, and at Menlo Park in the fall of 1877. Now he was on fire. After examining the Wallace-Farmer arc lights, he relates, "I determined to take up the search again. On my return home I started my usual course of collecting data. . . ."[7]

What made him feel in such fine fettle leaving Wallace's place, as he said shortly afterward, was that

I saw for the first time everything in practical operation. I saw *the thing had not gone so far but that I had a chance*. I saw that what had been done had never been made practically useful. The intense light had not been subdivided so that it could be brought into private houses. In all electric lights theretofore obtained the intensity of the light was very great, and the quantity (of units) very low. I came home and made experiments two nights in succession. I discovered the necessary secret, so simple that a bootblack might understand it. It suddenly came to me, like the secret of the speaking phonograph. It was real and no phantom. . . . The subdivision of light is all right. . . .

A reporter for one of the leading New York dailies had "shadowed" him to Wallace's at Ansonia and had obtained a startling interview with the inventor. With soaring imagination, Edison communicated to the reporter his vision of a central station for electric lighting that he would create for all New York, and from which a network of electric wires would extend, delivering current for small household lights, unlike the blinding arc lights made by Farmer and Brush. In some way (as yet unknown) the usage of electric light would be measured, that is, metered, and sold. He said he hoped to have his electric light invention ready in *six weeks!* He was building additional shops adjacent to the Menlo Park laboratory in which to carry out this large undertaking. Then he would erect posts along the roads there, connect lights with all the residences, and hold "a grand exhibition." [8]

He had an air of supreme confidence, with which he sought to imbue everyone around him (including interested capitalists). In truth, he had decided to enter the field rather late; other able inventors had begun long before him: Farmer, Brush, William E. Sawyer, Hiram Maxim, and in England, Joseph W. Swan and St. George Lane-Fox, were all then at work inventing electric lights. But who could move faster than Edison?

The plans he announced were, in breadth and originality of conception, far in advance of anything earlier attempted in this field. As he had lately become a national figure, newspaper reporters besieged him at every step, and he fired off many interviews, often indiscreet but sometimes quite revealing, so that we may follow the stages of his progress almost day by day in the newspapers, as well as in his laboratory notes and correspondence. His records at this period are also more complete and explicit, possibly a legal precaution.

During the visit with Wallace the intuition had come to him that he must somehow "subdivide" the intense light given by the arc light into many small, mild lights for domestic use, which could be controlled individually. At the same time the image of the central gashouse and its distributing system, of gas mains running to smaller branch pipes and leading into many dwelling places, had flashed into his mind. With the gas system a man could turn a single jet in one room on or off, or do so with a hundred jets. Why could not Edison do as much with an electrical distributing system having a central power station? In the case of electric current there would be far more difficult problems of distribution, those of resistance and consequent loss of pressure in the conductors; but Edison was confident that he could work them out quickly.

The possibility that big carbon arc lights could be subdivided into small ones, mild and safe enough to be installed in private houses and regulated independently of each other, had already been discussed by Edison and Barker in the course of their recent western journey. Arc lights ran as high as 3,000 candle power, but they gave an almost blinding glare, burned in open globes emitting noxious gases, and could only be employed high overhead, in streets or in high-ceilinged factories or shops. Moreover they used large amperages and were wired to the dynamo "in series," so that all the lamps had to be turned on or off together. Their inventors had not yet learned how to make smaller lights, nor how to connect them in specially adapted circuits so that they could be controlled individually. But how such subdivision of the current was to be achieved Barker, of course, did not know; though a good fellow, he was only, in Edison's estimation, a college professor. Edison, however, thanks to his telegraph work, had enormous experience in the engineering of electrical circuits. The subdivision of current for a circuit with multiple branches, so to speak, supplying a large number of outlets with small amounts of current for each, had never been done and was considered an "impossibility," or at least only theoretically possible, and prohibitive in cost. Moreover, the small lights would have to be entirely different in principle from those subject to the experiments of contemporary inventors. The kind of light he had in mind—at first, only vaguely, as an intuition—one that used very little current, had never been invented. So much the better! Edison was willing.

Seeing the Wallace-Farmer arc light apparatus with his own eyes had given him the inspiration. He saw what they had *not done*. Only two weeks after the whole idea had come to him, he cabled a European agent, who handled foreign patent negotiations for him, *"Have struck a bonanza on electric light—indefinite subdivision of light."* [9]

From the first his plan to devise a system of many small lights having the mildest of ordinary gas jets (8 candle power), was entirely different from that which the arc light inventors, such as Jablochkoff, Farmer, and Brush were working out. Edison was aiming to duplicate the gas-distributing industry—with electricity! Gaslighting in the past half century had reached the stature of a major industry in America, with annual revenues of 150 million dollars. It was centered in the cities, while three fourths of the population then lived in rural districts by the dim glow of oil lamps or candles. About 10 per cent of the gas business was in street lamps, now threatened by the electric arc light. Edison now proposed to replace the remainder (90 per cent) of the gaslighting facilities, those used for private and business illumination.

"If you can replace gaslights, you can easily make a great fortune," a New York newspaper reporter remarked to him when these plans were announced.

"I don't care so much about making my fortune," Edison replied tartly, "as I do for getting ahead of the other fellows." [10]

There was the man, very much himself. To have more money meant little; he had enough, he said, for his needs. But to stand as leader among the world's foremost inventors, to make again and again a great impact on society and industry—even to "change the world," if possible—meant everything.

First, he stated the problem for himself as clearly as he could, setting down his notes while ruminating in solitude:

Electricity versus Gas as General Illuminant

Object: E. to effect exact imitation of all done by gas, to replace lighting by gas by lighting by electricity. To improve the illumination to such an extent as to meet all requirements of natural, artificial and commercial conditions.
Previous inventions failed—necessities for commercial success and accomplishment by Edison. Edison's great effort—not to make a large light or a blinding light, but a small light having the mildness of gas.[11]

His attention is riveted, at the beginning, not so much upon the search for an improved type of incandescent light as upon the analysis of the social and economic environment for which his invention is intended. He studies the organization of the gaslight industry itself. Gas had its inconveniences and dangers. "So unpleasant . . . that in the new Madison Square theater every gas jet is ventilated by small tubes to carry away the products of combustion." But whatever is to replace gas must have "a general system of distribution—the only possible means of economical illumination."

Gathering together scores of volumes bearing on gas illumination, and all the back files of gas industry journals, he studies the industry's operation, habits, its seasonal curves of consumption, and even its geographical character. He draws charts and tables covering such information, then maps out in his mind a network of electric light lines for a whole city, making the shrewd judgment: "Poorest district for light best for power— thus evening up whole city." He sees that in slum districts, for example, there will be more demand for small motors by business firms. His calculations cover the cost of gas conversion from coal, and the comparable cost of converting coal and steampower into electrical energy with existing dynamos. An expert gas engineer, whose services Edison engaged at this time, observed that few men knew more about the world's gas business than Edison.

Edison had a *homo economicus* within him, a well-developed social and commercial sense—though he might be careless of money itself and was no bookkeeper of the John D. Rockefeller type. Before experimental work on the invention itself was under way, he had formed a clear notion of what its object must be, stated in economic terms. This pragmatic concept guided his search and determined the pattern of his inventive work, so that its result would be no "scientific toy," but a product useful to great masses of people everywhere.

By his initial costing and careful quantitative calculations of power and raw materials, such as machinery and copper for a whole system of light distribution, he was led to define exactly both the kind of light he sought and the kind of circuit he needed. Indeed, the scheme of a central power station with a network of thousands of small lighting units in many dwelling places was his ruling concept almost from the hour when he saw the crude Wallace-Farmer apparatus. It was well understood by him, from the start, that the development of a new and different circuit would be necessary in place of the series circuit used by the arc light systems. In order that the current might be "divided" and that his lights might each be used independently, Edison concluded that those small lights must be connected in a *parallel,* or *multiple,* circuit.

Experiments had recently been made in Europe with an improved parallel circuit that assumed the form of a ladder set down horizontally along the ground, with outlets connected at each "rung" and the electric current flowing along the "legs" of the "ladder" and through its "rungs." If one light were turned off, or broken, it would not affect the others on different "rungs"; whereas in the straight-line series, as with a string of beads, if one unit broke, all would go. The new circuit (but little tried as yet) would need much study and calculation.

Edison had a very strong feeling, as he began this investigation, that the existing techniques of electric lighting were inadequate and did not even take full advantage of available scientific knowledge. He saw, for example, that he would need a light so constructed that it used little current, one made on entirely different principles from those contrived by other inventors, which were, in almost all cases, of low resistance and consumed much current. His own plans had come to him at first only in vague outline; none of the details had been worked out in terms of actual ohms, volts and amperes. He would be striking off into unexplored ground—but surely great opportunities beckoned there. Hence his initial outburst of optimism, exuberant and, seemingly, excessive.

There were two main avenues which experimenters seeking to develop the electric light had followed: that leading to the big arc light, and that of the small incandescent unit, the enclosed glow lamp. Though the arc light had recently been brought to the commercial stage, Edison struck out for the incandescent light in a vacuum—the *ignis fatuus,* the will-o'-the-wisp, which so many inventors had pursued in vain during a half century. But then Edison's projected lighting circuit for popular domestic use demanded a small illuminating unit that would consume little current.

Progress in this field had been terribly slow. As far back as 1820, De La Rive, in France, had tried to make such a glow lamp in a vacuum enclosed in glass. Then De Moleyns, in 1841; after him the young American inventor J. W. Starr, in 1845, among others, had used metallic or carbon "burners" in more or less exhausted glass vessels or tubes, with indifferent results. In England, Joseph W. Swan, a skilled chemical scientist, had made glow lamps with platinum or carbon conductors over a period of twelve years, then had abandoned his experiments as failures in 1860. More recently, the Russian inventors, Konn and Lodyguine, had contrived promising models of incandescent lights in vacuum globes; and several Americans had done as much, notably William E. Sawyer, who enclosed carbon rods in a vessel of inert gas, such as nitrogen. All in all, as Edison found, the "state of the art" was still unsatisfactory: none of his predecessors had succeeded in making an incandescent lamp that would burn for more than a few moments; the variables they had encountered proved to be too baffling.

In making his decision to choose the incandescent light over the arc light he was "refusing a path that looked very promising," and "putting aside the technical advance that had brought the arc light to the commercial stage." [12] For the problems of the incandescent light were entirely different from those of the arc light; it would need a different type of dynamo and a different kind of circuit, not to speak of many new safety devices. Edison was, in fact, going against the stream, was undertaking a far more difficult investigation, than that of the adapters of arc lights.

Meanwhile it remained for him also to find the right vessel for his idea, an incandescing substance that would endure a fierce heat, yet would not fuse or melt and burn out within a few minutes—the very thing that scientists had been seeking for fifty years.

He naturally turned to carbon in his first experiments, since he was very familiar with that wonderful material and knew that it had a very high melting point. Strips of carbonized paper were tried as "burners," or partial conductors; they were made incandescent in the open air and quickly oxodized, merely to ascertain how much current was required. Then he used some glass jars, partially evacuated of air by means of a hand-worked pump, and managed to keep his carbon strips (a sixteenth of an inch in breadth) incandescent for "about eight minutes" before they went out. He was now inclined to accept the prevailing view that carbon was easily destructible and laid it aside for the time being.[13]

He seems also to have tried making a rapid *tour d'horizon* (over an already well-traveled area) so as to familiarize himself as quickly as possible with this new subject. For his incandescing substance he tested various infusible metals that others had long experimented with. Out of a whole group of refractory metals he chose platinum; to be sure, its melting point was somewhat lower than carbon's, being about 3,191 degrees Fahrenheit. On introducing a spiral of platinum wire into a globe partially exhausted of air, he was able to bring it to incandescence and achieve a brilliant light. However, at high heat the platinum burner quickly melted and the light went out. Therefore, after the first week or two of experimenting, he contrived a little shunting device by connecting a straight rod, also of platinum, to the burner; when the temperature rose too high the additional rod quickly expanded enough to short-circuit the burner and allow it to cool off. After the rod cooled off, it contracted and, within a fraction of a second, reopened the circuit. What he had was a lamp that blinked instead of going out entirely. Though there was nothing exactly new in this thermostatic regulator, and it worked unreliably, Edison quickly signed a first application for a patent on October 5, 1878.[14]

He then humped himself over his laboratory table and designed a second platinum-wire lamp with a more sensitive regulator, having a tiny diaphragm that responded to the heated air within the glass globe. After that he tried a third lamp using an iridium-platinum composition for his incandescing substance. These early experimental lamps played exasperating tricks on the inventor, who studied them for long hours, brooding, distracted, letting his cigar go out, lighting it, letting it go out again.

To be sure, he was learning a good deal as he went on. The earliest notes on the electric light experiments of September 11–26, 1878, show him already keenly aware of the importance of obtaining a higher vacuum, which he achieved by cementing the bottom of his leaky glass container. He was also trying to measure exactly the resistances of various incandescing materials in ohms and, in the early autumn of 1878, was beginning to make his first rough calculations of a multiple, or parallel, circuit—which would permit him to "subdivide" his electric current for many small units. It was going to be a long, hard job, he realized, requiring added equipment, a larger staff, and a good deal of money.

Grosvenor Lowrey had kept in close touch with him in those early weeks. He had promised that if Edison undertook to invent a practical electric light, he would approach the Western Union directors, who were his clients, for funds to finance Edison's research work. Lowrey, an informed patent attorney and also one of the leading corporation lawyers of New York, had fallen completely under the spell of the self-taught inventor and regarded him much as an ardent collector of paintings regards a great artist whose works he believes are destined to become immortal. On receiving assurances that Edison would actually undertake the work on the electric light, Lowrey began to form a syndicate of capitalists to back his inventive research. From the start of the lamp experiments, the lawyer went about buttonholing his well-heeled clients among the Vanderbilt-Morgan clan, assuring them that "Edison has discovered the means of giving us an electric light suitable for every day use, at vastly reduced cost as compared with gas." [15]

After the first brush with the new subject, Edison really had nothing of practical value; he had only a platinum burner with a shunt device that worked for about ten minutes. This did not prevent him from calling in the metropolitan newspaper reporters and declaring that he had "already discovered" how to turn electricity into a cheap and practical substitute for illuminating gas. After the sensation made by his phonograph the leading New York dailies, especially James Gordon Bennett's *Herald* and Charles A. Dana's *Sun,* made it their business to treat everything that Thomas A. Edison said or did as "copy." Knowing this, Edison initiated a full-scale press campaign aimed at assuring the public that the success of the electric light invention was already assured.

Was there ever such effrontery? Here the showman in the applied scientist certainly revealed himself. In numerous press interviews authorized by him and printed in the New York *Sun,* the New York *Herald,* and the New York *Tribune,* from mid-September to mid-October, he announced the early arrival of the age of electric light. In somewhat mystifying style he remarked:

Singularly enough I have obtained it [the light] through an entirely different process than that from which scientists have sought to secure it. They have all been working in the same groove. When it is known how I have accomplished my object everyone will wonder why they never thought of it. . . . I can produce a thousand—aye, ten thousand lights from one machine.

Together with light, he would transmit energy for power and heat, to cook food, to run an elevator, a sewing machine, or anything requiring a motor.[16] A few weeks after his first forecast, he admitted that while there would be "no difficulty about dividing up the electric current" he was still looking for a good "candle" which would give a pleasant light. He had quit experimenting with carbon and was now using platinum wire, though he still hoped for something better. He concluded his remarks to the press that day in a challenging tone: "I have let the other inventors get the start of me in this matter . . . but I believe I can catch up to them now." [17]

Those first Barnum-like trumpetings in the press made a sensation. But greater still was that which followed a public interview of October 18, wherein he solemnly declared that he had found just the kind of light he wanted. The reporter asked:

"Are you positive? . . . .
"There can be no doubt about it.
"Is it an electric light?
"An electric light and nothing else. . . . We simply turn the power of steam into electricity. The greater the steam power we obtain, the more electricity we get."

To be sure, the inventor now conceded, he needed time, perhaps several months, or a year, to "get the bugs out." Nevertheless he put on a show of his platinum-wire lamp. As a reporter described the scene, he turned on the Wallace-Farmer 8-horsepower dynamo and

touched the point of a wire on a small piece of metal near the window casing . . . there was flash of blinding white light. . . . "There is your steam power turned into an electric light," he said. Then the intense brightness disappeared, and the new light came on, cold and beautiful. . . . The strip of platinum that acted as a burner *did not burn.* It was incandescent . . . set in a gallows-like frame; but it glowed with the phosphorescent effulgence of the star Altair. A turn of the screw, and . . . the intense brightness was gone; the platinum shone with a mellow radiance through the small glass globe.[18]

But if the artful Edison had not turned off that screw, the light would have gone out by itself within a few minutes.

The boast made on this occasion that he would soon light up the entire downtown area of New York with 500,000 incandescent lamps, powered by a few steam dynamos, created excitement on both sides of the Atlantic. In London as well as New York, electrical scientists promptly stigmatized Edison's claims and half-revelations as bombast. In a leading scientific journal Professor Silvanus P. Thompson asserted that anyone who tried to invent an incandescent electric light was "doomed to failure"; that Edison's talk of subdivision of currents showed "the most airy ignorance of the fundamental principles both of electricity and dynamics." [19] In New York a rival electrical inventor, Sawyer, predicted that Edison would fail not only in his scheme of subdividing the electric light current but also in his attempts to make a burner of platinum—for Sawyer himself had proved that platinum was useless.

Nonetheless, the repute of the Wizard of Menlo Park was so formidable that his published claims created a fair-sized financial panic on both the London and New York stock exchanges. "Owing to the publication of Professor Edison's discovery of the distribution of electric light," a cabled dispatch from London reported, all gaslight securities within a few days had lost about 12 per cent of their value. It was now widely believed that, thanks to Edison's entrance into the field of electric lighting, the whole flourishing gaslight industry might soon enter into a decline.

Edison himself showed less concern than anyone else at the economic convulsions his advance advertising had provoked, or over the current attacks upon his capacity as an applied scientist. Was it part of wisdom to start out with so much publicity upon such a hazardous series of experiments? One of the inventor's closest associates of that period remarked, long afterward, "I have often thought that Edison got himself into trouble purposely, by premature publication . . . so that he would have a full incentive to get himself out of trouble." [20]

It was the forehanded Lowrey, however, who had encouraged Edison to make these premature public statements; and it was at the lawyer's advice that Edison maintained silence thereafter and gave orders that the Menlo Park Laboratory be kept closed to all visitors. [21]

The idea of a press campaign was not at all foolish under the circumstances. It was designed to kindle the hearts and loosen the purse strings of the Wall Street capitalists whom Lowrey now eagerly solicited in behalf of his talented protégé.

Grosvenor Lowrey was intelligent and handled Edison with much skill. A New Englander by birth, he had lived through a romantic and adventurous youth, having gone to Kansas to fight against slavery under the banner of John Brown; in the Civil War he had risen to be a major in the Union Army, and afterward had returned East to establish himself as one of the leading members of the New York Bar, with wide political and business connections.

On October 1 Lowrey wrote Edison that he had approached the son-in-law of W. H. Vanderbilt, Hamilton M. Twombly, and told him that the inventor "was willing to sell half of this invention [the electric light] for $150,000." Edison noted on the back of this letter, "All I want at present is to be provided with funds enough to push the light rapidly." [22]

On the following day Lowrey was closeted for an hour and a half with Twombly and W. H. Vanderbilt in the latter's great mansion on Fifth Avenue, discussing capital subscriptions and royalties to be paid Edison. That night he wrote Edison that he hoped to get for him a "clear $100,000" from the powerful group of Western Union directors. Lowrey also solicited the interest of the partners of Drexel, Morgan & Company. He urged Edison to be discreet, to avoid negotiating directly with anyone, or promising anything, without his (Lowrey's) counsel. To which Edison replied with a short message, "I shall agree to nothing, promise nothing and say nothing, leaving the whole matter to you." [23]

In heartening fashion Lowrey wrote the next day:

Give your whole mind to the light. I will see that not only do you get what you ought to have, but that every reasonable expectation of those you have spoken to is satisfied. I want the Western Union people to have first chance at this because both you and I know whom we are dealing with.[24]

Evidently the spate of sensational stories in the New York newspapers about Edison's new plans had helped the Western Union people to make up their minds. They had narrowly missed buying the telephone patent of Bell and now could not get it at any price. They would not pass over the chance of Edison's light. On October 12, 1878, Lowrey was able to report that he had a contract ready and would send Edison a first check for $30,000, which arrived a few days later.

The contract, drawn up by Lowrey between Edison and the syndicate of capitalists who were to finance his researches, was one of the most noteworthy in all the annals of American industry. In effect the Western Union people were buying rights, not in an existing invention, but in an unknown quantity, a promise. "Their money," Edison said, "was invested in confidence of my ability to bring it back again." Undoubtedly these financiers were taking risks with their capital, though if we read the roster of the initiating group it will be seen that for such grandees as Vanderbilt ("the richest man in America") and his associates it was "only peanuts." The original investing group, who became directors of the new company, figured prominently in the Social Register and the New York financial district: Vanderbilt, Twombly, President Norvin Green of Western Union, Eggisto Fabbri (a partner of J. P. Morgan), Lowrey, and also Tracy Edson and James Banker, both important capitalists and associated with Western Union. Each of these men, under a preliminary agreement of October 15, 1878, invested a few thousand dollars cash by subscribing altogether to five hundred shares of the proposed company's stock, amounting to $50,000, which was paid over to Edison in installments. The new company was to be known as the Edison Electric Light Company, and was originally to have 3,000 shares ($300,000 capital stock), of which 2,500 shares were to be given to Edison. For his part, he agreed to assign to the company all his inventions and improvements in the electric lighting field for a five-year period. If Edison's future inventions were successful, his stock would be worth a good deal; if he failed, it would be worthless. All that the backers risked was their original $50,000 cash—in return for which they had the chance of controlling all of Edison's patents in this field and of developing, by investment of added capital, a world-wide patent-holding and licensing company profiting from the inventor's patents in electric lighting.*

Behind the whole venture stood the figure of J. Pierpont Morgan, already the country's outstanding banker, who kept his name off the board of directors but had his partner Fabbri serving as director and treasurer of the new company. Its banker was also to be Drexel, Morgan and Company. At the time, Edison believed that the chances of ultimate success of his project were "vastly greater" because of the Morgan connection.[25]

To have gained support for such a novel undertaking from America's leading financiers was decidedly a feather in the cap for Thomas A. Edison's little all-purpose "invention-factory" at Menlo Park. He was simply given a blanket order to create an electric lighting and distributing system, and to develop a pilot plant, or model, of that system.

The launching of the Edison Electric Light Company by such sponsors is noteworthy also in that it inaugurates a phase of increasingly close relations between big business and technology in this country. It is significant that the leading innovators of incandescent electric lighting and arc lights, like Edison, were men who had first acquired their knowledge of electricity through work on telegraphic devices. At the same time it was the capitalists who had the experience of investing in the country's biggest electrical enterprise of that era, the Western Union Telegraph Company, who originally provided the funds for Edison's research and development work in electric lighting.

The closing decades of the nineteenth century witnessed the heyday of the practical inventor, or applied scientist; certainly the Yankee inventors were creating much new industry and wealth by their mechanical skills. Yet in the growing field of electrical engineering, during the 1870s, there was a considerable gap between the United States and Europe, where the Germans, Russians, French and British were already making fairly efficient dynamos and setting up brilliant arc lights in their great cities. America, before 1878, had almost nothing of the kind to show. By his new venture Edison hoped to win world leadership in this field.

* The articles of incorporation on November 15, 1878, stated: "The objects for which the said company is formed are to own, manufacture, operate and license the use of various apparatus used in producing light, heat and power by electricity." (Photostat in Edison Laboratory Archives.)

This whole period of the prime of his life, the twenty years spanning his thirties and forties, happened to coincide with a prosperous postwar cycle in which our capitalists showed a marked appetite for innovations and an unusually progressive spirit. At this time they seemed to grasp thoroughly the importance of investing in scientific research, and in the engineering of new steel furnaces, giant steam engines, locomotives, turbines, high-speed printing presses, and new electric generators and lighting devices. But instead of taking up an invention by chance, as in the past, the Drexel-Morgan and Western Union group of financiers, in the case of Edison's lighting project, were enlisting the services of a specialized research organization and its scientific captain in an effort to develop a new and revolutionary product.* On this occasion the group of capitalists backing Edison were showing real vision, though at other periods they were by no means consistently cordial to inventive research.

It is the lawyer Lowrey, however, who must be given chief credit for the original promotion of this enterprise. It was he who generated enthusiasm for the inventor's plans in high financial circles. Lowrey, in short, assumed that inventor-patron relationship which H. S. Hatfield said "can be as vital as the husband-wife relationship." [26] While the inventor may be obsessed with his one, overruling interest, and unable to follow good business procedure, or give heed to other practical difficulties affecting the realization of his scheme, the patron-capitalist—whether representing a government institution or a speculative group—is in a position to help bring the invention along to the stage of successful exploitation. Lowrey understood Edison's temperament and at the same time knew the money men very well. After a while the Wall Street group made it their practice to communicate with Edison only through Lowrey, while the inventor did likewise in dealing with them.

The influence of Lowrey showed itself even in the personnel selected for Edison's expanding staff. The inventor was thought to be careless or temperamental in his business procedure. At Lowrey's suggestion, S. L. Griffin, a junior executive at Western Union, was assigned to act as Edison's private secretary and oversee his business affairs. Francis Jehl,

---

* The Medicis of Florence subsidized engineers like Leonardo da Vinci, and later a scientist like Galileo, for the purpose of devising new military engines. In England and Europe, from the eighteenth century on, manufacturing capitalists sometimes helped inventors to devise or improve upon divers machines. On the other hand Samuel F. B. Morse was unable to gain the support of American capitalists and was only enabled to develop his telegraph and demonstrate it effectively, in 1844, through the grant of $8,000 by Congress. The support of the Morgan-Vanderbilt syndicate of a *research program* for Edison's laboratory—though small in scale—appears unique and marks a new stage in the relations of capital to technology.

a clerk in Lowrey's office, was also hired, in February, 1879, as a laboratory assistant, at the lawyer's recommendation. It happened that during two years at Menlo Park Jehl kept a private diary recording some of the remarkable events he witnessed there; it was to be the basis for his *Menlo Park Reminiscenses,* published fifty years later.

Not only were Lowrey and the Western Union group kept fully informed of Edison's movements, but his progress in experimental work, though ostensibly secret, seems also to have been reported in part to some of his enemies, particularly those connected with the gaslight industry. In November there were rumors in the New York press that the inventor was "ill" or "worn out" by his tremendous exertions; also that he had met with great disappointments. At once Lowrey issued denials of such reports, declaring that "Mr. Edison is in good health and excellent spirits . . . on the threshold of a new and wonderful development of electrical science." [27]

The relations of the inventor and his capitalists continued to arouse comment, sometimes wondering, sometimes invidious, not only in America but in Europe. Thus the Paris *Figaro* around this time published a rather fanciful article by its New York correspondent:

It should be understood that this astonishing Eddison [sic] does not belong to himself. He is the property of the telegraph company, which lodges him in New York at a superb hotel, keeps him on a luxurious footing and pays him a formidable salary, so as to be the one to know and profit by his discoveries.

This company has in the dwelling of Eddison men in its employ who do not quit him for a moment, at the table, on the street, in the laboratory; so that this wretched man, watched as never was a malefactor, cannot give a second's thought to his personal affairs without one of his guards saying, "Mr. Eddison, what are you thinking?"

The facts were quite otherwise. To keep Edison under control of the financiers as a "captive scientist" was no simple matter, as is shown by a clash that came at the very outset of his electric light campaign. As soon as the new project was reported, other inventors, in Europe as well as in America, had become unusually active in this field. In England, the accomplished Joseph W. Swan and St. George Lane-Fox pushed forward with their own experiments in vacuum lamps with carbon rods. Then, at the end of October, the American William E. Sawyer and his partner, Albon Man, claimed that they had "beaten" Edison in the race for the electric lamp when they applied for a patent on their model of a carbon-pencil light enclosed in a glass tube containing nitrogen.[28] This invention later proved to be unsatisfactory. But the mere reports of such developments, together with attacks on Edison, created a flutter of panic among his financial backers. At a directors' meeting of the Edison Company the suggestion was made that they join forces with Sawyer and Man by buying out their patents. Lowrey passed the suggestion on to one of his henchmen at Menlo Park.

The vigor of Edison's reaction seems to have given his backers pause, judging from a letter (marked "confidential") written by his new secretary, Griffin, to Lowrey.

Menlo Park
November 1, 1878

DEAR SIR:

. . . I spoke to Mr. Edison regarding the Sawyer-Man electric light, being careful not to say anything beyond what you told me. I was astonished at the manner in which Mr. Edison received the information. He was visibly agitated and said it was the old story, that is lack of confidence—the same experience he had had with the telephone, and in fact, all of his successful inventions, was being re-enacted! No combination, no consolidation for him. I do not feel at liberty to repeat all he said, but I do feel impelled to suggest respectfully that as little be said to him as possible with regard to the matter. He said that it was to be expected that everyone who had been working in this direction, or has any knowledge of the subject would immediately set up their claims, upon ascertaining that his [Edison's] system was likely to be perfected. All this he anticipated, but had no fears of the result, knowing that the *line he was developing was entirely original and out of the rut.*

I was careful to say to him that as far as you were concerned there was no lack of confidence. He will give you his own views in full, so I will abstain at present.

S. L. GRIFFIN [29]

After that there was no further talk of Edison working with Sawyer or any other inventor.

The whole project, of course, turned out to be much bigger and more difficult than anyone had foreseen. Indeed the alliance between the high-spirited inventor and his capitalists might have run even less smooth a course than it did, in the early, arduous years, were not the tactful and patient Lowrey always on hand.

3   One reason why Edison believed he would "get ahead of the other fellows" was that he had unbounded confidence in his laboratory, his superior scientific equipment, and his staff. All of this was rapidly expanded that autumn, when three sizable buildings were added to the original "tabernacle." One was a separate office and library, placed near the gate to his grounds; another, an engine

house of brick construction, was set in the rear, to contain two 80-horse-power steam engines; a third was a glass blower's shed.

One of the happiest effects of Lowrey's personal influence was the engagement of Francis R. Upton as chief scientific assistant and mathematician at Menlo Park Laboratory. Upton, a native of Massachusetts, was a tall black-bearded young man with distinguished manners. As if to compensate himself for his sense of inferiority in formal scientific knowledge, Edison jocularly nicknamed Upton "Culture"; and to put him in his place Edison played one of his typical scientific tricks on the "green" mathematician.

He brought out a pear-shaped glass bulb intended for lamp experiments—according to a story often repeated, with many variations—and gave it to Upton, asking him to calculate its cubic contents in centimeters. Upton drew the shape of the bulb exactly on paper, and got the equation of its lines, with which he was going to calculate its contents, when Edison again appeared and impatiently asked him for the results. The mathematician, after having worked for an hour or so, said he was about halfway through and would need more time. "Why," said Edison, "I would simply take that bulb, fill it with a liquid, and measure its volume directly." That is, he would pour the liquid contents of his bulb into a graduated cylinder for measuring volumes, and would get it in five minutes. Apparently Upton had not thought of that one, but only of obtaining the most precise measurements. He was taken aback. The story, which has been repeated in many different ways, is supposed to illustrate the contrast between the practical, "Edisonian" rule-of-thumb method and the mathematical scientists' different mode of attack on the same problem.[30]

This work, however, would require a good deal of precise calculating of electrical conductors, resistances and dynamo capacities from now on. Edison found Upton very useful, and strong where he himself was weak. "Any wrangler at Oxford would have been delighted to see Upton juggle with integral and differential equations," Jehl recalled. Edison would lead the way with his "intuitions," and Upton would do the checking and calculating. Upton, for his part, later described the astonishment he first felt at the inventor's "wonderful flow of ideas, which were very sharply defined, as can be seen by any of [Edison's] sketches, as he evidently always thinks in three dimensions." The university-educated scientist was also much surprised at Edison's clear understanding and unusual application of Ohm's and other electrical laws, which enabled him to choose correct voltage, lamp resistance, and conductor size. Edison's "guesses," though seemingly at variance with contemporary interpretation of scientific principle, usually turned out to be right.

I cannot imagine [Upton said later] why I did not see the elementary facts in 1878 and 1879 more clearly than I did. I came to Mr. Edison a trained man, with a year's experience at Helmholtz's laboratory ... a working knowledge of calculus and a mathematical turn of mind. Yet my eyes were blind in comparison with the eyes of today; and ... I want to say that I had *company!* [31]

Upton, nevertheless, proved indispensable to Edison in the second and very important stage of his search, which followed almost immediately upon his decision to use a multiple, or parallel, circuit. This concerned the *form* of incandescent lighting unit that was to be used in his proposed circuit.

Careful quantitative studies of those first, unsatisfactory lights he had tried, with carbon and platinum burners offering low resistance to the passage of current, had yielded estimates of how much copper wire conductor would be needed for a circuit of such lights, and of what requisite thickness, or cross section. After allowing for a given percentage of drop in voltage in the line at a given distance from his dynamo for a certain number of lights, he found that his circuit would need a fabulous amount of copper in order to connect up a few city blocks. Such a system of low-resistance lights was simply a commercial impossibility! Those first estimates, made with the help of Upton, had been for (low-resistance) incandescent lights consuming about ten amperes of current at ten volts, with a resistance of only one ohm.

It will be recalled that, after September 8, 1878, the idea had come to Edison—perhaps as a "flash of inspiration," perhaps by way of close reasoning and fresh insights—that he must contrive a light with a very *high resistance* to the current and using, therefore, but small quantities of it. Yet no other scientist or inventor had attempted this. There was his "secret" that was so "simple." Now he proceeded to test his idea. Reversing his course, he made estimates for a system of relatively high voltage, supplying lights of very high resistance that would use little current. At his order, Upton, in November and December of 1878, made calculations on the basis of a given number of lights of 100 ohms resistance, consuming only one ampere of current at 100 volts, over the same distance of line and with the same assumed loss of tension. What thickness and total quantity of copper main would be needed for such a circuit? The result was astonishing: only *one one-hundredth* of the weight of copper conductor would be required for such a system as compared with that of the low-resistance system. And copper was the most costly element involved—really the crux of the problem.

Thus Edison determined both what kind of electric distributing system and what form of incandescent lamp would serve his purpose—and set out to look for it; that is, to devise exactly such an incandescent burner as was needed. He knew that it must be one offering great resistance to the passage of the current, and, therefore, having a small radiating surface, or cross section.

It would seem that Edison even had to reeducate his physicist to a true understanding of Ohm's law of electrical resistance. Most electricians in 1879 did not thoroughly understand this fundamental formula for measuring the flow of electric current (though it was first made known in 1827) and scarcely were able to calculate in terms of volts and amperes.* [32] Most of the would-be inventors of electric lamps were hunting for some substance that would withstand great heat without melting. They had not yet conceived of a fundamental change in the form of the electrical circuit determining the components of the illuminant, as Edison projected it in his mind.

Up to now, the practical, or applied, scientists had not needed much electrical theory, and the theoreticians had had little contact with practical problems except in telegraphy and arc lighting. The work done at Menlo Park proved to be far in advance of that attempted by other inventors and helped solve key problems in applied electrical science.

Upton was surprised at Edison's unconventional attack on the problems of electrical distribution. But Edison pounded away at his idea that the low-resistance lights used both in arc lights and in existing glow lamps would not do. The people who were making these things were all wrong, he insisted. Once Upton had grasped Edison's application of Ohm's law of resistance, he was able to make the preliminary computations for the new high-resistance incandescent light adapted to a multiple circuit and using very little current.

When they looked at their figures on paper, the practical inventor and the mathematician were thrilled with hope. A new and strategic invention —thanks to Edison's original insights into the problems of electrical transmission, and his glimpses of new potentials—was now surely within

---

* The relationship between electromotive force, resistance, and current, discovered by the German physicist Georg Simon Ohm, goes as follows: The intensity of the electric current flowing in a conductor equals the electromotive force divided by the resistance. In electrical terms of today, it would read: The amperes in a circuit equal the voltage across it divided by the circuit's resistance in ohms.

reach. It was a great moment. What did it matter if the university scientists were almost all of them against Edison's idea; he had had a hunch all the while that they were wrong.

The verdict of all but a few scientific authorities was in condemnation of his announced schemes, which, they declared, violated the laws of conservation of energy. In England, a committee of Parliament, after having investigated the recent crash in gaslight securities and having obtained the advice of British scientists on the reports of Edison's projects, early in 1879 ruled that while these plans seemed "good enough for our transatlantic friends," they were "unworthy of the attention of practical or scientific men." [33] Adhering to the currently accepted interpretation of Ohm's law of electrical resistance, it was argued that if an electric light of 1,000 candles' luminosity were divided into ten smaller lights and connected by ten equal branches, each would carry not one tenth, but "one hundredth only of the original light." * [34]

These assertions, supported even by Lord Kelvin and John Tyndall, were right in their way, assuming the old conditions of a fixed *amount* of current which, divided among many lamps in parallel circuit, would result either in great losses of current, or would require impossibly large and costly copper conductors. Edison was canny enough not to speak in public of his plan for devising a stable, high-resistance incandescent light, which introduced a new factor into the situation. In the second place, he foresaw that with increasing dynamo efficiency, it was possible to increase current units and thus solve his problem of distribution. In an interview of October 20, 1878, he did reveal the fact that he considered the Wallace-Farmer dynamos he had at present of too low power, and "had an intention of constructing a machine of his own ... that would carry out his ideas more satisfactorily." (The dynamo that he was to design for his proposed system would be unlike the constant-current machines used for arc-lighting; it would be a constant-voltage dynamo, adequate for a parallel circuit of independent lights and supplying as much current as was needed.)

It is not surprising that Edison, the self-taught mechanic, showed greater resourcefulness than the Victorian-era scientists in applying Ohm's

---

* Sir William H. Preece, reasoning in the same manner, said in the course of a lecture before the Royal Institution in London (February 15, 1879):

"It is, however, easily shown ... that in a circuit where the electromotive force is constant, and we insert additional lamps, then when these lamps are joined in one circuit, i.e., in series, the light varies inversely as the square of the number of the lamps in circuit, and that joined up in multiple arc, the light diminishes as the cube of the number inserted. Hence a subdivision of the electric light is an absolute *ignis fatuus*."

law of resistance, which they held he contravened. Like Faraday, he had little mathematical knowledge but a profound experience in handling and using electrical currents; moreover, he had his "intuitions," his fresh insights.

Tyndall, though concluding that Edison's announced plans would be impracticable, had the wit to make some reservations in his adverse opinion because of his genuine respect for the American inventor:

Edison has the penetration to seize the relationship of facts and principles, and the art to reduce them to novel and concrete combinations. Hence, though he has accomplished nothing new in relation to the electric light, an adverse opinion as to his ability to solve the complicated problem . . . would be unwarranted. . . . Knowing something of the practical problem, I should certainly prefer seeing it in Mr. Edison's hands to having it in mine.[35]

One is astonished at how little was known then about the transmission of electrical energy. No incident shows more clearly than this electric light controversy the real interdependence of work in applied science and the theoretical studies of "pure" scientists; in this case the practical experimenters were gathering experience that would eventually serve to correct or modify the interpretations of electrical laws that the theoreticians had long replied upon.

As a modern commentator has summed up this old controversy,

Edison knew he wanted a small light which could be independently controlled. Independent control implied a parallel circuit where, with a constant-voltage generator, the current units could be multiplied as desired. But the increase in current could mean either heavy current loss in transmissions or excessive copper cost. To avoid these the light would have to be of high resistance. That conclusion is easily reached by an elementary application of Ohm's law. But in Edison's time it was an important achievement which placed him far ahead of the other incandescent-lighting inventors and scientists. . . .[36]

All that autumn, and through the winter of 1879 that followed, Edison studied the problems of high-resistance lamps and multiple circuits, as his notebooks show.*

* An example of his original deductions is seen in the Laboratory Notebooks for December, 1878:

If you have 100 lamps each of one-inch radiating surface and each of a resistance of one ohm, all connected in series, and to a battery which will keep them incandescent, then you can make 100 lamps of 100, 1,000, or 10,000 ohms resistance, arranging them (in multiple) so the combined resistance of the whole equals that of the 100 one-ohm lamps in series, and the result will be the same. But for general lighting the high-resistance lamp will be the best, not because it is more economical, but because it is impracticable to work in series, and all lamps given to customers must be in multiple arc. (Edison Laboratory Notebooks, Dec. 15, 1878, *Subdivision,* Edison Laboratory Archives.)

As he himself confessed, this investigation of electric lighting required a most arduous program of research: Edison and his staff made lengthy studies of the electrical resistances of various substances, and examined them also for their heat radiation, recording their specific heats. The effect of increasing or lowering resistance, and of changing voltage or amperage on such materials, and of using them in different forms, was also measured carefully.

Contrary to the prevailing impression, Edison was not primarily "the great tinkerer," but was remarkable rather for his power of observation, his imagination, and clear-cut reasoning faculty. The delicate mechanical tinkering was handled by Charles Batchelor, who was literally Edison's "hands"; Kruesi served as a superb machine maker; while John Ott did the drafting work after Edison's rough sketches. One day Ott drew attention to the inventor's hands; though usually soiled or acid stained, they were soft, white and beautifully formed—the hands of an artist, of a man who imagined things, not those of a workman or a craftsman.[37]

In the period 1878 to 1881, more and more university-trained scientists and engineers were added to the staff at Menlo Park: such were the electrical engineers, William S. Andrews (trained in England) and Charles L. Clarke, both of whom rose to high posts in the electrical world. There was also E. G. Acheson, the future inventor of carborundum, and Frank J. Sprague, afterward a leading inventor of traction motors. These men usually surpassed Edison in theoretical knowledge, but he was the conductor of the "symphony orchestra" at Menlo Park.

At an early stage of this campaign, Edison in a very revealing letter of November, 1878, admitted that all sorts of unexpected difficulties were turning up.

It has been just so in all my inventions. The first step is an intuition—and comes with a burst, *then* difficulties arise.—This thing gives out and then that— "Bugs"—as such little faults and difficulties are called—show themselves and months of anxious watching, study and labor are requisite before commercial success—or failure—is certainly reached . . . I have the right principle and am on the right track, but time, hard work and some good luck are necessary too. . . .[38]

On a later occasion, he also explained of his method that, after the intuition had come, he would settle into purely deductive labors. In this investigation, during several years, the role of chance was small, the accidental discoveries few.

I would construct a theory and work on its lines until I found it untenable, then it would be discarded and another theory evolved. This was the only possible way for me to work out the problem. . . .[39]

In January, 1879, Edison designed and completed his first *high-resistance* lamp, having a very thin spiral of platinum wire as its incandescing substance set in a globe that was as effective a vacuum as he could get with an ordinary air pump. A second, improved model of this lamp dated from April, 1879. The results so far were encouraging; those first lamps burned "an hour or two." [40] He then tackled the dual problem of getting a higher vacuum and also improving the incandescing element (with which inventors had struggled for more than a generation).

Here the conditions for success were rather narrowly defined. The incandescing substance (burner) must resist a tremendous heat before it could give light; if the heat were too great it would fuse or melt. Carbon had the highest melting point: 3,500 degrees C. But when he attempted to maintain a heat of no more than 1,500 to 1,700 C. in his first vacuum globes, the carbon incandescing substance tended to burn out around that level. He tried other materials, and in fact "everything": platinoiridium, boron, chromium, molybdenum, osmium—virtually every type of infusible metal. (He thought of tungsten, but could not use it with existing tools.) A thin wire of nickel seemed to offer promise, during a month of troublesome tests—then gave rise to an accident that almost ruined his eyesight. In his notebook for January 27, 1879, the entry occurs: "Owing to the enormous power of the light my eyes commenced to pain after seven hours' work, and I had to quit." On the next day he wrote: "Suffered the pains of hell with my eyes last night from 10 P.M. to 4 A.M. when got to sleep with a dose of morphine. Eyes getting better...." But he was mortified at losing a whole day recovering!

Giving up the troublesome nickel, he came back to platinum, which, though having a lower melting point than carbon, seemed to show a longer life when incandescent. After having tried inert gas in his globes, such as nitrogen, he resumed efforts to obtain a higher vacuum, which was a shrewd judgment for that time. In England, Sir William Crookes had lately made great progress toward very high vacuums by means of an improved type of pump, called the Sprengel pump, whose flow of mercury trapped air bubbles and expelled them to the outside atmosphere. On learning of this, Edison decided to get hold of one of the first of these new vacuum pumps to reach America, which was then at the laboratory of Princeton College. Upton was sent to Princeton by train and buggy to borrow it until Edison could obtain one of his own from England; on his return, late at night to Menlo Park, broken with fatigue, he found Edison waiting up for him, so ravished at the sight of the new instrument that he was resolved to try it out at once and kept his assistants pumping until dawn. Now a vacuum was obtained that came within one or two millimeters of full exhaustion of air; in a globe having such a high vacuum the thin platinum wire gave forth a brilliant light of 25 candle power and continued to do so for some time, whereas in the open air it melted at once when raised only to 4 candle power.

At this stage, in the late winter of 1879, Edison also made an important chemical discovery that would stand him in good stead. He noted, "I have discovered that many metals which have gas within their pores have a lower melting point than when free from such gas." [41] He had been observing the action of occluded gases within the glass bulb, absorbed within the metallic "burner" elements, then released when they were heated to incandescence, and he noticed their destructive effect. With the aid of the new Sprengel pump he devised a method of expelling these occluded gases from the burner element by sending a current through it and heating it, while air was being pumped out of the bulb. (The same process was devised independently by Joseph Swan, during his concurrent electric light experiments in England). The platinum wire, or coil, within the globe thereupon became extremely hard and achieved a greater resistance to high temperatures. Edison himself, reviewing his work later, felt that at this stage he "had made the first real steps toward the modern incandescent lamp." On discovering that the platinum had been made more infusible by driving out the occluded gases, he was induced to try for a still higher vacuum, on the supposition that this would make for still greater infusibility in the incandescing coil. He would find that this step was of the greatest importance in his later experiments, when he turned to a more rewarding substance than platinum. [42]

By now he had reason to congratulate himself, having made considerable progress beyond the stage which, as he knew, had been reached by others attempting to perfect an incandescent lamp—though it was also true that he seemed to be meeting with fresh difficulties at every step. On April 12, 1879, at any rate, he executed a new patent application for his first high-resistance platinum lamp having an improved vacuum (Patent No. 227,229).

In his exuberant fashion, he now indulged in a few more optimistic reports before newspaper men, aimed both to confound his detractors and to give heart to his financial supporters. He claimed that he had a "nearly perfect vacuum," and also described the working of his proposed multiple circuit, declaring that he had solved the problem of costly copper conductors. Platinum, he held, was just the thing. [43]

As the newspapers reported that Edison was sending mining prospectors to search throughout the Rocky Mountains for more abundant supplies of platinum, Professor Barker wrote warning him against the "indiscriminate examination of rocks"; Lowrey also grew worried and wrote begging him not to set off on wild hunts for platinum, as the world supply was known to be very meager.[44]

Though the tireless inventor had improved the melting action of his platinum coils, his money was now melting away more rapidly than he had expected, as he reported in disconsolate tones. The task he had undertaken looked much longer than was anticipated; moreover, to light up Menlo Park alone, within a half-mile radius, as he planned to do, would need some $18,000 worth of copper.[45]

The spirits of his financial sponsors, meanwhile, began to droop. Their brilliant inventor, through whom they hoped to win a patent monopoly over a new domestic lighting system replacing gaslight throughout the world, far from having completed anything tangible, was already hinting plainly that he needed more money. Meanwhile the first Brush arc lights, introduced in September, 1878, were already blazing over lower Broadway in New York, and more were being installed elsewhere with impressive effect, so that the bankers now began to have serious doubts whether Edison had pursued the right course in trying for an unproved lighting system.

In very friendly spirit, Lowrey reported to Edison that at a recent meeting of his financial backers at Morgan's office, some members of the group had expressed serious misgivings about the inventor's progress. Lowrey, however, had defended him stanchly, declaring that in such a venture "no great end could be obtained without considerable doubt and tribulation.... We must all stand by the inventor and the enterprise." Mr. Morgan, he confided, stood there listening, without saying anything. He was plainly a man "who was not easily frightened"; such brief remarks as he made indicated that "he was perfectly ready to go on." [46]

Lowrey urged that instead of trying to conceal any real troubles he faced, for fear that his sponsors might lose courage, the inventor should be as frank as possible. He also proposed that Edison come to New York to confer with Mr. Morgan. When the inventor indicated that he was too busy to leave Menlo Park, Lowrey offered to bring Morgan himself and his partner Fabbri out to Edison's laboratory for a tour of inspection, "to see the rubbish and rejects, so that they might form some idea of the actual operation and its present difficulties." But it was not until mid-April of 1879, after much pressure had been put upon him, that Edison agreed to hold a private demonstration at Menlo Park—for the benefit of his financial partners—of his high-resistance platinum lamp.

As Jehl relates:

They came to Menlo Park on a late afternoon train from New York. It was already dark, when they were conducted into the machine shop where we had several platinum lamps installed in series. . . . Mr. Morgan and Mr. Lowrey and the others stopped at the library for half an hour or so while our chief reported to them verbally on the results of his experiments so far.[47]

Menlo Park's scientific community had undergone a considerable physical expansion in the previous six months. About fifty men were still at work constructing the big engine house in back of the laboratory. In front of the laboratory, at the advice of the "stage manager" Lowrey, a neat brick building housing the office and library had been completed, with a clublike reception room equipped with the finest cherry-wood furniture, such as one saw then in Wall Street.

After the conference was over, the group crossed the yard to the laboratory, where the "boss" showed them pieces of platinum coil he was using for his lamps, pointed out the arrangements of lights on brackets along the walls, described the Gramme type of generator he hoped to install soon, as well as the voltage and current characteristics he was trying for. Then, the room having grown quite dark, he gave "Honest John" Kruesi the order to "turn on the juice slowly."

Today, I can see those lamps rising to a cherry-red, "like glow-bugs," one of the eye-witnesses wrote afterward, "and hear Mr. Edison saying: 'A little more juice' and the lamps began to glow. 'A little more' . . . and then one emits a light like a star, after which there is an eruption and a puff, and the machine shop is in total darkness. We knew instantly which lamp had failed, and Batchelor replaced that with a good one. The operation was repeated two or three times, with about the same results, after which the party went into the library to talk things over until it was time to catch the train for New York.[48]

The platinum coils consumed a lot of power for the light they gave, were costly and short-lived, and still fused; the Wallace-Farmer dynamos being used temporarily heated up badly. It was a transitional model of a lamp that Edison showed Morgan and the others, for he was bent on obtaining a much better vacuum. Moreover his parallel circuit seems not to have been tried yet, as Jehl indicates, for this could not be demonstrated until he had finished making a new type of constant-voltage dynamo (on which he had also been experimenting at the time).

Edison stated that the system was not yet "practical"; it was not the parallel circuit he envisaged. But one of the visiting magnates, Robert F. Cutting, had been learning something about the early carbon vacuum

lamps of Starr, demonstrated in 1845, and, in a tone of keen disappointment, remarked, "I have read Mr. Starr's book, and it seems to me it
would have been better to spend a few dollars for a copy of it and to begin
where he left off, rather than spend fifty thousand dollars coming independently to the same stopping point."

Edison tried to explain that the incandescent light would not be found
where Starr had searched and left off; that Starr had "passed over it. So
have I. That is why I want to go back over it again." [49]

But it was not easy to explain these things to a man of business. It
was a gloomy gathering that broke up on that cold, raw April evening
and returned to New York. All of Lowrey's abounding faith would be
needed to rally their spirits and persuade them to shell out more cash.
Some rumors of the disappointing demonstration of his platinum lamps
now leaked out; as a result, Edison Electric Light stock, which had risen
to $600 a share, fell sharply, while gaslight securities began to recover.

A leading New York daily chose this time to publish a scathing article
entitled, "What has Mr. Edison Discovered?" It reported that though the
impulsive young inventor had made many sweeping claims, "well-informed
electricians did not believe that Mr. Edison is even on the right line of
experiments." In the opinion of a rival inventor of much experience in
this field, W. E. Sawyer, the publication of Edison's latest lamp patent
revealed "nothing new"; and all his efforts were doomed to "final, necessary and ignominious failure." [50]

After the delegation of financiers had left, Jehl recalled, his chief walked
about the laboratory with a distracted air for a long time, his hands thrust
into his side pockets in a characteristic pose, his head down, his hair
falling over his face—"like Napoleon on the eve of a battle."

He had found out a good many things: how to make an improved
vacuum; how to raise the resistance of his incandescing coil. But time
was pressing. And he realized that there were fatal defects in his first
high-resistance light. Platinum, he saw, was really an obstacle in this
hunt, and he had expended enormous effort in struggling with it.[51]

Later he said that in trying to perfect some invention he often would
run up against "a granite wall a hundred feet high." If after many trials
he could not get over it, he would retreat and turn to something different.
Then some day something would be discovered, by himself or someone
else, that he saw would help him scale "at least a part of that wall." [52]

In adversity he could be highly philosophical. "Even if you gave much
time and labor to learning the hundreds of wrong ways of doing a
thing," he would say, it might lead, in the end, to the right way. But once

he became convinced that he had mistaken his path, as with the disappointing platinum light, then he could show great resolution and speed in retracing his steps.

"After that exhibition," Jehl said, "we had a general house-cleaning at the laboratory, and the metallic (i.e., platinum) lamps were stored away."[58]

chapter XI

The break-through

1

"The electric light has caused me the greatest amount of study and has required the most elaborate experiments," Edison said in recollection; "I was never myself discouraged, or inclined to be hopeless of success. I cannot say the same for all my associates." [1]

Thus, in 1905 he could review in unemphatic terms the years of storm through which he had passed a generation earlier. At the time, however, he was like a man possessed. After every reverse or disappointing trial, he worked on, day and night, compulsively. Was it pride that turned him into a demiurge? Was it some dim, long-buried sense of "guilt" that disposed him to endure any ordeal and made him toil while others slept?

One of Edison's favorite books, since his youth, was Victor Hugo's mythopoeic romance, *Toilers of the Sea.* Its hero Gilliatt, described by his neighbors as both "crafty" and laborious and believed by them to be in communication with supernatural powers, sets out alone to extricate a small steamship from a reef in the English Channel on which she has been stranded. In his long effort he must contend against the immeasurable forces of nature: the sea, the tides, the storm, to reach his goal. By his courage, his powers of endurance, and his mechanical skill, Gilliatt assumes for us the symbolic figure of *homo faber;* the steamship that he is bent on saving, of course, symbolizes Progress. So Edison too strove against the unknown elements of nature. Titanlike, he seemed in those days chained to his "cave" at Menlo Park, given over to superhuman exertions, so that in the end he might "steal the fire of the gods" with which to light the lamps of the world.

With characteristic optimism he had talked of completing this enterprise in "about six weeks." To be sure, he was already distinguished among inventors for the "lightninglike rapidity with which he seized upon an idea elaborated by someone else and, foreseeing its application before others, improved and adapted it to use." This time the weeks turned into months; a year passed, and he had not reached his goal.

Unfortunately, once he had made himself a favorite subject of the press his investigations could no longer be carried on behind closed doors; he was *afflicted* with the very publicity he had courted, for the world insisted on having bulletins periodically reporting his advances and retreats. After the favorable public interest first aroused by news of his electric light venture, a reaction of feeling prejudicial to his just fame set in during the spring of 1879. The influential gaslight interests undoubtedly inspired the publication of articles bitterly attacking him. It was said that

He incautiously raised expectations which he is anxious to fill, but which are taxing his energies . . . We hope that he will not drive on at this Herculean task, until, some day, despite . . . his extraordinary powers of endurance, he has sacrificed his health and broken down, on his work.[2]

In other articles it was charged that through want of education, and by his "feverish methods of research accompanied by propaganda," he had placed himself before the world as a "charlatan"; his reputation as a scientist was "smirched." One report had it that he lay almost at the point of death and that "a state of despair reigned at Menlo Park."[3]

To his intimate associates he seemed nothing daunted by public criticisms or misrepresentations, but continued patiently on the general course he had marked out for himself. As Upton relates:

His greatness was always clearly to be seen when difficulties arose. They always made him cheerful, and started him thinking; and very soon would come a line of suggestions which would not end until the difficulty was met, or found insurmountable.[4]

After his first quick onset against this elusive opponent, the incandescent light, he saw that the combat would be long. Unlike his earlier telegraphic and sonic investigations, this would require a very broad knowledge of physical and chemical science. In this area there was ever so much that remained unmeasured, hidden to the eye. For example, only recently, since about 1875, had chemists begun working in high vacua—an unknown, miraculous world. He himself described as "peculiar and unsatisfactory" the conditions under which this investigation must be pursued:

Just consider this: we have an almost infinitesimal filament heated to a degree which it is difficult to comprehend, and it is in a vacuum under conditions of which we are wholly ignorant. You cannot use your eyes to help you, and you really know nothing of what is going on in that tiny bulb. I speak without exaggeration when I say that I have constructed 3,000 different theories in connection with the electric light, each of them reasonable and apparently likely to be true. Yet in two cases only did my experiments prove the truth of my theory.[5]

In the second stage of this work, after April, 1879, despite the failures of numerous trials, Edison directed his efforts on a much broader front of the area "under siege." Added experience in this field made him more thorough in his methods. During the long train of incandescent-light experiments, it was observed that he seemed to become more "disciplined" and was "gradually converted into a scientific investigator." [6] In these researches he also employed instruments and means considered far beyond those possessed by most experimenters of his time.[7] For it was not only a light he sought, as he realized now; he must try to create "a whole new industry with all its ramifications." [8]

He now followed three main lines of investigation. First he worked out more thoroughly the scheme of electric current distribution he had conceived in 1878. The constant-amperage dynamo available then for arc lights must be adapted and redesigned, so that it was suitable for his new system requiring a constant-voltage current in a multiple circuit. Second, he directed a group of assistants in the key assignment of perfecting the pumping methods used in exhausting air from his lamp globes, so as to obtain a still higher vacuum. Third, another team, under his own watchful eye, carried out a long series of experiments in which about 1,600 different materials were tested for their worth as incandescent elements within his sealed vacuum globes.

His whole idea of the problem is shown (in his notebooks) to have changed in the course of seven or eight months. There are no more sketches of platinum lamps with thermostatic regulators; instead, his attention is devoted to the microscopic examination of materials tested under high heat, after the occluded gases in them were expelled.

To subdivide the electric current for numerous small lights in parallel, he would need a constant pressure dynamo, at an estimated 110 volts. Now the existing dynamos, of the Brush or Gramme type, were of too low efficiency and economy for his projected high-resistance lights. In 1877 a committee of scientists appointed by the Franklin Institute of Philadelphia had undertaken tests and measurements of several of the contemporary dynamo-electric machines. There existed then no precedent for a test of electrical machinery and the committee had to devise its own methods, which were carefully worked out. The Gramme type of dynamo was found to utilize from 38 to 41 per cent of the motive work produced, after deduction was made for friction and resistance of the air. The economy of the Brush dynamo was even lower, being estimated at 31 per cent; moreover these machines had strong induced eddy currents and often "ran wild." Such losses did not matter in a circuit of big, low-resistance arc lights. But Edison concluded that he must devise a dynamo of lower internal resistance and capable of converting mechanical energy into electric current with much higher efficiency.

Upton, describing Edison's independent mode of attack upon such problems, said long afterward:

I remember distinctly when Mr. Edison gave me the problem of placing a motor in circuit, in multiple arc, with a fixed resistance; and I . . . could find no prior solution. There was nothing I could find bearing on the counter-electro-motive force of the armature . . . and the resistance of the armature on the work given out by the armature. It was a wonderful experience to have problems given me by him, based on enormous experience in practical work and applying to new lines of progress. . . .[9]

Edison's notebooks for December, 1878, show that he had made rough sketches for a new dynamo generating constant voltage almost at the beginning of his electric light project. At that period, however, there was the widespread and fallacious notion, held by most of the electrical savants, that the internal resistance of a dynamo "must" be equal to its external resistance (or that of the circuit). Through study of primary battery circuits they had "proved" that they could attain only a maximum efficiency of 50 per cent. Edison reasoned, however, that while this rule held true for a primary battery, it did not apply to a properly constructed dynamo. As Jehl recalls, he said that

he did not intend to build up a system of distribution in which the external resistance would be equal to the internal resistance. He said he was just about going to do the *opposite;* he wanted a large external resistance and a low internal resistance (in the dynamo). He said he wanted to sell the energy outside the station and not waste it in the dynamo and the conductors, where it brought no profits. . . .[10]

Jehl, who was taught to carry out tests of resistances, remarks that the art of constructing dynamos was then as mysterious as air navigation. All electrical testing was in the embryonic stage. "There were no instruments for measuring volts and amperes directly; it was like a carpenter without his foot rule."[11]

The problem of a constant voltage dynamo was attacked with the usual Edisonian *élan*. Seeking to visualize every possible structural innovation for his dynamo armature, he had his men lay out numerous wooden dummies of drum armatures on the floor and wind wire around them, spurring them on in their task by laying wagers as to who would finish first.

After Edison had decided upon the form of winding, the type of magnets to be used, and the direction of the current, Upton made drawings and tables after which the real armatures were wound and attached to the commutator. Edison eventually worked out an armature made of thin laminated cores of sheet iron that showed less eddy currents and so produced less heat than the solid armature cores then used. When the new cores were run in an excited field, it was Upton who made the mathematical calculations and drew the final blueprints.

The self-effacing Upton may therefore be given principal credit for interpreting Edison's ideas and translating them into mathematical form. It was Edison, nevertheless, who *had the ideas*. Upton was frequently surprised by the accuracy of Edison's "guesses" and never considered that he himself was given insufficient credit for his work. A careful student of contemporary electrical science, he seems to have been conversant with, and to have guided himself by, some of the important critical writing of Dr. John Hopkinson, of England, on the structure of the Siemens dynamo. As Upton admits, the design of the Edison-Upton dynamo, as we may call it, was not essentially different in principle from that of the Siemens. Edison's main improvements seem to have been in dividing the formerly solid armature cores and the commutator into a far greater number of sections than had been formerly the practice. He also seems to have been the first to use mica in insulating the commutator sections from each other, a very effective method.

The new dynamo, though it was said to have incorporated a number of elementary mistakes, contained many admirable features for that period. With its great masses of iron and large, heavy wires, it stood in bold contrast to other dynamos having a more meager ferrous quantity. Its magnets seemed enormous in those pioneering days. When this machine was run at a certain constant speed, the voltage between its two armature brushes was approximately 110 and remained about constant, falling but slightly when increasing amounts of current were taken from the machine. The new dynamo's bipolar form, owning to its two upright columns, led to its being nicknamed Edison's "Long-waisted Mary Ann." *

Pictures of the Edison dynamo drawn by artists for the illustrated mag-

---

* Analyzing this old machine many years later, Edison's chief engineer, C. L. Clarke, while doing homage to "Mary Ann," recalls that the brush holders and rocker arms of the armature were "excessively crude"; that faults in armature construction still created fairly high resistance and resultant heat, which was prevented from escaping, so that current capacity was still greatly limited. Most of these faults of design were corrected during the two or three years of experimental trials that followed.

azines of 1879 show that the inventor and Upton had also contrived a ponderous dynamometer, set up in back of the Menlo Park engine house, with which they measured work output. Once Kruesi had completed the first going machine, Upton carefully checked the results. To his astonishment—and quite as Edison had "guessed"—the new dynamo, tested at full load, showed 90 per cent efficiency in converting mechanical energy (steam power) into electrical energy!

The irrepressible inventor also had some fine engineering ideas for driving his new generator. Others commonly used belting. To the devil with belts, he said; they were wicked wasters of mechanical energy. He calculated, at first, upon using an intricate system of countershafts, then worked his way toward direct coupling, or the direct connection of dynamos with steam engines, which was the plan generally adopted by others after him. After trying slow-speed steam engines as his mechanical force, operating at about 66 to 100 revolutions per minute, he ordered a Porter-Allen steam engine designed to run as high as 600 revolutions per minute. This was delivered to Menlo Park toward the end of 1880, and was directly shafted to his dynamos—dynamo and engine thus forming a self-contained generating unit, mounted together on the same iron bedplate.

Then at last he could make his long-awaited test—for we anticipate the order of events by some months. "All right! Open 'er up!" Edison yelled in his high voice. The engine clanked and shook, the steam pressure rose, and the dynamos revolved with a horrible din. The party watching the test at Menlo Park all thought it wise to retreat to the shelter of a brick wall adjacent to the engine house—when suddenly a steam pipe burst, and everything came to a stop. Now that, Edison remarked with philosophic calm, taught them something at any rate: the high-speed engine wouldn't work, and therefore would need to be replaced with something allowing of more safety and economy. The next year, he would manage better with 120-horsepower steam engines running at only 350 r.p.m.

In midsummer of 1879, however, Edison was as jubilant as a small boy over the new dynamo. As was usual with him, the world was soon told all about his "Faradic Machine"—so named in honor of his favorite scientist. It was described and pictured in the *Scientific American* of October 18, 1879, in a lead article written by Upton, though unsigned. Once more there was a great scoffing and ridiculing of Tom Edison's "absurd claims," which were attributed to sheer ignorance by certain suspicious savants. Edward Weston, a well-known inventor and metallurgist of Newark, who also made dynamos, wrote at the time that the claims for such a generator made it seem "more or less like a perpetual-motion machine." [12]

The hectoring of Edison by some of the leading American electrical experts, among them Dr. Henry Morton of Stevens Institute, now seems to us traceable to their own real ignorance of actual dynamo problems. Edison, on the other hand, as Francis Upton held, was opening up new paths. Reading Morton's positive predictions of failure for his whole enterprise, Edison grimly promised himself that, once he had it all running "sure-fire," he would erect at Menlo Park a little statue to his gloomy critic which would be eternally illuminated by an Edison lamp.

The engineering of an improved, constant-pressure dynamo during 1879 bore the same relation to the electric light system as the cheap production of gas from coal did to gaslighting. He had thus laid a firm foundation for his multiple circuit of small high-resistance lights—to which the most exhaustive studies were being devoted at the same time.

It must be remembered that, from the very outset of his work, Edison was guided by his overarching concept of a *whole electric distribution system* of which all the parts must be fitted into place. In contrast with other inventors, who searched only for some magical incandescing substance, he worked out all the supporting structure of his system: its power supply, conductors and circuit, and then came back to determine what kind of light would be demanded by it.

2   The alleged "mechanic" who was said to lack education was to be found reading scientific journals and institutional proceedings at all hours of the day or night. It was thus that he had learned about Sir William Crookes's achievements with high vacua, thanks to the Sprengel pump. The new opportunities to test the incandescent qualities of a broad variety of metals, rare earths, and carbons, under the hitherto unknown conditions of a vessel highly exhausted of oxygen, now allured Edison. (In England, Joseph Swan was also inspired by Crookes's high vacua to resume his experiments with the incandescent light after 1877.) Edison had his boys pumping away for dear life, until by August, 1879, he had all but one one-hundred-thousandth of an atmosphere expelled from his glass globe.

The globe itself, incidentally, was much improved by the inventor's own design, after he had brought to Menlo Park an artistic glass blower named Ludwig Boehm, who had been trained in the famous Geissler Works in Germany. Edison one day drew a sketch of a one-piece, all-glass globe, whose joint was completely sealed; and Boehm, late in April, 1879, working skillfully with hand and mouth, fashioned it in the small glass blower's shed back of the laboratory.

"There never has been a vacuum produced in this country that approached anywhere near the vacuum which is necessary for me," Edison wrote in his notebook. A hundred-thousandth part of an atmosphere was too much; the battle of the vacuum must go on unremittingly. Experience was gained during this constant struggle, and at last, after two months, he could say exultantly, "We succeeded in making a pump by which we obtained a vacuum of one-millionth part of an atmosphere." [13]

With growing excitement he realized that a key position had been won in the late summer of 1879. In his mind's eye he already saw what might be done with an extremely fine, yet highly resistant, incandescing substance under the conditions of such a high vacuum. His state of tension is reflected in the laboratory notebooks by some quite unscientific expletives, such as: "S——! Glass busted by Boehm!" [14]

In the later, much broader phases of his investigation we feel that Edison gradually converges upon the heart of the secret. He had his constant-voltage dynamo and a tight glass globe for his lamp with a high vacuum. All that remained for him, then, after his long coursing, was to discover an illuminant that would endure.

As we have seen, he had tormented himself over the platinum-wire burner until, as he himself admitted afterward, "It seemed as if it might make the search altogether vain." [15] But still, he had learned to raise the light of a slender platinum wire more than sixfold, as compared with its maximum light in open air, and within a globe that was still a poor vacuum. He had also discovered a way of expelling the occluded gases (or secreted oxygen) that still remained within the incandescing element and hastened its destruction. This had been done by keeping the burner aglow (with a charge of current) while the air was being pumped out. Tests then showed how greatly thereafter the platinum burner improved in hardness and candle power. Now he tried to solve his final problem —finding an illuminant that was superior to platinum—by using both his own interpretation of Ohm's law of electrical resistance and also Upton's mathematical tools.

It was in late August, or early September, after about a year's search, that he turned back to experimenting with carbon again as his illuminant, and this time for good. Carbon had the highest melting point (6,233 degrees F., or about 3,500 C.) and the highest resistance. However, those slender reeds of carbon he had tried earlier had been impossible to handle, as he now understood, because carbon in its porous state has such a marked propensity for absorbing (that is, occluding) gases. But once he had his higher vacuum and his method of eliminating these injurious gases, he saw that he might work even better with burners of carbon reeds than of platinum.

He was of course extremely familiar with the properties of carbon. In a shed in back of the laboratory there was a line of kerosene lamps always burning, and a laborer engaged in scraping the lampblack off the glass chimneys to make carbon cake.

"Oh, Mr. Edison, the lamps are smoking," visitors to Menlo Park sometimes warned him.

"Yes, I must remind them to turn the wicks down," he would answer banteringly. But he did nothing of the sort. He had the cakes of carbon brought to him and kneaded them with his fingers into fine reeds.

According to a romanticized account of this culminating experiment, given by a contemporary in 1879, he discovered the carbon filament by "chance":

Sitting one night in his laboratory . . . Edison began abstractedly rolling between his fingers a piece of compressed lampblack mixed with tar for use in his telephone. For several minutes his thoughts continued far away, his fingers in the meantime mechanically rolling over the little piece of tarred lampblack until it had become a slender filament. Happening to glance at it the idea occurred to him that it might give good results as a burner if made incandescent. A few minutes later the experiment was tried and, to the inventor's gratification, satisfactory, if not surprising, results were obtained. Further experiments were made with altered forms and compositions of the [carbon], each demonstrating that at last he was on the right track.[16]

A more rational, a more satisfying account of his process of discovery, an account supported by the documentary evidence of his own laboratory notebooks, indicates that Edison proceeded here according to the true methods of science. By a number of very intelligent choices, he had begun working for a higher vacuum instead of using an inert gas; then he had gone on to plan for a multiple circuit, a high-resistance burner, and the right generator for it. In the laboratory notebooks are recorded Upton's extended mathematical calculations of the nature of the voltage, current, and conductors that would be most effective with a multiple circuit system of distribution; they show also the determinations of how much resistance (ohms) the small light must provide, and how small an amount of current (amperage) it could suffice with. Thus Edison and Upton arrived at the conclusion that, given a 100-volt multiple circuit system, the resistance of his new lamps must be raised to about 200 ohms.[17]

The idea of reversing the contemporary practice of making incandescent lamps of low resistance (one or two ohms) was embodied in Edison's patent application of April, 1879, for his improved (second) high-resistance platinum light.

Not only must the illuminant be equal to this high resistance, but (in accordance with Edison's usage of Ohm's law) it must have a very small cross section, or radiating surface. "After considerable calculation," according to an engineer who served under Edison in early days, "Edison estimated that the carbon should be not over one sixty-fourth (or 15.6 thousandths) of an inch in diameter! He also estimated that the carbon filament should be about six inches long." [18]

It was a bold undertaking: raising the resistance of their illuminant about a hundredfold above that which contemporary technique used, and reducing it to the thinness of ordinary heavy sewing thread. Would so slender a reed of carbon bear rough usage and support those high temperatures, where thick rods had melted? By grace of the improved vacuum, and the new method of expelling gases secreted in the carbon, they hoped to succeed.

The quantitative calculations by Edison and his assistants on the precise structure of the filament and the lamp were "of outstanding merit," in the opinion of an English historian of modern invention who examined the work of the leading British competitor, Joseph W. Swan. By reducing the diameter of his filament by one half, say from one thirty-second to one sixty-fourth of an inch, Edison could make it incandescent with eight times less current.[19] Swan had an unsatisfactory lamp with its carbon "rod" about one sixth of an inch in diameter, or 10⅔ times as thick as Edison's; that meant a world of difference.

3   Through the summer months Edison and his staff worked at the tantalizing job of making some fine reeds of carbon lampblack mixed with tar behave properly in an incandescent lamp. His assistants kept kneading away at this puttylike substance for hours and hours. It seemed impossible to make fine threads of it, for the stuff crumbled, as an assistant complained one day.

"How long did you knead it?" Edison asked.

"More than an hour."

"Well just keep on for a few hours more and it will come out all right," said Edison.[20]

Before long they were able to make threadlike filaments as thin as seven one-thousandths of an inch. They then tested them. Systematically Edison investigated the relations between electrical resistance, shape, and heat radiation of these filaments, tests that required infinite pains and were time-consuming.

On October 7, 1879, he enters into his notebooks the report of twenty-four hours of testing: "A spiral made of burnt lampblack was even better than the Wallace (soft carbon) mixture." [21] This was indeed promising; the threads lasted an hour or two before they burned out. But it was not yet good enough.

In the later stages of this long campaign, as he felt himself approaching the goal, Edison drove his co-workers harder than ever; they held watches over current tests round the clock, one man taking a sleep of a few hours while another remained awake. Under the inspiration of the master, one of the laboratory assistants invented what was called a "corpse-reviver"; it was a sort of noise machine which would be set going with horrible effect to waken a comrade who had overslept.

Francis Upton said that Edison "could never understand the limitations of the strength of other men because his own mental and physical endurance seemed to be without limit." [22]

At this stage of his life Edison worked with minimum rest periods of three or four hours a day, his enormous recuperative powers helping to sustain him. He would doze off for a cat nap on a bench, or even under a table, with a resistance box for his pillow, but his assistants had orders to waken him if anything occurred that required his attention. When they did arouse him from his brief slumbers, he would be wide awake on the instant, ready to answer any questions. In response to a query he might refer his interlocutor to the exact page of some scientific lexicon, such as the *Dictionary of Solubilities*. His long memory for miscellaneous facts continued undiminished as the years went on.

To the original "Menlo Park team," the group of technicians and mechanics who knew him so well and worked with him so well, much credit must also be given for loyalty to Edison as well as for their varied skills and patient endurance. An egoist the inventor might well be, but his old associates came to love him as men who had shared dangers, fatigues, and triumphs together. He continued to behave much like the captain of some jolly pirate ship, who knew how to inspire his sea dogs to the utmost exertions. For one thing he was "always on deck" when the going was rough. Above all, he tried to keep his men on the alert and played on their emotions of pride and loyalty or set them to competing with one another to win bonuses for unusual effort.*

After they had suffered some heartbreaking failures and felt downcast,

---

* Henry Ford, the friend of Edison's old age, was most envious on hearing of the inventor's "inspirational" methods with his workers, but try as he would, Ford could never create a similar worker interest in the very different and impersonal environment of his assembly-line factories.

Edison would raise their spirits by starting the next morning on a new round of similar trials with the undimmed hope and joy of a guileless child. His irrepressible enthusiasm was infectious. At such moments he would say proudly, "The trouble with other inventors is that they try a few things and quit. *I never quit* until I get what I want!" [23]

On the other hand, when things were going badly and the old-timers saw by small signs that the Old Man was seriously troubled—showing himself choleric or extremely distracted—they would creep away as if overcome with emotions of guilt or fear, and set to work as if the devil were after them.

During the great hunt for the electric light, Edison's chief mechanical assistant, Charles Batchelor, particularly distinguished himself for his manual skill. After working for many hours to mount a tiny filament (according to Upton's account) he would keep his hand "as steady and his patience as unyielding at the end of those many hours as it was at the beginning. . . . The control of his fingers was marvelous, and his eyesight was sharp." Kruesi was equally able in his field—overseeing the production of machines or models by hand—and as tireless. Edison recognized the services of these two key workers, and those of Upton, by rewarding them with fractional shares of his own winnings, in cash or securities.

Tension among the laboratory force increased during the late summer days of 1879, as the several components of Edison's incandescent-lighting system were being perfected. But the "chief mucker" also knew how to divert their minds by serving some food and light wine after they had been working late at night, or by sitting down to swap stories with his men, or listen to music they made up for the squeaky tin-foil phonograph. One of the assistants on one occasion regaled his companions by improvising a parody of Gilbert and Sullivan's *H.M.S. Pinafore* in Edison's honor:

> For I am the Wizard of the Electric Light,
> And a wide-awake Wizard too. . . .[24]

Boehm, the glass blower, who loved to perform on a zither, would be called in from his shed to play German ditties or lullabies, or sad beer-hall melodies such as Edison was fond of—"The Heart Bowed Down," and "My Heart Is Sad with Dreaming."

Early in 1878, after the invention of the phonograph, a handsome organ had been installed in back of the laboratory's upper floor. It was presented to the inventor by Hilborne Roosevelt, a cousin of Theodore Roosevelt and a famous designer of such instruments. Sometimes Edison, deaf though he was, and without knowledge of music, would go to the organ and play chords or improvise melodies for himself.

Out of the period of the long hunt for the incandescent light many stories have come down to us illustrating what may be regarded either as Edison's unfailing sense of humor or his irrepressible vice of practical joking. For example, cranks used to force their way into the laboratory occasionally. One of this type, a man with bent back and snow-white hair, walked in softly one day with a sizable package under his arm and insisted on showing it to Edison. Laying it out on a bench, he disclosed a square contraption with some wooden cogwheels. Edison raised his large brows expressively, while everyone stopped and looked on.

"Mr. Edison, I have a machine here, which I should like you to look at; I have been working over it a good many years," the man said in a sad and gentle voice.

"What is it?"

"It is a perpetual-motion machine."

Edison's face was impassive. "Does it work?" he asked blandly.

"Well, it almost works," said the owner. "There is just one little point where it seems to stick. . . . I thought perhaps you might be able to tell me what to do."

"Yes," replied Edison firmly, "I can tell you what to do." The man's face brightened. "What you want to do is to provide the machine with a stomach. Then feed it good beefsteak and potatoes, and it will generate the energy it needs to make it work!"

The man gave a weak smile, repacked his machine and without a word left the room as softly as he had entered it.[25]

When it seemed, at certain dark moments, that the electric light would never be made to work, Edison would entertain his helpers with tales of new inventions to come; these were his "other irons in the fire" (some of them nonexistent) that would win fortune for all of them. Thus he had the thought of making a thermopile, which "Honest John" Kruesi was instructed to prepare for him in the form of metal bars to be subjected to heat. One of these alloys, as it chanced, was polished so brightly that it shone like gold. Studying it, Edison had a brilliant idea. Looking at Kruesi with an air of triumph (while winking to the other men nearby), he declared with great emphasis: "We *have* it this time, our troubles are over!" They could close up shop now, he added.

"What on earth has happened?" Kruesi asked in perplexity.

Edison remained silent, beaming. Then at length he exclaimed: "Why you never thought what you had in your hands, Kruesi; why, *it's solid gold!*"

"Gold!" cried the bewildered Kruesi.

"Yes," Edison said. He had had Dr. Haid, the chemist, give it the acid test and also determine its specific gravity. Kruesi could rest assured, it was "pure gold," and they could now "make it by the ton. Our worries are over," he cried, "and now we don't need the electric light."

Edison swore Kruesi to secrecy until he had had time to execute a patent for this "invention." Everyone of the other assistants listening tried to keep a straight face as long as possible, until one of them suddenly went off into convulsive laughter, and the joke was over.[26]

One day when the artistic glass blower Boehm was discouraged, after a series of mishaps with his pumps, someone remarked: "Could we not put the lamps in a balloon and send them up high enough to fill them with vacuum and then seal them up there?"

"Good idea," said Edison heartily; then he added with professional caution, "We'll have to take a patent on that, sure."

"But how could we seal them off, if there was no air to use in the blow pipe?" somebody else asked.

"That's always the way," said Edison with a long-drawn sigh; "that's always the way—no sooner does a man bring out a brilliant and practical idea, but some ignoramus must interfere and try to show some reason why the scheme is impractical. There's no chance for a real bright inventor nowadays!" [27]

4   The impersonal records of the laboratory notebooks for October, 1879, show Edison's mood of anticipation pervading the whole staff at this stage. He had been pushing on with hundreds of trials of extremely fine lamp filaments, so attenuated that no one could conceive of how they would stand up under terrific heat. The culminating experiments, verified by Upton, and published in December of the same year, are described in a contemporary account:

A spool of cotton thread lay on the table in the laboratory. [It was no accident, to be sure!] The inventor cut off a small piece, put it in a groove between two clamps of iron and placed the latter in the furnace. The satisfactory light obtained from the tarred lampblack had convinced him that filaments of carbon of a texture not previously used ... were the hidden agents to make a thorough success of incandescent lighting, and it was with this view he sought to test the carbon remains of a cotton thread.[28]

He had tried various methods for treating the threads before carbonizing them, but they would break. Finally he had packed them with powdered carbon in horseshoe form in an earthenware crucible which was then sealed with fire clay and heated at high temperatures. At last he had a firm, unbroken carbonized thread, or a *filament*, as the inventor himself called it. It was, however, very hard to handle, being then fastened by delicate clamps to the end of platinum lead-in wires. It was actually the ninth of a series of very fine incandescent filaments Edison had carefully constructed. The memorandum in the notebooks reads:

Oct. 21—No. 9 ordinary thread Coats Co. cord No. 29, came up to one-half candle and was put on 18 cells battery permanently at 1:30 A.M.

Afterward, Edison related, it was necessary to take it to the glass blower's house in order to seal it within a globe.

With the utmost precaution Batchelor took up the precious carbon, and I marched after him, as if guarding a mighty treasure. To our consternation, just as we reached the glass-blower's bench the wretched carbon broke. We turned back to the main laboratory and set to work again. It was late in the afternoon before we produced another carbon, which was broken by a jeweler's screw driver falling against it. But we turned back again and before nightfall the carbon was completed and inserted in the lamp. The bulb was exhausted of air and sealed, the current turned on, and the sight we had so long desired to see met our eyes.[29]

The trial of the No. 9 model of carbonized cotton filament was put on late during the night of October 21. The men looking on were thoroughly used to these things fizzling out. But the notebook entries for that night convey all the drama and the sense of triumphant resolution that the results meant for them:

No. 9 on from 1:30 A.M. till 3 P.M.—13½ hours and was then raised to 3 gas jets for one hour then cracked glass and busted.

That night there was no sleeping. The next day's entries for the performance of another No. 9 lamp tell us:

October 22: we made some very interesting experiments on straight carbon from cotton threads, so. We took a piece of 6 cord thread, #24, which is about 13 thousandths of an inch thickness, and, after fastening to platinum (lead-in wire) we carbonized in a closed chamber. We put it in a bulb and in vacuo; it had resistance of 113 ohms start and afterward went up to 140 ohms.[30]

The tales that were woven about that day of triumph (October 21–22), certainly a big day for all electrical science, featured a "death watch" of "forty hours" maintained by Edison and five of his associates; one of them, Upton, who felt ill, went home for a short nap and on returning at dawn found the others still awake; to his amazement the lamp was still glowing! The evidence of the records speaks only of an incandescent lamp that burned for more than half a day, that is, 13½ hours; but there is frequent mention of a lamp having burned forty hours—very possibly they wrote down no notes of this, in the excitement of the moment.

Contemporary accounts of this historic event, appearing soon afterward, told of how the inventor, after having been long misled, at last through "the happy discovery of the uses of a bit of cotton thread, had in a moment turned the whole current of the story into a fortunate channel. . . . The whole affair seemed like one of those little romances of science with which the road to great invention is strewn. . . ."

Edison was described as turning on the current for the perfected lamp:

Presto! a beautiful light greeted his eyes. He turns on more current, expecting the fragile filament to fuse; but no, the only change is a more brilliant light. He turns on more current and still more, but the delicate thread remains entire. [Edison said, "We sat and watched with anxiety growing into elation."] Then with characteristic impetuosity and, wondering and marveling at the strength of the little filament, he turns on the full power of his machine. . . . For a minute or more the tender thread seems to struggle with the intense heat—that would melt diamond itself—then at last it succumbs and all is darkness. The powerful current had broken it in twain, but not before it had emitted a light of several gas jets.[31]

As it went out the weary men waiting there jumped from their chairs and shouted and cheered with joy. Edison, one of them recalled, remained quiet, then said, "If it can burn that number of hours I know I can make it burn a hundred." Yet none seemed so completely astounded as were Edison, Batchelor, Kruesi, Upton, and the other workers at Menlo Park. They had become accustomed to labor without hope. "They never dreamed that their long months and years of hard work could be ended thus abruptly, and almost by accident. The suddenness of it takes their breath away." [32]

He took up the broken filament and examined it under the microscope, noting how hard it had become and how its very structure had changed. He knew at last that the high-resistance element he wanted must be tenacious, fibrous in structure, some form of *cellulose*. He would look for still better materials than cotton thread, which broke too readily.

We know that it was no accident that he turned to carbonizing that cotton thread and so *reducing* his burner to a filament, but that he arrived at this solution only after he and Upton had calculated the cross section and radiating surface they wanted for his high-resistance lights, connected in multiple circuit. Herein lay his boldness and originality. Now, he was at the crest, and on the right track; the rest of the going from here would be downhill.

The October 21, 1879, lamp gave but a feeble, reddish glow. (Turning one of the old models on again in a dark attic of the Smithsonian Institution, the writer could not but exclaim, "What a beautiful thing to have invented!") It was the best and most practical incandescent light contrived in more than fifty years of inventive effort, the future light of the world.

5   For once he tried to be discreet and keep his momentous discoveries a secret until he had improved upon his lamp filament. He wanted a more serviceable illuminant, and, since the cotton thread was a vegetable fiber, reasoned that some other vegetable substance of tenacious and fibrous character would provide the answer. His laboratory staff was then hurriedly put to work testing a long list of similar materials, among them bagging, baywood, boxwood, cedar shavings, celluloid, fishline, flax, cocoanut shell, hickory, plumbago, punk, twine. And what else might they try?

Edison racked his memory, then suddenly stared at the tough red beard of his old friend and mentor of Mt. Clemens, Michigan, J. U. Mackenzie, who had recently joined his staff as laboratory assistant. There, perhaps, was the answer to their prayers. They must, on the instant, cut off some hairs of Mackenzie's beard, carbonize them and raise them to incandescence! That tested out well enough, for a bit of elaborate horseplay. But in the end it was paper, in the form of tough, Bristol cardboard, that proved most enduring and infusible when carbonized and reduced to a hairlike filament. Edison was exultant when this filament burned for 170 hours, and swore that he would perfect his lamp so that it would withstand 400 to 1,000 hours of incandescence, before any news of it was to be published.

On November 1, 1879, he executed a patent application for a carbon filament lamp, which was quickly granted (January 27, 1880) as U.S. Patent No. 223,898. Its most significant passage was the declaration:

I have discovered that even a cotton thread, properly carbonized and placed in sealed glass bulbs, exhausted to one-millionth of an atmosphere, offers from one hundred to five hundred ohms' resistance to the passage of the current and that it is absolutely stable at a very high temperature.

The other specifications called for a distinctive one-piece, all-glass container with the conducting, or lead-in wires of platinum passing through the glass base and being clamped to the carbon filament, all the joints being sealed by fusing the glass. (Not long afterward the costly platinum lead-in wires were replaced with a metal alloy which, like platinum, had the same coefficient of expansion as glass.)

Here were the essential features of the basic Edison carbon filament lamp, in the form that was to be known to all the world during the next half century, and whose patents were to be sustained in all the courts. It was not the "first" electric light, nor even the first incandescent electric lamp. It was the first practical and economical electric light for universal domestic use. Excellent exploratory work in this field had been done before him by Americans and Europeans, including assorted Russian inventors. Swan (whom the English still consider the "prior inventor" in this case) had an operable incandescent lamp and a perfected vacuum; but his thick carbon rod was of the low-resistance type, using a hundred times more current than Edison's filament. In England only St. George Lane-Fox had experimented with a high-resistance burner, though his lamp functioned poorly. What Edison had accomplished was "a combination of old elements which produced a new thing." It opened the way to the electrification of men's dwellings throughout the world, and introduce the large-scale production and sale of electric power itself.*

It was early in November, 1879, that the worried capitalists who had supported Edison's venture learned the secret of his success. Two of J. P. Morgan's partners, Eggisto Fabbri (a man of musical tastes who greatly admired the phonograph) and Hood Wright, paid a quiet visit to Menlo Park a short time later, and found that Upton's house and Edison's were brilliantly illuminated by the new lamps.[33]

During the first thirteen months Edison had expended $42,869.21 on experimental work, not counting legal, patent, and other expenses, most of which had been met by the Edison Electric Light Company. Now he raised the question of further money advances for experimental and development work, so that he might complete a pilot light-and-power station at Menlo Park. Edison pointed out that he had spent out of

---

* Professor Abbot P. Usher, in his *History of Mechanical Invention* (1954), analyzing the mental processes of so-called inventive genius, likens Edison's return to experiment with carbon to James Watt's sudden solution of the problem of Newcomen's steam engine, during a Sunday afternoon walk when he improvised the condenser in his mind. Both were "acts of insight" resulting in an imaginative construction that answered the needs of a previously "unsatisfactory" condition, or problem, the formerly impractical incandescent light as well as the impractical steam engine.

A better clue to Edison's particular process of thought, however, is afforded us by his impulsive, almost boastful, statement in a newspaper interview, given at a very early stage of his experiments: "Now that I have a machine to make the electricity [generator], I can experiment as much as I please. I think ... there is where I can beat the other inventors, as I have so many facilities here for trying experiments." (New York *Tribune*, September 28, 1878.) His experimental research was far more thoroughgoing than his competitors', including even Swan. The decisive "flash of insight" may have been the decision to reduce his burner to a hairlike, filamentary cross section.

pocket more money than he had been given and pleaded for continued support. But the directors were stony. They were uncertain, as yet, about the future of his invention. Was it "only a laboratory toy," as one of them charged? Would it not need a great deal more work before it became marketable?

As before, Grosvenor Lowrey defended his protégé like a lion. To the doubters he boldly declared that this invention would be of "enormous value," exclaiming before the board of directors, "Edison is giving us the greatest return for capital that was ever offered—his talent, his knowledge, his health—while plenty of others [give] only capital." [34]

The future electric light and power industry, now aching to be born, would be valued, for its American section alone, at roundly 15 billion dollars at the time of the inventor's death. Yet there was prolonged haggling before another comparatively small sum was advanced to Edison, early in 1880, through loans by the original stockholders. And this was only obtained after Lowrey, prematurely, and over Edison's objections, made the secret of the electric lamp invention public. The lawyer foresaw that the news would startle the world, and that his fellow capitalists would be the more disposed to loosen their purse strings.

Rumors had been spreading about the ill-kept secret for several weeks. New Jersey neighbors told of brilliant lights blazing all night at Menlo Park; and railroad passengers between New York and Philadelphia also saw the bright lights, with astonishment, from their train windows. In Wall Street there was a flurry of speculation in Edison Electric stock, only a few shares of which were available as yet; the price rose, for a brief period, to a level of $3,500 a share—on the mere hopes or prospects of Edison's success.

Then came the front-page story in the New York *Herald,* on Sunday, December 21, 1879:

EDISON'S LIGHT

THE GREAT INVENTOR'S TRIUMPH
IN ELECTRICAL ILLUMINATION

———

A SCRAP OF PAPER

———

IT MAKES A LIGHT WITHOUT GAS
OR FLAME, CHEAPER THAN OIL

———

SUCCESS IN A COTTON THREAD

The exclusive story of the inventor's struggles for fourteen months, in a (more or less) secluded laboratory, was now told to the world *con amore* by Marshall Fox, who had written much of Edison before and won his confidence. For the daily press of the United States the detailed treatment of such an adventure in applied science as a feature story was something of an innovation—more than one member of the newspaper's own staff feared it might be a hoax! Its relative accuracy of detail was owing to the help furnished by Batchelor and Upton, whose drawings of dynamo and lamps illustrated the *Herald*'s Sunday supplement on Edison's light. The writer did his best to explain how this light was produced from a "tiny strip of paper that a breath would blow away"; why the paper filament did not burn up but became hard as granite; and how the light-without-flame could be ignited—*without a match*—when an electric current passed through it, giving "a bright, beautiful light, like the mellow sunset of an Italian autumn."

Acclaim for Edison (still mixed with some expression of incredulity) reverberated again on both sides of the Atlantic. The news "shook the scientific world to its foundations." [35] Scarcely two years before, thanks to his phonograph, Edison had gained international celebrity. Now his fame spread to the remotest corners of the earth. Menlo Park was again to be invaded by thousands of scientific pilgrims and plain curiosity seekers. But the new invasion was more troublesome than before due to a premature announcement that the inventor would have all the region of Menlo Park illuminated for New Year's Eve, 1880. Every Tom, Dick, and Harry wanted to see it; and Edison wasn't ready.

In the week following Christmas, 1879, hundreds upon hundreds of visitors made their way to the world-famous New Jersey hamlet from all over the northeastern part of the country, by train, carriage, or on foot. Edison hurried with his preparations as well as he could, but was forced to use his whole staff of sixty persons to handle the crowds. He could do no more than put on an improvised exhibition, with only one dynamo and a few dozen lights.

There were only two light globes attached to the entrance of the small library-and-office building by the gate; eight more were set out on wooden poles in the grounds outside the laboratory and along the road; while in the main laboratory building a miniature central station supplied a circuit of thirty lights connected in parallel.

Throngs of people, of all classes and every degree of scientific ignorance, had come to see "the light of the future" and to pay homage to its good "Wizard." The closing nights of the year 1879 actually turned into a sort of spontaneous mass festival, which reached its climax on New Year's Eve, when a mob of 3,000 sight-seers flooded the place. The crowds never seemed to tire of turning those lights on and off, as they moved slowly through the rooms of the laboratory. Nearly all who came acknowledged that they were satisfied that they had seen "progress on the march." [36]

They had also come to see Thomas A. Edison. The cry would go up, "There's Edison," and a rush would start toward him. As usual, he appeared in his working clothes, which were almost of a deliberate negligence, a white handkerchief at his throat in place of a cravat, and his vest half buttoned. He was by now well accustomed to appearing before admiring crowds in the guise of a modest, hard-working inventor. Among the visitors were bankers, manufacturers, scientists, journalists, artists making sketches, and farm hands whose questions ran from "How you got the red-hot hairpin into that bottle?" to inquiries about the cost of such lamps and their economy of operation, in terms of coal and horsepower.

What they saw here was but a token of what was in store for them in the near future, as the inventor promised the great throng of merry well-wishers who came on New Year's Eve. He was waiting for the completion of his new generator, he said, and intended to illuminate all the surroundings of Menlo Park, for a square mile, with 800 lights. After that he would light up the darkness of the neighboring towns, and even the cities of Newark and New York! His lamp would be sold for twenty-five cents, he predicted, and would cost but a few pennies a day to run. At the moment, the lamps were very costly to produce, for it took two men six hours to pump a high vacuum for his globes, and only a few score lamps were on hand.

In the crush a few minor accidents occurred. Several persons who ventured into the dynamo room, despite warnings not to do so, had their pocket watches magnetized. A well-dressed lady, who bent down to examine something on the floor near one of the generators, found all the hairpins leaving her head.

Not all who came were well-wishers; after the party was over, it was found that a vacuum pump had been smashed and that eight of the still rare electric lamps had been stolen. Then there were the representatives of the gaslight "monopoly" (as Edison called them) who after leaving, aired "a general feeling of disappointment," declaring that Edison had gulled the public by showing them only forty burners, none of which gave more light than an ordinary gas jet. There was also one disagreeable visitor, said to have been the rival electrical inventor, William E. Sawyer, who seemed wildly "inebriated," and who, while moving through the crowd, gave vent to his bitter disappointment, yelling that it was all a trick and shouting imprecations at Edison—until the crowd hushed him up. Edison also said in recollection:

I remember the visit of one expert, a well-known electrician, then representing a Baltimore gas company. Sixty of the men employed at the laboratory were used as watchers, each to keep an eye on a certain section of the exhibition and see there was no monkeying with it. This man had a length of insulated No. 10 wire around his sleeve and back, so that his hands would conceal the ends, and no one would know he had it. His idea, of course, was to put this across the ends of the supplying circuit and short-circuit the whole thing—put it all out of business without being detected. Then he could report how easily the electric light went out and a false impression would be conveyed to the public. He did not know that we had already worked out the safety fuse and that every little group of lights was protected independently. He slyly put this jumper in contact with the wires—and just four lamps went out in the section he tampered with. The watchers saw him do it, however, and got hold of him, and just led him out of the place with language that made the recording angels jump to their typewriters.[37]

The world's praise showering upon him seems to have been thoroughly enjoyed, for the moment, by Edison in his relaxed mood. One useful result of the wave of publicity at Christmastime was that the directors of the Edison Company now willingly "coughed up" another $57,568 for his proposed development work in the next twelve months. A less pleasing side effect was the hasty entrance of quite a number of imitators and "pirates" into the electric lighting field with merchandise very similar to his own, when he was still unready to market his product. Hiram Maxim, for example, with such patent applications as he could muster, began to offer incandescent lights of some sort for buildings early in 1880. Also upon the publication of Edison's second carbon filament patent, January 9, 1880, using carbonized paper, his inveterate adversary, William E. Sawyer—"that despicable puppy," Edison called him—entered a petition of interference at the United States Patent Office, claiming that he, Sawyer, had first used carbonized paper as an incandescing element and patented it a year earlier.[38]

In the face of this litigation, another filament material, in lieu of carbonized paper (which Edison considered not satisfactory enough) had to be quickly improvised.

He had won a great battle on October 21, 1879, yet the campaign was far from ended. The future commercial value of the incandescent light, he held, was now established beyond question; but it remained for him to put a complete central-station lighting system in operation on a large scale. This would mean facing many more problems, inventing numerous electrical appliances then unknown, and in fact developing a whole new art. In such development work Edison would be driven to efforts even more original than the work on the lamp itself. He was still at sea, so far as concerned the actual working-out of such a large-area light-and-power-distributing system. "Remember, nothing that's good works by itself, just to please you," he used to say; "you've got to *make* the damn thing work."

chapter XII

The Edison system

1 From the first, he had never ceased working over his larger plans for a complete electric light and power *system*. The earliest notebooks of the electric light campaign record his preliminary studies looking to the cheapened production of electric current; incidentally these were done in 1878, before Upton's arrival, with the help of the distinguished university scientists, Henry Rowland of Johns Hopkins and John Trowbridge of Harvard.[1]

Edison had promised to build up a system that would supply power and heat as well as light. For its components it would need many dynamos of the new type, underground distribution mains, safety fuses against short circuits, insulating materials that were still in the experimental stage, power switches, regulators for the generators, meters to measure current, and in fact a whole family of varied lighting fixtures, sockets, and service wiring for households. Without this system the miraculous incandescent lamp by itself was not enough; it was a scientific toy until it could be commercially exploited and made useful as one of many thousands of such units functioning within a complete lighting circuit.

That his ruling concept was not yet generally understood is suggested by the skeptical attitude of Professor Elihu Thomson, of Philadelphia, one of the ablest of the new crop of electrical engineers, who dropped in on Edison at Menlo Park early in January, 1880, and was most cordially received. Their only previous relationship had been Thomson's reading of a paper, four years ago, before the Franklin Institute, in which he attacked Edison's claims to have discovered a so-called "etheric force." At that time Thomson had won a little notice by assailing Thomas A. Edison. Now as a rival inventor he was hard at work, together with his partner, Professor Edwin J. Houston, organizing an arc-lighting business of his own.

Nevertheless, his host at Menlo Park freely exhibited the new incandescent lamp and even made him a present of one, inviting him to examine it for himself. On his return to Philadelphia, however, Thomson made a statement to the newspapers, that, as an electrical expert, he "did not think very highly of the Edison lamp and expected no great future for it." Not only was the light too dim, but setting up a parallel circuit of many such lights and supplying direct current for them through conducting mains and branch wires, he said, would be too costly and impractical; it would need, in fact, "all the copper in the world." [2]

How wrong Professor Thomson was would soon be made evident—when he himself (in 1885) turned about and "pirated" the Edison lamp and distributing system. In the meantime, his warning words, like Dr. Henry Morton's dire prophecies, frightened Edison's financial backers. The inventor himself was only spurred by them, for he obviously thrived on opposition and combat.

The truth of the matter was that the premature display of a few dozen lamps without a complete distributing system within which they were designed to operate was unconvincing, as Ed Johnson remarked. The whole plan for distributing current, the engineering and the economy of it, must first be tested on a fair scale, with a pilot system of several hundred lights. After a brief respite—during which it was said Edison slept for twenty-four hours one day—he was back at work.

The whole project was so novel that even some of his laboratory associates did not understand it until the master made it clear. Explaining his idea to Johnson one day, Edison ended by asking him, "Do you catch on with your intellectual grippers?"

"Holy Jesus," Johnson was said to have replied, "this is a great thing you have got onto now." [3]

In later years Edison claimed to have had his "comprehensive system of electric light distribution, analagous to gas," clearly formulated at the beginning of his undertaking as a "seven-point program": the parallel circuit; the durable, high-resistance light; the improved dynamo as a cheap source of electrical energy; the underground conductor network; the devices for maintenance of constant voltage by which current was made to reach distant lamps evenly; safety fuses and insulating materials; and, finally, lighting fixtures with keys to turn them on or off.

However, although Edison did have an over-all plan in mind, it was not at all so neatly mapped out in advance. The grand design of a central power station and spreading network of wire, like that of gas lines serving individual jets, was to be achieved only after some three years of trial and error. It was only then that Edison would be ready to march on New York. It was only then that the completed "Edison system"—which eventually impressed informed persons as being so comprehensive, so "beautifully conceived" in all its detail—became what he had hoped for: a giant distributing mechanism as practical and economical as that of gas.

He had first planned for a parallel circuit of very thick copper conducting mains, offering the least possible resistance to current, starting from his power source (the generator), then tapering gradually, as thin branches of wire, to the more distant lighting outlets in his circuit, so that the voltage would remain more or less constant and the "drop" moderate. It was the well-known "tree" circuit which, in fact, was what he used for his first impromptu exhibition during Christmas week of 1879 at Menlo Park. Under such a system the size of the two main conductors—starting at the "root of the tree," or generator—in order to serve a given number of lamps—must be as large as the sum of the separate branch conductors carrying the necessary current to each individual lamp and back. But he soon realized that to supply current for the lamps of only nine city blocks (8,640 in number) he would need, under the "tree" plan, 803,250 pounds of copper, costing $200,812 for this item alone! [4]

Through the winter of 1880 the "Wizard of Menlo Park" was furiously at work to eliminate the difficulties of the multiple-arc or parallel circuit that Elihu Thomson had touched upon. Since he always tried to analyze his problems visually, much of the work was done by laying out mains and service wires in the open fields of Menlo Park, in preparation for his future lighting plant. An expanded force of "one hundred earnest men," as he said later, toiled under his command through the winter and spring.

On January 28, 1880, Edison made application for a patent (No. 369,280) providing for an entire system of "multiple-arc distribution" from a number of generators through a circuit of metal wire; it had the basic form of the future Edison system and constituted part of the legal bedrock upon which he was later to defend the authenticity of his plan in the courts.

Then, late in the summer of 1880, he contrived an invention corollary to the parallel circuit, which he called the "feeder and main," a scheme that was as original as it was simple, and which, at one stroke, reduced the dimensions and cost of copper conductors to about *one eighth* of all previous estimates. By this device he also overcame the "drop" in the group of lamps that were most remote from the central station; in the ordinary parallel circuit these "dropped" almost a third below the candle power of the lights nearest the current source.*

* Under previous transmission methods, resistance in the conducting mains produced waste heat, progressively lowering the voltage available for lights at the

At this time many electrical scientists did not believe that electric lamps could be worked efficiently in parallel circuit. Then they were no less certain that such a circuit of high-resistance lights would be enormously expensive. This was the problem that Edison worked so hard to solve in 1880–1882, by trying new current and voltage characteristics and introducing new economies of distribution. In the one year, 1880, the indefatigable man applied for sixty patents, among them five covering auxiliary parts, six covering dynamos, thirty-two pertaining to improvements in incandescent lamps, and seven to advanced systems of current distribution. The last group of devices was scarcely exceeded in importance by the invention of his incandescent lamp. When one of them, the feeder-and-main system, was patented and demonstrated in England, Lord Kelvin, who was present at the demonstration, was asked why no one else had ever thought of it before, since it was so simple and yet eliminated the serious loss of pressure always expected in a parallel circuit. He is said to have replied, "The only answer I can think of is that no one else is Edison."

If not for such economies in material, and if not for the elimination of the former "drop" in voltage, the electric light system could not have hoped to compete with gaslight. But Edison was still not satisfied. About two years later, in November, 1882, he effected another break-through when he contrived the "three-wire" distributing system.

The new techniques involved in the "feeder-and-main" and "three-wire" systems, while less spectacular than the invention of the carbon filament lamp, signified tremendous strides toward making the electric lighting industry a going thing. The "three-wire system" was first tried out by Edison at the small-town lighting plant of Sunbury, Pennsylvania, in July, 1883. It was developed just too late to be used in the first New York City station opened nine months earlier, but was introduced there afterward.

Edison had been searching for improved balance and independence of

---

distant end of an ordinary parallel circuit. By the feeder-and-main arrangement, "feeder" wires transmitted current from the dynamo to short sections of main placed at points of dense light distribution (as at a given city block). A drop in voltage, say from 120 to 110 v., took place *only* in the feeder wire, while voltage in the main conductor sections, for local distributing circuits, was maintained at nearly constant voltage (for the lamps). Thus great economies in the amount of copper

conductor were achieved; while the formerly serious drop of voltage from lamp to lamp was reduced to no more than two or three out of 110 volts at the more distant outlets. (J. W. Howell and Henry Schroeder, *History of the Incandescent Lamp*, Philadelphia, 1941, pp. 70–72.) The feeder-and-main network cut down copper-wire cost per lamp by about 85 per cent, or from $200,000 to $30,000 for 8,640 lamps in an area of nine city blocks.

connection for his individual lamps, and consequently worked over new patterns of current distribution. In Sunbury, a sparsely populated town, using an overhead circuit on poles, he tried economizing by connecting lamps in parallel, but in groups of two on each "rung"; this saved wire but meant that if you turned out one light the second also went out. He brooded over this problem for a while. Previously the only way of having each light controlled independently was to have it connected individually to the two conductor wires. Out of a plan for using two sets of two-wire circuits, servicing a number of lights, Edison then evolved one with the same number of outlets, but with the two inner wires combined, that is, replaced by a central, or third ("neutral") wire paralleling the two outside (positive and negative) wires. This resulted in a perfect balance, since with individual lamps turned out in unequal proportion on the two sides of the circuit, enough current to restore the balance always flowed back through the third, or "neutral" wire in whichever direction it was needed. Three-wire distribution, as it happened, also brought about a splendid economy in copper: the wires could be cut down in dimension to one fourth of their previous cross section (under a two-wire system), making the total saving in copper conductors 62½ per cent.*

* The three-wire system, which soon came to be used on all electric light circuits (U.S. Patent No. 274,290, applied for on Nov. 27, 1882), consists of two 110-volt dynamos connected in series to give a potential in the main circuit of 220 volts. Two wires are connected to the outside wires of the dynamos, while the third or "neutral" wire is connected to the connection between the two dynamos. Thus the "neutral" wire serves as the outgoing conductor of one dynamo and the return conductor of the other. Two standard 110-volt lamps may be used on each individual lamp circuit. If a lamp on one side of the system, let us say the positive, is extinguished, the third wire compensates by serving as positive for the excess lamp on the other, or negative, side of the circuit. The same holds true when a lamp is turned off on the negative side, since the third wire acts as negative for the lamp in excess on the positive side. And thus perfect balance is preserved no matter where lights are turned on or off. In representing the parallel circuit we have earlier used the analogue of the "ladder," with its two "legs" as positive and negative conductors, and single lamps, operating independently, connected with the "rungs" between the "legs." Now the reader need only imagine two lamps connected on each "rung," which is bisected by a third "leg," the neutral wire, running between those two lamps. Actual use of the three-wire system became far more elaborate in practice.

Edison's first plans for a three-wire distributing system were worked out by rough calculations, with the help of his engineering aide, William S. Andrews. At almost the same time the British engineer and mathematician Dr. John Hopkinson worked out the same scheme, which he translated into precise formulas. Hopkinson, who was employed as a consultant by the Edison Electric Light Company, Ltd., of England, appears as joint patent owner (in England), with Edison, of the three-wire invention. This brilliant technician also redesigned and improved the Edison dynamo. (J. A. Fleming, *Fifty Years of Electricity*, p. 226.)

2   To return to the 1879 lamp filament of carbonized paper, it
never seemed to Edison more than a stopgap; its life span of
about three hundred hours was not long enough; many of the
lamps broke down; and two other inventors claimed prior
patents on incandescent burners of such material.

"Somewhere in God's mighty workshop," Edison is reported to have
said, "there is a dense woody growth, with fibers almost geometrically
parallel and no pith, from which we can make the filament the world
needs."

Now he began another dragnet operation in the hope of finding a
superior light-emitting substance, a cellulose of homogeneous structure.
In chemical research Edison habitually fell back on cut-and-try methods.
Nikola Tesla, the electrical inventor and scientist, who had an entirely
different temperament from Edison's, chanced to work for him for about
a year, in 1882–1883, and defined his method as follows: "If Edison had
a needle to find in a haystack, he would proceed at once with the diligence
of the bee to examine straw after straw until he found the object of his
search." To Tesla the time lost by Edison's method seemed deplorable. "I
was a sorry witness of such doings, knowing that a little theory and calculation would have saved him ninety per cent of his labor." [5] However, the
narrow limits of chemical knowledge at the time may have made empirical
methods indispensable.

This form of empirical investigation at periods resembled the process
of elimination as much as anything. It was, in fact, the system we follow
when we say to ourselves, before a locked door, "One of these keys will
spring that lock."

Now he turned again to his microscope; his laboratory assistants were
ordered to gather up mountainous piles of rubbish and lay them before
the master. Here was a medley of materials such as bark, rags, old
carpets, wild grasses, horses' hoofs, hides, cornstalks, and a thousand
other such materials—all of which were neatly dissected, examined under
the microscope by Edison himself, and the results duly entered into his
notebooks. In addition to all this, volumes on botany and lengthy cata-
logues of all sorts were pored over; agents were sent forth to make rounds
of wholesale drug houses and even museums, in order to comb out everything belonging to the vegetable kingdom.

In such cases Edison threw away the rule book. Working solely from
experiments and observations in the matter of the fiber hunt seemed to
him, in the words of Newton, "the best way of arguing which the Nature
of Things admits of." When it was all over, Edison was able to say, as if
with pride, "Before I got through I tested no fewer than 6,000 vegetable

growths, and ransacked the world for the most suitable filament material." [6]

On a day of sudden spring warmth, such as often comes in New Jersey in late April or early May, the master of Menlo Park picked up an old bamboo fan lying on his table to fan himself. Studying it, he cut away part of its binding rim, made of a long strip of bamboo, and placed it under his microscope. Very interesting, this fibrous structure seemed to him. He called Batchelor and ordered that it be carbonized in the crucible and promptly put to the incandescent test. The results were far more satisfying than those of any material tried during many months. Strips of the bamboo cane were workable and proved durable under intense heat, giving off a lovely reddish light for as long as twelve hundred hours. Now, at last, they had a truly rugged lamp filament—much superior to carbonized paper fiber—whose manufacture could be carried out on a large scale, for the supply was ample and costs low. Edison was thus enabled to bypass the interference suits that involved use of carbonized paper, entered by Sawyer in the United States Patent Office, and by Swan in England—an important consideration at the time.

In the second year of the electric light campaign, things seemed to be moving all too slowly, though Edison was making a whole series of strategic inventions that were to bring his system to completion. Some members of the Vanderbilt-Morgan syndicate were rather alarmed at the way Edison used up about $150,000 that year—for machinery, steam engines, dynamos, copper wire, and sundry materials—merely to test a pilot plant at Menlo Park. He was now shaping his first round lamp globes into longer pear-shaped vessels, employing the long horseshoe (later hairpin) filament of bamboo that would soon endear itself to millions of users throughout the world. He also improvised wooden screw sockets into which those lamps could be fitted, later making them of plaster of Paris.

Meanwhile, as the development work stretched out, Edison became restless and longed for more publicity, for which he often showed a healthy appetite. This was the heyday of Jules Verne's scientific fictions, which our inventor read with enjoyment. Voyages of exploration to the poles, to the bottom of the sea, to the jungles, or up in balloons, were then being vividly reported in all the Sunday supplements. Why should not Edison, though he could not himself afford the time to travel, send explorers around the world to search for the "most perfect" vegetable fiber? It was an idea that seems to us like a travesty of Edison's empirical method—exercised on a global scale—but one that caught the fancy of all the popular newspapers.

In the summer of 1880 the press announced that Edison had engaged a Mr. W. H. Moore of New Jersey to set off for Japan and China, and conduct systematic investigations of all types of bamboo or vegetable fiber to be found there. Another agent was sent on a similar mission to the West Indies and Central America, but he died of yellow fever upon his arrival in Cuba. A third man, the redoubtable mining prospector Frank McGowan, was dispatched from Menlo Park, at Edison's expense, to explore the upper Amazon, journeying by canoe to the wildest regions on earth. McGowan was gone fifteen months, the newspapers carefully following such reports of his fairly harrowing adventures as reached them. After having survived the hazards of life among the men and beasts of the Amazon, McGowan at last reached New York, was given a large sum of cash by Edison's secretary, and celebrated his safe arrival in Mouquin's Café. Then he walked off at night toward the West Side docks, never to be seen or heard of again—thus creating one of the great police mysteries of the eighties.

Undaunted, Edison afterward called on another far-traveled character, one James Ricalton, who possessed some learning in the natural sciences. "You like to travel, don't you," the inventor said to him abruptly. "I want a man to ransack all the tropical jungles of the East to find a better fiber for my lamps . . . in the palm or bamboo family. How would you like the job?" Ricalton agreed to go; Edison, then, in great secrecy, turned over to him a rare specimen of bamboo which he had long kept hidden under the sofa in his library at home. An explorer had once sent it to him from the Orient, the inventor explained, but had forgotten where he found it. Ricalton was to search the globe and locate the source. "You must leave at once," Edison said, and the other man agreed. Proceeding to the Far East by slow steamer, Ricalton plodded through the jungles of Indo-China, Ceylon, and India. Upon his return, exactly one year later, to the same dock in New Jersey from which he had departed, Edison briskly made his way through the crowd of reporters to take his hand. Smiling broadly, he greeted him with the words, "Did you *get* it?" [7] Ricalton had many varieties of fiber, but nothing remarkable.

The newspapers seemingly could never have enough of the derring-do of Edison's scientific emissaries, who, as "dauntless knights of civilization," braved all the menace of reptiles, deadly insects, and the poisoned arrows of Amazon natives, to discover a mysterious substance that could meet the requirements of the Edison incandescent lamp, "a material so precious that jealous nature had hidden it in her most secret fastnesses." The ill-fated McGowan and the steadier Ricalton were both likened to heroes of mythology, such as Jason and Siegfried. [8] Such stories, we may be sure, did no harm to the Edison enterprises or the Edison name, especially in the years when the actual installation of lighting stations proceeded with wearying slowness.

The curious thing was that in spite of all the newspaper talk of Edison spending "a hundred thousand" for exploration, his first agent, Moore, an experienced trader, had easily arranged for a dependable supply of good quality bamboo, which was regularly shipped from Japan at moderate cost. After 1880, a single Japanese plantation owner, who served the furniture trade of Tokyo, provided a few wagonloads of bamboo each year, which sufficed for millions of Edison lamps. The heroic searches of Ricalton in the equatorial regions had yielded nothing better and were, after all, unneeded. By the time the last of the Jules-Verne-type explorers had got back to New Jersey, in 1889, the whole technique of lamp manufacture had changed; carbonized bamboo was being abandoned, even by Edison, in favor of "squirted" cellulose, a product perfected by chemists such as Joseph Swan, who had stayed at home and pursued the methods of theoretical, as well as empirical, science.

3   Hope and pessimism alternated in the minds of Edison's financial sponsors during the transitional stage between the invention and its development. In a letter written at the time by Grosvenor Lowrey to his wife, he remarks sadly, "There are many steps yet to be taken before we can enter into the great profits which are to be the final results. . . . Manufacturing companies are to be formed, etc. etc." But after visiting the inventor and being exposed to his magnetic field, Lowrey's state of mind quickly changes.

Menlo Park, Ap. 20, 1880

Here we are again! An hour with Edison has restored the spirits which were lost, when I wrote you from the office today. Perhaps I had better *marry* him, since he cures me! The atmosphere of the place is conducive to invention, and now I have conceived that bright idea. I was quite blue—ultra-marine— this morning, now I radiate a pure white light of about 16 candle power. Can you see me by that light? Everything is progressing most satisfactorily here. The experiments are all, or nearly all done. We are now manufacturing lamps in great numbers to make an important and complete exhibition about August 1st, which will cover every point. . . .[9]

A new and enthusiastic supporter of Edison, who became an investor in Edison Electric Light early in 1880, was the railroad magnate Henry Villard. After having seen the first public exhibition of the incandescent lamp at Menlo Park, he encouraged the inventor by giving expression to the high hopes he entertained for the future of the new industry. Villard (born Heinrich Hilgard) a German immigrant of good education, on coming to America in his youth had won celebrity as a war correspondent during the Civil War. After the war he had become an expert stock market operator, acting as the representative of important German banking interests and directing their investments in American railroads and shipping lines in the West. In the summer of 1880 he formed his famous "blind pool" in the stock of the Northern Pacific Railroad, whereby he won command of America's second transcontinental system overnight.

A few months before that event, he had appeared at Menlo Park and invited Edison to install an independent lighting plant in the new *S.S. Columbia*, then under construction at Chester, Pennsylvania, for the Villard-controlled Oregon Railway & Navigation Company. Edison described Villard as a man of uncommonly "bold ideas" and of vaulting ambition. On his frequent journeys to Germany, Villard had had occasion to observe the notable progress being made in the field of electric power by inventors like Werner Siemens. About a year after first meeting Edison, Villard, by then one of the most spectacular financiers in America, was elected to the Board of Directors of the Edison Electric Light Company.

The contract for the *S.S. Columbia* required the installation of an independent lighting plant for a large steel ship of 3,200 tons, and 334 feet in length. It proved to be a most useful rehearsal for the larger central station work that was to follow. Four bipolar "Mary Ann" dynamos, of the so-called "Z" type, were placed in the engine room of the ship in tandem, and were driven by two steam engines. Connecting gear was still made up of belts, countershafts, and pulleys. A complete multiple circuit of 115 lamps using the carbonized paper filament was then wired to illuminate staterooms and saloons. Edison and Batchelor supervised all the work, which was finished in time for the departure of the *Columbia* in May, 1880, on its long voyage around Cape Horn to California. There it was to serve the Pacific Coast trade. The ship made a most brilliant display by night as it sailed down Delaware Bay.

This was the inventor's first commercial lighting plant of the "isolated" type, improved versions of which were later to be installed in hotels, stores, and private mansions. Many commentators raised the all-important question of safety at sea, asking whether the great white ship with its unproven lighting plant might not go up in flames, since fires were regularly reported on the new arc-light systems then being installed in cities. The wiring, by present standards, was crude; the insulation likewise. The dynamos were regulated only by an attendant, without instruments, depending on the brightness of the lamps. Yet the *Columbia* completed its two months' voyage, at fourteen knots, without adverse incident and arrived in San Francisco to report that not one of its lamps had given out during 415 hours of continuous service. Greatly heartened, Edison sent out a shipment of his new bamboo filament lamps to replace the older ones. Aside from that change, the *S.S. Columbia*'s original lighting plant functioned with entire satisfaction for fifteen years, up to 1895, when it was renovated.[10]

In the spring of 1880 the residents of Menlo Park, inured though they were to strange sights and sounds, saw something wholly unexpected. A gang of laborers began laying ties and rails out in the open fields adjacent to the laboratory to make a narrow-gauge railway track running out a third of a mile around a hill and then back to the starting point. Not long afterward, a little iron monster of a locomotive, about six feet long and four feet wide, was placed on the tracks. It was in fact, the first full-sized electric locomotive ever made in America, and consisted of a "Z"-type dynamo, laid sidewise on a four-wheeled truck and functioning as a motor; current was supplied to the two rails from the generating station in back of the laboratory normally used for the light circuit. The two dynamos there generated 75 amperes of current at 110 volts, permitting the small locomotive to develop about 11 horsepower, more or less. Electric current was drawn from the rails through the flanged metal rims of the locomotive wheels. A rather crude transmission mechanism made up of pulleys and friction wheels transmitted power to the driving axle.

Edison meantime tried to give the impression that he was merely "playing on the job" in this way, while waiting for new machinery for his large light plant to be completed. "Mr. Edison has, of late, been relieving his severer labors by the erection of an electric railway," commented a popular scientific journal.[11] But if it was sport, he entered into it with his usual verve. There had been experiments with toy electric trains in America as long ago as 1834, when the Vermont blacksmith Thomas Davenport made a miniature motor-driven locomotive powered by batteries. More recently, in 1879, Werner Siemens in Berlin had exhibited a full-size electric train consisting of three passenger cars driven at eight miles an hour, current being supplied through a third rail. Having given much thought to a variety of uses for his new dynamo and distributing system, Edison had resolved to construct an electric locomotive of his own device. From his notes and rough sketches, Upton, Clarke, and a mechanical engineer named Charles T. Hughes (especially engaged for this assignment), had worked out the blueprints; and Kruesi had built the engine in his machine shop. As an old railroader himself, Edison added a big searchlight to the head of the locomotive and a bell to warn off cows, while a small open-air passenger truck was coupled behind.

The day of the trial run came May 13, 1880; crowds of people from neighboring towns and newspaper reporters were on hand. Though Clarke, the engineer, had expressed great anxiety about the crude transmission arrangement, Edison gave the order to switch current off the laboratory line and onto the railway track, and bravely mounted the locomotive. The light-fingered Batchelor was at the throttle, and some twenty persons in addition crowded aboard the locomotive and truck. The doubting Clarke stayed behind.

"All aboard!" Edison cried in his high voice; "Fire her up!" The crowd hurrahed and waved handkerchiefs gaily, as Batchelor closed the switch. The little train started slowly, gathered speed, and bumped along on the rough track until it was brought to a halt at its terminal. But when the switch was reversed, in order to go back over the same course, Batchelor had to apply the lever with some force, so that the friction wheels burst under the sudden strain, and the locomotive was disabled. "We all walked back pushing the train with us," Jehl relates.[12]

Such was the first grown-up electric train in America; it was technically superior to that of Siemens in Berlin, and demonstrated good tractive power by Edison's dynamo-motor of low internal resistance. It was plain, however, that friction gears would not work, and some other transmission mechanism must be used. The next move was therefore the introduction of a belt connected with the armature shafting, as well as a countershaft and idler pulley, so that power was transmitted more evenly to the driving axle. The belt slipped a good deal and it was said the odor of "burnt armature" was grimly familiar during the tests. The motorman was forced to rely both on his judgment and his sense of smell. After further study, a number of resistance boxes in series were put on the locomotive and connected with the motor, so that the motorman could cut out one resistance box after another and thus bring up speed gradually. Within three weeks the locomotive was pulling well enough for Edison to demonstrate it before his financial backers, the directors of the Edison Electric Light Company.

Of these, Henry Villard proved the most enthusiastic supporter of the new project. He was, as he declares in his *Memoirs,* "a firm believer from the outset in electricity as a motive power for transportation...." He was also convinced that "the certain progress in the art ... would sooner or later lead to its substitution for steam, even in factories and on standard railroads." On numerous occasions Edison discussed with Villard plans for both light electrified trains—carrying wheat on the Northern Pacific's narrow-gauge feeder lines—and powerful machines that would operate efficiently on steep grades over the Rockies.

The first crude track at Menlo Park had some sharp curves and at least one descending grade dropping about sixty feet in a short distance. It needed courage to go joy-riding with Edison. In a letter describing the demonstration of June 5, 1880, Lowrey writes:

Calvin Goddard [Secretary of the E.E.L.C.] and I have spent a part of the day at Menlo, and all is glorious. I have ridden at forty miles an hour on Mr. Edison's electric railway—and we ran off the track. I protested at the rate of speed over the sharp curves, designed to show the power of the engine, but Edison said they had done it often. Finally, when the last trip was taken, I said I did not like it, but would go along. The train jumped the track on a short curve, throwing Kruesi, who was driving the engine, with his face down in the dirt, and another man in a comical somersault through some underbrush. Edison was off in a minute, jumping and laughing, and declaring it was a "daisy." Kruesi got up, his face bleeding and a good deal shaken; and I shall never forget the expression of voice and face in which he said, with some foreign accent: "Oh! Yes, pairfeckly safe!" Fortunately no other hurts were suffered, and in a few minutes we had the train on the track again.[13]

Villard, however, was so happy about the prospects of the Edison locomotive that in the summer of 1881 he advanced the inventor the sum of $40,000, to cover extended experiments with an improved engine. Henry Villard was then riding high; he had become the president of the Northern Pacific Railroad. He was soon to complete its transcontinental main line and himself would drive in its last "Golden Spike" during a famous ceremony held in Montana in 1883. Under an agreement with this newly arrived railroad king, dated September 14, 1881, Edison undertook to build a large electric locomotive that would be comparable in efficiency to a steam locomotive and capable of drawing ten tons of freight at sixty miles an hour.

Curiously enough, the ruling idea in Edison's mind seems to have been to replace the long-haul steam railroad train, and *not* the horsecar of the street railway—to which other engineers and inventors soon afterward turned as the best field for applying Edison's own ideas. He introduced substantial improvements in the second and enlarged model of his locomotive; the ponderous resistance boxes were replaced by copper-wire resistance coils connected in series with the armature, so that the motorman could plug them in and out easily to control starting of the machine. With its cowcatcher and headlight, the second model looked much like a steam locomotive, save for the absence of a smokestack. Three cars were attached to it: one an open car, covered with a striped awning and furnished with two park benches placed back to back; the second a flat freight car; and the third a covered "Pullman," with an electric braking device invented by Edison. This machine was reliable, well controlled, and ran at speeds of forty miles an hour over a track that was extended to two and a half miles in length. The inventor demonstrated it before several prominent railroad authorities, among them Frank Thomson, president of the Pennsylvania, for whom he offered to build a short line. Thomson's judgment was that the electrified train would never supplant the steam-driven train, and his adverse opinion "threw a wet blanket over my hopes," as Edison said.[14] Railroad men were notoriously conservative: when George Westinghouse first offered his air brake to Commodore Vanderbilt, the old magnate exclaimed impatiently that he had "no time to waste on fools."

There was still the possibility of using the electric locomotive for street-cars and for the new elevated railroad of New York. The prospect of eliminating so much noise, smoke, and dust would be greatly in favor of its adoption. The Sixth Avenue Elevated Railroad of New York, however, had already come under the control of Jay Gould, who would never be persuaded to rebuild or improve anything, so long as he operated that smoke-belching traction system to his own profit.

Edison's initial work on the electric locomotive was highly "empirical," but also very ingenious, as far as it went. Moreover, he heartily enjoyed these experiments, and liked nothing better than to drive his train at forty miles an hour over a light wooden trestle, seven feet high, spanning a pit near Menlo Park—while experienced railway men accompanying him shuddered with fear.

A whole combination of untoward circumstances brought his pioneering work in railway electrification almost to naught. The patents for this group of inventions had been assigned to the Edison Electric Light Company, which was supposed to promote the electric railway project, but the directors refused to invest any large sums in it. Then Henry Villard, who would have backed Edison, about a year or two later suffered financial disaster when the Northern Pacific Railroad, under his management, had to be thrown into receivership. In any case, Edison's electric railway patents, as it happened, turned out to be poorly drawn; interference suits were entered against him by a rival inventor, Stephen D. Field. The Edison Company directors eventually decided that it was advisable to join forces with Field and thus settle the patent suits. Edison took little part hereafter in the electric railway enterprise, which was managed by Field.

He had accomplished a notable work of pioneering in railway electrification and dealt with some very difficult problems of traction engineering. "Edison was perhaps nearer the verge of great electric railway possibilities than any other American," wrote Frank J. Sprague, who was probably the ablest inventor to enter this field after Edison pointed the way.[15] But Edison dropped it all. He made no effort to design a motor especially adapted to railway work; and his ordinary lighting dynamo was not made to withstand the heavy abuse of streetcar operation, with its constant stopping and reversing. Sprague, who was employed by Edison in 1882 and remained with him for years, did foresee the future of the street railways and developed a special traction motor to meet the need. Within four years, advanced work in this field was being carried on by Sprague and Charles J. Van Depoele, the inventor of the overhead trolley, while Edison practically abandoned the field.[16]

Nevertheless, the Electric Railway Company of America (a subsidiary of the Edison Company), incorporated in 1883, carried on development work based on the patents and inventions of both Edison and Field, and was chiefly instrumental in creating the first widespread public interest in electric railways. An Edison-Field locomotive, weighing three tons, twelve feet long and five feet wide, and with a speed of nine miles an hour, was completed in time for the Railway Exposition that opened in Chicago in June, 1883. There it was placed on a curved track a third of a mile long within the main exhibition building and, drawing a trailer car with passengers, operated smoothly for 118 hours, carrying in all 26,805 paying passengers. This modest though interesting demonstration aroused enormous excitement over electric railways, and intense speculation in them followed. "There was a tremendous rush into the electric railway field ... and an outburst of inventive activity. ..." [17] Edison, however, had no part in such exploitation, nor in the vast profits ensuing from it. For him the electric locomotive had been a salutary distraction, a beloved "hobby"— during the intervals when he waited for the completion of the big machines and boilers needed for his lighting system, or of lengthy tests and trials. In an interview given in the summer of 1884 he spoke as if the electric railway conceived at Menlo Park were already far from his mind. "I could not go on with it. I had too many other things to attend to, especially in connection with electric lighting." [18]

4   We must now turn back to observe Edison's strenuous preparations for a second "full-scale" demonstration of his incandescent-lighting system. The troubles, delays, unexpected breakdowns, and actual dangers met with in this undertaking would have been enough to dishearten any ordinary inventor, let alone one who had twenty other schemes under way at the same time, ranging from locomotives to talking dolls, and even patent medicines! Doubtless

the dignified financiers sponsoring the Edison Company were greatly taken aback, one day, to find the name of their favorite scientist, and his physiognomy, on the label of a bottled product named *Polyform,* advertised as having been concocted by the great Thomas Alva himself, and "guaranteed to cure sick headaches, neuralgia, and other nervous diseases." [19] He used to insist that it soothed a variety of his own aches, but was persuaded, after a while, to drop the business.

Before they would risk more money in exploiting Edison's lighting system the men financing the enterprise "desired that a complete operative system be installed on a suitable scale for testing its economy and uncovering weak spots," as Charles Clarke, then the chief engineer of the company, has testified.[20] The trouble was that Edison himself, eternally dissatisfied, kept changing his plans; he saw greater economies to be won by more powerful steam engines and dynamos gigantic for his time, and kept throwing out his old machines to order new ones. The summer passed and he was not yet ready for his second demonstration, as Lowrey had hoped he would be; nor would he be hurried, though many urged him to hasten the installation of his system in New York.

As the Christmas season approached, he was found "sunk in melancholy," according to a newspaper reporter who interviewed him, and it was an unusual condition for this sanguine man. He was then waiting for the arrival of new Porter-Allen steam engines and large Babcock-Wilcox boilers, whose shipment had been repeatedly delayed. Some of the newspapers slated him for "holding himself subservient to a Wall Street syndicate . . . while dallying with his Aladdin's lamp in Menlo Park."

Once Edison had shown the way, other inventors, who had long tried in vain to make practical incandescent lights, quickly revised their methods and appeared on the market with electric lamps which, in principle, were impossible to differentiate from Edison's. Though Grosvenor Lowrey had placed a Pinkerton detective among the workers at the Menlo Park shops, Hiram Maxim, one of Edison's rivals, had managed to lure away the expert glass blower Boehm, by offers of more money. A few months later Maxim's company was marketing incandescent lights identical with Edison's, save that instead of having a horseshoe form Maxim's carbon filament was a hooked cross. It was a "plain steal," Edison growled, and he threatened to sue. Meanwhile in England, Swan at last produced a durable incandescent burner made of carbonized thread. Thus Edison was to be involved in wearying lawsuits in two countries to protect his patents.

"Our system is not a guess nor a hazard," he proclaimed at this time. Unlike others, he had "no stock to peddle"; he would not move until his plan for domestic electric light distribution was complete and "figured down to the smallest item." [21]

To help accomplish this he mapped out an area in Menlo Park of about half a square mile surrounding the laboratory and, along the lines of imaginary streets, set out white wooden lampposts. The half-dozen neighboring houses all were wired, as was the group of laboratory buildings and shops, for the installation of 425 lamps of 16 candle power. Then trenches were dug for underground conduits, and conducting mains laid in them for a length totaling about eight miles. For insulation a hot tar compound was poured into the conduits, which at first were boxed in wood. Much leakage resulted; later the method of insulation was improved, when cheap iron pipe was used for conduits. Current was now supplied from a central power station in an annex of the laboratory, consisting of eleven "Z"-type dynamos (the "long-waisted Mary Anns"), one of which was used to excite the magnetic fields of the others. Dissatisfied with this numerous array of small dynamos and with their belting, Edison soon afterward had Clarke undertake the mechanical design of a large-capacity dynamo that was to be directly shafted to 120-horsepower Porter-Allen steam engines.

By Election Day, 1880, they were "almost ready." A stanch Republican, Edison gave orders that part of the system, the "street lamps" along the Pennsylvania Railroad tracks, should be illuminated temporarily that night, if Garfield were elected President.[22]

During one of the trial runs the directors of the Edison Company came, observed the big display of lights over Menlo Park, and inquired anxiously as to the economy of the system as compared with gas. The November–December, 1880, installations, however, were provisory, since the larger steam engines had not yet arrived.

On a raw day early in December one of the directors, Robert L. Cutting, came out to Menlo Park with a Very Important Person, Madame Sarah Bernhardt, the "divine" tragedienne, then making one of her tours of the United States. She had heard, of course, of the phonograph, longed to know "*le grand* Edison," and to have him record her beautiful voice.

The inventor's eyes lit up when he saw Madame Bernhardt, then at the height of her beauty and dramatic power. She proved to be curious about everything in his laboratory; swinging her long skirts, she followed close on his heels in a tour of the machine shops, the glass blower's house, and the dynamo room. For her benefit, Edison turned down all the lights, then opened them full force so that the lamp posts outside blazed like great stars, while Madame Bernhardt clapped her hands for joy. Still more marvelous to her was the little tin-foil phonograph, into which she recited passages from Hugo's *Hernani* and Racine's *Phèdre*. The usually shy and reserved inventor held her hand and explained everything with great eagerness, while Cutting translated back and forth.

"*C'est grand, c'est magnifique,*" she exclaimed on leaving. Edison for his part declared that Sarah Bernhardt, in his estimation, was "the greatest woman he had ever seen." That winter he went to New York to see her perform, though it was difficult for him to hear or understand.

Sarah Bernhardt was no less impressed by him. In her *Memoirs* she wrote not long afterward:

I looked at this man of medium height, and I thought of Napoleon I. There is certainly a great physical resemblance between the two men, and I am sure that one compartment of their brain would be found identical. Of course, I do not compare their genius. The one was destructive, the other creative.[23]

From France, she sent Edison two small, though amateurish, landscapes she herself had painted in oil, inscribing them to him as "the giver of light." [24]

A very different, less agreeable occasion was the visit of a delegation of New York City aldermen, headed by the Mayor, on a December evening two weeks later. The Edison Company, through Lowrey, was already making petition for a franchise to establish a central-station system in New York; an impressive reception was therefore arranged for the city officials. Edison made them a speech of welcome in one nervous sentence, then proceeded to demonstrate his model lighting system. He switched all the lights on, described his mechanism for an hour, then turned everything out save for a few dim globes, inviting the aldermen to follow him to the upper floor of the laboratory.

As they reached the upper floor it was in complete darkness. Edison suddenly switched on lights affixed to new and ornate chandeliers, which made a dazzling scene. The usually untidy place had been cleaned up and a long banquet table placed at one end, filled with luxurious food and wines catered by Delmonico's Restaurant. Waiters in black coats and white gloves served everything to perfection, ending up with champagne and cigars. It was Grosvenor Lowrey, an old hand at politics, who had planned this elaborate entertainment.

It seems, however, that the politicians were indifferent or bored. They were in no haste to introduce the so-called "light of the future" to their city. The gas company lobbyists, with whom they had had long and friendly commerce, had brought before them many persuasive counter-arguments. As for their ideas about progress, was it not said of these aldermen, when a new machine eliminating many manual street cleaners was first proposed to them, that they protested, "But those machines *don't have votes*." Now, when it came to granting the Edison company rights to lay electric conduits under the streets of New York, they suggested that a tax of one thousand dollars per mile of conduit might be in order. To Edison it seemed an outrageous charge, and inequitable as well, since no such tax had been imposed upon the gaslight companies.

After the repast, the aldermen unbent a good deal and in friendlier vein offered toasts and cheers for Edison, who appeared embarrassed and tongue-tied. The Tammany politicians left, however, without making any commitment. "All in all," a newspaper commented sardonically, "Mr. Edison's tests were a decided success, especially of his guests' capacity for champagne." [25]

That Christmas snow fell, and at last the central station at Menlo Park was ready, as hundreds of visitors arrived to see its promised illuminations. A contemporary relates:

In every direction stretched out long lines of electric lights, whose lustre made wide white circles on the white-clad earth. One could not tire of gazing at those starry lines. Edison did not bother about his overcoat as he walked in the open air. In an instant all was dark . . . in another instant the whole scene of fire and ice sprang into being again. "Eight miles of wire!" Edison exclaimed; and yet he observed, some people were not satisfied with his demonstration.[26]

In January, 1881, the new Porter-Allen steam engine arrived at last; Edison meanwhile had completed a larger model of his dynamo, with massive field magnets and even lower resistance in the armatures (Patent No. 242,898). Engine and dynamo were mounted together on the same bedplate to form a self-contained generating unit. The new "monster" machine was first tested on a winter night—and showed serious defects. For one thing, the armature developed high internal heat. Charles T. Porter, who had designed and built the steam engine, opened the throttle, watching the governor warily. The engine began to make a racket that sounded "like an immense drop-forge foundry with ten thousand hammers in operation." The link motion was so swift that it became a triangular-shaped blur. Edison, who held the watch during the test, recalled with gusto:

We set the machine up in the old shop that stood on top of one of those New Jersey shale hills. We opened her up, and when she got to about 300 revolutions, the whole hill shook under her. . . . After a good deal of trouble we ran her up to 700. Every time the connecting rod went up she tried to lift the whole hill with her.[27]

With his 120-horsepower generator, even though it was faulty, Edison could at last demonstrate how an abundant current supply was to be produced and prepare an economy trial. This was held on January 28, 1881, before the directors of the Edison Electric Light Company and a group of impartial technical experts. As a preliminary step, Edison had sent out fifteen men to canvass a selected district in the downtown area of New York and report on the probable consumption of current and the number of lights to be served. With such data, and the results of a careful test run of the full load of 425 lamps at Menlo Park, he could work out his cost accounts for a city central lighting station.

The test run was conducted for twelve hours. Results, carefully checked and computed, showed that after counting resistance loss in the armature from so-called "Foucault currents," and loss in the conducting mains, approximately six lamps were operated per dynamo horsepower. As they put in only a ton and a half of Lehigh egg coal to run the 425 lights for twelve hours, it was estimated that each lamp used 0.4 of a pound of coal per hour; adding labor and capital investment, it was determined that Edison's light could be operated at a fraction of one cent (about three mills) per hour, or approximately the cost of gaslight.

Edison was jubilant; even his backers looked somewhat hopeful. "After this," the inventor exclaimed to Clarke, "we will make electric light so cheap that only the rich will be able to burn candles." [28] In his joyous mood he wanted to publish the results to the whole world. But the wary Henry Villard, who had witnessed this test run, persuaded him that it would be best to "keep information from business rivals." The momentous economy report of 1881, as signed by witnesses, thus remained a secret for many years.

The relations between the inventor and his bankers, who were to provide the sinews of war for the Edison system, were at best those of an uneasy coalition between opposing interests and temperaments. To the men of capital Edison might be a "genius," as Lowrey urged, and a great "creative force" in himself, but he was still an eccentric, an impetuous character, whose ways of doing business were fantastic. Capitalists were traditionally wary of inventors who offered them new and untried products, or whole new industries having as yet no markets. The directors of the gaslight companies, for instance, were among the wealthiest men in the country, and yet with complete blindness, they rejected the whole idea of Edison's light and publicly ridiculed his schemes—when they might still have bought his patents cheaply.

Since the spring of 1880 Edison had pressed the directors of his company to furnish capital for the new manufacturing enterprises that would be auxiliary to his central-station lighting ventures. He foresaw that he would need factories in which to manufacture incandescent lamps, light fixtures, sockets, switches, fuses, fuse blocks, meters, voltage regulators, and dynamos. He wanted "millions" in order to do things in a big way. But the directors of the Edison Company were extremely loath to enter into the business of manufacturing electrical equipment. They considered that they had been brave enough in supplying the money for Edison's experimental work in exchange for control of his patents—and wanted none of the headaches, or limited profits, of manufacturing. The parent Edison company, in their opinion, had "the most valuable patent rights ever granted"; it was conceived as a patent-holding corporation, which would eventually do a handsome business in issuing licenses and earning royalties, without any of the trouble of manufacturing and selling.

After having perfected his incandescent light, Edison felt himself held back by his financial sponsors as he prepared to take up the next phase of development work, the lighting of great cities from central stations. "Wall Street could not see its way clear to finance a new and untried business," he said later in testifying at a patent suit in 1890.[29] "We were confronted by a stupendous obstacle. Nowhere in the world could we obtain any of the items or devices necessary for the exploitation of the system. The directors of the Edison Electric Light Company would not go into manufacturing. Thus forced to the wall, I was forced to go into the manufacturing business myself."

He would not trust others to manufacture the new appliances he would need at every step. To S. B. Eaton, a lawyer who was closely connected with the Morgan group and served as president of the Edison Company in early years, Edison declared, "If there are no factories to make my inventions, I will build the factories myself. Since capital is timid, I will raise and supply it. ... The issue is factories or death!" [30]

In the summer of 1880 he set up a factory for making incandescent lamps—the first in the world—in an old barn across the tracks from the Menlo Park laboratory. It was started on a shoestring capital of $10,000, almost all raised by himself, his associates Johnson, Batchelor and Upton, who invested small sums in it, being junior partners. At first about two hundred hand operations were needed in making these lamps, so that they cost $1.21 each; Edison, however, contracted to sell them to the Edison Electric Light Company for 80 cents. Upton was manager and worked with Edison to improve methods of production. Within a year they were employing 133 men, turning out 1,000 lamps a day, but "stacking up" half of them, since there were not enough users to buy them as yet.[31]

Again Edison dipped deeply into his resources; selling some of his

Edison Company shares or borrowing against them, he organized more factories in which electrical accessories were to be made. Early in 1881 Sigmund Bergmann, who had been making the first appliances used in Edison's pilot systems, went into equal partnership with Edison and Edward Johnson and opened a shop in New York which was to supply lamp sockets, switches, fuses, light fixtures, chemical meters, and other instruments, all devised by Edison. Bergmann, who proved to be an able manufacturer, had to expand his quarters within a year and employ three hundred men.

The next step for Edison would be to begin making dynamos, a difficult and unfamiliar business then. For this he would need huge factory space, much machinery and equipment, and a large labor force. According to Samuel Insull, the inventor's private secretary,

Edison had hoped to have associated with him his Wall Street friends, notably those connected with Drexel, Morgan & Company, and possibly Mr. Villard. He visited Wall Street to get them interested in his machine works . . . and growing impatient of his financial friends for delaying the matter, he decided to finance the Edison Machine Works himself.[32]

Ninety per cent of the original money was supplied by Edison, and the rest by Batchelor, both of them selling Edison Company shares, or borrowing on other collateral.

In the course of 1880 and 1881, Edison, by heavy borrowing, risked all the fortune he had accumulated during many years of inventive labor, in order to launch the manufacturing enterprises associated with an entirely new industry, that of electrical equipment. As an entrepreneur, he was doing the most constructive thing of all: carrying out the formation of new wealth, based on his own inventions. His Wall Street friends, however, remained skeptical and sometimes declared themselves bewildered by his unorthodox business methods.

One of them, the financier and society man Robert L. Cutting, happening to buy a few shares of stock in the Edison Lamp Company, discovered that its directors declared dividends whether earned or not, every Saturday night, during a game of poker. The dividends were doubtless small, but it seems that Edison nevertheless refused them. After two or three years had passed and costs of lamp production had been cut—as the inventor had foretold—someone proposed that extra compensation be paid to Edison, in view of the hard work and money he had put into the company. He said he would still prefer to wait, "but if you insist, send me some cigars." For years, thereafter, the Edison Lamp Company regularly sent Thomas A. Edison a box of cigars every Saturday night. When the firm was absorbed eventually by the General Electric Company, it was said the same arrangement was continued for a good while longer.[33]

Another important turn of events was the winning of a New York City franchise by the Edison parent company, in April, 1881, permitting it to lay underground mains (without tax) in the streets of New York, and distribute and sell electric light. The January, 1881, economy tests, moreover, were heartening to the leaders of great banking, railway and telegraph enterprises, so that they loosened their purse strings and supplied the inventor with fresh funds, about $80,000 in all, with which he would at last be able to begin building the central station for New York City. For this purpose another subsidiary corporation, called the Edison Electric Illuminating Company of New York, was organized (December 17, 1880) and additional capital was raised for it by the same financial group. This was the pygmy ancestor of the present-day Consolidated Edison Company.

During the winter of 1880–1881 the inventor kept rushing back and forth between New York and Menlo Park, attending to a multitude of affairs that were coming to a head. He saw that he would now have to establish himself in the great city in order to supervise the building of the first central station, upon which so much depended. In February, 1881, on a day that carried hints of early spring, he returned to Menlo Park from a highly satisfactory conference with his "money men," as he called them, and to his associates seemed to be "boiling over with enthusiasm and energy."

"Come on, Clarke," he cried, "pack up at once and come with me to New York. You have been appointed chief engineer of the company. We're going to begin business right away." [34]

chapter XIII

Pearl Street

1

When he left Menlo Park in February, 1881, Edison said, "My work here is done, my light is perfected. I'm now going into the *practical production of it.*" [1] He must supervise the installation of the lighting plant in New York himself; he must start the manufacturing shops that were to supply the equipment for the new electric light and power industry he was initiating. The man of inventions must now double as industrialist and administrator.

He thought he was leaving the seclusion of Menlo Park only temporarily; but the remove to New York marked another turning point in his life and the beginning of a long period of furious activity. In addition to the New York lighting plant, an important Edison showing of a model lighting station for the Paris Electrical Exposition was scheduled for the summer of 1881; an even larger model was planned for the Crystal Palace Exposition in London at the end of the same year. These foreign exhibitions were extremely important for the Edison Company's future international business; hence, some of the ablest members of the Menlo Park staff, Johnson, Batchelor, Hammer, Jehl, and others, were being sent to Europe to oversee that work; while the epoch-making laboratory, with its machinery and instruments removed to New York, would gradually fall into neglect.

For his headquarters and showroom, a large and ornate four-story mansion at 65 Fifth Avenue, just below Fourteenth Street, had been leased. It was near the center of what was then the fashionable residential and shopping district of the city; its elegantly furnished and spacious rooms gave an impression of dignified luxury appropriate to the growing fame of the inventor and his company. Here Edison promptly installed a small steam engine to run the generators for the house lighting plant, then hung up great numbers of his new incandescent lamps from chande-

251

liers, or "electroliers," as they were called, so that the place made a brilliant show up to the late hours of the night.

"We're up in the world now!" Edison exclaimed at this time to an old acquaintance. "I remember ten years ago—I had just come from Boston —I had to walk the streets of New York all night because I hadn't the price of a bed. And now think of it! I'm to occupy a whole house on Fifth Avenue." [2]

It was not long before everybody knew of Edison's arrival in town. Crowds stood outside the building in the hope of seeing him come and go. He was now rather full of face and figure, and usually dressed for business in a seedy Prince Albert coat and waistcoat, a top hat, and a white handkerchief tied around his neck in a careless knot.[3] In his office at the rear of the building Edison received many persons notable in New York's business and political life, among them James Gordon Bennett, owner of the *Herald,* William H. Vanderbilt and his wife, Robert L. Cutting, J. P. Morgan, Henry Villard, and Norvin Green, president of Western Union. After seeing the brave new lights at Edison's house, Vanderbilt and his wife ordered a similar electric plant for their palatial home on Fifth Avenue; thereupon J. P. Morgan ordered the same for his own residence on Madison Avenue. Christian Herter, the successful artist who decorated many of the homes of the American rich, including the Vanderbilts', had occasion to visit Edison and consult with him at this period. As he was a highly intellectual man, in the inventor's words, Edison loved to converse with Mr. Herter. "He was always railing against the rich people, for whom he did work, for their poor taste." [4]

Since Edison counted on a fairly long sojourn in the city, he had brought his wife and the three children to live near him in a hotel adjacent to Gramercy Park. He had not much time for his family, however, for he was promoting the Edison system night and day.

More than two years earlier he had written to a European agent: "It is my intention never to show or offer the electric light until it is so perfected in economical and all other aspects that I can feel sure of its instantaneous victory over gas." [5] To this pledge he remained true. When, about six months after his arrival in New York, a model of the Edison central-station lighting system was shown at the Paris Exposition, its completeness of conception made a profound impression on the foremost European electrical engineers of that era, among them Emil Rathenau, founder of the German General Electric Company, who purchased most of the Edison lighting patents for Germany. Rathenau recalled many years afterward:

The Edison system of lighting was as beautifully conceived down to the very last details, and as thoroughly worked out as if it had been tested for decades in various towns. Neither sockets, switches, fuses, lamp-holders, or any of other accessories were wanting; and the generation of the current, the regulation, the wiring with distributing-boxes, house-connections, meters, etc., all showed signs of astonishing skill.[6]

Just then, March 1, 1881, Samuel Insull arrived in the United States from his native England to take up employment as Edison's private secretary. After having worked as a clerk at the Edison office in London for two years, the twenty-one-year-old Insull was sent to Edison, on the urgent recommendation of Gouraud and Edward H. Johnson, as a youth with a real head for business. On the very day he disembarked from his ship, Insull found himself pressed into service immediately by his bustling employer, and with little ceremony. Johnson, in New York for a flying visit, was to sail for London at dawn to supervise the work on the model station intended for the Crystal Palace exhibition; and, as Insull relates, he was briefed on his duties by Edison, "who spoke with very great enthusiasm . . . of the work before him. What struck me above everything else was the wonderful intelligence and magnetism of his expression and the extreme brightness of his eyes." [7] All sorts of jobs were being done at the same time and with no little confusion. For example, as there were no workers experienced in electric lighting service, Edison had established a school on the top floor of 65 Fifth Avenue, to train some thirty men in the art.

The new secretary's first impression was that Edison "was engaged in a gigantic undertaking." He was fighting the gas companies; he was struggling with the difficulties of getting the machinery he needed; he was taking over some large works and told Insull he counted on employing 1,500 men within a few months.

Right after dinner Mr. Edison explained to me that it was necessary for him to start three or four manufacturing establishments to produce dynamos and lamps and underground conductors for his first district. He produced a wallet from his pocket, told me that he had $78,000 to his credit at Drexel, Morgan & Co. and asked me where he could get the balance.[8]

The English youth was rather startled at the idea of becoming a financier with so little preparation; but by the time Johnson departed for his steamer, at 4 A.M., Insull had gone over the books and made a list of the revenues Edison might anticipate from patent rights sold in various European countries, on the strength of which he might borrow additional money. Edison appeared highly resolved to pour everything he owned into the manufacturing end of the electric light business. It was that same morning, after a brief sleep, that he and Insull proceeded to 104–106 Goerck Street, in downtown New York, near the East River docks, where the old ironworks of the shipbuilder John Roach had recently become vacant. One of Insull's first tasks was to help Edison negotiate the purchase of the dilapidated Roach property. Here the driving John Kruesi was put in charge of several hundred employees and soon began turning out many small dynamos and some "Jumbos" as well. It was in grimy Goerck Street that Edison planted the acorn out of which the vast oak of the General Electric Company was to grow.

From that day Insull was Edison's financial factotum. It was the inventor's way to trust unreservedly the men of his entourage; and he found Insull to be not only devoted to him, but, as he put it, "as tireless as the tides." As to Edison's method of doing business, Insull has related:

I do not think I had any understanding with Edison when I first went with him as to my duties. I did whatever he told me to, and looked after all kinds of affairs, from buying his clothes to financing his business. I used to open all the correspondence and answer it all, sometimes signing Edison's name with my initials, sometimes signing my own name.... I held his power of attorney and signed his checks. It was seldom that Edison signed a letter or a check at that time.... Edison would make his own (marginal) notes on letters and I would be expected to clean up the correspondence with Edison's laconic notes as guide to the character of answer to make. It was a very common thing for Edison to write the word "Yes" or "No" and this would be all I would have.

Edison would suddenly disappear from 65 Fifth Avenue to rush out to Menlo Park and deal with production troubles at the lamp factory; or now he would be up all night working over new dynamos being constructed at Goerck Street; or again Insull would go looking for him, with urgent financial matters to be disposed of, and find him working in trenches below the streets, where he was engineering the underground mains for his first lighting district.

I never attempted to systematize Edison's business life [Insull continues]. Edison's whole method of work would upset the system of any business office. He cared not for the hours of the day or the days of the week. I ... would get at him whenever it suited his convenience. Sometimes he would not go over his mail for days at a time. I used more often to get at him at night than in the day time, as it left my days free to transact his affairs....[9]

In the summer of 1881, one would often have found Edison in the back parlor at 65 Fifth Avenue, sitting on a high stool placed before a large wall map, about twelve feet high by fifteen feet wide, which he studied incessantly; it represented in colored ink a chosen district of lower Manhattan. This he had selected as the area for his first city station. Canvassers had been sent from door to door throughout this quarter to ask people if they would change from gaslight to electricity, the cost being the same; every house using gas, and how much of it, was listed; the number of establishments using small engines that might be replaced with electric motors was also ascertained, as well as the places where stationary electric motors might be used for elevators or in hoistways. All these data were indicated on the map; and upon it also were drawn the lines of heavy conducting mains and feeders that were to be buried underground, as well as estimates of the current load needed at various points. The map, in fact, gave a clear picture (as Edison always liked to have it) of what the electric current demand would be by day and by night.

His first intention had been to cover a district of about a square mile, running south from Canal Street to Wall Street; but in the end, to play safe, he greatly reduced this area so that it ran north from Wall to Spruce and Ferry Streets, and from the East River to Nassau Street, about a half mile each way. In this district there was a fair-sized slice of the financial community, so that the opening of his lighting system would have a strong impact, he hoped, in high financial circles. The area offered striking contrasts; it included not only banks and offices, but slum streets having small factories and the tenements of the poor.

The power station he envisaged would be placed as nearly as possible in the center of this district, with its conductors in underground conduits running out in all directions to the outlets of consumers. After prowling all about the section, he determined upon a site 50 by 100 feet at 255–257 Pearl Street, which he purchased early in August, 1881.[10]

He knew nothing about the market values of real estate in New York, and, by going to the slums, had counted on obtaining liberal space for his station, as much as 200 by 200 feet.

So I picked out the worst dilapidated street there was, and found I could only get two buildings ... they wanted $75,000 for one and $80,000 for the other. Then I was compelled to change my plans and go upward where real estate was cheap. I cleared out the buildings ... and built my station of structural iron-work running it up high.[11]

His plans called for the installation of six of the new "Jumbo" dynamos. To save space, boilers were placed on the ground floor, while the dynamos, directly connected to Porter-Allen steam engines, were laid out on the second story, which was supported by heavy structural-iron girders on iron columns. Provision was made for a sufficient supply of city water and for the reception and disposal of coal.

Pearl Street had come down in the world since the days in the early nineteenth century when it was lined with fashionable taverns and thronged with fine carriages. There had even been a popular song about "Going to Live Uptown in Pearl Street." Now it was squalid, and downtown. Nor did the building at 255–257 add to its charms. It was a homely edifice, a sort of wrought-iron cage in the architectural style of the stations on the Third Avenue Elevated. Yet Pearl Street was making history again. For here was to be the cradle of the modern electric light and power industry.

A formidable task was the digging of trenches and laying of underground conduits. The other early electrical services, from telegraph and burglar-alarm systems to arc lights, all used overhead wires, either strung on poles or simply running from roof to roof. It was a cheap method; and, in the opinion of many experts, it avoided leakage of electric current into the earth. The city officials also tried to dissuade Edison from using underground mains. One of them said, "Some electricians wanted all the air, Edison only asked for the earth." He insisted, however, that his plan would eliminate all danger of shock or fire from short circuits. "Why, you don't lift water pipes and gas pipes on stilts," he protested.[12] Now that higher voltages were being used for supplying arc lights, there was a series of grisly accidents in which electric linemen were literally roasted to death, high up on poles, within sight of many passers-by in the streets. Though directors of his own company protested at the heavy additional expenditure for insulation and tubing, Edison was determined that there must be no accidents on his circuits; safe procedures must be established at the outset. His standards of insulation were soon afterward adopted in the first New York legislation imposing safety provisions for electric light and power.

At Menlo Park he and his staff had experimented with various insulating compounds, until, finally, a malodorous brew of hot asphaltum was cooked up which served their purpose. Iron tubing in twenty-foot sections was laid out in trenches; "gangs" of insulated wire or bare copper rod were run through it; then the hot tar compound was pumped in and allowed to harden. However, air bubbles often arose in the insulating compound, causing leakage and shocks to men and horses on the pavement above the conductors. For seven years Edison and John Kruesi, who, among other things, specialized in tubing, continued experiments for the improvement of insulation.[13]

The arduous breaking of earth for underground mains began in the late fall when frost was almost upon them. There were fourteen miles of trenches to be dug and much tubing to be laid in the district served, under the two-wire system first used. The illustrated newspapers of the time picture Edison himself in a battered stovepipe hat directing this work and personally overseeing the installation of all couplings and safety-catch boxes. The Irish laborers of that time, according to the reminiscences of John Kruesi, were mortally afraid of "the devils in the wires." This was one of the reasons why Kruesi, Batchelor, and Edison attended to the installation of junction boxes at every pipe length of twenty feet.[14]

At the start of the work, one of the Tammany men who served as Commissioner of Public Works threatened to hold up the whole business. He said, as Edison recalled:

"You are putting down these tubes. The Department of Public Works requires that you should have five inspectors to look after this work and that their salary shall be $5 a day, payable at the end of each week. Good morning." I went out very much crestfallen, thinking I would be delayed and harassed in the work which I was anxious to finish, and was doing night and day. We watched patiently for those inspectors to appear. The only appearance they made was to draw their pay Saturday afternoons.[15]

Sometimes the day's work ended so late that Edison and some of his helpers slept in the cellar below the unfinished power station in Pearl Street. Two of the men who aided him in testing the tubes, after sleeping in that cold and damp cellar that winter, became sick and died, while he himself, as he relates, was never affected. Kruesi recalled that on one night, he, Batchelor, and Edison, finding themselves belated, decided to sleep in a vacant shop where they happened to store their iron tubes. Kruesi related:

There was room on the floor for one, on a work bench for the other. They drew lots and Edison drew third; so he was reduced to lying down for the night on the iron tubes. I remember that he had on a very light-colored suit which, by morning, was marked with streaks of tar from shoulders to feet, for the warmth of his body had softened the tar.[16]

2   Meanwhile, having become dissatisfied with the small 8-horse-power dynamos used at Menlo Park, Edison had been experimenting with more powerful, direct-connected dynamos and steam engines, the second model of which he named the "Jumbo," after P. T. Barnum's famous circus elephant. This steam generator unit had an immense magnetic field, was bar-wound, and weighed twenty-seven tons in all, the armature alone weighing six tons.

Built especially for the Paris Electrical Exposition of 1881 at the machine works in Goerck Street, the new unit was of 200 horsepower, and capable of lighting 1,200 incandescent lamps, in contrast with the first 50-lamp dynamos at Menlo Park. Typically enough, this monster of an Edison dynamo was the largest built up to that time.

After it was completed, Edison relates, it was found that the voltage was too low. He had only a few weeks before the sailing date of the steamer in which it was to be transported to France. "I had to devise a way of raising the voltage without changing the machine, and the crankshaft of the engine broke and flew clear across the shop." By working night and day they managed to put in a new crankshaft. A final test was made on the last day; after which sixty workers, who had been given full written instructions in advance, took the machine apart, packed its components in 137 boxes, and moved it on a train of horse-drawn vans to the docks. By grace of a friendly Tammany leader, a fire-bell, and a police escort, the streets on their route were cleared, and the whole caravan reached the ship an hour ahead of time.[17]

At that time, in the summer of 1881, the only complete model of an incandescent-lighting station was at Menlo Park. Only a few Edison lamps had been shown in Europe by his ever zealous lieutenant, Edward H. Johnson, and these had been powered by batteries. The Paris Exposition that year was to feature the display of a variety of electric lighting systems, ranging from the arc lights of Jablochkoff and Charles Brush to the incandescent devices of Edison, Swan, Lane-Fox, and Hiram Maxim.

Of all these exhibits, however, Edison's was the only one showing an arrangement of incandescent lights within a complete electrical distributing system, including not only five hundred of his 16-candle-power lamps, but also the Jumbo generating unit, and all the novel appliances and controls of the Edison system. His lamps consumed only 0.7 amperes each, and showed an economy of 4.5 watts per candle. The Edison display made a powerful impression, particularly upon Europe's scientists, and easily took first honors.

Batchelor, who headed the Edison team that prepared his exposition in Paris, reported by cable: "Official list published today shows you in highest class in inventors. Swan, Lane-Fox and Maxim receive medals in the class below." [18]

In addition to receiving a little heap of gold medals and diplomas, Edison was awarded the ribbon of the Legion of Honor. The leading British inventor in his field, Joseph Swan, magnanimously cabled his congratulations to Edison.

One effect of this public demonstration was that it gave a strong impetus to the newly formed Société Continentale Edison of France, which controlled his lighting patents; under Batchelor this French branch soon launched a busy manufacturing unit of its own at Ivry-sur-Seine. European technicians who worked here—among them Nikola Tesla— learned enough to install central stations in other large cities. These engineers, in turn, made important contributions to the technique of electrical distribution, particularly in their pioneering work during the eighties on alternate, or high-tension, current engineering.

At about the same season Johnson, with William J. Hammer, had gone to London, where preparations were being made for another important exhibition, to be held at the Crystal Palace in January, 1882. For its opening, Johnson, the boomer of the Edison system, managed to bring together all sorts of notables, including the Prince of Wales, to witness the brilliant display of 213 "A" lamps on January 18, 1882. In highly theatrical fashion, he entertained the crowd with his electrical tricks. Enclosing an Edison bulb within a linen handkerchief he smashed it in his hands, thus demonstrating the nonincendiary character of the lamp. When some fuses blew out by accident, plunging the exhibit into darkness, he had his assistants replace the fuses within a few minutes and passed it all off as a funny trick.[19]

A leading London daily reported:

Mr. Edison's exhibition is the wonder of the show, and his representative [Johnson] is the prince of showmen. One feels that after an hour with Mr. Johnson that there is only one system and . . . there is but one Edison and Johnson is his prophet."[20]

The Edison Electric Lighting Company, Ltd., had been organized in the preceding year with British capital and soon began the construction of what was actually the first commercial incandescent-lighting station in the world, at the Holborn Viaduct. Sir William H. Preece, chief engineer for the Government's postal and telegraphic departments, had seen the Edison exposition at Paris, and now not only publicly retracted all that he had formerly said against Edison's lighting projects, but also befriended the British Edison Company by ordering lights for a large part of the great Post Office Building, which was within the power radius of the Holborn Viaduct station. Two newly completed Jumbo generators had been hurriedly shipped out from New York, and from the Holborn Viaduct station they supplied an installation of two thousand lights for commercial use in that quarter of London. An Edison underground main-and-feeder circuit was a notable feature of this installation.

The experience gained by Edison's lieutenants in London was fully reported to him before he had completed the first New York station. Luckily, Johnson had engaged the knowledgeable Dr. John Hopkinson as a consultant, together with J. Ambrose Fleming. Hopkinson went to work on the "bugs" in the first Jumbo dynamos and made carefully conceived proposals for improving their performance and lowering their internal heat, his exact calculations being forwarded to Edison. After Hopkinson's valuable work on the dynamo much of the mystery about internal heat and wasteful eddy currents was dissolved.*

The Edison Company, Ltd., began business with high hopes of lighting up all of England's streets and dwellings. At an early stage of affairs, it had entered suit for infringements of its patents against the Swan United Electric Light Company and soon obtained a preliminary court order for an accounting of lamps made and sold by Swan.[21] But late in 1882 Parliament passed an Electric Lighting Act empowering municipalities to buy up street and residential or business lighting systems, after a period of twenty years, at fair market value. The prospects of all private electric light and power companies, therefore, became quite dim, financially speaking. Edison's English counselors, therefore, advised that it would be best for his company to join forces with Swan's by an exchange of stock. Swan was a distinguished native son; he had strong financial support; and the nature of his old carbon lamp patents left the outcome of a suit somewhat in doubt.

In the end, Edison seemed disposed to accept the counsels of his attorney, Sir Richard Webster, and other reliable English associates favoring a merger; but a serious stumbling block to negotiations proved to be the proposal of the rival group that the new amalgamated company bear the name of Swan together with that of Edison. Writing his London attorneys in 1883, Edison stated firmly that he would consent to the merger only on condition that

the company shall be called the Edison Electric Light Company, Ltd., or at least shall be distinguished by my name *without the name of any other inventor in its title*. The Company may put forward the fact that it is the owner of the Edison, Swan and Hopkinson, or other patents.

The last condition is made solely upon business grounds of great importance to me, and not at all upon such feelings as might be imputed to me of wishing to distinguish myself as against Mr. Swan. . . . I leave out of view the fact that he and I are disputing the claim to the lamp. The property which this company

* Edison was much impressed by this young British scientist (who died early in life, in 1898); he purchased rights to several of Hopkinson's electrical patents, including his three-wire circuit developed in England simultaneously with Edison's in America.

is to own and operate is an *electric lighting system* to which I have contributed inventions as indispensable as the lamp, my title to which is not, as I understand, disputed anywhere. Moreover, I am under contract to give your company all my future inventions concerning electric light. . . . I remain in this country, and wherever possible, as large an owner as possible in my inventions. I have never parted with any of my holdings, except when compelled to. . . . I am bound by pride of reputation, by pride and interest in my work. You will hardly expect me to remain interested, to continue working to build up my new inventions and improvements for a business in which my identity has been lost. . . .[22]

With justice Edison asserts that his name has become synonymous throughout the world with incandescent electric lighting. That this should continue to be so was more important to the inventor than any profit or loss involved in this dispute. Nevertheless he was finally persuaded to permit the two companies to be merged as the Edison & Swan United Electric Company, Ltd. Edison's priority as inventor, incidentally, was clearly established several years later (1888) in British courts, in the course of a suit against other interests charged with infringing upon his lamp patents.

3   Back in New York Edison was running into problems such as could not have been anticipated in those apprentice years of electric lighting. For example, Mrs. Vanderbilt's "box-house" on Fifth Avenue and Forty-seventh Street had been wired for an independent generating plant, the wires being strung along gas pipes, as was then the fashion. The large picture gallery, as Edison relates, was lined with silk cloth interwoven with fine metallic thread. One rainy day two wires got crossed behind these tinsel hangings, which were soon afire. Edison at once ordered power shut off and the fire died out immediately. But Mrs. Vanderbilt "became hysterical" and on learning that there was a steam engine and boiler in the cellar, declared that she "could not live over a boiler. We had to take the whole installation out." [23]

On the other hand, Mr. J. P. Morgan, who also had an isolated plant installed in his home in the autumn of 1882, directed that boiler and steam engine be set out in a pit under the garden outside his house—to the great vexation of other residents of Murray Hill, who complained to the Fire Department that they were kept from sleeping by the ghastly clanking of the engine. But Morgan thoroughly enjoyed his electric lights, and his neighbors' threats of court suits disturbed him not at all. Even the outbreak of a little fire, as a result of a short circuit, which damaged his library walls and carpets, failed to discourage him.[24]

Time was passing, yet New York still waited for the opening of Edison's power station. He and his labor gangs down in the trenches had raced against the coming of winter, then halted and resumed the excavation work and laying of mains at a frantic pace in the spring of 1882. The job was not completed until June. Early in the next month the central station at Pearl Street was furnished with four great boilers, six Porter-Allen steam engines of 240 horsepower, and the first three of six improved Jumbo dynamos. In addition, switchboards and control instruments were installed in the station, and, on an upper floor, a bank of one thousand electric lamps with which to test the system. The direct-connected steam dynamo units were rated at 200 horsepower each, the armature rotating at 1,200 r.p.m., and the generators having current capacity for 1,200 lamps.[25]

The control instruments, however, were still fairly crude and were being changed from day to day. As there was as yet no central switchboard; individual switchboards regulated each dynamo. If voltage rose above the desired limit, a red lamp, set in the relay circuit from the dynamo, flashed on, and the attendant threw some resistance into the field of the machine. If voltage dropped, a blue lamp lit up and the attendant was warned to cut resistance.

As to measuring the current for sale to individual customers, Edison had been driven to invent still another ingenious appliance: a chemical, or electrolytic, meter consisting of a glass case with two small plates of zinc placed in a solution of zinc sulphate. The meter, on being connected with the house circuit, recorded the usage of current by removing molecules of zinc from the positive plate and depositing them on the negative plate. The plates were weighed once a month, their difference in weight representing the measure of power consumption. Despite some complaints it proved not inaccurate, though after a few years the simpler mechanical meter invented by Elihu Thomson was adopted.

On July 6, 1882, Edison, having fired up the boilers several days before, threw the switch for a trial run of the first Jumbo. Three days later, on a Sunday, when the business area was quiet, he connected a second Jumbo dynamo to the first, admitting that his heart was in his mouth as the engines engaged.

At first everything worked all right.... Then we started another engine and threw them in parallel. Of all the circuses since Adam was born, we had the worst then! One engine would stop and the other would run up to a thousand revolutions; and then they would see-saw.... When the circus commenced the gang that was standing around ran out precipitately, and I guess some of them kept running for a block or two.

One of those present has also related:

It was a terrifying experience as I didn't know what was going to happen. The engines and dynamos made a horrible racket, and the place seemed to be filled with sparks and flames of all colors. It was as if the gates of the infernal regions had suddenly opened.[26]

Edison recalled that he grabbed the throttle of one engine while Johnson, the only other person present who kept his wits, caught hold of the other, and thus they shut off the engines. This seesawing or "hunting" action of the dynamos, according to Clarke, was due to the ineffectual governors on the Porter-Allen engines, which alternately cut them off, then gave full steam. Vertical vibration of the whole wrought-iron structure carrying the generator units at Pearl Street was also assumed to be a contributing cause of the disorder. Edison and his aides then had to devise a makeshift arrangement by which a long shaft, placed along the wall of the station and connecting all the steam engine governors by means of pivoted rods and levers, secured torsional rigidity—tying the governors together, in effect. Thus if one engine went fast, all the engines were compelled to do "team-work," and so, Edison said, by straining the whole outfit, up to its elastic limit, in opposite directions, torsion was practically eliminated. It was hardly good enough. Clarke testifies that during the first few weeks of actual operation the station was run with only one generating machine.[27]

In this emergency Edison appealed to a talented engine designer, Gardiner Sims, for a steam engine of a new type having a mechanical governor operated by centrifugal force, whose functioning would be unaffected by torsional vibration. After considerable trouble, Sims produced two Armington-Sims engines that worked successfully in multiple; these were substituted for the Porter-Allen engines and corrected the trouble.

The première at Pearl Street, so long awaited, so oft-postponed (for reasons kept secret, in light of the July 8 fiasco), loomed up at last in the late summer of 1882. "I kept promising through the newspapers that the large central station in New York would be started at such and such a time," Edison confessed. "These promises were made more with a view to keeping up the courage of my stockholders, who naturally wanted to get rich faster than the nature of things permitted." [28]

He recapitulated the events of that day and described his own inner tension:

The Pearl Street station was the biggest and most responsible thing I had ever undertaken. It was a gigantic problem, with many ramifications. There was no parallel in the world. . . . All our apparatus, devices and parts were home-devised and home-made. Our men were completely new and without central-station experience. What might happen on turning a big current into the conductors under the streets of New York no one could say. . . . The gas companies were our bitter enemies in those days, keenly watching our every move and ready to pounce upon us at the slightest failure. Success meant world-wide adoption of our central-station plan. Failure meant loss of money and prestige and setting back of our enterprise. All I can remember of the events of that day is that I had been up most of the night rehearsing my men and going over every part of the system. . . . If I ever did any thinking in my life it was on that day.[29]

For once, he avoided any ballyhoo in connection with the official opening at Pearl Street, for he was haunted by the fear that something might happen to spoil the party. Steam was admitted to only one of the Jumbo dynamo-steam units in operation; and at 3 P.M. on Monday, September 4, 1882, Edison gave the order to John W. Lieb, chief electrician of the station, to pull the switch. The current flowed forth through the underground conductors and the Edison lights went on in the first district against the full sunlight of a summer afternoon, therefore giving a rather feeble effect. Only a small group of a dozen persons had gathered in the dynamo room, among them a few directors of the lighting company and some writers for the scientific press, who had come to wish the inventor well.

It was not only a quiet affair, it was anticlimactic, after so many ardors and alarms. Despite the tall talk of a central station that was to light 2,500 or even 10,000 buildings through one circuit, the Edison Electric Illuminating Company, as the operating subsidiary of the parent organization, opened for business with only eighty-five customers fully wired and a total load of only four hundred lamps.[30] The Pearl Street development costs up to this stage totaled about $600,000. The history-making power station limped along on only *one* dynamo, for Edison's makeshift control of the steam engine governors still worked poorly, the improved engines not being available until two months more had passed.

Somehow the metropolitan newspapers had got wind of the affair, and their representatives were on hand to cover it with modest little stories destined for the inner pages. "Most of the principal stores from Fulton and Nassau Street," the *World* reported casually, "were lighted by electricity for the first time." And the New York *Times* observed: "It was not until about 7 o'clock, when it began to be dark, that the electric light really made itself known and showed how bright and steady it was." The *Times* itself already boasted of fifty-two Edison lamps in its editorial and counting rooms—"soft, mellow, grateful to the eye; it seemed almost like writing by daylight." At the offices of Drexel, Morgan & Company in Broad Street, Edison himself appeared to turn on the lights—an imposing array of 106 lamps.

One newspaper (the *World*) described the inventor as being all dressed up for the occasion in a Prince Albert coat, white cravat, starched shirt front and a white, high-crowned derby. But another newspaper represented him as looking disheveled and collarless, with his white derby all stained with grease. Why this discrepancy? Both reports were, in fact, true. At one moment Edison was attired *de rigueur;* and then, in the twinkling of an eye, he was off, tearing through the small crowd outside the Pearl Street station to some point of trouble where an underground safety catch box had blown out.

The outlook for business seemed bleak too; the first customers were not even presented with bills during the early months of the trial. Edison remarked hopefully that it was all a novel enterprise, but that more buildings were being wired every day.

Edison's electric light and power business began in America's greatest city without fanfare and in the early years grew but slowly. The people of New York gave expression to admiration or dazzlement at the incandescent light, so evidently superior to the gas jet; but they were loath to adopt it during the decade that followed. For one thing Edison lamps cost a dollar apiece in the beginning—a high price. Moreover, there was always trouble of one sort or another, what with leaks from junction boxes under the streets, or, on occasion, a destructive fire in a house or shop that had just been wired by Edison's green mechanics. Of one such fire, in which a customer was killed, S. B. Eaton, president of the Edison Illuminating Company, ruefully remarked to the inventor that it was bad for business, that "the loss of life makes it all the worse," and that it was unfortunate that the newspapers had reported the grim tale fully.[31]

Only a few days after the central station had been opened, a policeman came rushing in, exclaiming that there was trouble at Ann and Nassau Streets and they must send a man up right away. Edison himself hurried over with another man and found a junction box leaking so that the moist surface of the street was passing powerful currents. With his gift of vivid recollection, Edison related:

When I arrived I saw a ragman with a dilapidated old horse come along the street, and a boy told him to go over to the other side of the road—which was the place where the current leaked. The moment the horse struck the electrified soil he stood straight up in the air, and then reared again; the crowd yelled, the policeman yelled, and the horse started to run away.... We got a gang of men, cut the current off ... and fixed the leak.[32]

"Look, the Edison lamps have been turned on," people said simply, at the start of the first city power station in America, and that was that. It was already commonplace to speak of Thomas A. Edison as America's "miracle maker" who was "pushing the whole world ahead"; his recent triumphs in Paris and London outshone the dull première in New York. By then more than a hundred "isolated," or independent, electric plants for residences, stores, and hotels were being supplied by the Edison companies, and men were already accustomed to the splotches of incandescent light that blazed up here and there in our relatively dark cities. Use advanced in little fits and starts.

It was only after many months and, indeed, years, had passed that men came to realize what Edison had wrought in that crude central station in dingy Pearl Street. The inventor himself, in the heat and dust of battle, doubtless had no time to evaluate the scope of his victory or its significance in history. It was, in fact, nothing less than the long-hoped-for "industrial marriage of steam and electric force," as his friend T. Comerford Martin summed it up. Almost unaware, the world began to move out of the age of James Watt into the Electrical Age.

More than fifty years had passed since Michael Faraday had discovered the mechanical production of induced electricity and had been possessed by a vision of the future electric power. It needed Edison, however, as well as Faraday, Ampère, Arago, and other scientific explorers, to make the Electrical Age.

"Scientists and inventors have more in common than in difference," James G. Crowther has written in discussing the relation of invention to science. A Faraday might discover far more new scientific facts than Edison, and in this their roles were different. "But the importance of Faraday's discoveries cannot be explained without reference to the work of Edison. . . . This is why Edison is truly a 'man of science.' " [33]

It was Edison who had finally *applied* the knowledge of electrical science that had been accumulating during those fifty years in a decisive form (his "system") and boldly imposed it upon the "new" commodity, electricity, which thereafter was introduced to practical usage on a large scale. Thenceforth, the mass production and sale of electric current was to be carried on in all the world's markets. His successful carbon filament lamp, taken together with his system of electrical distribution, constituted the key invention in this technological drama. More than any other individual, he gave impetus to the advance of this new art, creating new wealth immeasurable, new convenience and enjoyment, and a new tempo of life.

His campaign for his electric light and power system had consumed four years of the inventor's life, roughly from age thirty-one to thirty-five. He was now gray-headed, but still youthful in appearance, his round smooth-shaven face unmarked and unlined, and strong as ever. He seems also to have been so highly stimulated by his recent activities that he fairly proliferated important inventions of all sorts in these years and even, as will be seen, made one purely scientific discovery of very great moment.

Other men might have done as much as he, though they would have taken longer to do it; other masterful men would also move into his own field after him, to go forward again from the line where he would stop; but surely none could have exhibited a fiercer will to forge a great triumph, in the face of so many obstacles and dangers.

It will be remembered that Edison had been assailed as boastful or possibly "ignorant" by many scientific men four years earlier, when he first laid claim to having found some way of "subdividing" electric current and making an incandescent light for universal domestic use. On the opening day in Pearl Street the memory of these reproaches flashed through his mind. He could now say truthfully, with that pride in his workmanship that he never bothered to conceal, "I have accomplished all that I promised!" [34]

chapter XIV

"Science and dollars . . ."

1 At 65 Fifth Avenue, the New York "showroom" of the Edison Company, Edison was very much in the public eye during most of five years, meeting all and sundry who came to make inquiry about his system, explaining and demonstrating it. During this period the brownstone mansion also served as a busy social center, where financiers, politicians, artists, inventors, and "cranks," too, arrived in a constant stream to see and enjoy the "genius of electricity." One of the callers chanced to be the Hungarian Ede Reményi, one of the world's most celebrated violinists. Reményi was so greatly intrigued by the creator of the phonograph that on one occasion he played his violin for him all night long—"two thousand dollars' worth," as the calculating Edison figured it. But the most constant callers were journalists who came virtually every day to interview him. And these gentlemen of the fourth estate were most liberal smokers of the cigars Edison always kept in a box on his desk.

"I could not keep a cigar," the inventor used to complain of those days. Even if he locked them up in a desk, safe from newspaper men, his own associates would break it open to get at them. Thus he conceived the idea of teaching these people a lesson, and ordered a box of trick cigars—made of old paper and hair—which he left on his desk. Two months later the tradesman who had sold him these false cigars happened to come in and ask how their hoax had worked. The preoccupied Edison, however, confessed that he had forgotten all about the cigars and no longer remembered what had happened to them. They were all gone. "On coming to investigate, it appeared . . . that *I* had smoked them all!"

The first of the myth-making biographies of Edison had begun to appear in 1879; they were followed by a spate of articles, books, and pamphlets of the Samuel Smiles category, drawing lessons from his singular career. Edison was not a money lord, nor one of the new railroad kings, nor a soldier, nor a statesman; and yet (unusual in the case of a

268

scientific worker) he was widely accepted as one of the great men of the age. It was said that he typified "Yankee inventiveness" and the American's "materialistic optimism"; thanks to him, "Science was marching onward with giant strides." The whole world felt so much the stimulating power of his creativeness that men asked themselves, "What is there that cannot be? Where are the limits of human investigation? What next?" [1]

Was it possible that there were barriers, insurmountable to ordinary mortals, that might delay or resist the progress of the modern "king of inventors?" It was possible. In fact, the pity of it was that affairs were not going at all as well as appeared on the surface for the Edison Electric Illuminating Company, upon whose fortunes the inventor had staked everything he had. In its first years, the Pearl Street central station lost money. Its business grew steadily, to be sure, until it had 508 consumers using 10,164 lamps in 1884. But Edison had promised a whole chain of such light and power stations; yet years passed, and no other central station was built in New York, while the Pearl Street center remained limited in capacity.

The difficulty was that heavy investment in plant and machinery was needed in relation to output of electric energy and light. In this respect the electric light industry was wholly different from that of the telegraph, where some wire, batteries, and cheap sounders brought big revenues. The large capital needed for the rapid spread of the Edison system was simply not forthcoming. The directors of his company, having backed him on a limited scale, showed what the inventor called "the characteristic timidity of capital." There were many lines where men could make money rapidly; but in electric power the process seemed slow and costly.*

It seemed to Edison that he alone "could take hold and push the system," as he said in 1883. He added:

I have come to the conclusion that my system of lighting having been perfected should be promoted. ... It is all so complicated that I do not like trusting it to new and untried hands; *because science and dollars are so mixed up in it.*[2]

Now that his prototype of an electric power and light station had been produced, one would think that other and lesser men could be left to duplicate it elsewhere, rather than that his unique talents should be wasted in such routine work. Yet this was what he felt compelled to do.

"I'm going to be a business man," he told a friend in the summer of

* The declared policy of the Edison Electric Light Company directors, according to the annual report for 1882, was to hold patents and draw royalties therefrom, but not to go into the actual business of lighting. A year later, the directors still held that they must "go slow" in this new field, "until its practicability, economy and profitableness had been fully established." (E. E. L. Co. Bulletin, October 31, 1883.)

1883. "I'm a regular contractor now for electric lighting plants, and I'm going to take a long vacation in the matter of inventions."[3]

Though the newspapers might call him the "millionaire inventor," he described himself as "machine-rich and cash-poor." Most of his days were spent in managing his electrical equipment factories, overseeing the installing of new central stations himself, and promoting new Edison companies—which left him little leisure for experimentation. Most of Edison's workers then had still to learn their trade. And so he must have eyes everywhere; he must be literally in twenty places at the same time, lest some disastrous accident occur.

In 1882 and 1883 preparations were under way to establish central lighting stations in a dozen or more large cities, such as Boston, Chicago, Cincinnati, and Detroit. But to Edison's surprise it was the demand for small, independent, or "isolated," lighting plants, such as he had built two years ago for the *S.S. Columbia*, that provided the largest part of his company's business at the outset. Factories, department stores, hotels and even ranches called for such lighting installations in ever increasing volume. To meet this growing demand, the Edison Company for Isolated Lighting had been set up at the end of 1881, as a subsidiary of Edison Electric Light, a majority of its stock being allotted to the parent company. In addition to furnishing such plants for private concerns the Isolated Lighting company also installed them in many small towns where there was as yet no gaslight. To supply this demand a large stock of dynamos, conducting wires, lamps, and appliances of many kinds was needed.

The customers for isolated plants who flooded the Edison organization with their orders also needed credit, as a rule, in order to establish their local facilities. Edison therefore set up another subsidiary to expedite this business, naming it the T. A. Edison Construction Department. Young Samuel Insull was placed in charge here. The local capitalists in a small town would offer Edison's organization their notes and bonds, or sometimes stock in the local power company under construction (as well as lesser sums in cash); and central station equipment would be supplied to them by Edison's manufacturing companies. But such securities were not easy to convert into cash. As Insull relates, some of the plants were too hastily built; thus the Construction Department handling the business acquired the unsavory name of "Destruction Department." After a few years it was dissolved and its business turned over to another company in the Edison group.[4]

In the early period Edison insisted that everything of moment connected with his enterprises be brought to his personal attention by his assistants, and be subjected to his own scrutiny. It was impossible to handle the affairs of his sprawling organization in such wise; yet he attempted it, hastening from one place to another, going by train from Boston, to Louisville, to Chicago, and thence to a small town in Pennsylvania, for example, in one tour of inspection in 1883.

There was trouble-shooting without end. For the power station at Brockton, Massachusetts, a superintendent who was a former locomotive engineer had been selected because of his reputed steadiness of nerve. But when a smoldering fire started under the floor of the powerhouse as a result of a short circuit, he simply took off and ran. At another small station a "wire man" dropped an oil can between two conductors, and was astonished to find both oil and can melt away. Even a brilliant young engineer like Frank J. Sprague, on being put in charge of a small steam-dynamo unit at Sunbury, Pennsylvania, at first showed himself unfamiliar with the oiling mechanism of the new Armington-Sims steam engine. "I ... did not know how it worked," he related, "with the result that we soon burned up the Babbitt metal in the bearings and spent a good part of the night getting them in order." The next day Edison himself appeared at Sunbury and "there followed remarks that would not look well in print." [5]

Whether he promoted his system at the Fifth Avenue mansion, or hustled about to distant trouble spots, Edison was always without constraint or airs. On December 12, 1882, the Bijou was to open in Boston, the first American theater to use incandescent lights, and the inventor himself hastened there to supervise the installations. The Governor of Massachusetts and his military aides were on hand in full uniform at the première to hold a fete in Edison's honor in one of the theater boxes. The curtain duly rose on the operetta *Iolanthe;* but in the second act the lights went dark red and kept turning dim. Edison and Edward Johnson, who accompanied him, immediately rushed down to the cellar of the theater, where they discovered that a leak in the steam boiler had caused the fire to go down. Casting aside their swallowtail coats and silk hats, Edison and Johnson went to work at once shoveling coal into the fire so that steam pressure might be maintained. The ceremony and speechmaking scheduled for the entr'acte were held up indefinitely, much to Edison's satisfaction, for it was well known that he heartily disliked all such formalities.

A similar incident occurred a few months later, at the opening of the Southern Exposition in Louisville, Kentucky, for which Edison's company had arranged an important display of his (improved) electric locomotive and of a circuit of 4,600 lamps that was to create veritable "fountains of light." The city's Board of Trade had ordered a splendid banquet in his honor, to begin immediately upon the arrival of the great man by train from New York. The table was set, the toastmasters waited. No Edison. Wiring trouble had developed down in the generating plant below the exposition hall. On being informed of this, immediately upon his arrival at the gate, Edison had disappeared into the cellar to direct workmen trying to repair the power plant. One of the electricians present was H. M. Byllesby who, in later years, like Insull, was to become one of America's leading public utility magnates. To the young electricians it was "wonderful" to see Edison working by their side, all covered with grease and soot, while the sybaritic banquet prepared for him by Louisville's leading citizens grew cold. What they enjoyed most was "his attitude of caring nothing for what others thought of him, and paying no attention to frills and fancies." [6]

Back in New York, he might be working over the specifications for some new lighting plant, or snatching time to make some experiment in his little laboratory at Goerck Street. But glancing through the window for a moment and seeing the sky turn dark with an approaching storm reminded him at once that the Pearl Street station might be overloaded. In 1883 they had put the whole New York Stock Exchange, with its many lights, on that small station. Edison must telephone at once to the superintendent at the station and ask him about the load. "We are up to the muzzle and everything is running all right," was the reply. They had at that time only an "indexlike steam gauge, called an ampere-meter, to indicate the amount of current going out," Edison relates. When the sky grew so black that he could not see across the street, he telephoned again and learned that "Everything is red-hot and the ampere-meter has made seventeen revolutions. It's spittin' copper!" The Stock Exchange did go dark one of those days, and much of Wall Street with it, though Edison's emergency crew struggled to replace burned-out junction boxes as soon as possible.[7]

He was promoter and engineer in one; but there was also the Edison who was an irrepressible showman, a "histrionic" character, as his detractors put it. In any case, he would not have been Thomas A. Edison if he had not, betimes, indulged in horseplay or even downright clowning.

Niblo's Garden, an old music hall that was long the center of New York's night life, was a place he often enjoyed visiting. On one occasion the proprietor, who always showed him marked favor, conceived of a new pantomime, "a great Mimical Dramatic Ballet" entitled "Excelsior," which was to do honor to Edison's incandescent light. Through the co-operation of the inventor, a 55-volt dynamo, supplying five hundred small lamps, was obtained, and the drama of electricity was depicted. A high point in the affair was an electrically illuminated model of the recently completed Brooklyn Bridge; the concluding number featured a troupe of stout ballet dancers, Niblo's version of the Beef Trust, who were each to wave a wand with an Edison lamp at the tip of it. But what if the connections were broken by the animated movements of the girls? The great inventor himself was therefore called upon to supervise the installation of the lighting effects and check the wiring of the electrified costumes at the final rehearsals.

It was thus that the theatrical reviewer of one of the metropolitan dailies happened to observe Edison "moving about among the girls and adjusting their corsets," while inserting a little battery in the bosom of each member of the ballet. At a given signal, during the dance, electric lights flashed merrily from each girl's forehead, as well as from her wand. Niblo's program of 1883 proudly advertised "novel lighting effects by the Edison Electric Light Company under the personal direction of Mr. Thos. Edison." [8]

During his four years' sojourn in New York, Edison kept his usual late hours, and enjoyed a fair amount of night life, although more often than not this was of a quite serious order. After working in the impromptu laboratory at Goerck Street up to midnight, he would set off on a walk to some restaurant for a late supper. Often people stared at him, as his appearance was familiar to many. Since his city clothes were usually a sober black, others who did not recognize him lifted their hats, mistaking him for a priest. Often he would go to the home of Sigmund Bergmann, who lived nearby, and call him out, however late the hour: "Bergmann, get dressed and come down.... I want to speak to you on a matter." When Bergmann joined him they would repair to a German café on Second Avenue. While eating and drinking, Edison would make notes and draw rough sketches on the menu card; these were schemes for improved electrical appliances that he would turn over to Bergmann for production.[9]

2   He could pretend that he was giving up inventing for some years because of the pressure of business affairs. But he could no more stop observing things, or experimenting, than he could stop breathing. Under the stress of serious business troubles, and the nervous tension of life in the midst of New York, he seemed to grow not fatigued, but all the more stimulated; he fairly bubbled with new ideas.

The ten years from 1873 to 1883, or from age twenty-six to thirty-six, constitute the most productive phase of Edison's career. It is not unusual that scientific workers reach their peak in young manhood. The period that began with his multiplex telegraph, and then witnessed the arrival of the talking machine, ended with scores of incandescent-lighting inventions and new processes, one or two of which alone might have made the reputation of an ordinary inventor. In the one year 1882, he is recorded as having applied for 141 patents—this at the height of his intense activity in engineering and promoting the Edison system. Many of these patents covered new manufacturing processes for accessories, as well as refinements in his lamp, dynamo, and electrical circuits; others were entirely dissociated from his electric lighting work, reflecting some sudden shift of interest; such was a magnetic ore-separator patent, applied for on April 3, 1880, providing a new method for the extraction of low-grade iron ore. Still other contrivances were not even patented by him, but kept as trade secrets. More than one writer on the history of modern invention has remarked that Edison's inventive fertility during this period is without parallel.

The enormous creative activity of this decade is climaxed, in 1883, by a purely scientific discovery of an unknown, or hitherto unobserved, *effect*— as physicists call any fact of nature or group of natural phenomena not explainable or classifiable according to existing scientific theory. It was a discovery whose future importance none then could estimate, neither Edison, who came upon it "by accident," nor the scientists of his time, whose curiosity was drawn to it momentarily.

As was his habit, Edison was "looking for things," when he unexpectedly opened a door leading to the still unknown world of electronics, which was to become the familiar purlieu of the twentieth century's men of science. He did not understand the nature of the phenomenon he had discovered; but that it was a "peculiar" phenomenon and not in accord with existing knowledge, he did realize; he also grasped that it might conceivably be of importance. At that time men still thought in terms of molecules as the smallest unit of material; the *electrons* he was actually experimenting with were unknown and were to remain unknown for a good many years.

The discovery of the phenomenon later named by scientists "Edison effect" came as a by-product of the inventor's unceasing efforts to extend the life and efficiency of his incandescent lamp. This work went on intensively after he moved to New York, in the room he had equipped as a laboratory at the Goerck Street works.

He had noticed, as indicated by his laboratory notebook entries for February 13 and February 18, 1880, that his light bulbs became blackened, apparently through the collection of particles of carbon deposited on the inside of the glass bulb. As this accumulation of an opaque deposit shortened the life of the bulbs, he made patient efforts to determine what caused it. The deposit turned out to be particles or atoms of carbon that were evidently being discharged or "carried" (as he wrote) from the fine carbon filament, under high heat, to the inside surface of the globe.

On further study, Edison and his associates made two significant observations about this black deposit that so baffled them. In the first place, there was always, after a while, a narrow white streak on the inside of the glass bulb, a sort of negative "shadow" falling in the plane of the horseshoe-shaped carbon filament. It appeared that one leg of the filament obstructed the flight of atoms from the other leg so that not all of them reached the inside surface of the bulb; hence the shadow area in that line of direction. A second important observation Edison made was that the shadow was always cast by the leg of the filament connected to the positive side of the d-c circuit supplying the lamp. Therefore, he reasoned, the tiny particles of carbon were being shot off in straight lines from the negative leg of the filament.

These studies suffered long interruptions, while Edison pioneered the first central lighting stations; then, in the summer of 1882, we find him back on the same trail, puzzled but persistent. Was it electrical current that "carried," or deposited, those carbon particles? But how could this occur in such a high vacuum?

To be sure, it had been known, since experiments of two hundred years before, that the air in contact with red-hot metal had peculiar electrical properties as regards the dissipation of an electrical charge.[10] For years Edison had been aware that some sort of electrical discharge went on within his vacuous bulb, creating a blue glow at the stage when he sent a current through the carbon filament to eliminate its occluded gases. Now it occurred to him that the carbon particles "carried" from his electrically charged filament might themselves be charged, and therefore could perhaps be deflected, or collected, by means of an extra pole, or positively charged electrode, inserted inside the bulb. This idea is set forth and signed in a notebook entry for July 5, 1882; a first rough sketch by Edison of a two-element vacuum bulb was drawn by him, showing the added electrode, and his explanatory comment, *"Prevent electrical carrying."* The two-electrode bulb made after this sketch bears a striking resemblance to a diode, or two-element vacuum tube.

After some interruptions, extending for several months, he was back again at his laboratory table, in March, 1883, studying the problem of the discolored lamps. Perhaps the added electrode might serve to eliminate those black deposits? The experimental bulb that he began to test at this time had a platinum wire inserted vertically in the half-inch space between the legs of the horseshoe carbon filament. When the bulb was exhausted of air and sealed, and the added electrode connected to the positive side of the circuit, he found that a current flowed and gave a good deflection to the galvanometer; when it was connected to the negative side no current flowed. This was exciting: there was current passing through the vacuous space of the bulb, *without wires,* as he carefully noted. This lamp was tested during many experiments. Various metals were tried, such as tin foil, and in diverse forms, in his attempts to get more galvanometer deflection, until the Edison-effect lamp had a metallic plate as its second electrode, set between the legs of the carbon filament but not touching them.[11] (Here was the primitive vacuum tube that others, in the early 1900s, were to use for wireless telegraphy.)

Edison had discovered some of the fundamental facts about thermionic currents. They flowed through the vacuum of the lamp. We must have this trick patented, he tells himself, and so on November 15, 1883, he files application for a voltage-regulating device using the two-electrode bulb (Patent No. 307,031):

I have discovered that if a conducting substance is interposed anywhere in the vacuous space within the globe of an incandescent electric lamp, and said conducting substance is connected outside the lamp with one terminal, preferably the positive one of the incandescent conductor, a portion of the current will, when the lamp is in operation, pass through the shunt-circuit thus formed, which shunt includes a portion of the vacuous space within the lamp. The current I have found to be proportional to the degree of incandescence of the conductor or candle power of the lamp.

A report written at the time, at Edison's order, by one of the laboratory assistants, covering the work they had done on this lamp, or "tube," shows that the inventor realized he certainly had something strange there. "This is a recent discovery," the memorandum begins. It outlines the form and structure of the Edison-effect lamp and describes experiments made at different candle powers, so that at twenty-five candle power "a very powerful current" is achieved, one that is "perfectly continuous and capable of supplying a telegraph wire of 200 miles in length with current to work the instrument, notwithstanding that the globe is exhausted to a millionth of an atmosphere and the current must pass a space of at least half an inch." [12]

The Edison-effect lamp was, in fact, the first electronic instrument; its creation made Edison "the father of modern electronics." That it functioned through the generation and movement of free electrons in space none knew as yet. No one then dreamed of what it would mean for the future of radio, for long-distance communication without wires, for sound amplification, television, radar, and many other thermionic devices.

What Edison was concerned with was how to put this discovery to some practical or "commercial" use. His thought was that the Edison-effect lamp might be used as a sensitive indicator of changes in lamp filament voltages. He therefore had his faithful draftsman, John Ott, make a precise drawing of an Edison-effect lamp utilized in a circuit of ordinary incandescent lights, connected in parallel, so that it served as an indicator of change in voltage. It was not only the first instance of an electronic application, but it incorporated, at that early day, Edison's appreciation of a basic characteristic of the thermionic tube, its "immense responsiveness," as shown in the relatively high rate of variation of electron emission from a hot filament with variation of voltage of the filament.[13] Later inventors would find, long afterward, that the charges of electrons in a thermionic tube were so exceedingly small that they could regulate even the very rapid and irregular increases or decreases of current needed to reproduce music or the human voice with complete accuracy.

The Edison-effect lamp, mounted in a circuit as an indicator of incandescent-light voltages, was exhibited at the International Electrical Exposition held in Philadelphia in September, 1884. The inventor recalls that he also put a telegraphic sounder in circuit with the Edison-effect lamp, and it worked very well. Actually it did not operate efficiently as an instrument for indicating voltage changes as Edison had hoped; perhaps because the vacuum of the two-element bulb was not high and constant enough as yet. For various reasons he did not pursue these investigations. Many years later he explained his inaction by saying, "As I was overworked at the time in connection with the introduction of my electric light system I did not have time to continue the experiment." [14]

This first attempt at an electronic instrument aroused a serious, though passing, interest among a gathering of scientists who examined it in Philadelphia in 1884. Professor Edwin J. Houston, the associate of Elihu Thomson, who had ridiculed Edison's gropings with electromagnetic waves almost a decade earlier, now called attention to the extremely curious and mystifying "high vacuum phenomena observed by Mr. Edison," as something potentially important. "I am inclined to believe," he said in a quite prophetic passage, "that we may possibly have here a new source of electrical excitement." [15]

The learned British engineer Sir William H. Preece (also a repentant adversary) made it his business to visit America and arrived in Philadelphia to study the imposing display of Edison's recent electrical creations. The Edison-effect lamps aroused his curiosity more than anything else. He obtained several models and brought them back to England. In his subsequent paper on the subject, read before the Royal Society in 1885, Preece originated the term "Edison effect." As William D. Coolidge wrote in 1950, looking back on this episode, men were baffled because "they knew—or thought they knew—that no current could flow through a vacuum." They did not know, of course, that a hidden area of the natural world was now being opened to them.[16]

Ambrose Fleming, in London, also pursued lengthy experiments with an Edison-effect bulb which, as he afterward related, he desired to improve upon so that it could be used, as Edison had hoped, as a voltage-indicating instrument for generators. After a while he gave up these studies, scarcely comprehending what was under foot.

There followed a "dark period," or relapse of investigation in this field, such as often occurs in the history of science. Hertz's production and detection of electromagnetic waves still lay several years in the future. It was not until 1897 that J. J. Thomson, the British physicist, after intensive study of the Edison effect, demonstrated that it was caused by the emission of negative electricity, or *electrons,* passing from the hot element to the second, or cold, electrode inserted in the bulb.* Thus Edison's discovery, when finally accounted for, constituted an important step toward the unveiling of the electron.

Meanwhile, to return to the period of the Philadelphia Exposition in the autumn of 1884, Edison's attitude toward his own discovery was something like blank puzzlement, as suggested by some occasional correspondence he had then with interested scholars. A few months before the Philadelphia Exposition, he remarked in one of his outspoken interviews, this time with a writer for a technical journal, "In experimenting, I find a good many things I never looked for"; and he added that he had lately been running into a "lot of things I dared not touch," a possible reference to his experiments with the baffling two-element bulb.[17] He habitually assumed a pose of studied indifference to the work of theoretical scientists and mathematicians—eggheads to him!—revealing some sense of inferiority about his own lack of academic knowledge, as shown in the following note to one inquiring correspondent:

I have never had time to go into the aesthetic part of my work—never have—done very little with it. But it has, I am told a very important bearing on some

---

* At almost the same time the twenty-three-year-old amateur of physical science, Guglielmo Marconi, appeared with his first crude wireless telegraph.

laws now being formulated by the Bulged-headed fraternity of the Savanic World [*sic*]. I will send you a half-dozen lamps if you want to have a little amusement.[18]

Then he completely lost interest in those elusive, pulsating electrons and his thermionic tube. Thus he also lost a marvelous opportunity to play a leading role in the new electronic field of wireless telegraphy, and all the other technical developments based on free electrons.

3   Nevertheless, the problem of obtaining "distant effects" through nonclosed electrical circuits had held his attention for a good many years, and he studied it again after 1884. Had he not, when bringing forth his black box to exhibit the so-called "etheric force" in 1875, predicted that telegraphic communication would one day be carried on without the "useless encumbrance" of wires? In the 1840s Morse had sent electrical impulses, by induced current, for a mile or so through a body of water, or through several hundred feet of earth without wires. In the early eighties, William H. Preece in England, John Trowbridge, the Harvard University physicist, and Edison, among others, studied the same problem of inductive telegraphy. Lately an old friend of telegraph days, Ezra T. Gilliland, who had flourished a little as an engineer of telephonic devices, had begun to work with Edison over such experiments in the small laboratory at Goerck Street.

Edison had it in mind to devise an instrument by which a man traveling across the Western prairies could telegraph messages and receive replies while riding in a train. The plan involved the installation of a special telegraph line, strung on poles of car height and parallel to the track. Along the top of one or more railway carriages he would install an insulated metallic strip, connected in series with a telephone receiver and the secondary circuit of an induction coil through the wheels and rails to the earth. For transmitting, the system would include a battery, key, and high-pitched "buzzer" in the primary circuit, and a switch for changing from "send" to "receive." Duplicate apparatus, connected between wire and ground, would be installed at each wayside telegraph office. By such means Edison proposed to span by induction the 30 to 50 feet between metal strips on the cars and the telegraph wires and so communicate between moving cars and dispatchers' stations. Impulses at both ends would be received as musical buzzes of short and long duration.[19] Such was the space telegraph, or "grasshopper telegraph," as it was called. (Patent was applied for on May 14, 1885, granted as No. 465,971.)

Edison and Gilliland tried out their space telegraph on a small railway in Staten Island, New York, with very curious results. The operator reported that he could send messages all right when the train was going in one direction, but not when it was going in a contrary direction. In a spirit of raillery, Edison tells how he made a long list of suggestions, without avail; finally he asked the operator if he had any to make himself. "I received a reply that the only way he could propose was to put the island on a pivot so it could be turned around." [20]

After a while the bugs were eliminated and the system tried on another railroad, the Lehigh Valley, with considerable success. Then, according to Edison's droll but apocryphal account, the patents, which he held jointly with Gilliland and L. J. Phelps, of Western Union, were supposedly sold to a wealthy but highly eccentric individual who would neither promote the invention nor answer letters appealing to him to do so. He was a spiritualist and used it himself, no doubt, to listen to voices from another world!

A variant of this system was devised by Edison, using electrostatic induction to "broadcast" telegraph messages without wires, for distances of up to three miles. (At Menlo Park, in 1880, he had rigged up a similar inductive telegraph without wires that "sent" about 580 feet.) This was accomplished by means of masts a hundred feet tall, atop which he installed metallic plates. The height of his masts enabled him, presumably, to overcome a little of the curvature of the earth's surface and permitted signaling or telegraphing over the sea between distant points. His patent specifications also suggested that the masts of ships could be similarly utilized for signaling to other ships, and thus prevent collision in fogs.

Essentially, Edison had an electrostatic machine, creating no high-frequency oscillations, but using an arrangement for discharging an induction coil into the metal plate at the top of the mast. No aerial wires or antennae were involved. The electrostatic charge on the plate atop the sending mast induced a similar charge on the metal plate on the distant receiving mast, which sent a current through its own circuit and caused a click in the chalk-disk telephone receiver.*

"This was the forerunner of wireless telegraphy," Edison said ruefully many years later in recollection of his train telegraph and marine-aerial device. It is all very interesting and suggestive, but it is not so. Moreover, there is no evidence that his seagoing inductive telegraph was practically operative.

---

* The ship-to-shore, or ship-to-ship, inductive telegraph device is included in the same patent as the railway train telegraph, No. 465,971. Edison could not have sold it to a spiritualist, for it is recorded that he sold it to Marconi's radio company in 1904.

The inventor is also recorded as having said regretfully many years later:

What has always puzzled me since, is that I did not think of using the results of my experiments on "etheric force" that I made in 1875. I have never been able to understand how I came to overlook them. If I had made use of my own work I should have had long-distance telegraphy.[21]

His role in leading the way to the development of radio and electronic science has, however, been accepted by modern historians as significant. A year or two after he dropped his studies of the inductive telegraph, Heinrich Hertz, in Germany, using a condenser discharge, created and shot forth electromagnetic waves which were detected by tiny sparks in a resonator at the other side of a large room. The use of "radio waves," or electrical oscillations of high frequency, for wireless telegraphy now became a distinct possibility, as the young Marconi perceived several years after Hertz's experiment. The older methods of inductive telegraphy were thereupon abandoned.

After Hertz's world-renowned demonstration, Fleming was induced to examine again the possibilities of the Edison-effect lamp; in 1888 he finally devised a little metallic cylinder—instead of the plate formerly used—as the second electrode, which he inserted into the vacuum bulb to enclose the negative leg of the carbon filament. Now the improved Edison-effect tube could be used to *rectify*, or convert, an alternating or oscillating current into a unidirectional current.

Marconi's primitive wireless apparatus of 1895–1896 employed the so-called "coherer" of Edouard Branly to detect radio signals; it was a glass tube filled with loose iron filings (in a closed circuit with a galvanometer and battery). But some years afterward, in 1897, there came the important discoveries of J. J. Thomson concerning the physical properties and velocity of electronic charges; then Fleming, who began to work as a consultant for Marconi, finally conceived of the idea of using his refined Edison-effect vacuum tube, or "Fleming valve," as he called it, to detect high-frequency radio currents. It proved to be far more sensitive and reliable than the coherer; in consequence, after 1904, the Marconi Telegraph Company adopted what was really the Edison-Fleming vacuum tube as a detector of radio waves, obtaining a greatly increased range of communication.*

* At about the same time, in 1903, Marconi's company decided to buy the patent rights to Edison's 1885 space telegraph and aerial masts, lest they fall into the hands of adverse interests. Indeed such a group approached Edison with the same purpose; but suspecting that they were only bent on making trouble, he stipulated

Thus, after more than twenty years, Edison's old two-electrode tube, or Edison-effect lamp, came into its own as one of the basic elements of all electronic communication. Two years later, the Edison-Fleming device was to be improved still further by the American engineer Lee De Forest, whose strategic invention of the triode—with its platinum grid as a *third* electrode—gave the world a perfected vacuum tube that could amplify radio signals as well as detect them.

But it must be remembered that it was Edison who discovered what has been called by scientists the best source of thermionic emission, the current of electrons flowing through a highly exhausted vacuous space. Moreover he had had the wit to stop and demonstrate the unexplained phenomenon and, in the form of a first electronic device, to call the attention of scientists to it. On the other hand, experimenting with and studying free electrons was foreign to his type of mind. He could not *visualize* the minute electron, which, in the words of Sir Oliver Lodge, "stood in relation to the atom as a fly to a cathedral." He could not hope, with his own knowledge of science, which was wide though in some ways limited too, to measure and calculate waves of charged particles as did J. J. Thomson; or catch electrons "by ones, twos, and threes," as did Robert A. Millikan later on, and so measure the forces exerted upon them by an electric field.

Significantly enough, three of the most vital elements of radio telephony were actually in his possession around that time: the microphone, invented by him several years earlier; the Edison-effect bulb; and finally, the elevated aerial mast he was to contrive in 1885. Yet he passed them all by, and pursued other interests. Had he applied himself seriously to study of this new field, it is conceivable that, with his unrivaled inventive talent, he might have advanced radio communication by some fifteen years. But the history of science is full of such tremendous near misses. The brilliant French scientist Ampère, teetering on the edge of discovery of induced electricity ten years before Faraday, missed it completely.

No, the empire of radio was not to be Edison's—though his carbon filament lamp of 1879 and his two-element vacuum tube of 1883 were the direct ancestors of De Forest's marvelous radio tube of 1906. It has, in fact, been used, by way of demonstration, in modern radio receivers. Edison's empirical studies in electromagnetic, sonic, and incandescent-

---

that his patent be sold only to Marconi. Edison expressed a deep admiration for Marconi, who had shown the courage to carry out his grand design in the face of the greatest difficulties. Marconi, for his part, acknowledged that he had "derived encouragement from Edison's interest in my work." (Telegram, G. Marconi to T. A. Edison, Apr. 18, 1912, Edison Laboratory Archives.)

lighting phenomena swelled the stream of scientific knowledge, just as James Watt's practical inventive work on the early steam engine led the way to a wider knowledge of thermodynamics. But the old order of technical and scientific work belonging to the "heroic age" of inventors was to change drastically as the nineteenth century drew to its close; the interaction of invention and scientific activity was to assume a wholly different relationship, in which theoretical scientists, grouped in "teams" with technicians, around large research institutions, were to become preeminent in advancing the technology of radio and electronics and, later, of atomic energy, while the older type of individual or free-lance inventor played a lesser role. Paradoxically, it was Thomas A. Edison, the foremost of that older type, who himself invented the organized research laboratory which would tend to eliminate his own kind.

After he read of Hertz's famous experiments, Edison said to one of his intimates: "Well, I'm not a scientist. I'm an inventor. Faraday was a scientist. He didn't work for money.... Said he hadn't time to do so. But I do. I measure everything I do by the size of the silver dollar. If it don't come up to that standard then I know it's no good." [22] In 1884, incidentally, Edison could scarcely raise enough silver dollars to promote his already practicable electric light system—let alone experiment with unknown and, to him, incalculable free electrons.

It is, however, true that he continued to maintain his pose of *méfiance* toward the theoretical scientists and to snipe at mathematicians. After Upton had become absorbed in managerial work for Edison, he was succeeded, in 1887, as scientific consultant, by a truly distinguished mathematical physicist, Dr. Arthur E. Kennelly, of Anglo-Irish descent and education, then in his young manhood. Kennelly was patient and charming with Thomas A. Edison and showed a real appreciation of his qualities. One day Edison's private secretary, A. O. Tate, heard Kennelly, while Edison was in his office, roaring with laughter. When the inventor came out, Tate inquired what the fun was about. Edison exclaimed impatiently, "Oh these mathematicians make me tired! When you ask them to work out a sum they take a piece of paper, cover it with rows of A's, B's, and X's and Y's ... scatter of mess of flyspecks over them, and then give you an answer that's all wrong!"

Kennelly, however, was to play a notable role in twentieth-century science and one day, with his pencil and paper, was to discover—concomitantly with Oliver Heaviside—the ionosphere, since known to science as the Kennelly-Heaviside layer.

The same witness who quoted Edison on his acceptance of the silver dollar as his standard of scientific currency also goes on to remark (as did so many others) that he actually never seemed to care for the possession of money itself, nor even know how to hold on to it. "In the expenditure of money for experimentation... it made no difference to him what the cost might be." It was his religion of experiment, his passion for more exact knowledge and superior results, his extravagances in research, that used to frighten his own capitalist patrons. Though he was admittedly no "pure scientist," he was no "mere" mechanical inventor either.* His prodigious and watchful activities in so many branches of scientific work, as well as his "intuitions," contributed to the advancement of knowledge itself. It was because Edison was "always looking for things," as Charles F. Kettering said, that he inserted a straight wire into an incandescent light bulb and found he could pull a current of electricity out of a vacuum, "thus discovering an inexhaustible source of free electrons." W. D. Coolidge remarks that the greatest credit was due Edison for both discovering and publishing this new fact of science. "Other men have studied the Edison effect... applied the underlying principle, but it was Mr. Edison's work which opened the door to this whole field." [23] It was at any rate a scientific discovery of the first class, though only acknowledged as such rather tardily in scientific histories.

4   In those hurried years of the earlier half of his career Edison had virtually no private life as other men usually know it; the relaxed enjoyments of home, a sweet and personable wife, his young children, were things he tasted quickly and sparingly, as if in haste to be back in his "man's world." There, in a domain where women were excluded, where all were masculine and bearded fellows— but for the smooth-shaven "chief mucker"—and where his associates worked long and late and swore hearty oaths while they worked; there he was most himself and knew his greatest pleasures.

In his boyhood and youth he had seemed rather solitary; many able inventors have been rather lonely and difficult personalities. Edison, however, had grown accustomed to working with fairly large groups of people and keeping them attached to him through the sheer force of his per-

* In his memoirs, *Fifty Years of Electricity*, published in 1922, J. Ambrose Fleming gives all too little credit to his former employer for his path-breaking experiments. He tells us that "Edison in 1883 noticed the phenomenon known as 'The Edison Effect'; but he could not explain it; nor did he use it in any way." Neither did Fleming, who also dropped the Edison Effect, until Hertz's experiments years later roused him from his slumbers. Reading this sentence forty years after the event, the aged Edison bridled with wrath and wrote in the margin of Fleming's book, "Fleming's statement untrue and he knows it is untrue." For, in fact, Edison had patented his two-element vacuum bulb both in American and England as a device for indicating voltages.

sonality. He seemed to enjoy best of all the company of the veterans of his laboratory, who had marched in many a campaign with him, such as Batchelor, Upton, and the voluble Johnson. But in the middle years new favorites shone in the master's circle. The diligent young Insull became more and more his confidant, his *fidus Achates*. Then his friendship of earlier years with Ezra T. Gilliland, the former telegrapher of Cincinnati, was renewed after 1881. Gilliland, a sleek man of large frame, sporting a thick walrus mustache, was now his constant laboratory associate and most trusted friend—Damon to his Pythias, as Edison said. It was Gilliland who drew him a little toward the amenities of social life; Gilliland's wife, who was intelligent as well as pretty, became a close friend of the Edisons, and her parlor was one of the few that Edison cared to visit.

He clung, nonetheless, to his old-fashioned notions of women's mental inferiority. "It is very difficult to make women believe anything that is so," he used to pontificate. "Women as a class are inclined to be obstinate. They do not seem to want to get out of the beaten path." [24] Such opinions, or prejudices, were perhaps strengthened by his feeling about his first wife, Mary Stilwell Edison, a good and simple woman of limited education, who, as he fully realized, scarcely knew anything of what was going on in her husband's head.

Their married life might have been unhappy had Mary not been so gentle a person and had she not known how to reconcile herself to her lot. In the small hamlet of Menlo Park, though it had become world-famous, her own life was prevailingly of a "dull loneliness," as the other residents noted.[25] The Edisons lived in a substantial and comfortable home; there was a staff of two Negro servants and a coachman; a barn and stable with horses and carriages in the yard; a summer house on the big lawn. But Mary Edison, while seeing all too little of her husband, had only the wives of his married associates, Batchelor, Kruesi, and Upton, as her occasional companions. Thus the young woman was left much alone with her children. At first her sister Alice Stilwell lived with her and helped with the children; but after several years Alice married William Holzer, a foreman of the lamp factory, and moved to a home of her own.

It was fairly customary for Edison to work in his laboratory eighteen or more hours a day, but at the time of the electric light project, he never came home at all for weeks on end, according to a close neighbor, who saw him finally one morning "coming along the plank path to his house walking as though he were asleep." Some carpenters were at work in the house at the time, and though her husband would not have heard them, Mrs. Edison sent them away, saying, "He has gone into my spare room and rolled right over on the bed in all his dirt and grease, on my nice counterpane and pillow shams, but I don't care, as long as he gets rest and sleep." [26]

Since she was often alône at night, Mary Edison seems to have been terrified of burglars. As her daughter Marion recalled, "She would often sleep with a revolver under her pillow. One night my father forgot his key and, not wishing to waken the whole place, climbed up the trellis onto the porch roof to the bedroom window. Mother, thinking he was a burglar, almost shot him. She let out a scream which father heard, then he called to her, thus preventing a catastrophe." [27]

By the 1880s new honors and dignities clothed her. Was she not the consort of one of America's first citizens, whose fame overleaped the oceans? Yet she remained modest and self-effacing. Rarely did she appear at the laboratory, only eight hundred feet distant, where she knew her husband was engaged night and day. On the few occasions when she did come there the workmen regarded the beautiful young Mrs. Edison, now grown quite stout, with unconcealed admiration. "She was greatly beloved by the men in Edison's employ," relates W. K. L. Dickson. "They were proud of her—for she had been one of their own rank in the Newark shop and yet remained as gracious and friendly to them as ever." [28]

When he was minded to do so, the unpolished Edison could be most attentive and charming to women, as when Sarah Bernhardt visited him. Though often quite distrait with his wife, he was far from indifferent; indeed he felt a strong attachment to her. He had courted her passionately; she was (after his mother) the only woman in his life up to that time, his first romance, so far as we know. She had shared the lean years with him. Now he was a rich man, though with him ready money was always lacking and he could not often be extravagant to please her. But when a real windfall came, such as a payment for some important patent, he would bring Mary the most costly gifts. Thus, in her first year of residence in New York, she made an impressive appearance one day at a tea party—all bejeweled, dressed in a gown of the richest brocade, with a bodice of satin, the folds of her skirt looped up in a bustle and then flowing behind her in a long train. On such occasions she not only "dressed to kill," but saw to it that her growing daughter Marion was also elegantly turned out.

When the pace of work permitted relaxation, Edison would devote his Sundays, at least, to his family, sometimes taking them to a nearby beach. In his lighter moments he could be merry enough, bantering his wife and teasing the children unmercifully; but on those Sunday excursions, as

Marion recalled, Mary Edison was always "proudly happy" as she rode off with her husband by her side.

When at home on a Sunday, Edison's mind would go spinning along, still absorbed in the problems and experiments of the week. He might start by reading the newspapers, then the *Police Gazette,* but would end up poring through technical and scientific publications such as the *Transactions of the American Institute of Electrical Engineers.* At meals he would often say little to his family, eat quickly, and leave the table before the others. Decidedly it was not easy to have such a remarkable father.

The stories of his playing with his children seem to date mostly from the earlier years of his first marriage. By way of toys he would bring a batch of old alarm clocks for them to "experiment" with. Sitting down on the floor, he would take the clocks apart and put them together again, bidding his boys to do likewise, evidently hoping that, like himself, they would become fascinated by the mechanical arts. Disappointingly, his small sons, Tom junior and William Leslie, showed no budding passion for rusty alarm clocks; in fact Willie, the younger one, after having reached the considerable age of nine or ten, persisted in playing with a toy train. Angrily the father took the train away from him, declaring that he was "too old" for such things, while the boy wailed.[29] But as toys, old alarm clocks could not compare with the tin-foil phonograph and the talking doll that squeaked out Mother Goose rhymes, which he had brought them earlier, to their great delight.

Usually Edison's love of mischief—which on occasion showed a vein of playful cruelty—came out in his games with his children. One day he brought home a little glass toy in the form of a swan, made by his glass blower, Boehm, and invited one of the children to put the tail of the swan in his mouth and blow. The child was not amused when, from the neck of the swan, water sprayed all over his face, but to Edison that was fun. On another occasion, Mrs. Edison gave a large birthday party for her daughter Marion. The father himself put in an appearance this time and saw to it that the many electric lamps in his house were all functioning. But at the height of the party he slipped down to the cellar and opened the switch, plunging the whole house in total darkness.[30]

The children found their father not so much unkind as puzzling, and often "difficult." Marion, the eldest, being a girl, was not expected to become an inventor. She was tall, blond, pretty, and had her father's verve. "She seemed to be always dancing rather than walking," one of his associates said of her.[31] Her father tended to be more affectionate with her than with his sons, as fathers often are with their daughters. "I think I must have been my father's favorite," she recalls. When only ten or eleven she had her own pony and cart and used to drive all about the village at a fast clip; and she could be a daredevil on the electric railway too. Of the children only Marion was allowed to come into the laboratory, for she often brought her father his lunch at noon. At that hour he would usually be taking his cat nap.

He would have liked his eldest son and namesake, who resembled him physically, to be keen of mind and aggressive and energetic like himself. The boy, however, turned out to be rather delicate and sickly.

The holiday of the Fourth of July Edison always devoted to his children. He would rise at five in the morning, put a giant firecracker in a barrel out on the lawn and set it off, arousing the whole neighborhood. "He would become a child himself," one of his children recalls. "He would have us children run around barefoot and would throw those little Chinese firecrackers at our feet, enjoying himself hugely."

Sometimes he would put up a ten-foot pole and invite his own children and the neighbors' children to shinny up and win the coins he placed at the top. To the father's disappointment Tommy would always prove to be the weakest and the last at such games. On one occasion, the inventor thought of rubbing rosin on the poor boy's knees, so that he managed in the end to climb the pole—to his father's evident relief.

Tommy also had an unfortunate way of getting into scrapes. When he was about six he wanted above all things to play in the laboratory and machine shop, for the roaring engines and generators fascinated him. Fearing that the boy would hurt himself his father sternly ordered him to keep out of the building. Tommy disobeyed him, and was soon afterward seen by his father sneaking out of the back door of the machine shop. At once the inventor went to his newly installed telephone and called Mrs. Edison, ordering that the boy be thoroughly spanked. Then with one of his laboratory assistants, Edison climbed to the little balcony on the upper floor of the laboratory and looked toward his house to see what would happen. Tommy arrived at the gate and, to his great surprise, found his mother already there, waiting for him with a switch. "The poor boy could never understand how his mother, while at home, saw him in the shop." [32]

Edison's own father had been at best an indifferent sort of father, with little real understanding of his son. Of Thomas A. Edison his children recalled that he could be both warmly affectionate and playfully, unconsciously cruel, but that most of the time "he hardly ever saw us," or, "he never thought of us."

Unfortunately, his disappointment in his eldest son, and also in Willie, revealed itself. Their upbringing, he may have felt, was not of the best, possibly because of Mary's habitual overindulgence with her children, or her own lack of education. He could not be unaware that his wife was, in fact, somewhat self-indulgent. Her daughter remembered seeing her mother spend whole afternoons idly chatting with a woman friend and consuming an entire box of chocolates. It was bad for Mary Edison's figure—she became tremendously stout—and bad for her health.

Though not unforgiving, Edison could show a fierce temper, as when an assistant in the laboratory was found in error; then his face would turn black and become distorted with emotion. His family knew these moods too. There was, no doubt, a normal amount of anger and grief and forgiveness in his life with Mary. In the fragment of a diary he wrote in 1885—after she was so suddenly, tragically gone from him—he dropped some hints of his views on the problems of marriage. His daughter, Marion, though only twelve, had for the moment some literary ambitions.

Dot says she is going to write a novel, already started on. . . . Dot just read me outlines of the proposed novel. The basis seems to be a marriage under duress. I told her that in case of marriage to put in bucketfuls of misery. This would make it realistic.[33]

Mary Edison was delighted at last to leave the dull solitude of Menlo Park for New York in the winter of 1881. After stopping at a hotel for a while, the family settled into an apartment overlooking Gramercy Park, and Mary sent out cards inviting people to tea-and-champagne, though her husband, who detested such parties, never appeared at them. Marion was sent to a private school for girls; the Menlo Park house was still used as their summer home.

In the winter of 1883–1884 Edison, suffering from neuralgia, made a vacation journey to northern Florida with his wife and daughter. The climate of St. Augustine proved so beneficial that, thereafter, he went down to Florida every winter for a few weeks, these intervals of travel being immensely enjoyable to his family. To be sure, as soon as he had recovered from illness or fatigue, his mind would race back to the problems of the central station system; and from Palatka, Florida, he would fire off letters or telegrams by the hour to his factotum, Insull, such as: "Don't forget to have Tomlinson [his attorney] draw up contract with the engineers"; or "Let me know how the new pressure indicator works. . . . I think I have struck a way of utilizing the surplus power of our stations. . . ."[34]

After the winter journey to Florida in 1884, the Edisons returned to New York; the following summer, Mary and the children went back to the Menlo Park house, as usual. Edison then was much occupied by work in the city and remained there most of the time. In July Mary contracted typhoid fever, which at first did not alarm her family; but soon she was under the constant care of a doctor and her sister Alice. When suddenly she began to sink, Edison was called, and he hurried back from New York to her bedside. On the morning of August 9, 1884, his daughter Marion recalls, she was wakened by her father, who had been up all the night before. "I found him shaking with grief, weeping and sobbing so he could hardly tell me that mother had died in the night." [35]

From that day forward, he rarely came back to Menlo Park; it was as if he hated the place. Moreover, it was a time of much trouble and internal dissension in the Edison Electric Light Company. He was an agnostic, and so could find no solace in praying for Mary's soul. As was his wont, he buried himself in work, and thus stoically put out of mind the tragedy of his beautiful young wife. She was not yet thirty when she died.

The Menlo Park laboratory, its stores and apparatus removed, fell into disuse. For a while the village people used the "tabernacle" for their local entertainments and dances. Later the lower floor served as a cow-barn; but, when some years had passed, the historic building began to fall apart. A farmer discussing with a neighbor the site of Edison's memorable experiments was heard to remark, "It's a shame such a fine farm was allowed to go to ruin!"

With the passing of Mary Edison in 1884 Edison's way of life seemed to undergo a marked change. At thirty-seven he was no longer a Bohemian nor a hell-for-leather, free-lance inventor, but a man of substance, heading a big industrial enterprise that increased his wealth year by year. He was, in fact, a millionaire. By force of circumstances he assumed an outlook that was more worldly than before. When, not long afterward, he fell in love and married again it was with a woman wholly different from his simple and touching Mary, one more fitted to share and enjoy the high station he now held, and under whose inspiration he would live in a style very different from that of the past.

Nancy Elliott Edison, Thomas Edison's mother (left).
Thomas Alva Edison at the age of fourteen (above).
Samuel Edison, Jr., Thomas Edison's father (right).

In this house in Milan,
Ohio, built by his father,
Thomas Edison was born,
February 11, 1847.

The Weekly Herald,
written, printed, and
published by Thomas
Edison at the age of
fourteen in a baggage
car on the Grand Trunk
Railroad.

Madame Marie Roze "warbling a *scena* from an opera" into the phonograph, as pictured on the cover of *Frank Leslie's Magazine*, April 22, 1878.

The original phonograph, built at Edison's direction by John Kruesi on December 6, 1877. The recording surface was tin foil, wrapped around the brass cylinder, and embossed by a needle protruding from a flexible diaphragm in the mouthpiece.

Replica of the telephone developed by Edison in 1879, employing a carbon transmitter and a chalk receiver. The small crank on the receiver (right) had to be wound continuously while the instrument was in use.

Printing telegraph or stock ticker, with a keyboard of letters only, manufactured by Edison at his Newark factory around 1871.

In April, 1878, when he was thirty-one, Edison went to Washington to demonstrate the phonograph to the American Academy of Science, to members of Congress, and to President Rutherford Hayes. This picture was taken at the studio of the noted Civil War photographer Mathew Brady.

Mary Stilwell Edison, Edison's
first wife, around the time
of her marriage, December 25,
1871, when she was sixteen.

Thomas Edison at about
the age of thirty-four.

Edison's laboratory at Menlo Park in the winter of 1880–1881.
The main laboratory is in the center of the yard, with the machine
shop behind it, and the glass blower's shed set diagonally in
the angle between them. To the left is the carbon shed, where
the lampblack carbon was produced; in the foreground is the
library and office; to the right is the electric railway.

The second floor of the laboratory, early in 1880. In the photograph, Edison sits at left center wearing a cap; Francis Jehl, with knees crossed, sits reading a book at right center; at the far right is Edison's friend and early teacher of telegraphy, J. L. McKenzie. The organ in the background was given to Edison by Hilborne Roosevelt. The engraving is of the other end of the laboratory, and shows clearly the cabinets containing the instruments and chemicals along the walls, and the electric lights strung from the gas fixtures.

Edison carbonizing a paper
lamp filament, which is
enclosed in a metal
mold to keep it from
oxidizing in the furnace.

Pages from Edison's notebook when he was experimenting with
the "Edison effect," an outgrowth of electric-light research
which foreshadowed the development of the electron tube.

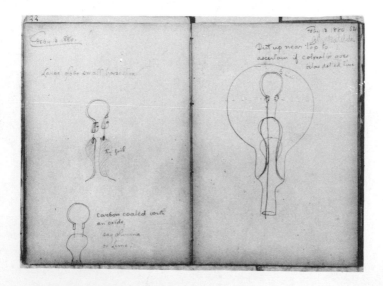

MACHINE SHOP.

The machine shop at Menlo Park in the fall of 1879. The Edison dynamo (the "Long-waisted Mary Ann") in the right foreground is attached by belts and pulleys to a dynamometer, also of Edison's design, which measures the power produced by the steam engines in the room behind. Two field magnets lie beside the dynamo, and at the far left are several field electromagnet cores. (From a drawing by Edison's draftsman Samuel D. Mott.)

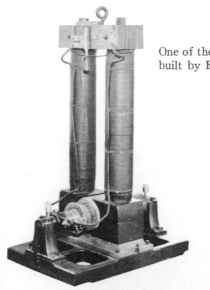

One of the "Long-waisted Mary Anns" built by Edison in 1879.

Replica of the first successful electric incandescent light, made October 21, 1879, with a filament of carbonized cotton thread. This model was made by Edison at the Festival of Light at Dearborn, October 21, 1929.

Edison's electric railway in 1880. The motor was a "Long-waisted Mary Ann" dynamo, laid on its side.

The improved 1882 model of the electric locomotive, with an enclosed cab and cowcatcher. It was capable of speeds up to 40 miles an hour.

Courtesy New York Historical Society

Much-publicized photograph of Edison, "as he appeared at 5 A.M. on June 16, 1888, after five days without sleep," working on the improved model of the phonograph. Actually the vigil lasted seventy-two hours.

Mina Miller Edison, Edison's second wife.

Edison's letter to Lewis Miller, asking his consent to Edison's marriage to Mina.

Cor Ave B & 17th St
New York Sept 30 1885

My Dear Sir

Some months since, as you are aware, I was introduced to your daughter Miss Mina. The friendship which ensued became admiration as I began to appreciate her gentleness and grace of manner, and her beauty and strength of mind.

That admiration has on my part ripend into love, and I have asked her to become my wife.. She has referred me to you, and our engagement needs but for its confirmation your consent.

I trust you will not accuse me of egotism when I say that my life and history and standing are so well known as to call for no statement concerning myself. My reputation is so far made that I recognize I must be judged by it for good or ill.

I need only add in conclusion that the step I have taken in asking your daughter to intrust her happiness into my keeping has been the result of mature deliberation, and with the full appreciation of the responsibility I have assumed, and the duty I have undertaken to fulfil

I do not deny that your answer will seriously affect my happiness, and I trust my suit may meet with your approval.

Very sincerely yours

Thomas A Edison

To Lewis Miller Esq
Akron
Ohio

A partial view of the iron mine and ore-processing plant at Ogdensburg, New Jersey, in the nineties. Photograph by Edison's assistant W. K. L. Dickson.

Edison in front of the office of the ore-processing plant.

The Giant Rolls, which used the momentum of two large rollers to break up ore-bearing rocks as large as pianos.

Strip kinetograph of 1889: the prototype
of the motion-picture camera.

The Peephole Kinetoscope, commercially manufactured
in 1894. Moving strips of film were viewed
through the magnifying eyepiece on the top.

Edison operating an early
motion-picture projector
in the library of his West
Orange laboratory (1897).

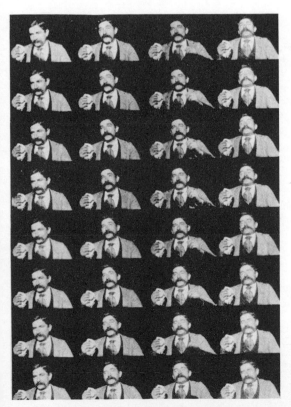

"The Record of a Sneeze," starring
Edison's assistant Fred P. Ott. The first
motion picture ever copyrighted (1894).

Model of the prefabricated concrete house invented by Edison in 1910.

A 1910 production model of the Edison phonograph, with a "Cygnet" horn.

The library at the West Orange laboratory. Edison's roll-top desk is in the center. The bronze statue to the left of the desk portrays Orpheus laying aside his lute for a phonograph record. The statue to the right, with the uplifted light bulb, is called *The Triumph of Light*.

Upstairs sitting room at Glenmont, Edison's
home in Llewellyn Park, as it appeared in 1907.

Edison on the lawn of Glenmont, around 1917.

Edison at about the age of seventy-eight.

chapter XV

The middle years: second marriage

1   Four years after Edison had perfected his carbon filament lamp there were only twelve city stations, on the model of Pearl Street, in operation; in all America only 150,000 of his lamps were in use. Much more rapid progress had been made in selling isolated lighting plants for factories, stores and hotels; by January, 1884, these numbered about two hundred, and used from a few dozen to three hundred lamps each. But all of it amounted to almost nothing compared with the huge gaslight industry which he had sworn to supplant. After the original Pearl Street station reached its full load limit, no more generators were added for several years; no annex was built until 1886, and added business was simply turned away.

The trouble was obviously due to the extreme conservatism of his financial backers. But how could he expand the business when others controlled its policy?

Try as he would, Edison could not stir things up without the approval of the holding company which owned his lighting patents. At the formation of the Edison Electric Light Company in 1878 he had been the largest stockholder, owning 2,500 of the original 3,000 shares. But two years later, when the project promised to be successful, the capital of the company was increased about threefold; the Morgan partners and Western Union directors subscribed to the bulk of the new capital issue, while Edison was forced to sell part of his original holdings to finance the lamp and generator factories. Thus, as soon as the business showed promise—even Pearl Street was earning a modest profit by 1884—he had lost control of the patent-holding company, which was the key controlling the whole enterprise.

"The importance of Morgan support to Edison's electric lighting project can hardly be overemphasized," remarks a historian of the electrical industry in America.[1] Edison himself had told people he expected great

291

benefits from this connection. The biographers of Pierpont Morgan (especially his son-in-law, Herbert Satterlee) have described him as enthusiastic for the electric light, and "intensely progressive."[2] He ordered one of the first isolated plants for his own home. Economists and social historians have cited Morgan's interest in and support of Edison's invention as an example of the innovative spirit that ruled in high capitalist circles after the Civil War.

Certainly the forehanded Morgan was interested in the potentialities of the Edison system from the beginning; it appealed to his imagination. One of his partners—at first Fabbri, and after his early death, C. H. Coster—regularly acted as treasurer of the Edison parent company and of its subsidiary, Edison Illuminating. If there was anything in this new project Morgan was going to keep his hand over it.

The situation cried out for an expansive strategy. But the banking group stolidly opposed Edison's enthusiasm and insisted that he "go slow." His three-wire invention of 1883, for example, offered great economies of distribution and, as he pointed out, made it possible to introduce lighting systems profitably even in small country towns where houses were far apart. But the directors firmly refused to finance such projects, and the inventor was forced to draw on his own capital to carry out the historic three-wire installation at Sunbury, Pennsylvania.

After Morgan had installed the Edison lighting plant in his redecorated house on Madison Avenue, in the autumn of 1883, he held a big reception for four hundred guests. One of them, Darius Ogden Mills, the famous gold mine operator and stock market plunger, was so impressed with all those brilliant new lights that on the following morning he walked into the office of Drexel, Morgan & Company and ordered the purchase of a thousand Edison shares. "Pierpont heard of this at once," and before Mills could go out the door, caught him and asked him what he knew about the Edison light.

"I know all about it," answered Mills.

"All right, we will take your order," said Pierpont, "and any other orders of the same kind, but I am going to put a condition on my partners with respect to such orders . . . *that for every share of Edison stock they buy for you they buy one for me.*"[3]

This incident has been cited as showing Morgan's enthusiasm for the Edison venture. What it suggests rather is that Morgan was serving notice that he would allow no one else to take control over this promising industry; and he was a most determined and formidable man. By the end of 1883, the meetings of the directors of the Edison Electric Light and of its two non-manufacturing subsidiaries, the Edison Illuminating of New York and Isolated Lighting, were regularly held in Morgan's office.

Major Sherburne B. Eaton, a prominent member of the New York bar who in 1882 succeeded Dr. Norvin Green of Western Union as president of Edison Electric Light, reflected the Morgan policies when he admitted frankly that the banking group was "loath to make use of money to push the enterprise." [4]

The new electrical industry intrigued Morgan; but he saw that it was as yet pretty small for him, and earned limited returns for the large capital it used. While refusing Edison some modest credits of a few thousand dollars, Morgan was raising millions to finance big railroad combinations like the New York Central and the Great Northern, which then yielded assured profits.

Of Edison's financial patrons only Henry Villard seemed to "believe in the light," as he said, with all his heart. Villard might have lent Edison powerful support at this stage of affairs, for he entertained grandiose ideas about the future Electrical Age. But in January, 1884, the newly arrived railroad king had a great fall; he had overextended things in expanding the Northern Pacific, exhausting its treasury, so that the big railway system was thrown into receivership. This affair caused a fair-sized panic in Wall Street and created much feeling against Mr. Villard, who was attacked in the press on the score of alleged mismanagement. Morgan, it was said, had always mistrusted Villard.[5] At all events he was *hors de combat*—so far as being able to help Edison in his fight—at least until he could accumulate another war chest.

A few years earlier the railroad promoter had contributed $40,000 to Edison's experiments with the electric locomotive. In Villard's hour of need, the inventor had returned this sum; he recalled later:

When Mr. Villard was all broken down and in a stupor caused by his disasters in connection with the Northern Pacific, Mrs. Villard called for me to come and cheer him up. It was very difficult to rouse him from his despair and apathy, but I talked about the electric light to him, and its development, and told him that it would help him win it all back and put him in his former position.[6]

That was exactly what Villard did, after a while, in trying a "comeback." He retreated to Europe, resumed his former connections with German banking interests, and persuaded them to invest large sums, under his direction, in the growing Edison companies in America, which brought in very handsome profits.

Edison's impressions of Villard, not only on that unhappy occasion in 1884, but also in later, more fortunate, years, was that he was humorless like Jay Gould. "He was a very aggressive man with big ideas, but I could never quite understand him." When Edison, with all his powers of invention, told his choicest stories, Villard "could not see a single point and scarcely laughed at all."

Many of the big money men of that time impressed other observers besides Edison as cold fish, who knew only the icy pleasures of the market place in which they perpetually swam. Was the acquisitive instinct, then, incompatible with a sense of humor? By contrast, a creative being like Edison seems to us full of the joy of life.

At all events, he was forced to give battle alone against the financial conservatives. He tried prodding them, though it helped little. From Florida, during the winter, he would dash off letters that crackled with impatience.

What is Eaton doing about putting agents all over the country to get the towns started? How about Louisville, Auburn, Haverstraw, Hudson & other towns? Any prospects? [7]

According to financial gossip in the newspapers there were now two parties in the Edison organization—

one slow and conservative, the other, including the inventor, energetic and willing to spend money for the sake of making money.... The result has been, as Mr. Edison himself put it, that his light has not been used on anything like the scale he might reasonably expect.[8]

In his view, it would be good business to set up local power stations in many small towns, show that they could be operated profitably, and then induce local capitalists to take them over. But it would need much capital to initiate such a spreading movement; and this the majority of Edison Company directors refused, holding that the local capitalists ought to raise funds themselves, in the first place, and pay for lighting plant equipment in cash, within thirty days.

At length Edison grew thoroughly fed up with the directors of his company, and especially with President Eaton, who seemed much more suited to the law than to industry. Samuel Insull described him as a pompous little man of military bearing, who strutted about his office at 65 Fifth Avenue, "holding forth like a great mogul"—but bringing in no business. Edison, however, had his champions on the Board of Directors, especially Edward Johnson, who protested at the inroads being made in their business by competitors who were infringing upon the company's incandescent lighting patents, and urged that vigorous legal action should be instituted against them. But the attitude of Eaton and Grosvenor Lowrey, who had lately joined the conservative group, was to avoid such action for the present, in accordance with the policy announced in the Directors' annual report of 1883:

The Edison patents, as a matter of law, not only endow our Company with a monopoly of incandescent lighting, but aside from the patents, our business has obtained such a start, one so far in advance of all competitors ... that the business ascendancy is of itself sufficient to give us a practical monopoly. The one or two ... competitors have thus far failed to make it worth while for our Company, in the opinion of your Board ... to go to the expense of bringing suits for infringement.... Our Company, from the enormous start and business advantage it has already acquired, will be able not only easily to keep at the head but also to maintain a practical business monopoly.... However we still have our patents in reserve, and are free to bring suit upon them at any time that the Company may think best.[9]

It must be remarked that Edison himself had a horror of being involved in lawsuits over patent rights, as a result of unpleasant experiences on the witness stand in earlier suits over telegraph inventions. Sawyer and Maxim had begun to infringe on his lamp patents two years earlier. There had also been a potential contest with Swan, in England, in 1880. But Edison had written then:

My views are strongly in favor of not sueing [sic] either Swan or Maxim. My reasons are the same as advanced over a year and a half ago, when it was urged with great persistence that the directors sue Maxim.[10]

He had little faith in the power of the courts to protect an inventor under the public patent laws. Such litigation usually became snarled up in a web of legal technicalities introduced by the opposing lawyers, so that the ends of justice were usually defeated. Meanwhile, an ingenious rival could easily find methods of copying an original invention by introducing insignificant variations which involved no change in principle. "A lawsuit is the suicide of time," Edison wrote in his diary of 1885.[11]

It was far better to depend on improving techniques and on "trade secrets," he maintained, than to lose time and money on lawsuits. If this position seems inconsistent with his own practice of applying for patents on the hundreds of devices or processes he invented, it must be remembered that commercial usage, no doubt, accounted for a good deal of that.

The Edison lamps, in early years, surpassed all others by a wide margin in tests conducted at various expositions of 1884 and 1885 to determine life, efficiency, and economy in number of light units per horsepower. More and more talented technicians were appearing in this field, however, and the policy of the Edison Company was perhaps too complacent. Patent infringers, though only a small cloud on the horizon in 1882–1884, would become a formidable threat a few years later.

There were other issues in dispute between Edison and his patrons. Though, in truth, he had some excellent ideas about business management, his methods were decidedly informal. For example, on his own initiative he engaged a new manager to oversee the first power station in New York, which had been running at a loss for the first year or two. Edison personally guaranteed that if the man brought the business at Pearl Street up to the point where it earned 5 per cent on its $600,000 capital, he would give him $10,000 out of his own pocket. "He took hold, performed the feat and I paid him the $10,000."

But President Eaton rightly objected that such expenditures, when not authorized by the official head of the company and its treasurer, were irregular, and the directors refused to honor them. As Edison remarked, "They said they 'were sorry'—that is, 'Wall Street sorry'—and refused to pay it. This shows what a nice, genial, generous lot of people they have over in Wall Street." [12]

There were also instances of Edison's laying claims against the E.E.L. Co. for out-of-pocket expenditures made in order to insure the success of their project. In one case he had authorized certain agents to carry on lobbying activities before the New Jersey legislature in behalf of a favorable electric lighting act. Reflecting the easy political morals of the time (and other times too) Edison promised certain legislators one thousand dollars "payable after the passing of the Act." [13]

But when it came to submitting clear records of such transactions to the executive officers of the company Edison usually did not have them. What Insull called "his habitual contempt for bookkeeping," and his informal or impulsive ways of doing business forbade keeping such accounts. It needed, in fact, much patient pleading by President Eaton before Edison was persuaded that he must make no expenditures without the authority of the parent company's officers.

A more serious issue, however, arose from a real conflict of interest within the group of Edison companies that had been so hastily thrown together. The manufacturing group were usually at loggerheads with the holding company, Edison Electric Light, which controlled the Edison patents, the New York power station, and the Isolated Lighting Company —in short, the patent-owning and utility-operating end of the business. Edison and his private partners—mainly Johnson, Batchelor, Upton and Bergmann—manufactured accessories: lamps, dynamos, motors, conducting mains, and other appliances needed for lighting plants using the Edison system. On these products royalties were usually paid to the parent company; and they were supplied under contract with the parent company.

For example, by an agreement of March 8, 1881, the Edison Lamp Company sold lamps to the parent company and its licensees for thirty-five cents each, the lamps selling at retail for one dollar each during the first few years. But as volume and production methods improved, and retail prices were reduced somewhat, this business became highly profitable for the lamp company, and the E.E.L. Co. urged that the wholesale contract price be lowered. Generally, the more profits Edison and his manufacturing partners made on lamps, dynamos, and other equipment, the more the holding and utility companies, in the view of the banking group, were disadvantaged. On the other hand, if the banking group resisted the expansion of the lighting business, the manufacturing companies were held back.

"I have always deplored the circumstances which placed you in double, and somewhat antagonistic, relations with the [Edison Electric Light] Company," one of the directors wrote him at this period.[14]

In the spring of 1884, C. H. Coster, partner of Morgan, and Grosvenor Lowrey, who also wore the "Morgan collar" these days, proposed that the holding company be given "a suitable interest in, and large influence over, the manufacturing business. . . ." It was further proposed that this be accomplished by the exchange of stock which in effect would give the holding company 40 per cent control of the Edison Machine Works, the Edison Lamp Company, and Bergmann and Company, makers of appliances, fixtures, and other "small deer." But Edison and his chief partners, Johnson, Batchelor, and Bergmann, were unwilling to part with so much of their ownership in the going manufacturing shops which gave them a means of livelihood, in exchange for the parent company shares which as yet offered no regular income. The inventor and his laboratory assistants had launched these manufacturing firms with their own limited means, as pioneering ventures in which the banking group at first had absolutely refused to invest money. Now that the lamp company (which lost heavily the first year) and the other electrical-equipment concerns showed themselves increasingly profitable, the Morgan group were quite willing to take stock in them.

Edison was so indignant at such proposals that at one point he threatened to terminate his contract with the parent company, declaring that he had no confidence in its executive officers' will or ability to carry out the expansion program he had in view. He also insisted that Major Eaton be supplanted as president by Edward H. Johnson.

A lively stockholders' fight flared up during the summer of 1884. Edison and Insull spent several weeks "working like Trojans getting proxies." Though the inventor was now a minority holder in his company, his glory was then at its zenith; a good many stockholders gave his party their proxies, one of them remarking to him that the company "seemed to have been managed with imbecility, or worse." [15]

A former friend and supporter, Professor Barker, however, wrote the inventor in a tone of anxiety: "I am sorry at this collision. If you win, your capitalists are alienated, and if they win you are dissatisfied." [16]

Insull worked hard for the Edison party, and he was beginning to show ambitions of his own. He seems to have regarded President Eaton as the big stumbling block. "There is no one more anxious after wealth than Sam Insull," he wrote to a friend in reporting the battle, "but there are times when revenge is sweeter than money." [17] When the day of the annual directors' meeting arrived, October 28, 1884, the Edison faction had won about half the stockholders' proxies, and, as Insull announced triumphantly, everything was settled to their satisfaction. Major Eaton was to be demoted to the position of corporation counsel; and one Eugene Crowell, an elderly and inactive figure, was to succeed him as nominal president of the Edison Company, while Edison's man, Edward H. Johnson, was to be executive vice-president. As Insull summed it up, "Drexel & Morgan have . . . in the face of this support which we have obtained come to the conclusion that the most graceful thing to do is to give Edison what he wants." [18]

The inventor was described as having been wrought up to a high pitch of excitement and belligerency during the conflict, but felt in better humor when it was over and settled, as Henry Villard urged, "in a spirit of mutual concession." However, a majority of the directors were now Edison's personal adherents. At this time Edison parted company with his friend and patron Lowrey, who was dropped from the board of directors for having allied himself with the conservative faction.

Pierpont Morgan, however, continued to watch over this young enterprise. One of his partners, J. Hood Wright, became a director; and in place of Villard, who resigned (after the Northern Pacific crash), C. H. Coster also was elected to the board of directors and made treasurer. The Morgan firm continued to be banker to Edison and his companies, and remained as financially conservative as ever, especially when the inventor tried to borrow money on his anticipations—no matter how alluring— instead of sound and sufficient collateral.

In the autumn of 1884, Manhattan residents were diverted by the sight of an evening parade down Fifth Avenue, the like of which had never been seen before. Several hundred men advanced in a hollow square formation, each caparisoned in a helmet surmounted by a little glow lamp; within the square a steam engine and Edison dynamo on wheels rolled along, supplying current through a cable connected by flexible wire to each of the marchers. At the head of the column rode a gaily uniformed commander on a war charger, bearing before him a baton tipped with light. The parade signalized the arrival of Ed Johnson in power. Another of his publicity schemes was the "Edison Darky," a dancing performer wired for light, who appeared at many expositions and fairs that season. Johnson was promoting the Edison system strenuously, and after his own circuslike fashion.

In more serious mood, on May 23, 1885, he issued a formal warning in behalf of Thomas A. Edison and his companies, that vigorous defensive measures would be taken against all infringers of Edison's patents for incandescent lamps. Suits had lately been instituted against the United States Electric Lighting Company (manufacturing the Maxim lamp), the Consolidated Electric Light Company (having the Sawyer-Man patents), and several other manufacturers. Potential customers were now guaranteed that all who made and sold incandescent lamps not authorized by the Edison company would be prosecuted and punished to the full extent of the law.[19] Hitherto no such policy had been made public, and competitors had entered the field in growing number. The first suits encountered some initial difficulties in the courts; appeals then dragged them out for almost four years longer, before the status of the contested patents was determined.

Meanwhile Johnson expanded the facilities of the Pearl Street station by adding an annex and then began building two "uptown" power stations, at Twenty-sixth Street and Thirty-ninth Street, which were completed in 1886 and 1887.

The autumn of 1885, in contrast with the preceding year of financial panic and depression, saw business humming for the Edison electrical manufacturing companies. After moving to a larger factory space at Harrison, New Jersey, the lamp company steadily reduced the number of manual operations and the labor time required. This factory was always Edison's "baby." With the help of Upton, its manager, and the engineer J. W. Howell, he worked out improved methods of pumping air quickly out of the vacuum bulbs and preparing carbonized bamboo filaments on a large scale. From an initial cost of $1.21 per lamp, when manufacture was started in 1880, the company lowered the production cost to about 30 cents in 1885, when it turned out as many as 139,000 lamps. Though not the first to think along these lines, Edison had very clear conceptions of mass production techniques to be contrived for this technically complex product. In the late eighties the lamp company's output at last approached a million lamps annually, as Edison had prophesied in the beginning, and costs per unit were down to 22 cents.*

Bergmann & Company, in which the inventor held a third interest, swiftly expanded its production of fixtures, sockets, and a whole variety of light electrical goods. Edison's old mechanic, Sigmund Bergmann, showed the commercial instinct to an unusual degree, and his concern proved to be highly profitable from the start. The bigger machine works, at Goerck Street, now employed eight hundred workers; it encountered more difficulty in its early years in producing heavy equipment and raising the larger capital needed.

The business of the Isolated Lighting Company also grew by leaps and bounds from 1885 on. Johnson sent out a whole team of salesmen, typified by the breezy Sidney Paine of Boston, who sold isolated lighting plants to nearly all the textile mills in New England. The new Edison lights going up in dreary factory towns always made a great sensation; and it was widely reported that millworkers felt more comfortable and became more efficient under their cheery glow.

Sometimes there were alarming accidents, however, as when a large hotel at Sunbury, Pennsylvania, which had just been wired for lights, suddenly began to emit huge sparks and vivid lightning flashes from all its fixtures during a storm. The insulation and wiring, tied to old gas outlets, had obviously leaked static electricity on a large scale. Edison's representatives were often nimble-witted and ready with some explanation, true or otherwise, to calm the spirits of those who were frightened by such fireworks. To the hotel manager, who demanded the next day that the wires be ripped out, one of these agents said, "You may not realize it, but your hotel was struck by lightning yesterday. If it hadn't been for us you'd be the proprietor this morning of ... a heap of ashes. Those sparks were the lightning being shunted into the ground on our wires."

"Well! ... We'll let the wires stay, of course," the proprietor conceded hastily.[20]

By October 1, 1886, the combined Edison organization, with assets approaching 10 million dollars, amounted to a big business for those days.

---

* In response to inquiries by R. G. Dun & Co., the credit-rating institution, Edison wrote in 1883 that his lamp company was not incorporated and had only $35,000 invested in it thus far, adding, "Don't think we owe $1,000. Hope nobody will ever give us credit." His share of its ownership amounted to about 80 per cent.

More than five hundred of its isolated lighting plants were in operation at various points throughout the country, using over 330,000 lamps. Central stations in large cities had risen from only twelve, in 1884, to a total of fifty-eight two years later. Similar facilities were being sold and installed in the great cities of Europe, South America and Japan. Now revenues flowed from all over the world, literally, to the man who had created the practical incandescent-lighting system. Millions were made happy by the usefulness and great convenience of the carbon filament lamp and blessed the name of Edison. His affairs were now so prosperous that at last he could look forward to relinquishing most of his purely business cares to other hands and going back to his laboratory table, like the incorrigible experimenter he was. Everything he touched, people said in those days, turned to gold. He could not but be aware of the broad change in his circumstances. It followed naturally that his tastes and interests, his whole way of life, changed gradually to suit his altered circumstances.

2   Six months had passed since the death of Mary Edison; more and more the widower of thirty-eight felt the difficulty of his position. He was left with three motherless children whom he could not properly look after; in any case, he had not the temperament to do it.

Marion, who attended a boarding school in New York for a while, was nearest to him and much in his company during the wifeless interval. At thirteen she was almost full-grown, and went out with him a good deal to theaters, restaurants, or on Sunday carriage drives. That he should want to see her often, that he should show an interest in her clothes or in the fanciful things she wrote in her diary was a delight to her.

"Why don't you come down to my shop to see me? You're not studying anyway," he would say, for he seemed lonely then. And she would leave her lessons and come to the machine works on the East Side, bringing him his favorite five-cent cigars. She tried to appear as grown-up as possible and be his companion in leisure moments.

My father's idea of my education was that I shouldn't have any [she has said]. Or, at any rate, that I should get it by reading everything, as he did, perhaps beginning with Gibbon's *Decline and Fall of the Roman Empire,* or Watts' *Encyclopedia.*[21]

The two boys, however, he saw much less of; Tom and Will, at this period, lived at Menlo Park with their Aunt Alice. Having formed strong prejudices against formal or academic education, Edison saw to it that his sons, as well as his daughter, had as little as possible of it. He would have liked the boys to acquire the training of a mechanic or an artisan, according to the principles expounded in Rousseau's *Émile,* which he had read with approval. The results in the long run were not such as to have gratified either Rousseau or Edison, America's own "child of nature." The sons by the first marriage now had no mother and knew their father very little.

In his friendships, it was observed, Edison was given to "sudden personal enthusiasms ... impatience and impulsiveness." [22] At times different favorites, like new satellites, revolved around his sun. Where Johnson had formerly reigned, the stout Gilliland now sat as the master's first "apostle." It was probably true, as was often said of Edison's intimates, that they often dreamed of becoming rich in his service; somehow, the Old Man would come through with a pot of gold for them all. But Johnson, nowadays, seemed more avid than any of the others to "pick up something on the side"—something more than the minor shares or bonuses Edison gave his associates—and joined with the young engineer Frank J. Sprague in forming the Sprague Electric Railway and Motor Company, a small business which grew out of the Edison enterprises.

Gilliland seemed different. The jovial man with the handle-bar mustaches was not only sympathetic, he was deferent to Edison; in experimental work he showed an intelligence attuned to the inventor's. A deep fellow, yet humorous, and always quick with a pun. Though Gilliland often came to New York to work with Edison, his home was then in Boston, for he had some profitable business of his own in telegraph and telephone patents. Then there was his accomplished and beautiful wife, like her husband a native of Ohio, presiding gracefully over their home, which Edison visited often in the winter of 1885 on his trips to New England. One of the new friends he often saw at the Gillilands was a handsome young lawyer named John Tomlinson, who soon became Edison's trusted personal attorney.

Under the civilizing influence of the Gillilands and Tomlinson the formerly unsocial man of the laboratories is found going out to buy a sixty-five-dollar coat (very dear then) and dressing for the evening. "For fear that Mrs. Gilliland might think I had an inexhaustible supply of dirty shirts, I put on one of those starched horrors procured for me by Tomlinson," he confesses. He was used to shuffling about in oversized boots, but now admits that he has bowed to fashion and purchased "premeditatedly tight shoes." They look nice, he concedes; but, he asks himself, "Is it not pure vanity, conceit and folly to suffer bodily pains that one's person may have graces [that are] the outcome of secret agony?"

He who was always formerly so pressed for time, night and day, now sits for hours on end in a drawing room with elegant but idle ladies and gentlemen, enjoying light talk, listening to music, and even playing parlor games! One of the games is "mind reading," which yields poor results. The ladies insist on talking about "love, Cupid, Apollo, Adonis, ideal persons. One of the ladies said she had never come across her ideal." And Edison quips that she had better "wait until the Second Advent."

A more rewarding pastime to which he was introduced, during a holiday with the Gillilands on the North Shore of Boston Bay, involved writing a diary. Each person in the party undertook to put down the truth, and only the truth, in his or her diary and then read it to the others. It is thanks to this that we have a precious fragment in Edison's own hand covering ten days of the summer of 1885 in which he sets forth his thoughts of the moment, his emotions, his inventive fantasies.[23]

Here are snatches of his daydreams, of his uncontrollable imaginings: gigantic demons "with eyes four hundred feet apart"; hideous crawling insects or monsters; a preposterous machine to be inserted into some unnaturally thin woman's joints to provide them with automatic lubrication, as with one of his steam generators. Then there are notes of jokes, more or less salty, such as Edison liked to produce on suitable occasions, and humorous speculations on recent developments in electrical science. The abandoned laboratory at Menlo Park, he learns, is now to be used for experiments in hatching chickens in an electric incubator. Edison exclaims to himself in mock indignation, "Just think of electricity employed to cheat a poor hen out of the pleasures of maternity. Machine-born chickens! What is home without a mother?"

His favorite relaxation, however, is to retreat to his room and get into bed with a miscellaneous assortment of books: Goethe's *Wilhelm Meister*; Rousseau's *The New Héloise*; Lavater's *Human Physiognomy*; *The Memoirs of Madame Récamier* ("I would like to meet such a woman"); Hawthorne's *English Notebooks* (too many ruined abbeys in them—"Perhaps I'm a literary barbarian?"); and finally, best of all, the *Encyclopaedia Britannica* ("To steady my nerves").

With his new friends he also goes off to one of Boston's old music halls, a form of entertainment he has always liked, sits in the "bald-head row," and enjoys a display of "the usual number of servant girls in tights." After the chorus has completed its long military maneuvers, the junoesque Lillian Russell makes her appearance. "Beautiful woman, beautiful voice," is Edison's appraisal. But with a touch of prudery he also remarks that the lady is said to have been married more than once.

Then there are sedate yachting and fishing parties in which he must join, the men clad in white duck, the ladies in long white dresses. But on one occasion the ocean spray blows hard, the ladies' long dresses are thoroughly soaked, and they turn back to shore without any fish.

What has become of the Thomas A. Edison we knew, the man who labored like a Titan, chewing tobacco and spitting on the floor? How could he pass whole days in such politely frivolous diversions? Was he corrupted, bewitched? Certainly the charm of the Gilliland home and its circle held him at this period and distracted his mind. His hostess, as he writes, would greet him "with a smile as sweet as the cherub that buzzed around the bedside of Raphael." But in truth there was another woman in the case, younger, even more beautiful, whose image would not leave his mind, as the diary reveals by many artless passages.

The thought that for all his fortune, he lacked such a home as the Gillilands', with its cheerful talk and laughter, and such a homemaker as its mistress to care for his children and himself, was present in his friends' minds as well as Edison's. He was certainly one of America's most eligible widowers. At news of the death of his wife, scores of unknown ladies wrote him letters expressing their pity at his bereavement and freely offering him their hearts and their fortunes. Sam Insull, in fact, was kept busy answering these fair correspondents, often sending them photographs of the inventor by way of solace.

The Gilliland couple, evidently with the tacit approval of Edison, now undertook to invite to their home a flock of marriageable females for the widower's inspection. By ones and twos they arrived in Boston from as far off as Ohio and Indiana during the winter and spring of 1885 and marched into the Gilliland parlor, while Edison examined them one after the other. The game grew more and more engrossing, though Edison remained coy.

"Come to Boston," he wired Sam Insull, who liked this sort of thing. "At Gill's house there is lots of pretty girls." [24] There were, for instance, "two Hoosier Venuses," one of them a blonde, like his late wife. Finally there appeared a striking brunette from Ohio, whom Mrs. Gilliland knew well, and whose visit was carefully planned. She was Mina Miller, the daughter of Lewis Miller, of Akron, Ohio, a wealthy manufacturer of farm tools. Mr. Miller was well known as an active and philanthropic churchman, for he was the cofounder—with Bishop John Vincent of the Methodist Church—of the Chautauqua Association. A colorful man in his own right, he had begun his career as a plasterer, educated himself to be a schoolteacher, and afterward became a successful inventor and manufacturer of agricultural implements. Samuel Smiles had made Mr. Miller's career the subject of one of his essay-sermons on self-help.

As was so often the case in American families of recent fortune, Lewis Miller's daughter Mina was sent to a finishing school in Boston, and then, for further improvement, on the "grand tour" of Europe, whence she had lately returned. She is described as having been in her youth "accomplished and serious, with a liking for charity and Sunday school work"—but also a veritable belle, with an imposing figure, "rich black hair and great dazzling eyes." [25]

The first meeting of Edison and Miss Miller took place in the early winter of 1885; by prearrangement the girl was visiting with Mrs. Gilliland, whom she had known well in Ohio, when the inventor arrived from New York. The Gillilands had told him all about her.

Edison was talking to several men in the parlor, when Gilliland, who had left the room, came back and said to him, "Mina Miller is here and she is going to play and sing for you." In a few minutes he returned accompanying the beautiful young woman, who swept into the drawing room like a great lady. It was well known that Edison would be annoyed when women, on first being presented to him, expressed wonderment and rapture at the honor, or paid him exaggerated compliments. The eighteen-year-old Miss Miller, however, greeted the great man with dignity, and, while he stared at her, returned his gaze with the utmost composure. On being asked to sing, she went at once to the piano and performed for the guests. She was far from being an accomplished musician, according to one witness, but what was impressive was the self-assurance and aplomb with which Miss Miller did everything.

The inventor, who was twice her age, on first exposure to Miss Miller's charms and virtues, felt, as he himself said, "staggered." Why this was so, just what was the nature of the electrochemical reaction that took place he could never explain. "Ask me nothing about women," he wrote on a later occasion, "I do not understand them. I do not try to." Not long after meeting Miss Miller, he sent an urgent telegram to Insull, in New York, asking that two photographs of himself be sent to Miss Miller in Boston.

Later that winter he was in Chicago, in connection with Edison Company business; the weather was bitter, he caught cold, and then became quite ill, staying in bed at his hotel for a week. Gilliland, who had accompanied him on this trip, thought his condition so serious that he sent for Edison's daughter Marion. When Edison arose from bed, he was still extremely weak. It was then agreed that a winter vacation in Florida was indicated if he would restore his strength quickly. Mrs. Gilliland joined them, and Edison proceeded with his party directly from Chicago to St. Augustine.

He would have returned North within a few days, but that someone spoke to him of the tropical region of southwest Florida, of the Everglades and the Keys, then still a frontier zone. When he was told of the jungles down there, he thought he would surely find vegetable fiber that would be perfect for his incandescent lamp. Off he went, with his two friends and his daughter, for a journey of some four hundred miles on a rickety train, which on one occasion jumped the tracks in the deep pine woods and could not be moved for two days. At length they reached Punta Rassa, on the Gulf Coast, a tiny harbor for the cattle trade, with an old iron lighthouse facing it across the bay from Sanibel Island. Here he was told of plantations along the Caloosahatchee River, near the village of Fort Myers, where tropical fruit and palm trees grew, and even bamboo sixty feet tall!

Bamboo! Had he not sent agents to the Far East and Brazil in search of it? In a small sloop the Edison party pushed eighteen miles up the river, lined with moss-covered live oaks, to the exotic paradise of Fort Myers, then a hamlet with one general store and a wretched inn. In this sun-drenched scene of royal palms, mangoes, and giant tropical flowers, Edison, who had felt himself near death in Chicago, revived wonderfully. After prowling about the place for two days he determined to take an option on a site of thirteen acres along the Caloosahatchee and build a winter home there for himself.[26]

Wherever he went that season, he spoke of the perfections of Mina Miller. So much so that, as he noticed, it made his daughter Marion fierce with jealousy. "She threatens to become an incipient Lucretia Borgia!" he wrote in his diary. On his return to New York, he found that the thought of Miss Miller permitted him no sleep. He wrote in his jocular manner, "Saw a lady who looked like Mina. Got thinking about Mina and came near being run over by a streetcar. If Mina interferes much more will have to take out an accident policy."

He had worked while others slept, and had become far more ill and exhausted than he realized. In the Floridian jungle Edison, like so many other weary travelers before and after him, had felt his strength and youth return. Like the suddenly awakened Faust, he felt he must now live in the sun and enjoy the love of a young maid. He decided that he must marry Mina Miller, bring his bride there, and build not only a winter home but a new laboratory to be placed in that fabulous setting of tropical flowers and palms. His beloved friend Gilliland must also live there and share his life. Gilliland enthusiastically agreed. Edison's diary shows him, in the summer of 1885, buzzing with romantic schemes for his castles in the Everglades; he even attempts a flight of poetic prose, then mocks at himself:

Studying plans for our Floridian bower ... within that charmed zone of beauty, where wafted from the table lands of the Orinoco and the dark Carib Sea, perfumed zephyrs forever kiss the gorgeous florae—RATS.[27]

Determinedly he wooed the young lady "via the post office," a novel experience for him. Though he was uncommonly busy again that year with new lighting plant installations and space telegraphs for railroads and ships, he made prodigious efforts to see her. Her home was in Ohio; but in the summer the Millers regularly migrated to Jamestown, New York, to attend the gathering of the Chautauqua Association. Though Edison was "not much for religion," as he said, he determined to visit that repair of evangelists in order to see her whom he now called "the Maid of Chautauqua."

On arriving at Lake Chautauqua and meeting one of many ministers assembled there, Edison is reported to have said, "I am afraid we do not belong to the same denomination." His outspoken irreverence posed something of a problem for his sweetheart's father. "My conscience seems to be oblivious of Sundays. It must be incrustated with a sort of irreligious tartar," he wrote.[28] It was useless for him to go to church, since he was so deaf. In fact he jested that if he chose any church at all it would be the Roman Catholic, "because there you paid your money, and the priest whose business it was, looked after the rest." [29]

Now, Mr. Lewis Miller had been so long and closely associated with Bishop Vincent, the head of the Chautauqua Association, that their children had virtually grown up together. There was a rival suitor on hand, a younger man than Edison, the bishop's son George, who later became a distinguished educator himself. George Vincent, in fact, was considered to be destined for Mina since boyhood—until the middle-aged but impassioned inventor appeared upon the scene. It was Edison who was at her side during a crowded afternoon steamer excursion on Lake Chautauqua. The throng surrounded them closely, they could not be alone; and he could not hear her save when he was very close to her. But the inventor was once more equal to the emergency. As he related:

My later courtship was carried on by telegraph. I taught the lady of my heart the Morse code, and when she could both send and receive we got along much better than we could have with spoken words, by tapping our remarks to one another on our hands.

On another occasion soon afterward, he managed to take her out alone with him in a small rowboat. Mina Miller feigned resistance to her magnetic and ardent lover for a while, though with steadily weakening resolve.

By dint of much persuasive effort he then gained her parents' consent to have her join him in a long excursion by carriage through the White Mountains of New Hampshire, the Gilliland couple acting as her chaperones, while his daughter also accompanied them. It was during this drive through the mountains that Marion saw her father tap out his proposal on Miss Miller's hand, as he too testified:

I asked her thus in Morse code if she would marry me. The word 'yes' is an easy one to send by telegraphic signals, and she sent it. If she had been obliged to speak she might have found it much harder. Nobody knew anything about our many long conversations. . . . If we had spoken words others would have heard them. We could use pet names without the least embarrassment, although there were three other people in the carriage.[30]

Mina's consent was conditional upon her parents' approval. Therefore Edison, upon his return to New York on September 30, sent off a formal application for her hand to Mr. Miller, which he evidently composed by dint of much labor and thought. Part of the letter, done in his manuscript hand, sounds as if written with the help of his secretaries (who may have drawn upon contemporary manuals of etiquette); but part is unmistakably and inimitably his own.

I trust you will not accuse me of egotism when I say that my life and history and standing are so well known as to call for no statement concerning myself. My reputation is so far made that I recognize I must be judged by it for good or ill. . . .

Edison had made his proposal after mature deliberation; and he declared in conclusion that he had learned to love Mina, and that his future happiness depended upon Mr. Miller's answer. The reply brought assurance of such happiness.[31]

The news of Edison's forthcoming marriage to Mina Miller, the daughter of the Akron philanthropist, was soon out. In the autumn of 1885 it was perceived by his associates that the inventor was making extraordinary preparations for this new phase of his life. He was busy with plans for building a winter home in Fort Myers, Florida, and having lumber shipped for his projected laboratory. In order to land there and unload building material and equipment a pier extending some distance into the river must also be constructed. Since it was not easy to reach Fort Myers overland, he would go by boat; and soon he advertised for a sixty-five-foot steam yacht with two staterooms (though he gave up the idea when he saw what it would cost).

To his agents in Florida he sent instructions covering every exigency:

We will erect two dwellings [one was for the Gillilands] on the riverfront and place the laboratory and dwellings for workmen on the other side of the street. ... Our buildings are being made in Maine and will be loaded aboard ship at Boston. We will send four of our employees to superintend the work....[32]

Another matter that engrossed him that autumn was the purchase of a home magnificent enough to be the setting for such a bride as was Mina, and for his own improved station in life. The simple house at Menlo Park was out of the question now. He made two hurried visits to Akron to see her (though he never seemed fully at his ease in the parsonical circle of the Millers) and saw her again in New York when she visited the city in December, just before their marriage. He had given her the option of living in a big town house in the city, or in the country, and she had expressed a preference for a country house.

On a day when the snow covered the ground he drove with her to suburban West Orange to see a large estate he considered purchasing. It was called "Glenmont," and had been built on a high ridge by a former millionaire merchant of New York at the staggering cost of $200,000. It was not a house, but a veritable castle of brick and wood in a style loosely called American Romantic, with spacious rooms, numerous outbuildings and greenhouses, and broad, landscaped gardens. Glenmont apparently suited Miss Miller's ideas exactly. There was the further advantage that it could be purchased at a fourth of its original cost, since the improvident owner, a notorious absconder, had skipped the country with all his movable wealth, leaving the disposal of his house to defrauded creditors.

All this, and the bride's jewels too, meant spending money like water, yet nothing daunted the inventor. Though not yet forty in 1886, he saw himself as one who was rising to the status of America's ruling industrial barons, one who stood at the head of a great industry he himself had created. Henceforth, all that he undertook must be planned on the grand scale; his mode of life itself was to be transformed. There was to be not only the great mansion in suburban New Jersey, but also a vastly enlarged new laboratory to replace Menlo Park, furnished with the finest equipment in the world, and to be situated on the acres he had purchased in West Orange, within a half mile of his future home and his new wife.

In February, 1886, Edison's friends gave him a last, very jovial bachelor's dinner at Delmonico's Restaurant in New York. Ed Johnson,

Batchelor, Bergmann, Gardiner Sims, and others were present; hearty toasts were drunk, the party breaking up with the understanding that its members were to meet again on the twenty-fourth of the month, at the wedding ceremony in Akron.

All northern Ohio seemed to be agog over the event, as Edison arrived there to be wedded to an acknowledged belle of the region. From the entrance of the big Miller house in Akron Mr. Miller literally rolled out a red carpet extending several hundred feet to a knoll on the grounds that overlooked the city. By every train crowds of guests arrived, the spanking Miller carriages with their high-stepping horses meeting and bearing them away swiftly to the Miller home. The big front parlor was ablaze with flowers and resounded with the music of an orchestra; a whole corps of waiters had been brought from Chicago to serve the wedding lunch. The ceremony was performed under an arch of roses by a leading minister of the Methodist Church. Later that afternoon the couple drove around Akron, cheered by crowds along their way, and had their picture taken before boarding the train for Florida in the early evening.

The winter home in Fort Myers was not ready for them; conditions, at first, must have been rugged. But nothing whatever was heard from Edison for three weeks; he seemed to have vanished from sight like one enchanted. A secretary said:

We have written to him; we have telegraphed him. We get no response.... We ask him questions requiring his immediate attention and that is the last of it. We are running the concern without him. He ignores the telegraph and despises the mail.[33]

In April they came back from Florida and Edison brought Mina to their new home in West Orange, that sprawling, red-painted chateau with its many gables, carved balconies, and stained-glass windows over which the second Mrs. Edison was to reign for many years as proud chatelaine. In character Glenmont was similar to those spacious residences of mixed architectural styles erected by the new rich of the 1870s and 1880s in Pittsburgh, Cleveland, Chicago, and along the Hudson River in New York—homes signalizing the status and power of the princes of oil, pork, iron, coal, railroads, and (now) electricity. It had a porte-cochere and a broad vestibule, vast living rooms, rich chandeliers, wide staircases, heavy red-damask-covered furniture, and masses of bad statuary, oil paintings by the yard, and bric-a-brac collected by its former owner. To be sure, an isolated lighting plant was promptly installed in it; and Mina Edison was to make it as bright as possible with flowers and pictures. To her this place was the appropriate setting for genius.

In speaking of the house at the time Edison said that he had intended to look for a country home costing no more than $20,000.

"But when I entered this I was paralyzed. To think that it was possible to buy a place like this, which a man with taste for art and a talent for decoration had put ten years of enthusiastic study and effort into—too enthusiastic in fact —the idea fairly turned my head and I snapped it up. It is a great deal too nice for me, but it isn't half nice enough for my little wife here," and he laid his hand on the arm of the beautiful young woman at his side. "So that secures the fitness of things." [34]

Many years later a famous European writer called at Glenmont to interview Edison and, in the course of their conversation, likened him to Faust.

"Faust!" exclaimed the inventor; then drawing his arm affectionately around his wife, he added with a twinkle in his eyes, "Ah, yes, Faust, and she is my Margaret." [35] But Mina Edison was, in reality, very different from the peasant maiden of Goethe's drama. Her predecessor, Mary Edison, in her innocence and simplicity, more nearly suited the part.

Some pains had been taken with Mina's upbringing so that she might be fit to play her part in good society. She was a young woman of indubitable strength of character and will, and entered upon the adventure of marriage with full awareness that her mate was one of the most noted of living men and of the duties she owed him. There was a good deal in her husband that she intended, nevertheless, to improve—if she could possibly do so. Under her sway his rough exterior would be gradually smoothed a little here and there, though it was a slow and difficult process. No power on earth could really turn this man into a tame parlor lion; to Mina's regret he continued to shun the kind of social life she would have enjoyed.

"You have no idea," she said on one occasion in later years, "what it means to be married to a great man."

Life itself, the changing focus of his interest, his very different preoccupations, nowadays worked to bring about a change in his ideas and in his aspirations.

In his middle years the worker risen from the ranks of the poor was the lord of Glenmont and its handsome acres and possessed of a good and beautiful young wife of proper breeding. His ambitions nowadays were unlike those of the typical man of science and bore a resemblance rather to those of the Carnegies and Rockefellers, the ruthless captains of industry who were his contemporaries. He spoke, at about this time, to J. Hood Wright, a partner of Morgan, of his purpose of building up a whole series of new industries. To A. O. Tate, who became his private secretary after Insull's promotion to an executive post, he also remarked, while gazing out of the windows of Glenmont, "Do you see that valley?"

"Yes, it's a beautiful valley," the other replied.

"Well, I'm going to make it more beautiful. I'm going to dot it with factories." [36]

chapter XVI

The Edison laboratory, West Orange

1 During the boom times of the late 1880s in America, the frontier was vanishing, the settled territory of the country was expanding, the whole economy was growing at an unprecedented rate. Was there ever anything like the newly completed Brooklyn Bridge, or the young skyscrapers in Chicago, or Mr. Carnegie's steelworks in Pittsburgh? Like a good American Mr. Edison, too, felt the urge to do things on a truly continental scale.

At this period, in 1886, the Edison Machine Works, which had been undergoing a sort of forced-draft expansion, and suffering, moreover, from labor troubles as well as growing pains, was moved from its location in New York City to vastly larger quarters at Schenectady, New York. One of Edison's agents had found some locomotive works in this old Dutch community, on a tract along the Mohawk River near the Erie Canal, these works being then unoccupied because the several owners had quarreled and gone out of business. Schenectady's Chamber of Commerce proved eager to invite in the Edison enterprise and helped to arrange for the sale of the property for $37,500, a sum far below its cost. The great presses and machine tools were then transported to Schenectady, and within a year all the Edison generator, motor, tubing, and insulation work was being carried on there by some eight hundred hired hands. A year later, the scientific journalist, T. Comerford Martin, described the Schenectady works as "a vast establishment of noble machine shops ... where the prosaic and marvelous jostle each other ... one of the greatest exemplifications of the power of American inventive genius." [1] The Schenectady works grew with the rapidity of a Western mining boom town, until they fairly overflowed with busy artisans and machines of every kind. Some referred to them as "the cathedral shops" because of their lofty ceilings which permitted the use of great cranes overhead. [2]

The next year there was the new laboratory to be built at a site in

West Orange's quiet Main Street, a half mile from Edison's new home. His plans for this were on a scale that completely dwarfed the wooden "tabernacle" of Menlo Park. The new conception, the altered scale of his ideas at this period may perhaps be illustrated by an incident that occurred some years later, when a noted French inventor of electrical devices, Edouard Belin, came to visit Edison at the West Orange institution. After showing the visitor about the place, Edison, during a friendly discussion, advised him to center his inventive labors on products that would come into universal use, so that he could "make a lot of money out of them." And what would he do with the money, the other asked, since he had enough for his needs? "You could build a fine laboratory, where you could have everything you desire and be able to hire the greatest technical experts as your assistants ... a place in which you could invent anything that came into your head, regardless of expense."

The French inventor mentioned that he had a small establishment and was content with it. "I already have a rather nice laboratory," he said.

"But if you had unlimited money," Edison insisted, "you could build a laboratory ten times as big ... and you could do ten times as big things." [3]

Such was the Edison of the middle years, after his marriage and the acquisition of Glenmont. He believed that with a larger scale of operations he could more easily make "useful things that every man, woman and child in the world wants," and would buy "at a price they can afford to pay."

The West Orange laboratory buildings, when completed after more than a year of construction work, aggregated ten times the size of the humble establishment at Menlo Park. They constituted then the largest and most complete private research laboratory in the world. In a notebook which he kept while construction was still in progress, Edison described his whole project with enthusiasm:

I will have the best equipped & largest Laboratory extant, and the facilities incomparably superior to any other *for rapid & cheap development of an invention* & working it up into Commercial shape with models patterns & special machinery. In fact there is no similar institution in Existence. We do our own castings forgings Can build anything from a lady's watch to a Locomotive. ... Inventions that formerly took months & cost large sums can now be done 2 or 3 days with very small expense, as I shall carry a stock of almost every conceivable material. [4]

The main building was of brick construction, about 250 feet long, three stories high, and having 60,000 square feet of floor space. It contained large machine shops, an engine room, glass-blowing and pumping rooms, chemical and photographic departments, rooms for electrical testing, stockrooms, and in its main wing a spacious office and library. The library, a great hall with a thirty-foot-high ceiling and two tiers of galleries rising from its main floor, contained 10,000 volumes. With its stores of minerals and chemicals in glass cases it was a museum and sample room in one. At the same time, four more long, one-story structures were laid out adjacent to the main building and at right angles to it, the whole group making a yard, or quadrangle, enclosed by a high picket fence with a guard posted at the entrance gate. The work here was now carried on with a fair degree of secrecy, visitors being admitted only on invitation.

Since everything about Edison was "prodigious" to the newspapers of the time, the contents of his new laboratory were reported in the usual superlatives, as embracing

eight thousand kinds of chemicals, every kind of screw made, every size of needle, every kind of cord or wire, hair of humans, horses, hogs, cows, rabbits, goats, minx, camels . . . silk in every texture, cocoons, various kinds of hoofs, sharks' teeth, deer horns, tortoise shell . . . cork, resin, varnish and oil, ostrich feathers, a peacock's tail, jet, amber, rubber, all ores, metals. Of one jobber [so the inventor gaily told the reporter] he "ordered everything from an elephant's hide to the eyeballs of a United States Senator." [5]

Menlo Park, within its small compass, had provided the first center for organized "industrial research" in the world. For seven years it epitomized the romance and the ardors of pioneering in new fields of applied science. The new laboratory, a whole community of buildings dedicated to applied science, was huge indeed by comparison. It was not fated, however, that it should outshine the small place at Menlo Park in the quality and brilliance of inventive work performed.

For one thing, a staff of forty-five to sixty persons assisted the inventor at West Orange. They ranged from scientific specialists, like Dr. Arthur E. Kennelly and the well-known chemical consultant J. W. Aylsworth, down to draftsmen, mechanical assistants, "bottle-monkeys," and unskilled laborers. Edison still kept a tight rein on everything being done in his laboratory, but inevitably the work had to be departmentalized; various persons or teams were constantly engaged in making tests or trials and writing detailed reports on different experiments going on concurrently. Edison therefore had to give an increasing part of his time to administrative work as the director of his staff.

At Menlo Park he was able to see at a glance what every one of his dozen assistants was doing in the one long room. Now, every morning after going through his mail, he would have the men in charge of different experiments come into the library one by one and report on the work in progress. He was always able, in a remarkable degree, to shift his attention quickly from one line of investigation to another that was entirely unrelated, asking questions rapidly, giving instructions, and making decisions as rapidly. It was also his habit, at least once a day, to make a tour of the different departments and with his own eyes study the work being done. Nevertheless, he was forced to depend more than before on specialists, such as Aylsworth in chemical studies, or J. W. Howell in electric-lamp research. These were very able men, who did fine things. But they were not Edison.

His own passion for experimentation never flagged. From conferences or supervisory work on matters pursued by his assistants, he often would escape to an upper room of the main building, a sparsely furnished place littered with instruments and materials of all sorts. Here he would remain alone, or with at most one mechanical assistant, engaged in lengthy studies of the favorite subject of the moment. Most of the time, however, he was to be found seated at his huge desk in the center of his library, near its great marble fireplace. After some years, with the addition of many sculptures, medallions, and paintings of the inventor, as well as mahogany and yellow-pine furnishings, the library acquired the cluttered character of most late-Victorian interiors; but it was always deeply stamped with Edison's personality. One day someone planted ivy outside the laboratory walls; it climbed to the roof, spreading everywhere about the whole group of buildings, so that the place came to resemble an abbey.

The strategic importance of Edison's original model of the private research center, as handmaiden to technology, was quickly grasped by the masters of some of our large industrial corporations. In 1881, the inventor Alexander Bell, for example, on receiving a prize of $10,000 in honor of his telephone work, established a small general research laboratory of his own in Washington, D.C., in imitation of Menlo Park. Out of this first center grew up the vast Bell System Laboratories. Others also imitated Edison in pursuing invention systematically; the early ones were the Thomson-Houston Company and George Westinghouse, who were already competing in the electrical equipment field. In the 1890s and prior to World War I, however, research work in the corporate laboratories was largely subordinated to the owning company's special needs; and despite the advantages in large staffs and elaborate instruments, their accomplishments in those years remained somewhat unimpressive. Until almost the turn of the century the Edison Laboratory continued to be unique among private industrial laboratories, as a center of "general research on inventions," with a varied program that might include anything from electric locomotives to phonograph needles.

2   One of his major preoccupations in the spring of 1887 was with renewed experimental work on his own phonograph. For nearly ten years the primitive tin-foil phonograph had remained only a scientific toy, now forgotten even by the entertainment parlors and country fairs that had once exhibited it. It was a beautiful concept, nonetheless, that of the preservation and reproduction of sound, and one that continued to fascinate men. In such a case, the original inventor having apparently neglected to improve his invention, it was inevitable that other skillful hands would take it up.

At the laboratory of Alexander Bell, recently established in Washington, his cousin Chichester Bell, a chemist, together with Charles Sumner Tainter, an able technician, devoted themselves for several years to the task of perfecting Edison's phonograph. In June, 1885, they applied for a patent on a talking machine which they named the "graphophone." In place of tin foil, whose impressions were faint and easily effaced, they introduced wax as a recording substance, which they imposed on cardboard and mounted on a cylinder. Their recording stylus, like Edison's, moved vertically up and down in the groove, but instead of indenting or embossing its vibrations, "engraved," that is, cut them into the waxed record, which they claimed as a new process; and whereas Edison's stylus was rigid, theirs was loosely mounted so that it floated along without scraping too much. After receiving the award of their patent in May, 1886, Bell and Tainter decided to approach Edison privately and seek to enlist his cooperation in promoting the improved talking machine as a joint undertaking. Theirs was still a crude affair, but it sounded distinctly better than the tin-foil instrument; they also contrived to turn the cylinder at a constant speed, either with a treadle or a motor, which was more reliable than Edison's hand crank.

Edison's secretary, Tate, describes the proposals they brought:

They said that they fully recognized the fact that Mr. Edison was the real inventor of the "talking machine"; that their work was merely the projection and refinement of his ideas, and that they now wanted to place the whole matter in his hands and turn over their work to him without any public announcements that would indicate the creation of conflicting interests. They were prepared to bear the costs of all experimental work . . . and provide all the capital essential for its exploitation. And for all this they asked to be accorded a one-half interest in the enterprise. . . . Though they had tried to differentiate their instrument from his by calling it the Graphophone, if Mr. Edison would join them they would drop this name and revert to the original designation.[6]

Reasonable though this proposition seemed on the surface, it made Edison furiously angry. He declared roundly that he would not deal with Bell and Tainter on any terms, and called them "specalators" who were bent on stealing his invention. A model of their machine had been left with him, and aside from their improved stylus, he found nothing new. It was astonishing to him that the Patent Office had awarded them a patent.

In truth the model of the "graphophone" of 1885, when examined today in the Smithsonian Institution, shows a startling resemblance to Edison's phonograph of 1877. When, several months later, Colonel Gouraud cabled Edison from London, saying that he had been invited to head a British agency for the Graphophone Company, and inquiring if Edison approved, the father of the phonograph dictated a thunderous dispatch in reply:

Have nothing to do with them. They are bunch pirates. . . . Have started improving phonograph. Edison.[7]

He had been literally "shocked into action," as Tate observed. The phonograph was his "baby." After having neglected his paternal duties toward it, owing to the pressure of so many other, larger affairs, he went back to his laboratory table again to consecrate long hours through most of 1887 to the phonograph. It might have been profitable for him to join forces with the Bell group, which now had ample resources. They, on the other hand, would be driven to change their talking machine drastically in order to avoid infringement on his patent rights. But Edison was nothing if not competitive in the arena of invention. He would fight them to the death. Like that imperious old railroad magnate, Commodore Vanderbilt, in a similar emergency, he vowed that he would not bother to sue his adversaries, but would *ruin* them—by making a better machine.

For most of two years experiment with the phonograph was given priority in the laboratory over many other affairs claiming his attention. And he was happier thus, as the engineers phrase it, with "his belly at the drawing table again"; for there were limits to his interest in purely commercial and industrial affairs, or in just "running wires" for one new lighting plant after another in different towns. In any case the light and power business was growing by itself in charge of his lieutenants, with Sam Insull, the sharpest of them, handling the business at Schenectady.

As was his practice, Edison began by establishing clearly in his own mind a concept of what the popular need and use of an improved phonograph would be, planning his line of development accordingly. The principal use of his invention, in his judgment, would still be as a business machine for dictating letters and commercial records. The "craze" of 1878 for recordings of fragments of popular music or vaudeville humor seemed to have passed off; and in any case, he disliked that business. Success would depend on accurate recording and faithful reproduction; he must find, principally, an improved material for his records, something much better than tin foil. He foresaw also that facilities for duplicating records in large volume would be of the utmost importance; it would also be necessary to find some means of re-using records if the machine were to be made practical for business dictation.

Having mapped out his campaign, he took steps to reorganize the old Speaking Phonograph Company of 1878, inviting its few shareholders to exchange their stock for shares in a new corporation which would take over the patents of the old concern. His new favorite, Ezra Gilliland, worked with him in experiments on the new machine. Gilliland told Edison that the new phonograph would be a splendid commercial affair some day, if a practical man were to promote it and give it all his time. He then pressed Edison to appoint him as the new company's general sales agent, or executive manager. In a most generous spirit, and largely because he felt Gilliland was his trusted friend, the inventor granted him this key position, and signed a contract to this effect, which was executed by Gilliland's friend, Tomlinson. This well-spoken young patent lawyer had become Edison's personal attorney in 1884, in place of Lowrey. Afterward when trouble arose people would remember that Tomlinson was a man who "never seemed to be looking at you directly when you spoke to him," perhaps because of a slight cast in one eye.[8]

In the many clauses of his basic phonograph patent of 1878 Edison had enumerated various materials that could be used for records, waxed substances among them. As a substitute for the old tin-foil sheet he now devised a hollow cylinder of a prepared wax compound, with walls about a quarter of an inch thick. The composition of this wax-compound record, or "phonogram," as it was then called, demanded infinite pains and hundreds of trials, the chemist Aylsworth furnishing Edison more help than any other in this work. Wax permitted closer grooving. In place of the old recording needle Edison then introduced a much harder cutting tool, which made "hills and valleys" at the vibration of the diaphragm; and for reproduction he employed, after a while, a somewhat blunt sapphire stylus. In the old machine there had been crude adjustment screws to hold the needle in place; now he devised a quite ingenious "floating weight" or flexible pickup mechanism that held the stylus down with a light touch. His recording process was so skillfully adjusted that grooves in the wax record were only one thousandth of an inch deep. Thus persons putting his machine to use for business dictation could shave their records down with a cutting blade provided for that purpose and re-use their records many times.

The idea of the floating stylus, it has been observed by a later commentator, was in all likelihood borrowed from Sumner Tainter's original device for his "graphophone." Edison may have felt that, inasmuch as his adversaries were trying to take over his basic invention, turn about was fair play. In fact, Bell and Tainter soon dropped their waxed cardboard record and again appropriated one of Edison's useful ideas, that of the solid wax record.[9]

A more original piece of mechanical invention involved a new process for duplicating records in large volume. The great trouble was that the wax mold used for the original, or "master" record, was a nonconductor and could not be electroplated. After many years of study, Edison and his assistants (in 1903) worked out an ingenious duplicating process which he called "vacuous deposit." Under this system the original record mold of wax was placed in a vacuum chamber, with a piece of gold leaf suspended on either side of it; high tension electricity was then discharged between the two gold leaf electrodes, while the record mold revolved. The electric charges vaporized the gold leaf and deposited it on the record mold in a film of tissuelike thinness. Upon this gold film, which faithfully represented all the original record grooves, a heavier deposit of another metal was then electroplated, so that—once the original mold was taken out—you had a very solid and durable negative mold, with which almost any number of positive wax copies could be duplicated.*

In testing results on the improved phonograph Edison had the great problem of trying to hear distinctly despite his deafness, which was total in his left ear. He would place his right ear—whose hearing was limited —in contact with the instrument itself or at its horn; or at times he would bite into the horn with his teeth, thus permitting the sound vibrations to be carried through the bones of his head to his hearing nerve. "It takes a deaf man to hear," he would say grimly. Batchelor and Gilliland also served as his ears, by reporting results under different conditions.

---

* Edison was certainly the original discoverer of the phonographic art. Tainter, said to have been the more active inventor of the Bell-Tainter partnership, showed fine technical insight and introduced important refinements in the original Edison apparatus, without departing from its basic principles. His best contribution, in the opinion of modern students, was the flexible pickup, which Edison imitated. Edison, in turn, added more refinements, so that by 1888 it was hard to tell where one inventor began and where the others left off. The phonograph, by then, became an affair of engineering and technology, and remained so until the introduction of electronic devices in the 1920s marked another "break-through."

The laboratory notebooks show him compiling a long list of "complaints" against his evolving phonograph:

Crackling sounds in addition to continuous scraping due, either to blow-holes, or particles of wax not brushed off—poor recording—uneven tracking—dulling of the recorder point—breaking of glass diaphragm—knocking sound: chips in wax cylinder—humming sound, due to motor.

In painstaking fashion he worked to eliminate these defects one by one, inventing various new stratagems and devices by which the first simple talking machine became more responsive and lifelike. Whereas the original discovery of the phonograph principle had been accomplished in a short time, a few weeks at most, the process of refinement in the second phase stretched out tantalizingly into months and even years. All these changes in detail and in mechanical design were incorporated in a series of approximately a hundred more patents covering his later phonograph models.

After about six months of hard work, in October, 1887, Edison felt he was far enough along the road to hurry off some of his customary advance bulletins for the press, serving notice to all the world that he had never really abandoned his "favorite invention" and that his perfected phonograph would be making its debut almost any day now. Though forced to throw overboard everything for the electric business, he had borne the phonograph for ten years "more or less constantly in my mind." In this case, the publicity was shrewdly timed to forestall, or at least to embarrass his adversaries, Bell and Tainter, who were reported to have made real progress in improving their own machine. But the snowy winter went by and further press notices that the new Edison phonograph would be ready "within two weeks" proved once more to be premature.[10] Though the rival group had a year's head start, they too had met with much difficulty in making their machine foolproof and also in promoting it, very likely because potential investors or customers were holding back to see what new miracle America's Wizard might yet produce.

In the early spring of 1888 Edison thought he had improved his machine to a point that would permit him to raise capital and launch it as an industry. A delegation of financiers was therefore invited to visit West Orange and inspect the new phonograph, rumors of which had already whetted their financial appetites. In the group were Jesse W. Seligman, of the important New York banking firm of J. & W. Seligman, as well as D. O. Mills and Thomas Cochran. Unknown to the inventor, his assistant John Ott at the last moment had substituted for the original apparatus what he considered a superior diaphragm, recently perfected in the shop;

he had also changed the recording stylus for a finer instrument, but neglected to change the old reproducing stylus, which was of the broad or blunt variety.

All unaware, Edison in his library opened the demonstration by making a speech into the machine in the low tones used in dictation, announcing the serious purposes he had in mind for the phonograph and describing its splendid business prospects. Then he switched diaphragms and needles and began to play it back. But the reproducing needle proved to be too broad to enter the groove that had just been made and merely scraped along its edges, so that the phonograph gave forth only a continuous hissing sound—which sounded positively derisive. "Edison looked astounded and perplexed. He examined the instrument carefully but could find no defect. He was bewildered," as a witness of the scene wrote. The same machines had been working quite adequately an hour earlier. The group of financiers, after waiting about, politely made their departure, saying they would return when he had located the trouble; but they never came back.[11]

Meanwhile, in April, news came that Bell and Tainter had finally sold out their graphophone patent to a millionaire manufacturer named Jesse W. Lippincott, and that he was going to promote the sale of the machine with great vigor. Edison now spurred everyone in his laboratory to the utmost exertions. More than money, as we have seen, he loved to win a stiff "horse race" for the honor of an invention.

As the spring of 1888 advanced, the pace of activities at West Orange increased. Edison and Gilliland worked together over the development of a small motor that was to turn the phonograph cylinder at a constant speed. A governor and a flywheel for the regulation of speed were also devised.

In testing his approved apparatus, Edison was perhaps the first to make records by famous musical artists. For all his bad hearing and want of musical education, he had very decided opinions on this art. He detested Wagner, but loved Beethoven. Emmy Destinn, to his mind, had "no voice," and Caruso he considered an indifferent singer. The tremolos of Italian tenors were profoundly annoying to him, and coloratura sopranos no less so. What he feared was that such tones would make his recording stylus jump over the ridge of the record groove. "I hate catcalls," he said, referring to sudden changes of volume, "they are apt to call up the unknown." Other emphatic remarks in his notes assailed the "quaver" in the Italian dramatic style, which he was determined to teach singers to renounce. An operatic singer, in his view, was not required to "yell at the villain to stop murdering Leonora, because the people who listen to the phonograph can't see the villain murdering her anyway and are not holding their breath." He also wrestled with the problem created by orchestral accompaniment, which seemed to him to be "booming the cannon around the cry of a singing mouse." [12]

The boy prodigy, Josef Hofmann, then twelve, came and played the piano before Edison. After him came Hans von Bülow, the most famous pianist of his era, to perform a brief Chopin mazurka, and to listen through the tubes as it was played back. Many years later, Edison, recollecting the audition, claimed that he had actually accused Bülow of having played a false note! When the artist protested heatedly, Edison cried out to his assistant, "Put on the wax," and they played the record over again. When Bülow heard the alleged wrong note he "fainted dead away." Edison then threw water in his face; whereupon the famous executant, reviving, left the laboratory without a word—"and that was the last I ever heard of the great von Bülow." [13] Conceivably the pianist may have been moved by his sense of outrage at the poor reproduction of his playing. Though the talking machines of 1888 showed a considerable technical advance compared with the first phonograph, they had serious limitations and encompassed but a small part of the tonal spectrum.

Henry M. Stanley, of African fame, a sternly religious little man whom God had brought safely through the jungle, also visited the Edison Laboratory about this time and spoke a few grave words into the talking machine. Then he turned to the inventor and said, "Mr. Edison, if it were possible for you to hear the voice of any man . . . known in the history of the world, whose voice would you prefer to hear?"

"Napoleon's," replied Edison without hesitation.

"No, no, I should like to hear the voice of our Saviour," said Stanley.

"Oh, well," laughed Edison, "you know, *I like a hustler!*" [14]

But to return to the War of the Phonograph, affairs approached a crisis by June, 1888. The American Graphophone Company, under Lippincott, was already offering its wares on the market; the Edison organization made public charges that its patent had been infringed upon and announced that it would soon be ready to put its own superior machine on sale.

At this point, Edison apparently had some afterthoughts about his "perfected" phonograph. Calling together his veteran assistants, he informed them that they were expected to stay on the job with him day and night until the new machine had been brought up to certain standards he had set for it. Somehow, news was soon circulated in the metropolitan press of the herculean exertions being attempted at West Orange, and of the last wild charge of the Edison team upon the sound-producing problems of the phonograph. Although reporters were barred at the gate and told that no one could see him, not even his wife, the newspapers during five days carried regular bulletins describing the great inventor's "frenzy" and the "orgy of toil" endured by his associates locked within the Edison Laboratory.

This famous endurance run was commemorated by a large photograph of Edison, said to have been taken "as he appeared at 5 A.M. on June 16, 1888, after five days without sleep." It pictured him sprawling over his table, while listening to his finished phonograph through earphones, in an attitude of great fatigue, his head supported by one hand, his hair tousled, save for the long cowlick that fell habitually over his brow. The set of his jaw was thought to resemble the expression attributed to the most celebrated of modern military conquerors at the end of a victorious battle. An oil painting made after this photograph, in which the artist consciously tried to represent him as "the Napoleon of inventors," was soon afterward widely distributed as an advertising poster by the Edison Phonograph Company.

Never before or afterward did Edison's capacity to dramatize his career in the public mind, and to generate myths about himself, work to better effect. Before the coming of the famous American public relations counselors (of the type of Ivy Lee) Edison was a master of free publicity, received as much of it as he could possibly use, and, whatever he might say to the contrary, remained avid for it, generally avoiding spending money for paid advertising space as long as he could.*

The "five-day vigil" was a slight case of exaggeration. The laboratory notebooks do, however, furnish evidence of a "stretch of seventy-two hours" ending June 16, long enough by any standards.

On May 31, 1888, at a time when the inventor still seemed overborne by the problem of the new phonograph, excitement of a different sort centered at Glenmont, for on that day, the first of the three children Edison was to have by his second wife was born. It was a girl and was named Madeleine. Edison had interrupted his work to await the arrival of the child, remained for a time at his wife's bedside, then hurried back

---

* One of the inventor's younger sons, Charles Edison, called attention to his father's dislike of paid newspaper advertising as superfluous in his case. The inventor was probably unconscious of his own desire for publicity, and no doubt thought of himself as a man of retiring disposition. When his British agent, Gouraud, undertook to advertise the 1888 phonograph in flamboyant style, Edison sent him a memorandum, protesting, "I don't propose to be Barnumized," and requested that he [Gouraud] talk phonography and not Edison." (Memorandum, March 2, 1888, Edison Laboratory Archives.)

to his laboratory. A few days later he had gone into his all-night sessions with the talking machine.

By now Mina Edison was more fully aware of what kind of great man she had married. Accustomed to a cheerful home, in which she played a leading part in a large family of eleven children, and to a busy social life, she must now spend her days and nights much alone in the big mansion on the hill at Llewellyn Park. There were few diversions, no games, and almost never any social gatherings here; only the servants, and some of the Miller relatives visiting her occasionally. She would be better off when she had more children of her own. The children by Edison's first marriage did not enjoy staying at Glenmont. Marion, only a few years younger than her stepmother, was at a difficult age and attended boarding school or traveled in Europe with a governess. The two boys after a while spent much of their time in the home of their aunt, Mrs. Holzer, at Menlo Park, or at the farm of Uncle Pitt Edison in Michigan, when they were not away at school.

As Mina Edison said in later years, the conditions of her marriage constituted a great challenge. With her deeply imbued sense of duty she recognized that her husband's career always came first. Such efforts as she made to soften or control him soon found their limits, for as he had once said, no woman could ever "manage" him. Evening after evening she would wait in vain for him to come home to dinner, then send him some warm food by the coachman.

A more resolute woman than his first wife, Mina is described as having forced her way into the laboratory and faced Edison down, on one occasion, when she discovered that the tray of tempting food she had sent him was untouched. "Now, Thomas Edison, you have just got to sit down and eat your dinner!" she exclaimed. She was loyal, and he, despite his teasing habit, was most affectionate with her. But when they were grown old, she did not shrink from admitting publicly that he was "difficult." In a very apt phrase she once called him "my impatient patient one."

Around the time of the coming of the new daughter and the new phonograph, Colonel Gouraud, Edison's English impresario, arrived in West Orange for a business conference with the inventor. Shortly after he returned to London, he received from Edison, in place of the letter or cablegram he had expected, a dictated "phonogram," reporting with great satisfaction the safe arrival of both "babies":

> In my Laboratory in Orange, N.J.
> June 16, 1888, 3 o'clock A.M.

Friend Gouraud:
This is my first mailing phonogram. It will go to you in the regular U.S. mail

via North German Lloyd steamer *Eider*. I send you by Mr. Hamilton a new phonograph, the first one of the new model which has just left my hands.

It has been put together very hurriedly and is not finished, as you will see. I have sent you a quantity of experimental phonogram blanks, and music by every mail leaving here . . .

Mrs. Edison and the baby [Madeleine] are doing well. The baby's articulation is quite loud enough but a trifle indistinct; it can be improved but it is not bad for a first experiment.

With kind regards,

Yours, EDISON [15]

In England, at the time, the father of the phonograph was probably the best known American in civilian life. At his offices in London, now called Edison House, Gouraud gave demonstrations of the new apparatus that made a tremendous impression. These would begin with an introductory speech by Edison himself followed by records giving the silvery voice of Mr. William Gladstone, the Prime Minister; of Robert Browning reciting his verses (but forgetting his lines); and of Sir Arthur Sullivan showing his wit—but also, in serious vein, prophesying the horrors of mass entertainment:

I am astonished and somewhat terrified at the results of this evening's experiments—astonished at the wonderful power you have developed, and terrified at the thought that so much hideous and bad music may be put on record forever! [16]

Edison said, "I don't want the phonograph sold for amusement purposes. It is not a toy. I want it sold for business purposes only." Nevertheless it was the entertainment side of his invention that the world wanted most. When this was perceived, the Edison organization later sold many music records, but did nothing to raise the standard of the music it chose to reproduce. Others applied the machine to the distribution of grand opera airs and vocal or instrumental music by serious artists. When, after several years, Edison learned that such things were being attempted in Germany, he at first refused to believe it.

The wax records of the new phonograph ran all of two minutes in those days, and could carry the volume of a brass band, or the warblings of a mockingbird. A typical picture of the Phonographic Age, often appearing in illustrated newspapers, was of two bearded gentlemen, each holding one tube of a talking machine to his ear, while gazing at the other with a rapturous expression. They were listening to "Home, Sweet Home."

3   The circle of Edison's old laboratory associates was evidently much agitated at the promotion of his new favorites, Gilliland and Tomlinson, to a dominant position in the reorganized phonograph company. The minority shareholders had been asked to exchange their shares for stock in the new company, but Edward H. Johnson and several others protested at the decision to place its management in the hands of Gilliland. "It is not a matter of money, but of wounded pride," wrote Johnson. Edison, he said, was forsaking those who had shown their loyalty to him through long years and was giving his trust to persons who, for their selfish advantage, kept close watch over him, spied on him, and even intercepted messages sent to him by his old partners.[17] In the end, Edison paid Johnson and several others cash to induce them to turn in their stock. He ascribed Johnson's complaints to jealousy of Gilliland.

He himself felt a growing resentment at Johnson, who, having recently become the president of the Edison Electric Light Company, seemed bent on feathering his own nest as much as possible. In 1884 Frank J. Sprague and Johnson had set up, on a shoestring, the Sprague Electric Railway and Motor Company, in which Edison had no interest, for he had his own Electric Railway Company. Sprague had designed an excellent stationary motor used principally for elevators; his partner Johnson, through his connection with the Edison organization, pushed the sales of the Sprague motors, which were manufactured by the Edison Machine Works in Schenectady, as subcontractor. Then Sprague entered the traction field by inventing a motor specially designed for streetcars. Here he competed, in some measure, with the Edison-Field company, which, in contrast with Sprague's, was then almost moribund. By his successful installation of a modern electric streetcar line in Richmond, Virginia, in the autumn of 1887, Sprague set the whole pattern for large-scale electric traction development—a field which Edison had tried to cultivate earlier. In temperament, Edison and the brilliant younger man proved to be incompatible. Sprague had the impression that Edison, at about this period, ceased to be progressive in matters of electrical science; while Edison held that the other tended to create dissension in the once "happy family" of his associates. At length, in 1889, the first Sprague company was bought out and merged with the Edison manufacturing organization, Sprague departing to organize another successful electrical manufacturing company of his own. Johnson also departed not long afterward. Edison said at the time that the whole Sprague-Johnson venture into motors and traction had been "a galling thorn among all the boys, and as I told you, would break up the old association." [18]

In Ezra Gilliland, however, Edison saw the devoted business manager and promoter he, as an inventor, needed so badly. With the help of this trusted friend, he hoped that he might be permitted to give up many of his present business cares and go back to purely inventive research.

Capital was always desperately needed for the promotion of the Edison inventions. Gilliland undertook to raise the money needed for the production of the new phonograph, and promised to protect Edison's interests in any transactions entered upon. One day in the spring of 1888 the bland, paunchy man turned up with a veritable angel, the same Jesse W. Lippincott of Pittsburgh who had recently bought the Bell and Tainter patent rights for $200,000. Formerly a successful manufacturer of glass tumblers, Lippincott had sold out his business for a cool million to one of those "trusts" that were then taking over almost every going industry in America, from barbed wire to mortuary caskets. The phonograph appealed to him as something new, in which he could put his money to work again. But on learning that there were prospects of a lawsuit with Edison, he had concluded that it would be better to combine forces peacefully than to go to war. Why not a Phonograph Trust? A proper venture in those days before the passage of the Sherman Act of 1890.

Impressed, no doubt, by the widely publicized "five-day vigil" at the Edison Laboratory, Lippincott approached Gilliland with an offer to buy the Edison phonograph rights in America. The inventor allowed Gilliland and Tomlinson to negotiate as his authorized representatives with the Pittsburgh entrepreneur.

About ten years before, during the 1870s, Edison had had a disastrous experience as a result of having put his trust completely in a certain patent lawyer of Washington, to whom he had sent papers covering numerous patent applications for minor telegraphic devices. When some time had gone by, he discovered that the lawyer had sold the numerous patent applications for small sums of cash to other parties, executing the patents in their names; then the man had disappeared, and it was hard to track him down. As Edison told the story long years afterward, the lawyer had been driven to commit these frauds because he was in desperate money straits, owing to the serious illness of his wife; and so Edison had done nothing about the matter. With equal innocence of heart he gave his trust entirely to Gilliland, the kind friend through whom he had first met Mina Miller and whose charming wife was Mrs. Edison's closest friend.

The new phonograph had come of age, in its inventor's view; it was to create a large new industry. Hence he hoped to receive in return for exclusive patent rights something like a million dollars at the least. But

Gilliland and Tomlinson, after repeated conferences alone with the "angel," reported that the best they could get from Lippincott was $500,000, as well as a contract for Edison to manufacture the machines exclusively for the new company. For ten days the inventor, somewhat disappointed, considered the matter. In the end he did as they urged.[19]

And what of Gilliland's agreement with Edison for an exclusive selling agency? Gilliland and Tomlinson said that they had arranged to assign this contract also to Lippincott in return for a payment of "only $50,000."

The thought evidently passed through Edison's mind that his devoted friend and personal attorney were making a pretty good thing out of his phonograph invention. But more than one of his close associates, like Johnson, had become passing rich through working with him; and he had no objection to that. The contracts between Edison and Lippincott were drawn up by Tomlinson and signed that summer, while a separate contract for the American sales agency passed between Lippincott and Gilliland and Tomlinson, the actual terms of which Edison was not told about. Moreover, the two men asked Lippincott not to reveal them to Edison for fear that he would think they were getting too much.[20]

Final arrangements, and payments of securities and cash, were completed on August 31, 1888. The next day Gilliland and Tomlinson were to leave for Europe (with their families) to negotiate the sale of Edison's phonograph patent rights in Europe. Just before departing, Tomlinson brazenly appealed to Edison for cash with which to meet the expense of his journey, pretending that he had worked hard on the big Lippincott deal for but small fees. Edison kindly handed over $7,000 more. Thus, as it turned out, he had even paid for Tomlinson's getaway. It was the last he ever saw of that precious pair.

Ten days later, Jesse Lippincott, the would-be founder of a phonograph trust, the North American Phonograph Company, had some second thoughts about his dealings with Gilliland and Tomlinson. He came to the Edison Laboratory and made a complete revelation of how both he and the inventor had been gulled. On the understanding that these trusted agents of Edison were covertly helping him, Lippincott, to buy the Edison phonograph patents cheaply, or at half of what the inventor thought they were worth, Lippincott had made a secret bargain by which he agreed to pay Gilliland the enormous commission of $250,000 in cash and securities for the Gilliland sales agency contract. That was half of what Edison himself was to receive; and yet Gilliland had never put a penny into the enterprise and had never sold anything as yet. Moreover, the lawyer, Tomlinson, had made a bargain with Gilliland by which he was to receive a third of the other swindler's share.[21]

When Edison learned of all this, he was heartsick; he realized he had been deceived and betrayed by his friend Gilliland and by his own personal attorney. As he said at the time, it was his custom to trust his intimate associates, and this habit of his was well known. When the affair was reported in the newspapers, a while later, it was rather widely remarked that the inventor "had no pretensions to business sagacity." [22] But Lippincott was no wiser. It was only after thinking over the matter that he decided the two men had acted in fraudulent fashion, and thought that he might somehow avoid paying them the rest of the large sums he had promised them by making a clean breast of his secret dealings. His own hands, however, were scarcely clean.

The lawyer in the case, by acting as attorney for both sides of a contract and deceiving Edison, his original client, had behaved in so unethical a manner that it could have led to his disbarment. His connection with the Edison companies would be terminated, his reputation would be ruined. As for Gilliland, Edison sent him a cablegram:

I have just learned that you have made a certain trade with Lippincott of a nature known to me. I have this day abrogated your contract and notified Mr. Lippincott of the fact and that he pay any further sum at his own risk. Since you have been so underhanded I shall demand refunding of all money paid you and I do not desire you to exhibit phonographs in Europe.[23]

Usually Edison laughed off his heaviest reverses, but this time, as his wife and eldest daughter Marion perceived, he was deeply wounded. He had truly cared for Gilliland as a friend; now, he said, he "would never trust anyone again." In his later years he had many close business and scientific associates and old retainers whom he was evidently fond of, but he allowed himself no intimate friendships—save that with Henry Ford, the Croesus of the automobile world.

When Gilliland and Tomlinson refused to consider repaying any part of the money, suit was instituted against them, in January, 1889, and the wretched affair became public, much to Edison's regret. The defendants were able to show, however, that they had full authorization to act as Edison's agents and that a breach of ethics, rather than provable fraud, was involved. A demurrer was entered by their attorneys, and litigation was dropped by Edison's lawyers.[24] It was, after all, only a minor scandal in a period notorious for business corruption on a large scale.

Wealth sometimes dripped from the great inventor's hands into those of men with superior cunning. At any rate, Edison's phonograph was reborn after ten years; with its clear, if thin, voice, it created renewed wonder and amusement for the public of the gay nineties. Being under contract to supply Lippincott's North American Phonograph Company with many thousands of phonographs, Edison erected a large factory for this purpose adjacent to his laboratory at West Orange and employed hundreds of workers. Soon afterward he engaged more factory space in a nearby town for the production of wooden phonograph cabinets. Within a few years the well-loved phonograph was the source of employment for many thousands of workers, in Edison's and in other plants. Its rapid growth as a large new industry, however, was attended with an incredible amount of business and legal trouble.

The inept Lippincott, who headed the abortive talking-machine trust in its early years, soon showed himself wanting both in commercial sense and in vision. Edison had contemplated selling phonographs outright at $85 to $100 apiece; Lippincott, in imitation of the successful Bell Telephone Company, followed the plan of issuing licenses to local dealers in cities and territories and, through them, renting the machines at $40 to $60 a year. The results, however, were extremely disappointing.[25]

The early business phonographs, moreover, with their two diaphragms for recording and reproducing, often broke down; their stylus points grew dull, and their motors proved balky and had to be replaced, after several years, with clock springs. The short, two-minute cylinder records had a capacity for only about four hundred words, which was inconvenient. To stop the machines for correction was well-nigh impossible. Stenographers in those days were usually male, and they tried to boycott the phonograph, as a machine that threatened their livelihood.

Although there was a growing interest and trade in phonographs, Lippincott seems to have managed things at the distributing end so poorly that Edison formed a very low opinion of him. On the occasion of some difference between them, he addressed a sharp note to Lippincott saying, "I shall send a civil engineer with a theodolite to see if your head is really level." [26]

With but few exceptions local distributors who had taken up licenses to sell the phonograph as a business machine soon went out of business. One of the exceptions was the Columbia Phonograph Company of Washington, D.C., which began by providing a good repair service and sold many instruments to government bureaus. In addition to that, this company, the forerunner of the present-day Columbia Broadcasting Company, was among the first to undertake the sale of duplicated music records for entertainment purposes.

The great success of the phonograph was to be not as a business machine but in the entertainment field, when, toward 1892, it was introduced into popular nickelodeons as a coin-in-the-slot machine, grinding out light music and comic dialogues for the multitudes. The brass bands, the music-hall tenors, and sopranos resounding in thousands of the tinny 1888 models offered nothing to be admired as art, but at all events they pointed the way to the future, to the rightful role of the phonograph as purveyor of music to the masses.

After a while the unfortunate Lippincott found himself unable to meet his business obligations, and at the same time became seriously ill, suffering a stroke of paralysis. His largest creditor was Edison, to whom he owed money for thousands of phonographs that had been delivered and not paid for—these debts, however, being secured by collateral, the controlling stock of the North American Phonograph Company. By the autumn of 1893, a time of profound business depression, Edison, as both principal creditor and largest stockholder of the tottering North American Phonograph Company, was its de facto head. While working constantly at the technical improvement of his machine, he also shaped plans to re-build the whole enterprise. To this end he had a survey made of the whole phonograph trade, which showed its condition to be perfectly chaotic. While some local firms, the Columbia company especially, had already built up a mass market in duplicated popular music records, North American itself had never produced a single record that was fit to sell.[27]

During the 1890s rival inventors also appeared on the scene, a short time before Edison's basic patent was due to expire; the most skillful of these was Emile Berliner, who had developed the flat-disk record, and a stylus making a lateral cut, instead of the vertical or "hill-and-dale" cut of the Edison and the Bell-Tainter instruments. Berliner's machine, known as the "gramophone" in England, was the ancestor of the Victor Talking Machine, whose trade-mark was the faithful little terrier listening to "his master's voice."

Edison's new plans, as they matured, called for the development of a simplified phonograph at a popular price, designed primarily for musical entertainment. He also contemplated making and selling popular music records. But before launching such a program he decided to put the North American Phonograph Company into friendly receivership, so that he could take back full rights in his invention. When he did this, in 1894, he was obliged to take on the liabilities of Lippincott's ruined phonograph "empire" and thus was involved in suits of various kinds directed at Lippincott's company, during which, by court order, he was prohibited from selling phonographs in the United States for a period of about three years. (His growing export sales, however, were unaffected.)

Henceforth, he resolved to keep the phonograph business in his own hands. When the unhappy interlude of the North American Company's receivership ended, he would reenter the field with an instrument whose sound recording was vastly superior to the second model's, and one that was remarkably cheap as well.

4
In pursuing the story of the phonograph and bringing it to a convenient stopping place we have, perforce, run ahead of important inventive work by Edison going on simultaneously in other fields. It is breath-taking even to attempt to follow him as he moves restlessly from one type of investigation to others wholly different. In 1888, for example, one would find him patenting a new incandescent lamp, in which the carbonized bamboo filament is replaced with "squirted" cellulose, making it both more efficient and more economical. At the next moment we find him at work on a revolutionary type of mining machinery. Then there is a secret dark room, upstairs in the laboratory building, where only two chosen assistants are permitted to work with the master. Doubtless there are fabulous goings-on in that room, which none as yet will disclose. Thus, even in what is for him a relatively unproductive period, full of business headaches, the Edison Laboratory at West Orange fairly boils with its varied activities.

A group of scientists arriving to visit him in the spring of 1889 find him, to their alarm, with head and face swathed in bandages. It is nothing serious, he explains; simply that a crucible happened to explode while he stood close by. The press reported the affair under the headline:

EDISON BURNED BUT BUSY [28]

In the winter and spring of 1889 he had felt more tired than ever. His young wife, attempting to minister to his health, tried for a long time to persuade him to rest. Then at last, to her great delight, he agreed to take a long holiday, and in fact to go to Europe with her.

The vacation trip turned into a triumphal tour. They sailed on the French liner *La Bourgogne* on August 3, 1889, their first destination being Paris. Putting the Gillilands and Lippincotts and all the growing pains of the electrical industry as well, out of his mind, Edison once more was as carefree as a boy. He was to visit the great Universal Exposition in France, see the new Eiffel Tower and all the contrivances of the nineteenth-century inventors, including his own electrical, telegraph, telephone, and phonograph inventions.

Nothing could have made Mina Edison happier, after the years of relative seclusion in her country house. She was a young matron of only twenty-two, in her full beauty, and she loved to shine in society. In Paris she would share in all the public honors that the world nowadays lavished on her husband wherever he made his appearance. On their very arrival at the dock in Le Havre an official delegation representing the French Republic was on hand to meet them; reporters from newspapers of many European countries were there to interview them, and crowds cheered them on their way to the train. In Paris there were more and bigger crowds, and more government dignitaries to bid them welcome.

A crew of Edison men, headed by William J. Hammer, had arrived several months earlier to prepare a display of Edison products that would cover a whole acre of the Fair grounds. Here a complete central lighting station had been erected, and was surmounted by one of Hammer's brilliant electric signs, made up of colored bulbs representing the flags of both France and the United States. There were also "fountains of light." Indeed, the Edison system provided a thrilling picture of the world's future illumination. But the longest lines of people, an estimated 30,000, were drawn to the new phonographs, which ranked only second to the Eiffel Tower as an attraction.

The fêtes for Edison seemed unending and often wearied him, though he observed that his wife apparently could never have her fill of them. Shortly after his arrival, King Humbert of Italy named him a Grand Officer of the Crown of Italy, and sent him impressive decorations at the same time, thus, as the newspapers said, "making Mr. Edison a Count and his wife a Countess." [29] The next day the inventor and his wife were received at the Elysée Palace by President Sadi Carnot, who, in a formal ceremony, decorated Edison with a red sash and named him a Commander of the French Legion of Honor. At a banquet given him by the City of Paris, and at another by the newspaper *Figaro,* the official speakers paid homage to him as "the man who tamed the lightning." *"Edison est un roi de la république intellectuelle; l'humanité entière lui est reconnaissant!"* it was said.

Well he might be a "king of the intellectual republic" to whom the entire world was indebted; and presidents, (Bonapartist) dukes and duchesses, and even Oriental sultans might do him honor; but Edison could neither hear nor understand what they were all saying.

"Dinners, dinners, dinners," he said in recollection, "but in spite of them all they did not get me to speak." Refusing to speak in public, he had the American Ambassador, Whitelaw Reid, reply for him when a retort courteous was in order.

During the ten days in Paris the American inventor, with his big head, expressive features, and youthful bearing, became familiar to the people in the streets, who besieged the entrance of his hotel in the Place Vendôme to see him—among them being many would-be inventors, desperately anxious to consult him on their own discoveries. No American in civil life, saving perhaps Buffalo Bill, of the Wild West Show (then playing in Paris), was more renowned in Europe.

At the Opera House, as he entered a box—the guest of the President of the Republic—the orchestra played "The Star-Spangled Banner," while the audience cheered and called on him to speak. He seemed overcome with emotion, merely rose to his feet, bowed, and sat down.[30]

After I had been there an hour [Edison relates] the manager came around and asked me to go underneath the stage, as they were putting on a ballet of 300 girls, the finest ballet in Europe. It seems there is a little hole on the stage with a hood over it, in which the prompter sits...I was given the position in the prompter's seat and saw the whole ballet at close range.[31]

Marion Edison, who had been studying at Geneva in the charge of a friend of Mrs. Edison, joined her parents in Paris. She recalls that her father disliked the crowds and social life, but that her stepmother insisted that he come out with her to meet all the distinguished people who wanted to see him. "Oh they make me sick to my stomach!" he would exclaim on returning from those long dinners.

It was plain to see that the youthful Mrs. Edison and her high-spirited stepdaughter of seventeen agreed in "everything save their opinions." It is not unusual for a young wife to find difficulty in asserting her authority over the grown-up children of a middle-aged widower, and, at that, children who have been reared in a way quite different from her own upbringing. Marion was clearly "jealous" of the beautiful young woman who had taken not only her mother's place, but as she may have felt, her own place as well in her father's affections. But, then, Edison himself used to indulge in little pleasantries about his wife's taste for "socials" and for pomp and circumstance. For example, he would refuse to appear in public wearing his impressive decorations, such as the red sash of the Legion of Honor, saying, "I could not stand for that.... My wife had me wear the little red button, but when I saw Americans coming, I would slip it out of my lapel, as I thought they would jolly me for wearing it." [32]

More agreeable to him was his visit to the Eiffel Tower, and the luncheon high in the sky, of which he partook as the guest of the famous engineer, Alexandre Gustave Eiffel, who had designed the great tower. Then Buffalo Bill, Colonel William F. Cody himself, was in Paris with his Wild West Show encamped in its outskirts, which, as a popular attraction, rivaled the Exposition itself. Edison could think of no finer diversion than to go there, eat an American "grubstake" breakfast of pork and beans with the famous Indian scout, and ride at top speed in a "Deadwood coach" around the enclosure, together with Senator Chauncey M. Depew and other dignitaries, while a "savage horde of howling redskins" pursued them fiercely and poured volleys of blank shot at them.[33]

In more serious mood, Edison dutifully toured the Louvre with his wife and saw its old treasures, which he did not enjoy. Though he possessed a great imagination in practicing his own mechanical arts, he said, "To my mind the Old Masters are not art, their value is in their scarcity and in the vanity of men with lots of money." On the other hand he did enjoy the painters of the nineteenth century and the great Impressionists whose works were then consigned to the Luxembourg Palace.

A noteworthy event was his visit, at the invitation of Louis Pasteur, to the famous Pasteur Institute. Edison's own notes show that he deeply respected the inspiring French scientist and considered his "germ" theory to be one of the greatest discoveries of all time. He had a good talk with Pasteur, then watched how he inoculated crowds of persons, hour after hour, with his marvelous vaccine. There was hidden drama here too; for example, a handsome boy was brought to him too late, after being infected with hydrophobia. "He will be dead in six days," Pasteur whispered to Edison, after the boy had been duly inoculated and had left.

How different, how contrasting were the careers of these two men: Pasteur, preeminently the scientific discoverer; Edison "the most famous inventor of his time." To be sure, certain discoveries of Pasteur were of immense commercial benefit to all agriculture and to the production of wine. A good many years before, in 1865, Napoleon III, in the course of an audience with the scientist, had expressed surprise that Pasteur had never used any of his discoveries as a source of commercial profit. Pasteur had replied that a true scientist would consider he "lowered himself by doing so." To make money by his discoveries, "a man of pure science would complicate his life and risk paralyzing his inventive faculties. . . ."[34] Pasteur, in fact, deeply enjoyed seeing the immediate, practical effect of his ideas on men and things, but nothing must be permitted to draw him away from experimentation.

From Paris the Edison party went on to Berlin, and the inventor met Hermann von Helmholtz, one of the greatest and most imaginative physical scientists of his century. Like Pasteur, Helmholtz showed esteem for Edison because his resourcefulness in applied science opened new avenues to scientific knowledge itself. But there could not have been much meeting of minds with Helmholtz, who, in any case, spoke no English. With the foremost German electrical inventor Werner von Siemens, who did understand English, communication was more possible, and more worldly.

Together they traveled from Berlin to Heidelberg, in a private railway compartment, to attend a scientific convention there, Edison regaling Siemens all the while with his fund of American stories.

Late in September he arrived in London, which he had last visited in 1872 as an obscure young inventor of telegraph devices. Now the tycoons of the electrical industry were most eager to talk with him and entertain him in their big country houses. In London he had occasion to inspect the Edison central station at Holborn Viaduct, which was run with direct-current generators, at 110 volts. But the dawn of alternating-current usage had come; Gaulard and Gibbs converters had already been invented several years earlier, opening the way to the distribution of more powerful currents. Edison also met S. Z. Ferranti, who was just then building a gigantic a-c dynamo designed to send currents of ten to fifteen thousand volts over long distances into the central district of London for electric lighting. The Edison d-c power station had a radius of only a mile and a half at most. On being asked for his opinion of these innovations, Edison declared that he deemed direct current the only safe medium for distribution of power, and held Ferranti's ideas to be "too ambitious." Nevertheless heavy current engineering was now advancing with great rapidity— though the man who had done so much, by his practical work, to stimulate progress in electrodynamics would have no part in it.

On the return voyage, Edison was thoroughly relaxed, having traveled and loafed for two months on end. He was a good sailor, and his humor was as unfailing and "aggressive" as usual. During the Channel crossing to England, a rough passage, when his family and all the passengers became ill and were forced to go below deck, he stationed himself in the saloon and there enjoyed one of his foul cigars. Whenever any of the wretched passengers came up for air, Edison, as he admitted, deliberately "would give a big puff and they would go away and begin again."

Immediately on his arrival in New York (October 6, 1889) he hurriedly crossed the harbor in a launch to the Jersey shore, then rode by carriage to the laboratory at West Orange. He was burning to know the results of certain secret experiments that had been going on during his absence behind the locked door of Room 5. But Batchelor and W. K. L. Dickson had meanwhile put up a small new building in the laboratory "compound" for these new purposes. Proudly Dickson showed his chief into the "studio" and prepared to demonstrate what had been accomplished, in accordance with Edison's instructions, during his stay in Europe.

Edison sat down; the big room was darkened; Dickson went to a bulky apparatus at the back of the room that resembled a large optical lantern.

Next to it, and attached to it, was a phonograph. Against the wall, facing Edison, there was a projection screen. Dickson turned a crank, and a vague flickering image of Dickson himself "stepped out on the screen, raised his hat and smiled, while uttering the words of greeting: 'Good morning, Mr. Edison, glad to see you back. I hope you are satisfied with the *Kineto-phonograph.*'" [35]

So Dickson wrote in 1894, not long after the event. The development of the first operable motion picture camera—invented but not yet patented by Edison, and hence unknown to the world—was well under way. Connected with it, though rather rudely synchronized, was a model of his improved phonograph. Thus the first motion pictures were not silent, but were talking pictures.

chapter XVII

Formation of General Electric

1   In urging that men of science should beware of mixing business with invention, Louis Pasteur might well have taken Thomas A. Edison as his object lesson, had he known the pattern of the American inventor's life. Edison would have loved to pursue his fascinating experiments in the motion picture with undivided attention, rather than intermittently. This work, however, advanced far more slowly than his earlier inventions.

"Dollars and science were so much mixed up" in his career, as he himself suggested. For twenty years, aside from his inventive activities, he had been involved in one commercial venture after another. By 1888 he was actually one of America's ranking industrialists, employing between 2,000 and 3,000 workers. With so many on his payroll, he could not escape labor troubles.

In his relations with his workers he had the attitude of the honest, old-fashioned entrepreneur. Since he himself had a puritanical horror of idleness and was content to work eighteen hours a day, he reasoned that his hired hands should be willing to hustle for eleven hours. But the period of 1884 to 1887 was a time of marked labor unrest, when nearly one million workers entered the ranks of the Knights of Labor, America's first full-fledged labor federation, to fight for the union shop, the ten-hour day, and higher wages.

Edison paid average wages, plus incentive payments for extra effort, in the days when his shops were small; but he could never abide labor unions. As a small employer he had made his men like him and had known how to keep them interested in the work. Nowadays the factory system's division of labor characterized his electrical shops, where hundreds of workers were employed in place of the few dozens he had had earlier, and no man made more than a fraction of any product.

When Edison had started lamp manufacturing, one of the most important processes, the sealing of the filament into the globe, required several months of training before the men were expert at it. "When we got to the point where we employed eighty of those experts they formed a union," he related, "and knowing that it was impossible to manufacture lamps without them, they became very insolent." The manager, for example, wanted to fire an objectionable employee, but the union opposed this, and the manager told Edison conditions were becoming intolerable. "So I started in to see if it were not possible to do that operation by machinery. . . . I then went secretly to work and made thirty of those machines." After his labor-saving machines were ready to be installed, the trouble-making individual was fired. "Then the union went out," Edison says. "It has been out ever since." [1] Thus he used invention as a social weapon with which to discipline labor and lower manufacturing costs, just as his fellow captains of industry were doing. To his workers, Edison's humor sometimes seemed rather grim.

A similar dispute had led to the removal of the entire Edison Machine Works from New York to Schenectady in 1886. As he tells the story:

It seems I had rather a socialistic strain in me; and I raised the pay of the workmen twenty-five cents [a day]. . . . But the men thought they would try coercion and get a little more, as we were considered easy marks. Whereupon they struck at a time that was critical. However we were short of money for payroll, and we concluded . . . it would give us a couple of weeks to catch up. So when the men went out they appointed a committee to meet us; but for two weeks they could not find us. Finally they said they would like to go back. We said all right. . . . When they went back to the Goerck Street shops they found them empty of machinery. It was quite a novelty to the men not to be able to find us when they wanted to, and they didn't relish it at all.[2]

At the time when the Edison Machine Works was transferred to Schenectady, Edison relates, he and Insull, his financial lieutenant, were carrying fearful burdens. They had enormous orders and great difficulty in meeting payrolls and buying supplies. The lighting and heavy electrical equipment industries were expanding so rapidly that their founder, paradoxically, often felt himself on the verge of ruin, for want of capital. Even the conservative Eaton, the company's lawyer, now remarked, "The danger is not that we won't have enough business, but won't be able to grow fast enough to handle it." [3]

In 1886 Edison had found himself with orders on hand requiring $200,000 worth of copper and no cash to buy it with. Fortunately the head of a large copper-producing firm allowed him consignments on credit. But there were other pressing needs. He found it hard to borrow money of his company's regular bankers. As he wrote Henry Villard, the bankers were merciless when credit was tight. "You said the way to raise money was to get loans. I once borrowed $20,000 from Drexel, Morgan & Company, and they called it suddenly when I was in a desperate condition carrying the shops, and I had to pay $4,000 to raise the $20,000." This was an interest rate of 20 per cent! [4]

On a later occasion, when he was developing his magnetic ore concentrator as a new venture, he approached James A. Stillman, president of the National City Bank, for loans and discounts. But Stillman coolly turned him down. Was it not good banking to help a man in this way, the inventor asked? "What you want is a partner," Stillman replied.[5] Edison's opinion of Wall Street banking practices was understandably very low.

In supplying equipment for city central stations and independent generating plants, the Edison manufacturing companies were often obliged to accept notes from local capitalists, and even shares of stock, in part payment. Thus the machine works and the lamp company ran through their working capital and encountered great difficulty in meeting their own bills. By 1891, according to Insull's records, the Edison manufacturing units were loaded with about $4,000,000 in the securities of various local power companies, collateral on which it was not easy or safe to borrow money in those days.

Insull, however, became adept in obtaining credit or in having it extended for the machine works. His motto seemed to be: "Never pay cash if you can buy on credit." Insull would insist on credit for a year or more even when it came to buying himself a twenty-dollar suit of clothes. At times Edison would fall on him with urgent demands for cash for some experimental project, and Insull would manage somehow to raise the money, though such extended borrowings spelled danger eventually for the manufacturing companies.

Competition against Edison in the electrical manufacturing field seemed at first a cloud no bigger than a man's hand. Then, toward 1885, it became ever more threatening. The makers of arc lights, such as Brush, began producing incandescent lamps and lighting plants like Edison's, but under license contracts with Maxim or with Sawyer and Man. The newly formed Thomson-Houston Company of Lynn, Massachusetts, appeared the strongest of these competitors, also operating under license arrangements with the corporate owners of the Maxim or Sawyer incandescent lights, known respectively as the United States Electric and the Consolidated Electric companies. The Edison Company, after long delays, finally opened suit in 1885 for patent infringement against the two companies owning the Maxim and Sawyer-Man patents. Surprisingly enough it met with an initial setback, in Federal court at St. Louis, when Edison's claims to priority were disallowed on somewhat narrow grounds. In the following year an appeal was taken to a higher court, and the fight was pressed with great vigor by the Edison lawyers. But these proceedings dragged for three years longer. Meanwhile every electrical company in America "jumped the Edison claims," with Thomson-Houston in the lead.[6]

Only a few years earlier, Elihu Thomson had predicted utter failure for Edison's lighting schemes; now he not only decided to imitate him, but considered, like others in his field, that it was ethical to do so—on the assumption that Edison's patents would never be upheld. As an apologist for Elihu Thomson has written, naïvely enough, he began manufacturing incandescent lamps "that were to all intents identical with Edison's, except that they were protected by the patents of Sawyer and Man."[7]

Still another important competitor was George Westinghouse, of Pittsburgh, who bought control of United States Electric, with its Sawyer-Man patents, and in 1886 entered the incandescent-lighting field. Westinghouse had known success both as a mechanical inventor and as a manufacturer, and attracted some able electrical engineers to his staff. Though he used a corporate disguise, he was for seven years the principal defendant in a series of bitterly fought lawsuits waged against his subsidiary companies by the Edison interest.

A man of large vision, daring to the point of recklessness in his enterprises, Westinghouse resembled Thomas A. Edison a good deal, and was of the same age. He was determined that his organization should be the most "progressive" in the field, as it was, in grasping for great opportunities in the new mechanisms of alternating-current distribution. However, the circumstances under which he began to compete with Edison seem morally indefensible even for that era of industrial privateers.

In a pamphlet giving warning to patent violators, issued in 1887 by Edward Johnson as president of the Edison Electric Light Company, we are told how numerous capitalists, engineers, and inventors visited Edison at Menlo Park, when his lamp was first invented, how they were shown everything in a trusting spirit, and how they went off saying to themselves that Edison was a "damn fool" for revealing his process. Among these early visitors had been George Westinghouse, then known principally as an inventor and manufacturer of railway air brakes. Several years later he announced with remarkable impudence that he was simply going to take advantage of Edison's pioneering work and manufacture and sell incandescent-lighting systems at lower cost than the Edison companies. A Westinghouse advertisement is quoted as saying baldly:

We regard it as fortunate that we have deferred entering the electrical field until the present moment. Having thus profited by the public experience of others, we enter ourselves for competition, hampered by a minimum of expense for experimental outlay.... In short, our organization is free, in large measure, of the load with which [other] electrical enterprises seem to be encumbered. The fruit of this ... we propose to share with the customer.[8]

Open warfare for the control of patent monopolies in key industries was typical of the 1880s and 1890s, when the leading corporations in those fields had not yet been catalyzed into trusts or cartels. Though the Westinghouse Electrical & Manufacturing Company appeared as the main antagonist of the Edison organization, the Thomson-Houston company, no less an infringer of patents than Westinghouse, by 1889 was already second to the Edison group in business volume. Under shrewd business leadership this company built and supplied 870 central lighting stations for big cities and towns within five years of its entrance into the incandescent-lighting field. That was more than the Edison company had achieved, though it held the leading position in the manufacture of lamps and other electrical equipment, and in selling isolated plants.

Between them Thomson-Houston and Westinghouse had bought up virtually all the non-Edison patents in their industry—now combining forces against the bigger Edison organization, now entering into conflict with each other. Thus it was really a confused three-sided struggle that centered about the famous "carbon filament suit," which the Edison Electric Light Company stubbornly directed against its two main adversaries in the years between 1885 and 1892. What the directors and attorneys of the Edison Company sought, by a legal action conducted at prodigious cost, was to establish once and for all the priority of Edison's key invention and thus remove the threat of the "patent pirates."

2   So long as the courts were not yet finished with all the technical-legal angles of this complex case and there was no clear-cut decision, Edison's chief rivals were free to produce and sell carbon filament lamps by the thousand and install lighting stations of their own manufacture. Edison and Johnson therefore conceived the idea of attacking their principal antagonist, Westinghouse, on another ground, where he would be more vulnerable than on the legal side.

A year after organizing the new Westinghouse Electrical & Manufacturing Company (incorporated early in 1885), Westinghouse was well launched upon a pioneering venture of his own in the electrical field: building and installing generating and distributing equipment that was, in part, radically different from the Edison system, since it was adapted to high-tension alternating current of 1,000 volts or more. For this Westinghouse supplied a-c generators, and newly invented transformers that stepped down high-potential currents to 100 or even 50 volts so that the system would be safe for domestic use. Alternating current offered danger of heavy shock unless insulation and transformers functioned perfectly. Now the Edison Company, which had been advertising its direct-current installations, using 110 volts, as entirely safe, determined to attack the exponents of alternating current by creating alarm among the public about the dangers of the Westinghouse installations.

In taking this course, Edison was swayed by passion rather than wisdom; he could be a good hater, and the thought that Westinghouse and others whom he considered no better than interlopers were forcing their way into his own vineyards filled him with rage. On first hearing that Westinghouse was bent on going into electrical manufacturing in a big way, Edison said some hard things about his new rival, to the effect that he knew nothing about the business and ought to stick to railway air brakes. Those words reached the bold-handed Pittsburgh industrialist and made him all the more resolved to distinguish himself in the electrical field —which he had first entered several years earlier, by making electrical switching and signaling devices for trains.

Whatever Edison might feel about this matter, his own d-c central stations were seriously limited in capacity, their conducting mains having a radius of at most two miles from the power supply. With the cheaper high-voltage current under the a-c system the radius of power stations could be greatly extended and large economies in conductors achieved.

The advance into alternating-current engineering had been under way since 1883, when Lucien Gaulard and John D. Gibbs had patented their a-c transformer in England; with this first inadequate instrument they converted high-potential current to a low potential. George Westinghouse saw the future possibilities of their invention—a subject of great disputation at the time—and purchased their American patent rights. He saw that, if successfully applied, it would permit either hydroelectric plants or steam generating plants to be placed at favorable locations for water-power or coal supply, and would make it possible to send currents of high voltage over long transmission lines for many miles, and cheaply, then, with the a-c transformer, step them down again to low voltages at the point of consumption. (This could not be done with Edison's a-c system.) William A. Stanley, an able American electrical inventor whom Westinghouse engaged as a consultant, managed to bring about a marked improvement in the Gaulard and Gibbs transformer by the autumn of

1885; thanks to the Stanley patents, Westinghouse felt sure that he could enter the electric lighting and power industry with a decided advantage over the Edison system.

At his home town of Great Barrington, Massachusetts, Stanley had built a small a-c power station which delivered current at 500 volts, then stepped down the voltage and distributed it to dwellings and shops, by means of his transformers, at a safe 50 volts. Following the success of this experiment Westinghouse set to work in earnest to produce and sell economical a-c long-distance power installations.

Similar opportunities to introduce a-c transformers and a-c systems had been offered to Edison at about the same time. German engineers, working on an Edison power station in Budapest, had devised an improved transformer, even before Stanley, and had transmitted high-potential currents without significant loss over a distance formerly thought to be far beyond the limits of the Edison mains.

The "Z.B.D. System" (so named after its inventors, Zipernowsky, Bláthy, and Déri) was patented in the United States and its rights were offered to Edison. In 1886, Francis Upton went to Europe, inspected the Z.B.D. transformers, and recommended that American rights be purchased. An option was taken for $5,000 on a total asking price of $20,000; but it was allowed to lapse three years later, and the Z.B.D. transformer was permitted to pass into other hands on Edison's insistence that it be rejected.[9]

In the early stages of the art, back in 1880, he had firmly opted for d-c at 110 volts as the safest form of electrical distribution—all the way from the powerhouse to the ultimate consumer. This had been an intelligent decision at the time, for it undoubtedly served to prevent fires and accidents that might have discouraged adoption of his system. It must be noted that the leading European electrical scientists, Lord Kelvin and Werner von Siemens, like Franklin Pope and Elihu Thomson in America, also warned against the use of a-c systems. Mistrust of the Stanley, or Z.B.D., devices arose from the supposition that, in the event of a transformer being short-circuited or a part of the high-tension system being accidentally grounded, the customer turning on a light in his home might be instantly electrocuted. Many disastrous fires and electrocutions of linemen had occurred in recent years after the arc-light systems had begun using a-c circuits of about 2,000 volts. Yet by 1887 Westinghouse and Thomson-Houston were boasting they would be able to use high-potential currents safely at 5,000 and 10,000 volts! In the opinion of conservative authorities, the development of a-c systems, with their "alternators," or a-c dynamos, and their transformers, must await the perfection of grounding and insulating techniques to eliminate all danger. Until then the Edison plan, with its low-voltage direct current, was held to be the only safe one.

In November, 1886, Edison received a report from the engineers of Siemens and Halske of Berlin, indicating that the Z.B.D. System and its transformer gave much trouble, was costly to install, needed a great deal of development, and was dangerous to work with, as compared with the Edison-Siemens system. Commenting on this report, Edison wrote E. H. Johnson:

Just as certain as death Westinghouse will kill a customer within 6 months after he puts in a system of any size. He has got a new thing and it will require a great deal of experimenting to get it working practically. It will never be free from danger....

None of his plans worry me in the least; only thing that disturbs me is that Westinghouse is a great man for flooding the country with agents and travelers. He is ubiquitous and will form numerous companies before we know anything about it.[10]

In afterthought, he adds:

If they use great insulation on their converters, their capacity and economy will diminish as the square of the insulation. I cannot for the life of me see how alternating current high-pressure mains—which in large cities can never stop—could be repaired. Whereas the main of the direct current would not produce death if a man received shock ... in a wet place.

Nevertheless all the younger engineers in Europe and America were buzzing with excitement over the new a-c distributing systems. Among them were Hopkinson, Kapp, and Fleming in London; in America, Nikola Tesla, the brilliant young Serbian immigrant—after quitting Edison's employ—invented the all-important polyphase (a-c) system, patent rights to which he promptly sold to Westinghouse. The young men were dreaming already of harnessing the power of Niagara Falls with gigantic a-c generators and producing electrical energy in immense volume.

Alas, how the once intrepid Edison had changed! Was it like him to shrink from an undertaking that would still require "a great deal of experimenting," or even offered risk to life itself? He was, however, as firm as the proverbial "stubborn Dutchman" in his belief that his rivals, Westinghouse and Elihu Thomson—who soon changed his mind and also took up the a-c system—were not only wrong, but would bring down the whole electric light and power industry in ruins. And so he began the "war of the currents," with his characteristic enthusiasm, denouncing the advocates of a-c and doing everything to arouse the public mind against them.*

A bitter press campaign now ensued, in which both sides hurled propaganda petards at each other. The big laboratory at West Orange was the principal source of "scientific" evidence purportedly exposing all those who were making and selling a-c systems to the public. There, on any day in 1887, one might have found Edison and his assistants occupied in certain cruel and lugubrious experiments: the electrocution of stray cats and dogs by means of high-tension currents. In the presence of newspaper reporters and other invited guests, Edison and Batchelor would edge a little dog onto a sheet of tin to which were attached wires from an a-c generator supplying current at 1,000 volts.

In one of those sadistic "experiments," Batchelor, while trying to hold a puppy in the "chair," by accident received a fearful shock himself and "had the awful memory of body and soul being wrenched asunder . . . the sensations of an immense rough file thrust through the quivering fibers of the body." Though badly shaken up, he recovered in a day or two, it was said, "with no visible injury, *except in the memory of the victim.*" [11]

The feline and canine pets of the West Orange neighborhood were purchased from eager schoolboys at twenty-five cents each and executed in such numbers that the local animal population stood in danger of being decimated. Public statements and pamphlets were also issued, such as Johnson's red-lettered "Warning!" of February, 1888, citing many press reports of fatalities experienced in theaters and factories, or along power lines where high-tension arc-light systems were in use. The pamphlet also

---

* The Edison d-c generator employed the mechanism of the commutator to convert alternating current into unidirectional current, i.e. "direct current." It functioned best at a low constant voltage, of 110 to 220; otherwise the commutator attached to the dynamo armature heated up badly. Thus the transmission radius was restricted to a mile or two, unless very large copper conductors were to be used—at fearful cost—which made long-distance transmission impracticable.

The "alternator," or a-c generator, eliminated the commutator and generated current of high voltage which could be transmitted very cheaply over long distances. Then, a transformer, consisting of two unconnected coils wound about an iron core, one coil having many turns of wire and the other fewer turns, came into play, to "step down" the high voltage a-c, at points of distribution and domestic use, from 500 or 5,000 volts to 100 or 50 volts. (The transformer was also often used, in the reverse sense to "step up" the voltage of the alternator to enable economical transmission over still longer distances.)

As Kenneth M. Swezey, a modern writer on electrical engineering, has aptly written, "the Edison d-c stations had to remain small and practically in a customer's backyard. Alternating currents however, could be generated in large bulk and transmitted many miles." (*Electrical Engineering*, September, 1956, p. 787.)

called attention to the alleged low moral character of the "patent pirates," Westinghouse and Thomson-Houston, who were bent on introducing those hazardous currents into the American home.

Westinghouse was so angry at these exaggerated and even false "warnings" that he considered bringing suit for libel against the Edison company. From the Westinghouse camp, at any rate, public statements and brochures were issued in reply, in which great pains were taken to prove the Edison propaganda misleading, inasmuch as the Stanley transformers were being placed at points along the street mains outside of buildings, within which the current was received only at low voltages, as safely as with the d-c system.

In the next round of the battle, a certain "Professor" H. P. Brown, formerly a laboratory assistant at West Orange, and evidently employed by Ed Johnson, carried on some rather ghastly experiments at public gathering places or lecture halls in New York, by which he made propaganda for the passage of a law in New York State providing for the electrocution of criminals sentenced to death—by charges of alternating, or "Westinghouse," current. In executing a number of large dogs with an a-c generator, Brown sought to demonstrate that death in such form was "instantaneous, painless and humane," and "left no disfiguring marks." The public was stirred up to a rare pitch of excitement by the lobbying of both factions before the New York State Legislature, which in the autumn of 1888 finally adopted a statute providing for the use of the "electric chair" in place of hanging as a means of capital punishment. Brown, engaged as a consultant by the state authorities, ostentatiously purchased three Westinghouse alternators as the most suitable for dispatching condemned criminals by "electrocution." Indeed, before that new word was widely adopted, some persons were so unkind as to suggest that the term *"to Westinghouse"* be employed in describing the new mode of capital punishment.[12]

In the public press, George Westinghouse tried to answer the false accusations directed against his a-c system in reasoned fashion, arguing that out of thirty deaths known to have been definitely caused in recent years by electrical-distribution accidents, sixteen took place in connection with d-c circuits, and none where Westinghouse equipment had been used. It seemed to him inconceivable that anyone could claim that direct current, at similarly high voltages, was any "safer" than alternating current. Edison also published a magazine article under his name defending his claims that the Westinghouse plan meant murder. He declared, "I have not failed to seek practical demonstration. . . . I have taken life—not human life—in the belief that the end justified the means." When his own company had considered purchasing patent rights for a European alternating system, he "had succeeded in inducing them not to offer this system to the public, nor will they do so without my consent." [13]

Despite the strong protests of Westinghouse at the grim use to which his a-c generators were to be put, the New York prison authorities, after long delays, finally proceeded to execute a condemned murder, a certain William Kemmler, in the "chair," at a supposedly secret ceremony held on August 6, 1890. The electric charge was reported to have been too weak, and the miserable work had to be done over again, becoming "an awful spectacle, far worse than hanging." [14]

That Edison could have lent himself to a publicity campaign intended to persuade the public that the a-c system (now universally used) was fit only for mad dogs or criminals seems strange to us now. That he should have permitted his adherents, and especially that "research expert" Brown, to make misleading statements, was no less strange. The whole dreadful controversy can be attributed only to an extreme bitterness of feeling toward his opponents that completely overbalanced his judgment.

At about that time, Edward Dean Adams, a very knowledgeable Boston financier who was a member of the Edison Company's board of directors and a friend of Westinghouse, tried to bring the two men together, urging Edison to come with him to Pittsburgh and see what was being done at the Westinghouse plant with the engineering of a-c alternators and transformers. But Edison refused, sending Adams a scorching telegram:

Am very well aware of his [Westinghouse's] resources and plant, and his methods of doing business are lately such that the man has gone crazy over sudden accession of wealth or something unknown to me and is flying a kite that will land him in the mud sooner or later.[15]

Some of his associates, who had a better understanding of the scientific principles involved, pleaded with him to reconsider his rejection of the new system; among these, it was said, were Frank Sprague and Dr. Kennelly.[16] But Edison, according to E. H. Johnson, remained "granitic" in his conviction that the a-c system would destroy the reputation for safety he had tried to build up for his industry. It was perhaps Edison's greatest blunder (as he himself would admit twenty years afterward).*

Edison's own pioneering work in electrical distribution had stimulated a growing demand for industrial current; thus, he himself created in great measure the urgent need for alternating-current development. Factories everywhere cried for heavy electric current for big motors to replace

* In 1908, Edison met William Stanley's son George, who was visiting at Glenmont as a friend of Charles Edison. "Oh, by the way," the inventor said to young Stanley, "tell your father I was wrong."

steam engines, which only a-c generators could furnish. Edison's company was actually in the best position to take up the new techniques. Yet, on the threshold of these great events, he seemingly closed his mind and would go no further toward the new phase of "industrial power" at high voltage and low cost—which his weak d-c generators could never supply.

It was as if he could not really visualize the inherently complex electro-dynamic problems of heavy-current engineering, as he had failed to visualize radio waves when he was almost upon them. Tesla, Stanley, and Westinghouse, on the other hand, saw that the a-c transformer was "the key to a new door in electrical progress." For the era of giant electric power a modern breed of engineers, armed with the calculus, men like Tesla, Charles P. Steinmetz, and B. G. Lamme would lead the way along high-voltage transmission lines. The initiative and vision of Westing-house, among others, would soon create the heavy electrical equipment of the new power age. To the Westinghouse organization would go the glory of harnessing and distributing the monstrous hydroelectric power of Niagara Falls—which would have been impossible by Edison's methods.

"In 1879 Edison was a bold and courageous innovator. In 1889 he was a cautious and conservative defender of the status quo," a modern commentator has said.[17] He stubbornly resisted the advance to a-c systems, much as the big gaslight companies earlier had rejected his incandescent electric light, and for much the same reasons, no doubt: the fear that all the effort, equipment, and capital invested in the old system would quickly be made obsolete by the new. Yet it is in the very nature of even the greatest scientific work that it is transitory, because it is "chained to the course of progress." *

In 1888, the largest of the Edison manufacturing companies, the machine works at Schenectady, was so busy that it was reported the workers were literally standing on one another's feet. The need for reorganization of the sprawling electrical manufacturing units was urgent; the need for fresh capital was even more urgent.

About two years earlier, some effort at reorganization had been made when the United Edison Manufacturing Company was set up, with offices at 65 Fifth Avenue, New York. Its function was primarily to co-

---

* "In science each of us knows that what he has accomplished will be antiquated in ten, twenty, fifty years. That is the fate to which science is subjected; it is the very meaning of *scientific work*...." (Max Weber, "The Destiny of the Scientist," in *Essays in Sociology*, Oxford, 1946.)

ordinate the selling and installation of lighting plants for the three companies that were making their main components, Edison Lamp, Edison Machine Works, and Bergmann's—which, however, retained their separate corporate individuality while sharing control of the United. The parent company, Edison Electric Light, remained a sort of paper organization, but held the Edison patents and licensed them to the local light and power companies that were being established all about the country. In these local utilities, like the big ones in New York, Boston and Chicago—though they used Edison's name as a synonym for electric light—the inventor himself had no material interest. He had also sold out nearly all his shares in the holding company (the E.E.L. Co.) in order to start manufacturing.

Friction within this loose combination continued, principally between the manufacturing group—which made the dynamos, conductors, instruments, and lamps—and the patent-holding company, whose directors were mainly financiers. The parent company remained limited in its profits; the manufacturing companies, which had originally been licensed by it, were growing and thriving, but were under constraint, being unable to get sufficient capital from the conservative bankers who controlled the E.E.L. Co. In one of his vivid phrases, Edison, writing to Henry Villard, spoke freely of the "leaden collar of the Edison Light Co." around his neck, hindering him from doing a better job in developing his industry.[18]

Henry Villard, after two years' absence in Europe, had returned to the scenes of his former triumphs, and, as often happens with talented speculators or gamblers, had made a pretty comeback. This he accomplished largely by centering his attention, as his friend Edison had advised, on the young electrical industry. Acting as agent for important financial groups in Germany, he turned their investments toward the electrical field instead of railroads. In Europe he had lately become closely associated with the directors of Siemens & Halske and with Emil Rathenau, head of the German General Electric Company. He had always been fascinated by Edison, but perceiving the difficulties the Edison electrical industries encountered in their expansion and their friction with the banking group in the parent company, he bided his time for an opportunity to "move in," with the backing of his German capitalists.

In the parent company the Morgan group were well entrenched, controlled the patents and the sale of licenses, and watched over the whole operation. But it was through the manufacturing companies that Edison and his lieutenants, Insull, Batchelor, Upton, and Bergmann, served as operating men and earned their living. For two years Villard had been maturing a plan for a grand consolidation; in the spring of 1888 he proposed the amalgamation of the several companies as the Edison General Electric Company. His scheme was to buy in the "electrical shops" at what he considered a fair value. Knowing that he could not operate without the support of the Morgan group, he sought to combine forces with them.[19]

As Sam Insull shrewdly remarked at an early stage of the negotiations, the Villard and Morgan group wanted to take in the electrical shops at a lower valuation while giving their own shares in the patent-holding company a higher value in terms of the new, amalgamated company's securities. They therefore tended to reject any proposals that would make "too good a deal for the shops," that is for Edison and his manufacturing partners.[20]

At this time, however, as earlier mentioned, the Edison manufacturing organization was loaded up with about four million dollars' worth of the securities of local lighting companies, "frozen assets" accepted in partial payment for equipment. Orders coming in at Schenectady and the other shops were far in excess of the capital available to handle the business, as Edison said later, "and both Mr. Insull and I were afraid we might get into trouble for lack of money. . . . Therefore, when Mr. Henry Villard and his syndicate offered to buy out the shops we concluded it was better to be sure than to be sorry." [21]

While Edison was working to build up the company that was eventually to become one of the world's largest industrial corporations—General Electric—he was often obsessed by the fear that he might be forced into bankruptcy. The necessity of meeting a weekly payroll for 2,000 to 3,000 workers, week by week, harrowed the mind of one who was, after all, predominantly contemplative in temperament. At times "he was embarrassed in getting enough money to pay even for his household expenses," according to the reminiscences of Insull. One day he finally said to Insull, "This looks pretty bad. I think I could go back and earn my living as a telegraph operator. Do you think, Sammy, that you could . . . earn your living as a stenographer?" [22] The proposals of Villard and his syndicate were therefore received by Edison with a deep sense of relief from intolerable burdens.

After a good deal of haggling, Villard worked out a final agreement by which the Edison Electric Light Company, the parent company, was to be taken into the combination to be known as the Edison General Electric Company for $3,500,000 (par value) in new stock. The various "shops" and all their assets, the lamp works, the machine works at Schenectady, and Bergmann & Company, were to have their shares exchanged for $2,158,333 in new stock and $1,166,667 in cash—a total of about

$3,300,000. Of this sum paid for the factories Edison himself, as principal owner, realized over half, or approximately $1,750,000, in stock and cash. Sprague Electric Railway and two other small units, in which Edison had an insignificant share, were also absorbed at the same time.

The original Drexel-Morgan group, on their total investment of about $779,600 in Edison's idea, thus, within a few years, realized over 350 per cent profit, or $2,700,000. The Morgan banking group also had separate investments in the profitable Edison Illuminating Company of New York (now Consolidated Edison), which had become a separate utility company, as had other big-city lighting companies established in Boston, Brooklyn, and elsewhere.

The three main Edison shops together had gross sales revenues of over $7,000,000 a year in 1889 and combined annual profits of over $700,000. Within the next two years they were to expand their annual business at a rapid pace. Sigmund Bergmann, Edison's partner in the appliance business, came out with a million dollars, went to Germany, and there founded one of the largest electrical concerns in Europe. Johnson also retired from the new firm to live in a huge mansion in Connecticut. Edison's friends had been made rich by their connection with him. He himself was now relieved of his most pressing business cares and owned some two million dollars in cash and securities, earned mainly through his manufacturing shops.

The new Edison General Electric was described in the metropolitan press as one of the biggest consolidations ever carried out under the laws of New Jersey and as constituting an "electrical trust." [23] The Edison G.E., on being consolidated, soon issued about four million dollars in new stock, to obtain fresh capital, part of which was bought by the Morgan banking syndicate, but most of which was taken up by the Deutsche Bank of Berlin, for the account of the Siemens & Halske firm —for whom Villard acted as American representative. Villard, who had "grandiose ideas" of a world cartel (!) said at the time that "the pooling of the brains of Edison" and those of "the eminent scientist and discoverer Dr. Werner Siemens," would provide the Edison G.E. with a virtual monopoly in electrical inventiveness and manufacturing resources.[24] After the whole deal was completed, Villard was elected president of Edison G.E.

Edison, owning a minority share of about 10 per cent of the stock, still was regarded as the company's inventive brain; he was also one of its directors. But now only three of his old associates, Insull, Batchelor, and Upton, remained with him on the board; the rest were Villard's and Morgan's people.

European capital then still played a large part in financing credit-hungry American railroads and industries. Morgan, as an international banker, often combined with British, French and German interests. But, through his partner, Coster, who was still treasurer of the company, he continued to keep a sharp watch over this company. A proposed contract by which Villard hoped to bind Edison's inventive services to the company for a long period of years—for an additional $500,000—was rejected as impracticable, through the influence of Coster.[25] Edison's voice in the great concern that had arisen out of his inventions grew perceptibly weaker; he seldom attended directors' meetings. Morgan, however, continued to buy shares or add to his control. Villard writes Edison:

Morgan has been after me for some time to make up a block of 5,000 shares of E.G.E. [from] among the larger holders, to be sold at 92½. . . . I told Morgan I would consult you. I think it is the Vanderbilt interest. . . . I am willing to contribute 2,000 shares. Will you give 1,000 shares? If you do I think we can get the balance from the others.[26]

When Pierpont made such requests, even Villard dared not refuse him.

4   It will be recalled that Edison was in an unusually carefree mood during his trip to Europe, in the summer of 1889, for it came just after he had sold out, as he thought, most of his business troubles to others. "I have been under a desperate strain for money for 22 years," he told Henry Villard in 1890, "and when I sold out, one of the greatest inducements was the sum of cash received, so as to free my mind from financial stress, and thus enable me to go ahead in the technical field."[27] Meanwhile, the Edison G.E. lawyers, like those of the predecessor company, still marched from court to court in the long legal fight to enforce its full patent rights to the carbon filament lamp and the Edison electric distributing system.

In the course of the "Seven Years' War" over the carbon filament lamp patent, the outcome of the contest seemed extremely uncertain during most of this period. Many experts believed, rightly, that Edison was not actually the *first* to have invented the incandescent electric light and that this could be proved historically. Then it was widely believed that the description in Edison's basic 1879 patent did not give sufficient or accurate enough information by which a mechanic could make such a lamp. On this point alone the Edison lawyers had met with some reverses in German and British courts, though they quickly regained ground in subsequent appeals.

Edison himself seemed to have lost all faith in patent litigation. Virtually everything he ever devised, every idea of his, was disputed by claimants who seemed to spring up on every hand as soon as he executed a patent. If a rival group were wealthy enough they could always dredge up some unknown with prior claims to Edison's inventions, tying him up in the courts for five years. As Edison put it:

This class of men has risen in every field where my work has produced anything of commercial importance. I remember one fellow who swore he invented one of my machines before I did. We found out where he had it built, and the records of the machine shop showed it was made a long time after I had put a machine on the market. He simply copied my mechanism, and rusted it up . . . I lost the German patent on the carbon telephone through the insert of a comma which entirely changed the interpretation of the patent. Another foreign patent was lost because the patent office in that country discovered that something similar had been used in Egypt in 2000 B.C.—not exactly the same device, but something nearly enough like it to defeat my patent.[28]

There was no justice in the patent laws, Edison concluded. The issuance of a patent was "simply an invitation to a lawsuit" as well as to artful imitators. Thus, on being asked if he would approve of still another suit over patent claims, he wrote bitterly, "Say I have lost all faith in patents, judges & everything else relating to patents. Don't care if the whole system was squelched." [29]

After the Edison lawyers had met with an initial reverse in the suit against a Westinghouse subsidiary for infringement of Edison patents, Westinghouse rashly decided to press his advantage by instituting a suit himself against the Edison Electric Light Company, for infringement of the Sawyer-Man lamp patent, which he controlled.

The suit in equity of Consolidated Electric Light (controlled by Westinghouse) against McKeesport Light Company (an Edison subsidiary) ran three years, but marked a definite turn in the tide of this wasting legal warfare. On October 4, 1889, Justice Bradley of the United States Circuit Court at Pittsburgh handed down a clear-cut decision upholding Edison's claim to priority over Sawyer and Man. True, these inventors had made an electric lamp that would burn for a few minutes; but it was a failure in every practical sense, and totally without *commercial* value.

It seems [Justice Bradley's decision reads] that they were following a wrong principle, the principle of small resistance in an incandescing conductor, and a strong current of electricity; and that the great discovery in the art was that of adopting a high resistance in the conductor with a small illuminating surface and a corresponding diminution in the strength of current. This was accomplished by Edison . . . and was really the grand discovery in the art of electric lighting,

without which it could not have come into general use in houses and cities. . . .
But for this discovery electric lighting would never have become a fact.[30]

Justice Bradley, moreover, noted that following the patenting of the
Edison invention "there was an entire change of base on the part of Sawyer & Man, giving their product a different direction and purpose from
its original form." *

By 1890, after the decision of Judge Bradley, the two chief adversaries
of the Edison Company were fairly desperate. Their legal advisers could
read the handwriting on the wall and already urged a compromise settlement with the Edison interests. Nevertheless, Thomson-Houston and
Westinghouse made appeal to a higher Federal court; the long-drawn-out
legal contest thus reached its culminating phase in the suit of Edison
et al. against the United States Electric Light Company, in the United
States Circuit Court of the Southern District of New York, in 1890 and
1891.

Today the lengthy conflicts over patents of half a century ago seem to
us grotesque. We think no longer in terms of the individual inventor's
"prior" claims, but regard scientific invention as the product of the total
culture of a society, and often the creation of many hands, of many
seekers after knowledge, employed in great corporate or institutional
laboratories. However, in the period after the Civil War, before the coming of great trusts and their patent-pooling plans, combat between industrial giants assumed a legal as well as an economic form. The love of
glory, the ego drive of individual inventors also counted for much in the
persistence of the long patent contests. Inventors are often stormy
characters. For example, William Sawyer, an able electrical inventor
in his time, often competed with Edison. Distracted by Edison's victories
over him, he became chronically ill, took to drink, and, in an altercation
with a man in his boardinghouse in New York, in 1883, shot the other
man. Poor Sawyer was then convicted of manslaughter and died while
waiting to begin his prison sentence.

The final act of the carbon filament lamp drama took place in New
York in the spring of 1890, before Federal Judge William A. Wallace.
The defendants, facing the prospect of being driven from the electrical
field, engaged eminent counsel, headed by General Samuel Duncan, of

* The decision of Justice Bradley, and
later of Justice William Wallace, paralleled
very closely the finding of a high English
court in a suit against interests owning
the European lamp patents of Joseph
Swan. The whole case turned upon the invention of Edison's attenuated "filament."
Had there been no other difference in the
lamps, this would have been sufficient; for,
as the English judge remarked, the very
word "filament" had been unknown before this.

the New York Bar, and a whole staff of scientific experts. Richard N. Dyer, well-known patent attorney, together with Clarence Seward and Grosvenor Lowrey, directed the Edison G.E.'s legal staff. Edison himself was brought to the witness stand and confronted the court and public, inwardly very shy, outwardly crusty. By now he detested these trial hearings, vital though they might be for his corporation. The opposing lawyers might have borne hard upon Edison, but they proved generally indulgent, for all the world esteemed him as one of the modern heroes of science.

Nevertheless, the shrewd Duncan, in cross-examination on June 19, 1890, deliberately tried to expose the inventor's ignorance of theoretical science. What, he asked, was Edison's knowledge of Ohm's law, at the time when he first experimented with the incandescent light? Edison admittedly candidly that he did not fully understand everything about Ohm's law in 1878. Such knowledge, he added, "would prevent me from experimenting."

Q. Why should a knowledge of Ohm's law prevent you from experimenting?

A. Because I would try to figure it out mathematically, and I have had a great many mathematicians employed by me for the last ten years, and they have all been dead failures.

Q. Upton included?

A. Yes sir, in his mathematical capacity.

But did not the laws of electrical science show Edison the way to the high resistance burner? Did not Upton teach him? the lawyer asked.

A. I don't think so. The mathematics always seems to come after the experiments—not before.[31]

It was at this period that Edison, expressing his views on the preeminent role of applied scientists, coined the expression, "I can hire mathematicians, but they can't hire me!" (The same sentiment was to be repeated in a court, many years later, by Edison's epigon, Henry Ford, in his remarks on historians.)

Fortunately for the plaintiff, however, the testimony of Francis Upton, on many technical points, proved to be highly important. During many days of cross-examination the mathematician remained unshaken under every attack.

The defense still maintained that Edison's original patent specifications would be impossible to follow in attempting to construct an incandescent lamp. But one of the Menlo Park veterans, John W. Howell, when challenged to do this, constructed exactly such a lamp as the 1879 model (of cotton thread) in the presence of court officers, and kept it alight for 600 hours. This piece of technical evidence was considered to have been the *coup de grâce* for the defendants.

On July 14, 1891, Judge Wallace ruled that the priority of Edison's carbon filament lamp patent over others was complete. Like the highest British court, he held that the difference between Edison's very attenuated filament and the various carbon "rods" of others—between a cross section of one sixty-fourth of an inch and a cross section of one sixteenth of an inch—was of enormous practical significance. Edison alone had reached the goal which so many had sought. But this great legal victory, as it turned out, was virtually barren. It had been won at a cost of about two million dollars to the company; Westinghouse was said to have been made almost insolvent. And by the time the court's decision could be enforced, only about two years remained before the life of the lamp patent expired. Moreover, Edison no longer owned the Edison General Electric. Well could he say, "My electric light inventions have brought me no profits, only forty years of litigation."

5   Victory in the courts for the Edison lamp patent created panic among the electrical-goods manufacturers, who (after the denial of their final appeals) were to have their lamp factories closed by injunction in 1892. To be sure, they would be able to re-open for business when Edison's patents in America ran out in 1894; but stoppage would cause them severe losses. Westinghouse at once made strenuous efforts to develop a non-infringing lamp, using an old idea of a "stoppered" base and a different type of filament. Its vacuum was somewhat inferior to the sealed-glass Edison globe and had a shorter life, but Westinghouse managed to produce many thousands of these stoppered lamps in time to fulfill his lighting contract for the Chicago World's Fair of 1893, though at heavy cost.

The position of the Thomson-Houston Company was also endangered. Under the aggressive management of Charles A. Coffin, a former shoe salesman, it had attained a volume of business almost as large as that of the Edison General Electric. Coffin believed in expansion at all costs; he bought up competitors one by one, and thus entered different lines of electrical business. One of these was the electric railway firm of Charles J. Van Depoele, the inventor of the overhead trolley; another was the important arc-lighting business of Charles F. Brush, of Cleveland, purchased outright for $3,000,000 in 1889. Coffin, moreover, vigorously promoted the sale of Thomson-Houston equipment by accepting large amounts of stocks and bonds in local electric lighting companies instead of cash. The Edison organization, which had been obliged to use the same financial methods, had found itself overloaded with such paper, and had cut down on such credits. But Thomson-Houston, whose directors were among the leading financiers of Boston's State Street—including Henry Lee Higginson, F. L. Ames, and T. Jefferson Coolidge—had large financial resources. It formed a subsidiary company which took over from the Thomson-Houston treasury the notes and stocks of local utilities and sold debentures to the public secured by this collateral, thus providing the manufacturing concern with ample cash.

The engineering talent of Thomson-Houston, though imitative, was skillful enough. After producing an excellent imitation of the Edison incandescent-lighting system, they later obtained valuable transformer patents, developed "alternators" after Elihu Thomson's designs, and, in competition with the Westinghouse company, began to sell power on high-voltage lines. Somewhat alarmed by these developments, President Villard of the Edison G.E. wrote Edison inquiring whether they might not do well to take up the high-potential current program. With his letter he enclosed the translation of a German scientific article reporting recent advances in a-c systems in Switzerland and Germany. Edison replied in his most stubborn vein, "The use of alternating current instead of direct current is unworthy of practical men."

He was still America's "wizard of electricity." But the Thomson-Houston, Westinghouse, and other American engineers were showing themselves more progressive in experiments with high-voltage currents. At this period, the mathematical physicist Charles P. Steinmetz became the principal scientific aide of Thomson-Houston.

Nevertheless, that rising organization faced the problem of having much of their business cut off by the lamp patent decision. Would central stations using Thomson-Houston lamps be closed down? Or would manufacturers and users of lamps other than Edison's be forced to pay heavy license fees to the Edison G.E.? Since Thomson-Houston's wealthy stockholders had had long and friendly ties with the Morgan group in the Edison G.E., would it not be better, men in both factions reflected, to have the two companies combine forces rather than go on fighting lawsuits eternally?

Henry Villard had the idea of a new consolidation as soon as he had completed his earlier one in 1889. Before the great carbon filament suit had got under way in the courts, he suggested to Edison that a combination with Thomson-Houston might end their troubles with "fierce competition and low prices." But Edison had other views. He answered:

The Thomson-Houston and the Edison Company can no more control the price than the tides. The Company with the best and cheapest machinery will do the business, patents or no patents. Fact is, Mr. Villard, that all electric machinery is entirely too high now. These high prices hurt the business. *With the leaden collar of the Edison Electric Light Co. around me, I have never been able to show what can be done.* The ground of cheapening has scarcely been scratched. Let us break the leaden collar and you will see a brainy competition that will show them what real competition is.

Some of the inventor's ideas about running the business were based on deep common sense and were much less trustbound, so to speak, than Villard's or even Coffin's. Instead of avoiding competition, there should be more competition, Edison urged; instead of maintaining high prices, they "must go down 50 to 75 per cent lower than now ... and we will make a great profit. ..."

At the very idea of combining with Thomson-Houston, Edison bridled. He continues:

The statement that they ask no favor from the Edison Company might be met by the fact that, having boldly appropriated and infringed upon every patent we use, there is very little left to favor them with, except our business, which they are now after.

Why should we divide? Why hire Thomson-Houston Company to do our business ...? The more I figure the "benefits" of a coalition the more worried I am that you may be induced to enter into one.... The more I figure, the more fatal it seems to me to the Edison G.E., being the dominant electrical company in this country, not only in prestige but in profit....

If you make the coalition my usefulness as an inventor is gone. My services wouldn't be worth a penny. I can only invent under powerful incentives. No competition means no invention. It's the same with the men I have around me. *It's not money they want but a chance for their ambition to grow.*[32]

However, Villard continued to negotiate secretly with the "enemy," as Edison called them. As early as May, 1890, rumors of an impending merger by an exchange of stock were rife among Thomson-Houston officials. In February, 1891, the superintendent of their factory at Lynn received a telephone message from President Charles A. Coffin to the effect that he must watch for the arrival at the factory of "... Mr. Henry Villard, President of the Edison General Electric Company. Mr. Villard is to be shown through our plant.... Please see to it that *his identity does not become generally known.*"[33]

The largest company in the field might still proudly bear the name of its founder: *Edison* General Electric. Edison had, nevertheless, little to say about its actual management. The improved incandescent lamp, for example, after 1888 was constructed with a "squirted" cellulose filament (instead of bamboo), but showed uneven qualities because of imperfect producing methods. This was still Edison's specialty, and he would have liked to hunt down the "bugs." But now, it seems, it was necessary that he *ask permission* of his lamp factory's management (!) before he could send an inspector into the plant who would report to him directly.[34]

Edison now felt himself only a cog in a gigantic corporate machine. "I feel that it is about time to retire from the light business," he said to Villard in 1890, "and devote myself to things more pleasant, where the strain and worry is not so great."

Meanwhile the whispers of an approaching coalition with one or the other of his big rivals continued to spread. If they were true, he remarks to Villard,

then it is clear that my usefulness is gone.... Viewing it from this light you will see how impossible it is for me to spur my mind, under the shadow of possible future affiliations with competitors, to be entered into for financial reasons ... I would now ask you not to oppose my gradual retirement from the lighting business, which will enable me to enter into fresh and congenial fields of work.[35]

The restless Edison now panted for fresh pastures, for new subjects to which he could apply his unique inventive talents in his own way, without the weight of a large organization bearing down as a yoke upon him. It seemed to him that to "play" with his secret "kinetographic camera" or his improved phonograph, or that elephantine machine he had lately designed for the magnetic separation of iron ore would be an endless delight to his mind. But always to be borrowing more money to make more money in stocks, as Mr. Villard advised, was surely no life for your true inventor.

"Electric lights are too old for me," he remarked jocularly at this period. It was no mere jest. He had generated a revolution in the usage of electric current for light. But the new movement to build up giant electrical power, flexible power to replace the ponderous steam engines, he would take no part in. It was now to be directed by the pupils of Maxwell, Helmholtz, and that little-known American, Willard Gibbs— in short, by engineers having the formal training of the mathematical physicist.

Edison was no mere mechanical inventor, but it was true that he lacked the power of abstraction. In his heart he sensed already that he could not go forward with confidence, and play the role of a leader in the new ground of electrodynamics.

One day, early in 1892, according to his private secretary, Edison was brooding in his library when a laboratory assistant came in and made some inquiry about electrical matters, to which he replied frankly that he did not know the answer. Brusquely he added that Dr. Arthur Kennelly should be consulted. "He knows far more about [electricity] than I do. In fact, I've come to the conclusion that I never did know anything about it." His words were tinged with deep bitterness.[36]

There were more miseries in the money market after the great Baring Brothers bank in London failed in the autumn of 1890. That indefatigable young financial expert, Samuel Insull, to be sure, managed to scrape up money day by day. "Buy on credit," he would say. It was thus that he managed by 1891 to accumulate a floating debt of 3.5 million dollars for the Edison G.E. It was thus also that, thirty years later, in the booming 1920s, he would operate again with "other people's money" to build up his billion-dollar Middle West Utilities Company pyramid, which, unhappily, fell in 1929 with a crash heard around the world.

During that early season of money strain in 1891, there were repeated rumors that the Edison organization was "in trouble" because of its load of debt. Henry Villard was said by the rumormongers to be headed for a fall once again. (Financial gossip in Wall Street, even when false, is always significant in that it is deliberately circulated. Thus, during disturbed market periods, powerful financial interests in Boston as well as New York were secretly buying the Edison G.E. stock.) It was also said in the newspapers now that Morgan was dissatisfied with Villard's management of affairs. Villard was brilliant, had large ideas, and at times a big speculative following. "In short he dazzles. But J. P. Morgan is a hard man to dazzle," said one contemporary.[37]

Meanwhile, in actual operations, the company was growing by leaps and bounds, its gross revenues rising within two years by 50 per cent, or from 7 to 10.9 million dollars. Insull and Kruesi were forced to make huge additions to the factory space at Schenectady. At the same time competition brought sharp declines in prices at the end of that year. Lamp prices, for example, are recorded as having fallen from a retail price of $1.00 in 1886 to 50 cents in 1891, and to 44 cents in the following year.[38]

The only solution seemed to be another big merger. By the autumn of 1891 secret conferences between representatives of both companies were held frequently and soon approached their decisive phase; Edison and his old associates, who still held managerial positions in his company, could do little but hope for fair terms. Coffin, boss of Thomson-Houston, however, had the repute of a hard man in a trade.

One of the last conferences between the directors of the two big companies was held privately in Boston at the residence of Hamilton M. Twombly, a member of the Vanderbilt family, who represented their large holdings in Edison company stock. Morgan had asked Twombly to go over the books and the plants of Thomson-Houston and was advised that it was urgent that the company be purchased and combined with Edison G.E. At the Boston meeting, to Morgan's surprise, it appeared that Coffin and his associates at Thomson-Houston refused to sell out. C. W. Dean, another Thomson-Houston executive, is reported to have said at the time, "We don't think much of the way the Edison company has been managed." [39]

When Twombly told all this to Morgan, the crusty banker said, according to an account handed down by Elihu Thomson, "Well, send them down here to talk to me." The tough-minded Coffin came to 23 Wall Street and showed Morgan enough figures to prove that his own company was the more efficient, that it netted about 50 per cent more profit on its capital than the Edison G.E. Morgan knew his financial arithmetic very well, and promptly changed his mind. If Thomson-Houston would not sell out, then he, Morgan, would sell Edison G.E. to them. Coffin persuaded him to finance, not a purchase, but a consolidation.[40] To Morgan it made little difference so long as it all resulted in a big trustification for which he would be the banker.

Edison G.E. was thus sold over the inventor's head. H. M. Twombly, D. O. Mills, and Morgan negotiated the sale with a Thomson-Houston committee consisting of F. L. Ames, T. J. Coolidge, and Henry Lee Higginson—State Street men with whom Morgan often collaborated in big investment deals. To Thomson-Houston, as to the financiers controlling the Edison organization, these arrangements promised to be most satisfying. There would be an end to the wasteful legal contests over patents and a great pooling of technical and manufacturing facilities—in fact, most of the engineering and managerial brains in the electrical industry. Certain electric railway and streetcar patents, those of Sprague and Van Depoele, formerly divided between each company, were needed by both. The Edison organization was ahead in incandescent lighting and central and isolated station equipment; Thomson-Houston was strong in arc-light systems for street lamps, and far in front of Edison G.E. in a-c power installations. From a technical point of view it was rational that the confusion caused by competing central stations, sometimes in the same cities, one with d-c and the other with a-c power installations, should be resolved. Only Westinghouse would be left as an independent hereafter, controlling no more than a fourth of the electrical business.

It was Edison's firm conviction that the main motive for the consolidation was "to do away with competition," which had caused sharp price cutting.[41] But at forty-five he already appeared an old-fashioned American entrepreneur, who believed in competition as an agency of progress and, like the Populists, feared both the coming of trusts and the strategic power of great banking groups in the money market. When he had started his electrical "shops" ten years before, the Morgans would not lend him a penny for them; now that they were grown up to the status of a large new industry, Morgan was determined to put the company Edison had founded into a trust under his financial leadership.

Once the terms of consolidation were agreed upon, in mid-December of 1891, Morgan seems to have given Villard his walking papers. Charles A. Coffin was to rule over the electrical "empire" until his death in 1926. The newspapers reported that Villard had been given to understand from "an authoritative source . . . that a courteous resignation would be courteously received. This understanding coming from the firm of Drexel, Morgan would of course have great weight with Mr. Villard." [42] Villard's German associates had sold out their holdings of Edison G.E. at a handsome profit of about 200 per cent. Samuel Insull was offered a minor executive post under Coffin, but chose to resign and take command of a rising electric utility company in Chicago.

Of the new capital of 35 million dollars to be issued for the amalgamated company, Thomson-Houston was to be alloted 18.4 million in the new stock, or the controlling share, and the Edison G.E. only 15 million in exchange for its own securities. The reason given for these harsh terms was that the Edison G.E. showed a floating debt of 3.5 million which the new company would have to assume. In other words, the smaller company actually swallowed the bigger one.

The new company, by the agreement drawn up on April 15, 1892, was to be called simply the "General Electric Company," and the name of Edison was thus to be removed from its banner. He was known as the "father of electric lighting"; his name had been a great trade name, widely used by city central stations selling current for light. That it should now be removed from the title of the organization he had founded seemed not only unduly hard, or ungrateful; it reflected clearly the new order and the will of Coffin and his associates, who were to form a majority on the board of directors. Most of the Edison management men, among them several of his old laboratory assistants, were now dropped by the new administration.

Up to the last hour, Edison hoped that the Edison Lamp Company would in some manner be treated differently from the other units and be kept under his control. But news of the transfer of all the Edison manufacturing units to the new company, without exception, suddenly reached him from a New York newspaper office in advance of its publication.

"I had never before seen him change color," his secretary, Tate, said afterward. "His complexion was naturally pale, but following my announcement it turned white as my collar." Edison said only, "Send for Insull." [43]

There were some stormy scenes of "crimination and recrimination" with Villard and Insull, who had been trusted to represent Edison's interests during the negotiations. But it is doubtful that they could have succeeded in having his name retained in the new company's title, or in keeping the management of the lamp factory for him.

"Mr. Insull was bemoaning the consolidation, when Mr. Villard turned upon him with the sharp words: 'If it had not been for you, Insull, this move would not have been necessary.' " The implication was that Insull, as the financial manager, had overborrowed.* [44]

MR. EDISON FROZEN OUT HE WAS NOT PRACTICAL ENOUGH FOR THE WAYS OF WALL STREET ran the headlines of a leading New York daily. He remained silent about what had happened; but he issued a brief public denial that he had been "mistreated" or "gulled." In reality, he had begun to retire from his company in 1889, when he reduced his ownership of its stock to about 10 per cent. After 1892 he unwisely sold the rest of it. He was elected a director of the General Electric Company; but he hardly ever attended any of its board meetings and soon resigned. His secretary, Tate, remarks:

Something had died in Edison's heart. . . . His pride had been wounded. He had a deep-seated, enduring pride in his name. And this name had been violated, torn from the title of the great industry created by his genius through years of intensive planning and unremitting toil. [45]

* Because of the notoriety attending Samuel Insull's spectacular rise and fall in the 1920s, it has been hinted by some writers that he may have deceived Edison; but there is no evidence of this; and the consolidation resulted in no benefit to him, since he was forced out of his high post in the Edison organization.

In memoirs written in 1934, shortly before his death, Samuel Insull declares that Edison "insisted on dropping the name of Edison when General Electric was formed." Insull favored the consolidation with Thomson-Houston as being the best solution financially; but Edison was unhappy over it, knowing that he would have no voice in the company's councils in the future. (*Samuel Insull's Autobiography*, manuscript, quoted by permission of Samuel Insull, Jr.) The Edison family tradition—though the inventor spoke little of the General Electric affair—holds to the view that he deeply resented the removal of his name. (Charles Edison to Author.)

One day, several weeks after the merger of the two companies into the General Electric "trust" had taken place, Edison, in his parlor in the Llewellyn Park mansion, speaking with a companion, declared with evident bitterness that he was done forever with electric lighting. "I am going to do something now," he said, "so different and so much bigger than anything I've ever done before that people will forget my name ever was connected with anything electrical."

chapter XVIII

Breaking rocks

1   For two years prior to the consolidation of G.E. Edison had been, in truth, very far away from things electrical. During those long-drawn-out negotiations over the exchange of stocks or notes, of money and power, he was never among the conferees. One would not have seen him very often in the Edison Laboratory, nor at the lamp works (which had been his pride), nor at the other shops; nor was he absorbed in experiment in his old way. Instead, current report placed him in a remote and forbidding region of the New Jersey highlands, those low mountains in the northwest corner of the state overlooking the Delaware River valley. There he was engaged in a most unusual engineering operation, which was quite literally of the earth-shaking, or, at least, earth-moving variety. By 1892 the G.E. affair was behind him; he was already well advanced in this new undertaking that was to be so different and "so much bigger" than anything he had attempted before.

Occasionally a curious journalist—or one of the many Edison idolators—wandered out to the village of Ogdensburg, New Jersey, some sixty miles west of Newark, to learn what the great electrical inventor was really doing. It was rumored that he was carrying on some sort of "revolution" in mining; he was devising new machinery and processes for extracting valuable metals from lean ores, which some people imagined he was going to apply to the recovery of gold itself!

Those who pursued Edison to the wilderness scene of his new enterprise journeyed past Lake Hopatcong by a narrow-gauge branch line of the Jersey Central, up the steep spurs of the Musconetcong Mountain, through an area of old worked-out iron mines over which forest had grown up again, and up to the 1,200-foot level near Ogdensburg. Here an industrial village called "Edison" was coming into being, its center a

367

nest of tall, red-painted mills that already boomed and clanked with their multifarious activities.

Edison had gathered together some enormous pieces of machinery, in order to take down whole mountains. Here was the largest steam shovel in America, the very one that recently had done the excavating for the Chicago drainage canal; here was a giant traveling crane with a span of 210 feet; here were miles of metal-and-wood conveyors, and an arrangement of huge rock-crushing machines. As a visiting engineer described the scene:

On all sides the roar and whistle of machinery, the whir of conveyors, and choking white dust.... The workmen look like millers, so coated do their clothes become with the flying white particles, and everyone wears a patent mask with pig-like snout. Big wheels revolve in the engine-houses; big dynamos transmit their heavy currents through overhead wires to the various parts of the plant. Little narrow-gauge locomotives puff their way in and out between the buildings ... with shrieking and whistling wheels and brakes.

This is "Edison," the mining village. Edison, the man, is always seen patrolling the ground of battle in his familiar linen duster and his great country hat. Usually you can find him watching the steam-shovel. The workmen say that he is always somewhere near it, as if it fascinates him. Follow the water-pipe through the cut, and you will find him, one is advised.[1]

Thus the young would-be inventor Thomas A. Robins, riding up the open gondolas of the branch line, finds his idol one day in 1891 at the scene of excavation. Edison's strong, round face looks happy. None of the earsplitting uproar bothers him for a moment. He is absorbed in everything going on; but he gives you his broad smile when you come to him. A whistle blows suddenly, and Edison shouts, "Look out, they are going to blast!" He pushes his visitor around the corner of a shed, and crouches down with him in the shelter of its wall.

The steam shovel has cleared off a ledge, preparatory to the blasting, the inventor himself having first staked out the ground precisely with a heavy magnetic "dipping" needle and with diamond drills. There is the roar of the dynamite charge, and segments of rock weighing five or six tons and about the size of a piano break off and are gathered by the shovel into a narrow-gauge train, from which they are moved by skips into the jaws of a huge crusher, to be reduced to stones, then pellets, and finally powder.

"We are making a Yosemite of our own here," he yells above the unearthly din. "We will soon have one of the biggest artificial canyons in the world."

How wonderful is a steam shovel! It is like a huge primordial beast.

"Look at this fellow," Edison exclaims in delight. "Wouldn't you think he is alive? Always seems to me like one of those old-time monsters or dragons we read about in children's books. I like to sit and watch it." [2]

The steam shovel is eating its way through a section of the mountain toward the group of mills about 5,000 feet distant. It will take about a year to reach its destination, and it will carry about 600,000 tons of rock. Edison's imagination for mechanical engineering has run away with itself. Everything here is planned and carried out on a colossal scale, for the success of the operation in all this "breaking of big rocks into little ones," depends on the machinery's crushing-capacity of 6,000 tons a day, said to surpass that of all the stamp mills in California at that date.

At noon the work stops, and Edison invites his visitor to take "hash" with him and several officers and foremen in charge of the concentrating mill. At a ramshackle cottage nearby, which he calls the "White House," a plate of lukewarm greasy food is served up, for there are no luxuries available here. There is little or no conversation. After lunch the inventor goes to sleep in his chair for about an hour, at the end of which a somewhat idiotic-looking fellow named Friday comes and wakes him by shaking him roughly.

"What does Friday do?" Edison is asked.

"Nothing. I have him with me because he never falls asleep. That's what I pay him for—just to keep me awake." [3]

He is on his feet, wide-awake, and proceeds amiably enough to show his visitor all about the works.

It is indeed a fabulous, a Titan's world Edison has made here for himself. Among his huge jaw crushers, among the highly ingenious earthmoving machines and the two miles of conveyors, he moves about explaining everything freely, good-humoredly, more than ever the child of nature.

The blasting and crushing was not all fun, however. There were many accidents, many breakages. On one of his first visits Robins saw a workman being carried out on a litter, with both legs broken. "Edison looked on grimly and said nothing." He was ever on the move to repair breakdowns himself, sharing all the risks of his laborers, five of whom were killed during the years of these operations. At night, in the early days, before living accommodations were provided, he used to sleep in his clothes in the warm oven room below the separating mill, with a heap of nut coal as his bed. He would often be up during the night to inspect the processing of his concentrate. For years he would remain here all week, returning home to Llewellyn Park only after the whistle blew on Saturdays, grimy and weary.

After the formation of General Electric he had lost his place at the head of America's electrical industry, but he was still a very rich man for those days. He might have rested on his laurels and grown even richer by doing nothing. Instead he was now pushing into the mainstream of American industrial life; he was bent on winning a commanding position in a new field, that of the country's vital iron and steel trade, by virtue of his new ore-milling process which would recapture economically the pure iron of exhausted mines in the Appalachian region.

Men believed at the time that iron ore was growing scarce. Edison declared confidently that in the acreage he owned or had leased, some 19,000 acres in the vicinity of Ogdensburg, New Jersey, alone, he had "over two hundred million tons of low-grade iron ore," enough to keep the United States supplied for seventy years.

Iron ore was in growing demand; but this new supply, and greater deposits still, of billions of tons of magnetite, could only be obtained by Edison's process.

2   The idea of this bold undertaking had come to the inventor's mind more than ten years earlier, when he was buying much iron for his first dynamos and found it high-priced. The iron mines along the East coast were virtually exhausted, and heavy transport charges raised the cost of ore shipped to Eastern mills from northern Michigan. One day in 1882, during a fishing trip with his laboratory assistants, he had landed on the shore of Long Island. There, on what should have been a white beach he noticed immense deposits of black sand "spreading for miles" and amounting to "hundreds of thousands of tons." Taking a magnet from his pocket he perceived how the particles of black sand swarmed toward it like tiny ants. They were particles of magnetite iron, the ancient "lodestone" eroded from Connecticut's shore of primal rock, finely divided by the action of the sea, and carried by currents across Long Island Sound. Examining his specimen of black sand on his return to Menlo Park, he found that it had a pure iron content of about 20 per cent.

"My first thought was that it would be a very easy matter to concentrate this, and I found I could sell the stuff at a good price," he said afterward. In 1882 he had William H. Meadowcroft, then employed as secretary to President Eaton of the Edison Electric Light Company, go out and put up a small concentrating plant for him on the beach at Quogue, where he would use strong electric magnets to separate the iron particles from the sand. Somewhat similar methods were in use, during the eighties, though in a small way, in Sweden and elsewhere in the United States.

Edison designed a magnetic separator of his own. It consisted of a hopper, shaped like an inverted cone, with a group of magnets below; by causing a thin stream of the magnetite particles to fall past the magnets, the iron particles would be deflected by a partition into one bin, while the waste sand would fall straight down into another bin.* But just before the apparatus could be put to work, Edison said, a tremendous storm came up, and every bit of that black sand went out to sea.[4]

By the late 1880s iron ore of Bessemer quality was in greater demand than ever, though found abundantly only in the Great Lakes region. Much occupied at the time with the electromagnets of his generators, Edison became convinced that he could use them on a large scale to separate the low-iron-content ores still found in the Eastern states.

In 1889, when he came into a large sum of money (with the formation of the Edison G.E.), he had a team of prospectors make an extensive survey of the magnetic iron ore deposits, or magnetite, to be found in the chain of the Appalachian Mountains running from Southern Canada to the Great Smokies in North Carolina. They used the magnetic dipping needle to guide them, and also managed some diamond-drilling in the terrain they covered, a strip twenty-five miles in width by some five hundred in length. Enormous masses of low-grade ore, in the form of magnetite rock, were located, one of the most promising regions being that along the New Jersey-Pennsylvania border where Edison bought and leased a large acreage.

Thus he had come to center his operations in Ogdensburg, New Jersey. The place, as it happened, had been named after one of the family of English Puritans who had settled in New Jersey two centuries ago, and of whom Thomas A. Edison himself (through his great-grandmother) was a direct descendant. The rich ores had long ago been taken from the old workings; only the lean magnetite rock was there to be processed.

The inventor had always been fascinated by the problems of mining engineering, to which he now gave long study. Extensive preparations were necessary; the ordering and manufacture of heavy machinery con-

---

* In a paper written by Edison in collaboration with the mining engineer, John Birkinbine, and read before the American Institute of Engineering in 1889, the separating apparatus was described as a "unipolar, non-contact electric separator," having no moving parts, and being therefore quite different from other separating devices of that time. It consisted mainly of a hopper at the top, an array of magnets beneath the hopper, and divided bins placed below, which were arranged to catch the separated concentrates and tailings in different compartments.

sumed much time. Two years passed before Edison was ready to begin operations on the "Ogden Baby," as he called his novel enterprise.

The site, a high ridge in the New Jersey highlands, was a bleak wilderness. Teams of men who had worked at quarrying or construction were brought in, until they numbered about 145. But there were no living quarters here, and it was hard to keep them. Thus Edison was led to expand his plans until they included eventually the construction of a whole village of fifty wooden frame cottages, with running water and electric light. Another difficulty was that the men were used to hard drink and went on sprees of Saturday nights. By setting up a company store that sold good beer at low prices, Edison weaned them away from strong liquor.

Tests of the magnetite rock near the site of the old Ogden mines showed its content as being about 18 to 20 per cent pure iron. Edison saw that everything would depend on the economy of his operations; he would have to organize things so that rock and earth in the thousands of tons could be "treated" each day, in contrast with the few hundred tons handled in rich ore fields elsewhere. Dynamite was very costly; therefore he planned to use it sparingly, and grind up huge chunks of rock by mechanical means instead of blasting. To one of his technical men Edison wrote in 1890, "In view of the extent in which I am going into mining. ... I want to learn the business from actual experience." [5]

The job to be done was essentially one of quarrying, rather than mining. Experienced mining engineers assured him that machines for crushing rock of the size he proposed to handle could never carry such a load. Their mathematical tables of materials and stresses showed that the shock of big boulders moving upon them would tear apart any driving engines connected with the crushing machines—would, in fact, "crush the crushing machine." [6]

That was all Thomas A. Edison needed to hear. He went to work with some of his own laboratory mechanics at West Orange and designed what he called the Giant Rolls, a crushing machine of 100 horsepower, driven by a Corliss engine. It was an adaptation of the familiar Cornish rolls long used in quarrying operations. Edison said, "I shall bring the rollers up to a high speed, and then just before the boulder is dropped in, cut the engine loose." *

He arranged his two Giant Rolls so that they revolved in opposite

---

* Edison's Giant Rolls were made of chilled iron plate, with knobbed faces to expedite the crushing of rock; they used momentum largely instead of steam power and it was claimed showed a friction loss of only 16 per cent.

directions; rock segments lifted by skips were dropped between them; the rolls were speeded up to 700 r.p.m., but just as they were to crash against the rocks, power was disengaged by an ingenious friction clutch mechanism. While the engine was freed from shock, the rocks crashed into each other at high momentum. Under this treatment those big rocks were usually reduced to about the size of a man's head, and dropped through an opening to a conveyor line below the crusher that carried them off to the intermediate rolls; from this stage they were passed on, as pellets, to another crushing device called the "three-high rolls," which reduced them to powder. The whole process was carried on quite rapidly so that a rock of six tons or so would be reduced to fist-sized stone in half a minute.[7]

A complex of conveyors in an endless chain moved the ore in baskets to a dryer, then rushed it off to the "refining mill," that is, the ore separator. This was a towering structure that formed the heart of the whole system. Hoisted to the top of this building by an elevator, the powdered rock fell through several screens in a fine curtain of sand past a line of 480 large magnets, which ranged gradually from weak to full deflecting power. The nonferrous sand fell straight downward to be carried off and sold as useful building material. The iron ore was deflected by the magnets into partitioned bins, and then moved on to a last processing stage.

Edison soon discovered that iron oxide in powdered form could not be smelted in blast furnaces, for without a "binder" much of it would be blown away as dust while being transported and handled. He therefore moved the "pure" ore by conveyors into a big oven chamber, where it was given a final treatment: mixed with a sticky binding material, it was molded under heavy pressure into porous cakes, or briquettes; these were baked, and then shipped off in open railway cars.

Thus the inventor lavished a considerable mechanical art upon his ore-separating project in the hope of bringing about profitable operations at prevailing prices in the East Coast market. He also kept improving his conveyor line, introducing rubber belting, with the help of Thomas A. Robins, who sold it to him. Robins was inspired by Edison to perfect and promote rubber conveyor belting, and thus became one of the key figures in the transition to assembly-line methods in industry. Henry Ford, the great engineer of automobile mass production, on being shown the site and the plans of the Ogdensburg project many years later, declared that it encompassed the most complete automatic conveyor system of material handling that had been designed up to that period, 1892–1897. Ford also recalled that it was his reading of articles on the Edison project of the 1890s (rather than the earlier example of the Chicago packinghouse conveyors) that inspired the idea of his first "beltline" system for the model-T Ford.[8]

At a time when Eastern smelters were closing down their mills for want of economical iron ore, Edison stood ready to provide concentrate taken from billions of tons of magnetite rock located only seventy-five miles, on the average, from the iron mills at Eastern seaports. Thus the iron and steel industry of the Atlantic Coast cities might still be saved from its disadvantageous position in regard to transport costs.

Nothing, however, went according to plan in this operation. There were constant breakages in the crushing machinery. A workman might carelessly drop a sledge hammer into the intermediate crusher, which would prove to be an irreducible object, and all the works would stop until Edison himself came running up to find the cause of trouble. Other laborers would sometimes drop boulders of indigestible size into the Giant Rolls, as if to challenge its magnificent slugging power— "Let's see if it will crush *that!*" they would say. Dan Smith, the expert head rigger, who had charge of the steam shovel and traveling crane, said later that the "Old Man" not only worked, ate, and slept with his men, but "would never send a man anywhere he wouldn't go himself." [9]

One day, the big dryer tower became choked up. Though Edison was warned not to go in, he and Walter S. Mallory, the plant superintendent, crept in at the bottom of the shaft through a manhole, to examine the situation. Suddenly a mass of about sixteen tons of ore within the eighty-foot tower was dislodged and slid down on them. He emerged, with Mallory, none the worse for wear. Telling of the incident later, he remarked that, anyway, the men knew he and Mallory were inside and "had a great time digging us out."

In 1892 a first trial run produced a few carloads of ore, one of which Henry C. Frick tested at the Carnegie mills in Pittsburgh. He reported that the first briquettes were of uncertain quality; what did not sift away along the car tracks blew away up the chimney of his blast furnace.

Edison went to work in earnest to perfect the "mix" of his iron ore briquettes, so that their binding material would be effective, resistant to water, and cheap. According to Mallory he carried on "several thousand experiments" to this end, and designed an entirely new briquetting machinery and plant. The great depression of 1893, however, intervened at this point; prices fell, and few orders came for iron ore of any kind. "The mill is not now running," he reports in the spring of that grim year. "The Ogden Baby is sick." [10]

This interlude of depression and stoppage—during three or four years —was used by Edison to tear down whole buildings and reconstruct the plant with a view to more efficient operation. After the victory of the Republicans over Bryan in the Presidential election of 1896, business conditions improved rapidly; by applying to many East Coast iron mills, Edison finally received some orders, including a large one from the Bethlehem Steel Company. Its manager, John Fritz, said to him at the time, "Mr. Edison, you are doing a good thing for the Eastern furnaces. ...I am willing to help you." [11]

In 1897 and 1898 the village of Edison, now populated by four hundred workmen, once more thundered with activity. In newspapers and magazines the herculean operations at Edison, New Jersey, were now reported to be a complete "success." An admirer wrote him:

After your tremendous feat of the iron discovery and successful production, I see you are going into gold, in which the world knows you will be equally successful. ...You are...alone of your kind—necromancer, alchemist, warlock, wizard.[12]

This correspondent was not alone in believing that Edison's inventive mind was to be applied to the recapture of gold from the tailings of exhausted mines. In any case, it was said that the ore-separating scheme was "an enterprise bound, sooner or later, to revolutionize the world"— by solving the problems of the iron industry. It was widely known that Edison suffered real privations, for he labored for long hours in those bleak Jersey highlands. But another of his admirers wrote to assure him that, like the ancient conqueror Tamerlane, who had once in the vicissitudes of war been forced to hide in a cave, Edison too would emerge one day from his mountain refuge to conquer the world! [13]

Sometimes crowds of visitors came to see the ore-separating works and to stare at the great man. When their numbers became too great, he would hide in some shack and go on with his job. On one occasion his Irish foreman, Dan Smith, came and begged him to come outside before a crowd "and show them what genius looks like," if only for a few minutes, so that they might be persuaded to go away. Edison was just then wearing a workman's cap that he had bought in the local store; his familiar linen duster was torn and stained a good deal, not only with dirt and grease, but with the juice of chewing tobacco he was always spitting through his mask. He was at his untidy best. But he came out of his hiding place and, a good mimic, paced up and down gravely before the gaping crowd, hands thrust deep in pockets, head lowered as if buried in thought—then suddenly, comically, made a dash back into his cabin.[14]

The myth of Edison as a magician was a long time dying. In 1898 the

Hearst newspapers, for example, published a serialized "scientific romance" entitled "Edison Conquers Mars," which was read by millions. It pictured a war of the future between invaders from our neighboring planet and the men of Earth, who are commanded, naturally, by "General" Thomas A. Edison. In the nick of time, he invents a new flying machine which directs the power of a "disintegrator" upon the hideous enemy.[15] Edison made vigorous protests at the use of his name in such a penny dreadful, but to the minds of simple beings everywhere there were no limits to his capabilities.

3   In reality the big ore-separating plant, during its entire life of ten years, was a great white elephant; it was kept in operation only intermittently, for brief intervals of trial runs. For years Edison nursed the project along, carrying out repairs and introducing improvements here and there. It was only at the end of 1897, when he finally received the big order from Bethlehem Steel, that he was able to plan for full, large-scale operation of his plant, and could determine even what its costs would be under conditions of volume production of 1,000 to 1,500 tons of briquettes a day. But by then, when he had brought his costs to a reasonable figure, it was too late!

The grim truth was that Edison was losing thousands of dollars on each day's shipment of his concentrate. The gargantuan Ogden Baby was simply eating up his resources. Though he had greatly improved his crushing and briquetting processes and at last had some chance of winning a market in the East, the continual fall of iron ore prices doomed his venture. Certain technical writers, who did not believe in Mr. Hearst's newspaper fictions, were already beginning to guess the truth and to speak of this affair, by 1898, as "Edison's Folly." The magazine *Iron Age* remarked that expert mining engineers were aghast at Edison's methods of ore treatment. It was indeed a grandiose engineering project for those days and was not to be imitated again until the time of World War II, with its recurrent scarcities in raw materials.

He had hoped to operate profitably when the market price for iron ore at Eastern points was at a level of $6.00 to $7.50 a ton; but he was able to get from Eastern mills only $3.50 to $4.00, after 1893. An experienced mining engineer, recommended by Henry Villard, came up to Ogdensburg in 1898, studied the problem, and advised Edison that his only hope resided in a very large-scale processing of 10,000 tons of earth and rock each day.

This was why, after having hung on in this losing venture for nearly a decade, Edison made elaborate preparations for a whole year, and then, in the autumn of 1898, began to work his plant at full capacity on the order from Bethlehem Steel for 10,000 tons. That autumn and during the winter that followed he remained buried in the wretched hamlet named after him, utterly absorbed in the operational difficulties of carrying a whole mountainside, through an ever-widening cut, down to his separating mill.

In November his wife, Mina, visiting her parents in Ohio, wrote him a letter bearing news from the outside world: a beloved sister of hers had just died. Edison in reply addressed her as "Darling Wife," also "Billie," earnestly commiserating with her upon her loss. "The thing we call Nature seems very cruel at times," he observed. In fact it was being extremely unkind to him just now, as he explained:

This is the first chance I have had to answer, I scarcely get any sleep as everything has to be attended to by me. We have a great Blizzard . . . as you have doubtless learned from the newspapers. The snow has drifted in great piles at the mills, everything has been blocked here as well as frozen up, and so a great number of my men could not reach the Mills from their homes. . . . For 4 days we have not run the mills. Today I got the crushing plant started, tomorrow the other mills will start . . . 400 men are here and all depends on myself. To attend the funeral in Ohio it would be necessary to shut down at a most critical time technically and financially. Had I an intelligent assistant I could have come. . . . I expect that until the whole thing is systematized and the men well trained, it will be impossible to find a man.[16]

He became so absorbed in the crushing of rocks that he never got home for the holiday season. His ever-thoughtful wife, however, sent him a large cake shaped somewhat in the form of Edison's ore mill.

"What, is it Christmas, already?" exclaimed the distracted inventor. For him it meant only that work would have to stop for a few days while his men were given time off—and there were so many difficulties still to be overcome. He was making his first big production run at last in the heavy winter of 1899; everything depended on its outcome.

The faithful Charles Batchelor, who had also retired recently from General Electric, was one of the few of the old guard who worked with him on this enterprise; he constantly inspected the works and reported to Edison what the men were saying and doing. One day he brought him a newspaper clipping about recent developments in the Mesabi Range in Minnesota. In that region, where the most fabulous high-grade iron deposits in the world had been discovered some years before, John D. Rockefeller and W. H. Oliver, of Pittsburgh, had recently bought the bulk of the acreage, and arranged for large-scale extraction of the ore at open-face mines, and for its economical transport in special ore trains and then in huge ore vessels plying the Great Lakes. Anticipating a torrential flow of Bessemer iron, the market price had lately fallen to $2.65 a ton f.o.b. Cleveland.

On reading this dispatch, Edison burst out laughing. He had sunk about two million dollars in his ore-separating venture, and his company was in debt to the tune of several hundred thousands more. "Well, we might as well blow the whistle and close up shop," he exclaimed.

Not only had the price gone down too much for him; but it seems that the magnetite rock at the old Ogdensburg mines, which at one end (where he began operations) showed 20 per cent iron content, had, by 1898, run down to about 12 or 10 per cent. This meant that he must crush twice as much rock in order to extract the same amount of ore, now worth only half of what he had hoped to get for it.*

The bountiful Mesabi Range deposits, enough for most of America's needs over the next sixty years, gave a tremendous geographical advantage to the Midwestern steel mills over the older ones in the East. It meant that Edison's New Jersey and Pennsylvania Concentrating Works was *kaput*.

He did not, however, halt the disastrous operations of that steam-belching, rock-devouring Frankenstein, but, during several weeks completed his last big run, filling the order contracted for—at great loss. He was still a man of science, and had to learn for himself what the cost accounting for his iron ore would be at full capacity. Those monster machines went pounding and screaming on for several weeks, while he watched knowing that it was all in vain. Some day plants like his for the reduction of low-grade ore might be needed (as they are now). At last he stopped to take his bearings.

In those last months there were all sorts of evil omens. A whole stock-house, containing four workmen, suddenly collapsed and sank into the undermined ground, killing its occupants. A skilled mechanic, repairing one of the crushing machines, was caught in its iron grip and was mangled to death. This accident, it was said, "greatly unnerved Mr. Edison, who discontinued the work a short time later...." [17]

He had given fully five years of his life over the last decade to this heartbreaking struggle, while virtually all his other interests were neglected. At the beginning and end of that period two more sons had been born to him: Charles, in 1890, a handsome infant who resembled his father; and Theodore, the child of his late years, who came into the world

---

* According to a geologist's estimate of these operations at Ogdensburg, made for New Jersey authorities, 75,206 tons of magnetite rock were quarried in the early part of 1899 to produce 10,000 tons of briquettes of low-grade ore. (William S. Bayley, Geological Survey of New Jersey, Trenton, 1910.)

in 1898. Edison's own father, Samuel Edison, had died in 1896 at the ripe old age of ninety-two. Meanwhile, Tom junior, the eldest son, pursued an erratic career; he had made an unfortunate marriage, and had got into financial straits from which he had to be rescued with the aid of papa's checkbook. Finally, the large, athletic Will, during the Spanish-American War, at twenty years of age, had enlisted for service and come down with yellow fever. It all sounded like the lot of joy and sorrow befalling the average American family. But during all those years the baby that gave Edison the greatest concern was the "Ogden Baby."

It was an unmitigated disaster, the worst he had ever faced; he had lost not only those years of his life but the entire fortune accumulated by his inventive labors. Yesterday a millionaire several times over, today he was "busted." And yet he told T. Comerford Martin, at about the time he stopped ore-milling: "I never felt better in my life than during the five years I worked here. Hard work, nothing to divert my thoughts, clear air, simple food made life very pleasant." [18] He still had his wits and his laboratory, at any rate.

One day in 1899 he rode back on the train to Ogdensburg with Walter S. Mallory, for the last time, to order the dismantling of the mills and the disposal of the machinery. Where another might have been in a state of despair, he discussed his situation and his prospects in the most jovial spirit, as if he felt no pain. Though now fifty-three he remarked gaily enough that he could always find a job as a telegrapher at $75 a month. Then, in more serious vein, he told Mallory that he had already arranged to put all that costly earth-moving and crushing equipment, and the technical knowledge he had gained, to good use: the establishment of a modern portland cement works according to his own design. He hoped in this way to settle the debt of about $300,000 the ore-milling concern was left with, adding: "No company I was ever connected with has ever failed to pay off its creditors." This was one case where he had permitted no one else to share his doubtful rewards. It was like him to waste not a moment in vain regrets, to forget the defeat, and with coldly rational courage apply himself at once to new undertakings.

About a year or two after they had dismantled the Ogdensburg plant, Mallory happened to call his attention to newspaper articles of the current boom in Wall Street, during which General Electric stock had risen to $330 a share. Edison asked, "If I hadn't sold any of mine what would it be worth today?" After some calculation Mallory replied, "About four and a quarter million dollars." Edison was silent for a moment, plucking at his right eyebrow, as was his habit when lost in thought, then looked up, smiled broadly, and said, "Well, it's all gone, but we had a hell of a good time spending it!"

chapter XIX

Motion pictures

1
At certain periods, after having been much away from his laboratory, Edison would say, "I am tired of *industrial science*,"
and would promise to "rest himself" by taking up "pure
science."

Nature, he said, was often full of surprises, and it was fun to come
upon them unexpectedly. In his laboratory notebooks there were observations of hundreds of curious, unexplained phenomena which would be
most fascinating to investigate—especially things chemical and metal-
lurgical.[1]

Though industrial science, that is, applied science, was really as the
breath of his life, it was highly characteristic of our many-sided and
restless inventor that he should allow himself sudden flights toward the
unknown, the fantastic, or the "impossible."

In 1890, a young writer named George Parsons Lathrop (son-in-law of
Nathaniel Hawthorne), in preparing a biographical study of Edison for
Harper's Magazine, had been permitted to see some of the inventor's
notebooks and in his article had given hints of Edison's ideas about future
scientific discoveries. This was the heyday of Jules Verne's scientific
fictions. The literature of the age romanticized its inventors; and the inventors, in their turn, were inspired by the spirit of romance. Edison, for
example, had read Jules Verne with enjoyment since the early 1880s. What
was an invention but a romantic vision, up to the point where it was made
practical?

Lathrop, at any rate, proposed to Edison that they collaborate in writing a novel having as its hero a scientist not unlike himself. Edison was to
furnish the ideas for the inventions of future times, Lathrop the fictional
intrigue, or plot. The title chosen for the work was "Progress"; and a
publisher offered to promote it in newspaper-serial and book form. In the
380

summer of 1890, Edison began to make notes and jottings on the imaginary scientific appliances of 1940–1950.

The manuscript notes for "Progress" consist of reveries or visions of things to come that fall within lines of direction already projected by contemporary scientific research. Edison prophesies a great advance in synthetic chemistry, biochemistry, and optics; artificial diamonds would become commodities of common use; the components of food, such as proteins, would be made by chemical processes; marvelous new plastic materials such as "artificial mother-of-pearl" would be contrived by chemical engineers and used to decorate the walls of palatial dwellings; all sorts of objects would be "plated" in some new and beautiful manner. (Could he have been dreaming of chromium plate?)

The future mode of transportation would be aerial navigation; and heavier-than-air vehicles would be developed to the point where journeys to Mars were entirely feasible. The spirit of "Progress" would also be introduced into warfare, which, in future times "would be conducted by dropping dynamite from balloons."

Would man himself be improved? Edison, as a convinced Darwinian, anticipated a large advance in knowledge of eugenics, through experiments in the breeding and interbreeding of higher anthropoid apes. His hero-scientist founds a community on the upper Amazon in which, after eleven generations, the apes are made capable of conversing in English. "The 80th generation would equal in intelligence and personal beauty the Bushman tribe of Africa. (Lathrop you can enlarge on this yourself.)" [2]

Alas, these literary exercises were brusquely interrupted, and Edison was off to the Ogdensburg works to crush rocks again. To his despair, Lathrop was unable to persuade Edison to come down from his hills and continue with their scheme.

About five years later, in November, 1895, came the remarkable scientific discovery by Professor W. K. Roentgen, the Netherlands physicist, of the so-called X-rays emanating from a cathode tube; that their radiations could pass through the flesh and muscle tissue (though not the bones) of living animals and so make a shadow picture of them on a photographic plate aroused tremendous interest among the scientific public. The value of such an instrument for surgery was grasped at once; soon many investigators were at work devising fluorescent lamps and fluoroscopes, Edison one of the first among them.

Professor Michael Pupin, of Columbia University, engaged his interest by appealing to him at the time for help in discovering the most effective fluoroscopic chemicals. During a lull of activities at the ore mill, Edison went back to his laboratory and with his assistants tested crystals made out of about eight thousand different chemical combinations. Within a few weeks he sent the Columbia University scientist a fluoroscope with a tungstate-of-calcium screen—with which Pupin was enabled to make a clear shadowgraph of a man's hand that was filled with shotgun pellets. Guided by the Edison fluoroscope a surgeon promptly performed the first X-ray operation in America, with complete success.

Pupin wrote Edison in acknowledgment:

Messrs Alysworth & Jackson sent me one of your fluoroscopes with your and their compliments. It is a beautiful instrument and it is of invaluable service in the new line of work opened up by Roentgen's discovery. I took the liberty of showing its marvelous power at my public lectures; and the public enjoyed it more than anything else that I had to show; this is partly due to your great popularity.... My experiments gave very encouraging results even with the poor screens that I have been able to make myself. With your very excellent screens the results will undoubtedly be very much better. The field of utility is enormous. ... Your success will be greeted with delight by all scientific men.[3]

Edison replied that he was working up a variety of X-ray tubes and was donating a number of them to hospitals where surgeons had quickly taken up the new X-ray techniques; he also sent them to various scientists, whose investigations, he noted, were "out of my line."

Soon the newspapers, with an excitement rivaling that caused by the incandescent lamp of 1879, were reporting Edison's progress in this new work day by day. He was going to photograph the human brain, it was said, and ever so many other formerly inscrutable or impenetrable objects. What could he not do with the miraculous new instrument? Among the hundreds of inquiring letters that came to him at this time was one urging him "to make an X-ray apparatus for playing against Faro bank." The correspondent promised faithfully that he would share his profits with the inventor and besought him to keep the whole matter secret.[4]

While popular excitement over the X-ray ran high, Edison made a number of fluoroscopes, in the form of a box with a peephole at one end, exhibiting them at the Electrical Exposition held in 1896 at the Grand Central Palace in New York. Here thousands of persons came to peer through Edison's fluorescent box and examine the shadows of their hands or other limbs, held between the vacuum tube and the screen. Some of them wanted particularly to have their brains examined, in order to learn, as one said, "if there is much in it." It was at any rate the first public display of X-ray action in America, or perhaps in the world, and provided Edison with much distraction from the troubles of "industrial science."

Before Edison could devise any commercial usage for the new X-ray machines, it was noticed at his laboratory that the primary or original

X-rays were probably highly injurious. The fact that secondary and tertiary rays in a room where a powerful tube was being operated could give cumulative burns was not perceived. This was soon brought home to Edison when an assistant, Clarence Dally, was poisonously affected, so that his flesh became ulcerated, and his hair fell out. Several years later, after numerous amputations, Dally died. Edison had been experimenting with X-ray fluorescent lamps for lighting, but when Dally's unhappy condition became apparent, Edison concluded that such a device "would not be a popular light" and so dropped it.

In the early stages of this work, he was by no means the only one to proceed without caution; many accidents occurred in hospitals and laboratories. Edison himself suffered from severe eye trouble which, at the time, was attributed by doctors to exposure to X-rays—though he recovered completely. In the 1890s little was known about radioactivity; X-ray machines were not made really safe until some fifteen years later, when operators began to use protective lead screens and "selectors," and W. D. Coolidge's hot cathode tube was incorporated in X-ray machines.

Other scientific "distractions" during the long years of rock-crushing were provided by intermittent development work on the phonograph. There was a great boom toward 1896 in the demand for low-priced phonographs and for cylinder records of music. Despite the competition of the Columbia machine and the new Berliner "gramophone," Edison's export sales in England, France, and Germany expanded rapidly and helped in part to repair his broken fortunes. After 1898 he was free of court restrictions and was again able to sell the Edison phonograph, equipped with an economical clockwork spring, in the United States. Work on the acoustical improvements for a large "Concert Phonograph" was also carried on by his assistants, under the inventor's periodic supervision.

But surpassing all these activities in interest for Edison during the middle period of his life was his secret "toy," the motion picture camera. This whole subject was both intriguing and difficult; he could see no practical commercial future for it and yet could not bring himself to drop it. If the balky and jerky kinetic camera could ever be brought to perfection, it might prove as "miraculous" as his phonograph and perhaps even save him from the ruin that overhung his affairs for years after 1893.

Our last glimpse of the motion picture camera was in the locked room at Orange, when Edison had returned in October, 1889, from his triumphs at the Paris Exposition, to be shown by his assistants a "kineto-phonograph" that actually functioned. In giving the story of a man who lived so many lives at the same time, we must use the film flash-back device (born of Edison's own invention) and return to the secret labors of an earlier decade.

2   For a century or more men had used a variety of mechanical tricks with pictures in order to give the illusion of objects or living creatures in movement. Edison himself wrote that the germ of his idea came to him through acquaintance with the zoetrope, or wheel-of-life device, invented in the 1830s, which used the phenomenon of persistence of vision to create, for a few brief moments, the illusion of pictures in motion. In this scientific toy a series of drawings of objects in successive stages of movement were placed on the inside surface of a slotted cylinder or wheel, which, being turned on its axis, appeared to show those objects in movement to an observer looking through the slots.

By the 1870s the same effect was obtained even more impressively through use of photographs instead of handmade drawings. The most notable of these photographic experiments in demonstrating the persistence of vision were by E. J. Marey in France, and by Eadweard Muybridge in America. By an arrangement of a series of cameras along a race track, Muybridge took instantaneous photographs of horses in full gallop, the shutter of his cameras being actuated by wires or strings contacted by the horses. Using sensitive wet plates, already available, Muybridge was enabled to present a short cycle of movement in a dozen or so pictures. These he projected through a magic lantern, by means of a revolving device much like the wheel-of-life.

In the course of a tour of the country, giving illustrated lectures, Muybridge visited Edison in 1886 and showed him his pictures of horses, dogs, birds, and other animals in motion. Marey's experiments in the analysis of movement, which were done with a single camera at the rate of twelve pictures to a second, by having a series of dry plates revolving on a wheel, seem to represent an even more advanced idea of pictures of motion. Both men's work constituted a significant, but limited approach to the subject; in both cases, the effect of a rapid succession of pictures of a single cycle of motion, lasting a second or so, was that of a central object, say a moving horse, placed on a treadmill against a background that was racing past the viewer.

Edison's mind, it has been said, was "a great reservoir of miscellaneous facts." He informed himself upon all the recent advances in the photographic art. He had recorded the movement of sound. Could not the same thing be done in some way with the motion of objects or living creatures, for the sight? As long ago as 1864, a French amateur of science, L. A.

Ducos, had actually worked out on paper the idea of a chain or film of photographs unrolling and, by the persistence of vision, giving the appearance of recorded movement. But he had never gone beyond the stage of a hypothesis. The pioneering efforts of Muybridge and Marey were highly significant, for the eye "saw" motion for a fraction of a second; they left Edison, however, unsatisfied. Yet their technical discoveries, their demonstrations of a scientific fact (persistence of vision), set the stage, so to speak, for Edison. He was aroused enough by what had already been attempted to go to work in their field and try to solve their problems himself.

In the autumn of 1887 the new laboratory at West Orange was completed. He was then extremely preoccupied with his improved wax-cylinder phonograph; but his mind was filled with all sorts of additional schemes. He had then revealed his idea of trying to make a camera that would record motion effectively to one of his assistants, the young Englishman named W. K. L. Dickson, who some years before had come all the way from London to seek work under Edison. It happened that Dickson was an impassioned amateur of photography, and often made photographic studies in connection with various experiments. As he remembered it, Edison one day engaged him in conversation in the yard outside the laboratory, disclosing a scheme for making a machine like a phonograph that would make pictures of objects in motion to the accompaniment of sound and voices.[5]

Shortly afterward, in December, 1887, Dickson was sent to the Scovill Manufacturing Company's headquarters in New York to purchase special photographic equipment. In Number 5, an upper room of the new laboratory building, space was set aside for experiments with new photographic apparatus that Edison was to contrive. At the beginning of 1888, while his chief was putting the finishing touches on the improved wax-record phonograph, Dickson set to work making many tiny microphotographs, on dry plates, of persons and objects in motion. This series of photographs, as small as $\frac{1}{16}$ inch square, or even smaller, when taken all together covered a short cycle of movement. But Edison had set himself the further problem of using the instantaneous camera to record movements in a continuous stream, and showing or projecting them with the effect of persistence of vision.

It is evident that he began with a fairly visual idea of the kind of mechanism he would use. This is clearly shown by the first document giving evidence of these early experiments—his caveat Number 110, written October 8, 1888, and filed with the Patent Office—which first describes such an apparatus:

I am experimenting upon an instrument which does for the eye what the phonograph does for the ear, which is the recording and reproduction of things in motion, and in such a form as to be both cheap, practical and convenient. This apparatus I call a Kinetoscope, "moving view." ... The invention consists in photographing continuously a series of pictures occurring at intervals ... and photographing these series of pictures in a continuous spiral on a cylinder or plate in the same manner as sound is recorded on a phonograph.

He goes on to explain that when the picture was taken, the cylinder would be held at rest; then it would be rotated for a single step, halted again, and another exposure made. What he describes includes a photographically sensitized plaster cylinder geared to a mechanical movement that automatically rotates it and operates a camera shutter. A feed screw simultaneously shifts the cylinder lengthwise. The result is a spiral of pinhead-size pictures made directly on the coated drum. He remarks significantly enough, in this caveat, *"A continuous strip could be used, but there are many mechanical difficulties in the way."*

Thus, after the first ten months of experimentation, during which he had studied the nature of a camera and viewing apparatus arranged both to take and view little pictures in motion, he had a clear, though general, idea of what he wanted, if not of how to get it. He had learned that while his phonograph must be run continuously, a camera record of pictures of motion would have to be run intermittently, with a stop-and-go action, at some speed still to be determined, so that persistence of vision would come into play for the viewer.

In this first search, carried on behind locked doors, Edison followed his own line of march, ignoring the findings of other men who had studied such phenomena. But at this stage, and for a long time thereafter, the idea of his phonographic mechanism controlled his thinking.

He had first fixed on about forty pictures a second as the rate necessary to get a perfect record of motion. Later that number would be reduced, as experiment showed was advisable. His cylinder machines gave him trouble, because the microscope used in viewing could not be focused evenly on the curved pictures. Also, the first photographs were too small and had to be increased in size from $\frac{1}{16}$ to $\frac{1}{4}$ inch to show good detail. It was tedious, even frustrating work.

The first cylinder apparatus made some sort of moving pictures. A fragment of one of the early models, still in existence at the Edison Laboratory, has a sensitized sheet of celluloid containing many diminutive pictures attached to a metal cylinder. The pictures show John Ott wrapped in a white sheet, waving his arms, and "making a monkey of myself," as Ott said. The pictures were meant to be viewed directly, through a magnifying lens, and raised pins encircling one end of the cylinder apparently closed the circuit of an induction coil to illuminate the pictures intermittently with an electric spark.[6]

The pictures were exceedingly crude, especially when magnified. But Edison was hopeful now and reported to his patent lawyer that he was "getting results." [7]

It was one of his special characteristics that he had, as has been said, an instinctive knowledge and discernment of materials. From the start he disliked the rigid and fragile glass plates then used in still photography, and experimented with new and improved photographic materials. When, early in 1889, he obtained from John Carbutt of New York, the pioneer of dry-plate making in America, some heavy sheets of celluloid coated with a photographic emulsion, his work took a long step forward. This tough, pliant sheet, about fifteen inches long, he could wrap around his big cylinder; he could also take the quarter-inch pictures in far greater number. At this stage he had a record of living movement running as long as five seconds.

In the late spring of 1889, Edison and Dickson, on the basis of their findings thus far, came to the decision that they must abandon the cylinder and contrive some radically different apparatus to provide the necessary stop-and-go movement for the camera shutter, while at the same time feeding a strip or tape of celluloid across the focal plane of the camera. Therefore, they cut up the Carbutt celluloid sheets into narrow strips which were cemented together end to end. The pictures of "slices of motion" could now be made still larger, with ¾-inch-wide film and ½-inch frame, and could be run to a greater number of exposures, covering movements more extended in time. The coming of transparent celluloid film was, therefore, decisive in Edison's search.

Further improvement in celluloid film was then made by George Eastman of Rochester, New York, the inventor of the rapid-action "Kodak." His film was still tougher than Carbutt's, yet light and flexible, so that it could be handled in rolls. On hearing of the new Eastman film, Edison promptly dispatched Dickson to New York to get a sample. Under Edison's prodding, Eastman later produced specimens of celluloid, for his special use, in fair-sized lengths of fifty feet, instead of short pieces. When Edison was shown these long strips, according to Dickson, his smile was "seraphic." He exclaimed, "That's it—we've got it—now work like hell!"

In the summer of 1889 he designed a camera mechanism for advancing, that is, feeding the roll of film forward at a given rate of speed. The strip of film was drawn through rollers sideways across the focal plane of the camera and rewound automatically; perforations on one side of the film were engaged by a sprocket wheel attached to a main shaft that was revolved by hand or motor. An escapement or Geneva mechanism propelled the sprocket and moved the film intermittently, one step at a time, now forward, now resting; when the film stopped, a revolving shutter, also geared to the main shaft, rotated for one step, so that the exposure was made in the proper time relation. With this mechanism Edison obtained a series of photographs which, moving at a given speed, gave a pretty good illusion of motion.

The first machines of the early summer of 1889 were balky; the teeth of the sprocket wheel tore the strip film at the perforations, and the first Eastman film was coarse-grained. Edison was obliged to leave for Paris early in August, it will be recalled, but gave his assistants explicit instructions about further improvements to be sought. During Edison's absence, they managed to complete pictures that lasted all of twelve seconds. When Edison returned from Europe on October 6, Dickson was bursting with pride as he showed him what had been done thus far.

According to Dickson's statements, Edison himself overcame the many mechanical difficulties he had foreseen in the way of the development of the strip-film kinetograph at the time of his first (1888) caveat. He drew on the same mechanical ingenuity he had shown in his youth, when he first came to New York and devised his clever automatic-telegraph instruments and stock printers. After his motion picture invention became known to the public, everyone exclaimed over the *simplicity* of his apparatus—the strip-film kinetograph which solved both basic problems of the motion picture art: taking pictures of motion and exhibiting them. The same machine was used at first for viewing the pictures directly through a magnifying lens.

Edison's first intention was to make sound pictures; he would synchronize the movement of the reel of film with a phonograph driven by the same motor powering the camera. There was to be singing or music accompanying a dance, or the sound of a voice as the film unrolled. "The establishment of harmonious relations between kinetoscope and phonograph," Dickson wrote later, "was a harrowing experience, and would have broken the spirit of inventors less inured to hardship and discouragement than Edison." [8]

With the thought of making the apparatus "commercial," he also devised a peep-show mechanism for exhibiting positive prints made from his kinetograph negatives, which he named the "Kinetoscope." This consisted of a cabinet of substantial size, containing batteries and a motor that turned the strip film on a spool bank and operated a light. In front of the cabinet was an eyepiece and lens; by looking through this the viewer saw the film as it moved along through a rotating screen with an aperture permitting him to glimpse only one picture at a time. Persistence of vision did the rest, giving the illusion of more or less lifelike motion.[9]

In the year that followed, Edison had his assistants build a better and larger motion picture camera for taking pictures one inch wide and ¾ inch high, on film 1⅜ inch in width, to allow space for perforations. This ponderous machine, which was in use between 1890 and 1894, is the true father of all modern motion picture cameras. The width of its film, 35 mm., is still standard today.

At this period of his life, in middle age, Edison worked much more slowly at his inventive tasks than in the past. Several different experimental projects were generally going forward at the same time, a team of men under his direction being assigned to each of them. In other words, he had the weight of a much larger research organization upon him at Orange than in the Menlo Park days. Four men, for example, worked on the motion picture job in the early stages, while it was still more or less a secret. Edison was just then leaving the electrical manufacturing industry, was experiencing much difficulty with the phonograph business, and was already deeply involved in his nerve-wracking ore-separating work. Thus he did not even file for a patent on his basic 1889 kinetograph and kinetoscope until the end of July, 1891 (Nos. 493,426, and 589,168); and those patent applications were, as it happened, incomplete and faulty. Two of his photographic assistants, not long afterward, left his employ to work for others who entered into competition with Edison in the new industry created by his invention. These competitors fought him in the courts to contest his patent claims, in the course of which litigation it was alleged that one of his assistants, Dickson, performed a large part of the inventive work on the motion picture, and that, anyway, there was "nothing new" in Edison's invention that had not been tried previously by other men. In short, his work on the motion picture camera was more disputed and disparaged than almost anything he ever did.

Marey and Muybridge were indeed his scientific precursors, whose work had the element of discovery of a mechanical principle. Edison himself, in 1894, made handsome acknowledgment to them as his forerunners. His own contribution to the art, nevertheless, was highly strategic and creative. Once again we find him entering a field of investigation in which many persons had preceded him and where the general scientific principles were already understood. But with his coming things begin to move. The art and industry of the motion picture at last is born.

The distinction between his work and that of his forerunners lay in his original use of a *single point of view* from the one, fixed camera's eye, so that in his pictures a man could be seen walking from one end of a room to the other, while its walls and the whole background image remained still—instead of rolling past our sight as in the sliding stage-sets of the old theater. Edison's great trick was the adaptation of the instantaneous camera and the flexible Eastman celluloid film, so that strip film (instead of plates) could be fed across the focal plane of the camera. There was indeed "nothing new" in the several materials and mechanisms he utilized.* But once again, we have an instance of what many have regarded as Edison's preeminent gift—"the ability so to adapt or combine ideas or materials already existing as to effect results at once distinctively new and thoroughly practical." [10]

The cumbersome Edison motion picture camera of 1889–1890 was the first machine that photographed objects in motion effectively. We need only view one of these early models of the strip-film kinetograph, in the museum of the Edison Laboratory National Monument at West Orange, and compare it with its predecessors, to realize this truth. (The English, to be sure, give "priority" to one L. A. A. Le Prince, who in 1886 patented a tape-reeling camera of multiple lenses, which, however, is known to have been completely inoperable.) A further significant contribution was the little black box he called the Kinetoscope, providing the world with a first exhibiting mechanism that used positive film in rapid movement, seen through an eyehole. It offered the first motion picture *show* in history. As Terry Ramsaye, historian of the motion picture, has said, the Edison peep-show is the "inescapable link" between past and present in this field. "There is not and has never been any motion picture film machine . . . that is not descended by traceable steps from the Kinetoscope." [11]

3   For a period of almost two years a good deal of mystery surrounded the new apparatus and the first small wooden frame photographic building (erected at a cost of only $516) in which it was locked up. After his betrayal by Gilliland, Edison, the once openhearted country boy, became fairly secretive. Only a few trusted and favored friends, such as his patent lawyer Richard N. Dyer, were permitted to witness private exhibitions of the new piece of "wizardry."

* Dr. Henry Morton, testifying as a scientific consultant in the suit of Edison against Mutascope Company (1900), was asked if it were not true that the inventor had merely taken advantage of new photographic materials becoming available to him. Mr. Morton's reply was: "No. When Edison began his experimenting there was no material available; it was his demands that helped create such material."

In one of the first authoritative articles written on the motion picture machine after news of the invention became public, Dickson relates:

On exhibition evenings the projecting-room, which is situated in the upper story of the photographic department, is hung with black, in order to prevent any reflection from the circle of light emanating from the screen at the other end, the projector being placed behind a curtain, also of black, and provided with a single peep-hole for the accommodation of the lens. The effect of these somber draperies, and the weird accompanying monotone of the electric motor attached to the projector, are horribly impressive, and one's sense of the supernatural is heightened when a figure suddenly springs into his path, acting and talking then mysteriously vanishing.[12]

There was a good deal of whispered gossip, nevertheless; and in June, 1891, the first vague newspaper accounts of the apparatus were published.[13] Edison thought it wise to file for a patent on an "Apparatus for Exhibiting Photographs of Moving Objects" (No. 493,426) on August 24. In his first caveat on the primitive kinetograph of 1888 he had referred to the possibility of using a projecting apparatus and a white screen; but in his patent application of 1891 that idea was dropped. Those early photographic exposures usually glinted and jerked about a good deal; when magnified and projected on a screen the effect was even poorer visually. It seemed to Edison that pictures of motion were under much better control in his little peep-show box, which might be used, like the phonograph, as a coin machine for popular entertainment.

The omission of the projector-and-screen device would prove later to be unlucky for him. He committed another and even more costly error by failing to take out foreign patent applications for his motion picture camera in England and Europe. As a matter of routine his lawyer advised him to file for such foreign patents. But when told that it would cost $150, he is reported to have rejected the proposal, saying casually, "It isn't worth it." * [14]

These two legal errors opened the door wide to competition by many rival inventors, borrowers, infringers, and plain pirates. Once a model or even a drawing of Edison's motion picture camera was available in Europe, anyone could imitate it and have it patented in his own name and offered in the United States market as well as that of Europe. On the other hand, while Edison was then greatly harassed by other prob-

---

* Some thirty years later, Edison told Terry Ramsaye, who, in 1924, was compiling his two-volume history of the motion picture: "The reason of this is because the more patents I took abroad the more I lost." (Manuscript notes of Edison attached to Ramsaye's proofs, in the Baker Library Collection, Harvard University.)

lems and could not give his undivided attention to the motion picture project, his imitators, as it happened, were able to introduce improvements upon his ideas and eventually to advance the motion picture art beyond the point where he had stopped. It is a curious aspect of the patent law problem that imitation and even infringement often breed technical progress.

As late as 1893, in a friendly note to the aged Muybridge, Edison wrote:

I have constructed a little instrument which I call a Kinetoscope, with a nickel and slot attachment. Some 25 have been made, but am very doubtful if there is any commercial feature in it, and fear that they will not even earn their cost. These Zoetropic devices are of too sentimental a character to get the public to invest in.[15]

He testified later that for several years his efforts to interest capitalists in promoting this machine proved vain.

By 1893, however, some enterprising men had begun to see promise in the queer little peep-show box. Thomas Lombard, who promoted the sale of coin machine phonographs for Edison's company, happened to be admitted to one of the private exhibitions at West Orange, and strongly urged the inventor to let him present the kinetoscope to the public at the approaching Chicago Exposition. There were some amusing "shorts" on hand, showing John Ott, his walrus mustache waving about, doing an Arabian "skirt dance," and of the same pioneer motion picture star going through all the phases of a prolonged sneeze—accompanied by some poorly timed sound effects on the phonograph. But Edison was unable to get this coin machine apparatus ready for the Chicago fair.

After having neglected the motion picture invention—in favor of low-grade iron ore and the phonograph—Edison's interest in the affair finally began to wax. On February 1, 1893, a strange new building, dedicated to the new art, soon reared its ugly form within the Edison Laboratory "compound." It was something the like of which had never before been seen and would probably never again be seen either in this world or the next—a wooden structure of irregular oblong shape, with a sharply sloping roof hinged at one edge so that half of it could be raised to admit sunlight. The whole building, fifty feet in length, was mounted on a pivot, like a revolving bridge, and could be swung around slowly to follow the changing position of the sun and admit its full glare to the interior. The walls of this jerry-built affair (total cost under seven hundred dollars) were covered with black tar paper; the stage at one end of the single large room was also draped in black, so that the whole décor was funereal in the extreme. Such was the first building constructed especially as a *motion picture studio,* officially called the Kinetographic Theater, but affectionally referred to by its staff as the "Black Maria."

In this somber edifice a spate of pygmy motion pictures was "ground out" in 1893 and 1894. The performing artists were drawn, as Dickson relates, from every walk of life and "from many a phase of animal existence" as well. They strutted for a brief moment of glory—the shorts now ran about a minute and a fraction—in the full glare of that sunlit stage, against a background so dark that the pictures were quite sharply defined. Among the early performers was "Gentleman Jim" Corbett, pugilistic idol of America, who provided a realistic exhibition of his art. As Corbett related, many years afterward:

> The Black Maria certainly did look like an old-fashioned police patrol wagon. We hadn't been inside very long before most of us would have preferred a police patrol at that—for the little moveable studio was the hottest, most cramped place I have ever known.
> The camera in use could take only an average run of 1 minute 20 seconds. The pictures . . . were only about six inches high and you looked at them through binoculars. But you could distinguish all the details of the action.[16]

A third-class pugilist from Newark had originally been hired to serve as Corbett's sparring partner, though without being informed of the name of his opponent. But when he reached West Orange, the lugubrious stage setting as well as the prospect of contending with Gentleman Jim so affected his emotions that he ran all the way back to Newark. However, a replacement was found for him at Edison's usual low rates, about ten dollars for ordinary performers and fifty dollars for celebrities, and the historic film was completed. Other performers were the famous strong man Sandow; Japanese dancers; French ballet girls; "Buffalo Bill," with accompanying Indians, recording a first "Western"; also acrobats, knife throwers, and divers fowl recruited for cockfights.

In one of his frivolous moments Edison went off to Daly's Theatre in New York and persuaded its celebrated Gaiety Girls and their leading lady, the beautiful Mae Lucas, to come to his studio and dance before his kinetograph. After such efforts it was an easy step to film short comedy skits from the vaudeville stage, as well as picturesque renditions of an organ-grinder performing with his monkey, or of a patient in a dental parlor reacting to treatment with laughing gas.

The camera's eye was also pointed outside the windows of the Edison Laboratory to take realistic views of everyday life along Valley Road, West Orange. These might be called the first documentaries. One sequence is a truly lifelike study (dating from 1897) of Edison himself working in his laboratory, while one assistant after another seems to come racing in to give his report and receive lightninglike orders. The inventor cups his ears, barks out instructions, and hurries on with his work until the next man comes running in with other troubles for him. (The pictures, having been taken at 16 exposures per second, and being nowadays shown at 24 a second, are speeded up.) Then one day the quiet of West Orange is shattered by the thunder of battle, as uniformed horsemen, simulating the British and the Boers in Africa, go charging over the empty lots outside the Laboratory—to the vast amusement of the eternal American boy in Thomas A. Edison.

In the early stages of the art, however, it was the exhibition of prize-fighting contests that was to create a first popular vogue for Edison's pictures in motion and bring him sudden fortune again. Early in 1894, Thomas Lombard, who had been so enthusiastic about the peep-show, turned up with a young capitalist named Norman C. Raff, recently arrived in the East from gold-prospecting exploits in the Rockies. Edison's former Wall Street friends, such as Villard, would not go near his motion picture invention; Edison himself at the outset seemed a bit ashamed of its apparently delinquent character. But Raff and Frank Gammon, an associate whom he soon introduced, were another breed—speculators, backers of race track entries or threatrical shows. They saw the skits and prize fights, and in 1894, after organizing the Kinetoscope Company, entered into a contract with Edison to purchase a large number of his peep-show boxes for two hundred dollars each and exhibit them in "Kinetoscope Parlors," furnished with film and coin machine boxes.

The first of these parlors was opened on lower Broadway in New York, on April 14, 1894, and was billed as "The Wizard's Latest Invention." There had been a few illustrated lectures previously; but this was the first public showing and created something like a riot. Crowds on Broadway waited all day and far into the night in long queues to see the pictures that lived and moved for ninety seconds through the eyeholes of five kinetoscopes. Soon there were similar parlors opened in Chicago, Baltimore, Atlantic City, San Francisco, and other centers, which were also besieged by crowds. The exploitation of the motion picture was thus begun as a form of mass entertainment, following the commercial pattern of the earlier phonograph parlors.

On one day in the summer of 1894 a gory ten-round boxing match was staged in the Black Maria between two professional gentlemen named Mike Leonard and Jack Cushing. It ran to a thousand feet of film and ended in a real-life knockout of Mr. Cushing. When this picture was shown, in six reels and through six different kinetoscopes, by one of Raff and Gammon's licensees in New York, it drew almost riotous crowds for many days on end, and the police had to be called in to keep order.[17] The motion picture, even in the diminutive films of the Edison peep-show box, had conquered the great masses. The impact that this new entertainment medium was to have upon the minds and lives of hundreds of millions throughout the world was something unforeseen and unimagined— perhaps surpassing the effect of almost all other nineteenth-century inventions, by Edison or anyone else.

4   Edison had shown the insight of the born inventor in his mechanical work on the motion picture in 1888–1889. After he had perfected his kinetoscope he permitted himself some prophecies as if referring to the distant future:

The kinetoscope is only a small model illustrating the present stage of progress, but with each succeeding month new possibilities are brought forth. I believe that in coming years, by my own work and that of Dickson, Marey, Muybridge, and others who will doubtless enter the field, grand opera can be given at the Metropolitan Opera House at New York without any material change from the original, and with artists and musicians long since dead.[18]

But for all his clairvoyance, he could not foresee the great developments that were to take place within a comparatively few years. Nor did anyone else really foresee them. Meanwhile he halted his efforts to develop the peep-show machine into something superior, projecting life-sized motion pictures. There is no doubt that he and Dickson attempted in 1889 to project pictures upon a wall screen five feet square. But the effect was of a flickering blur; the problem of magnifying the picture and illuminating the screen remained technically difficult, compared with the operation of stationary stereopticon slides shown with an optical lantern. The peep-show box with its rotating shutter gave, by comparison, very clear flashes of vision. When the first backers of his motion picture work urged Edison, in 1895, to give them an apparatus projecting film on a screen, he at first refused, then casually assigned the task to one of his machine shop men. He remarked at the time that if he could give his whole attention to the problem for a week he would "get it going." But he never had the time, after his first fine thrust at things kinetographic.

Meanwhile, two brothers named Otway and Grey Latham, who had been showing the Leonard-Cushing and other boxing films with much profit, under license from the Kinetoscope Company, had realized how awkward it was to have the crowds of viewers move from one box to another to see a minute's record of the fight in each. If they could manage to project the whole motion picture on a sheet against the wall, one of the Lathams remarked, "there would be a fortune in it." In other words the motion picture ought to be taken out of Edison's little box and shown in theaters where people could sit and watch the entire thing without interruptions.

When the idea was broached to Edison, however, he said firmly no. The great inventor was certainly playing the stubborn Dutchman again. He had said earlier that the flat disk, or plate phonograph record, would never amount to anything, and failed to patent his own version of it, so that others afterward, with the help of Emile Berliner, were to make millions out of the flat record. Now, when approached by Norman Raff, his business associate, to develop a screen projection camera, he asserted that he was going to stick to the peep-show box at all costs; "If we make this screen machine that you are asking for, it will spoil everything." At the moment they were selling a lot of peep-show boxes at a good profit; but if they put out a screen projection machine, Edison estimated about ten of them would take care of the demand for the entire country! At that rate there would be no profit in it. "It was better," he advised, "not to kill the goose that laid the golden eggs." [19]

He continued in this course, even after he learned that the Latham brothers, with the aid of their father, a professor of chemistry, had begun secretly working out a screen projection apparatus of their own in a small laboratory in New York. This is a turning point in the history of the new entertainment industry. It must be remembered that, at the time, Edison held potential control of all the processes of motion picture making. He had invested $24,118, over five years, in the kinetoscope's development, was winning it back quickly, but clung to his narrow view of the industry's future prospect. In truth the "craze" for the motion picture peep-show, after two years, showed signs of dying down.

Now began a chaotic race among inventors of all sorts—as well as mechanics, plumbers, and even steam fitters—to perfect a screen projecting apparatus for Edison's films. In France, Marey and Louis Lumière were soon developing excellent motion picture devices of all sorts; and in England R. W. Paul likewise. Their machines, appearing toward 1896, could be used in America in conjunction with a screen and a projector. There would be cutthroat competition and lawsuits on every hand; in the end, Edison by his great blunder in 1894 was to lose what was later reckoned to be a "billion-dollar monopoly." (We speak, here, from hindsight.)

In 1895 he also lost his ablest photographic assistant, William K. L.

Dickson, a thin young Englishman with waxed mustache and goatee and an artistic temperament, who for nigh on fifteen years had shown an almost filial attachment to his chief, and served him cheerfully for thirty dollars a week. Next to Edison, he knew probably more than anyone else about the mechanism of the motion picture camera and kinetoscope. While much credit is owing to him as Edison's principal aide in these investigations, it would be absurd to say, as some have done, that on the motion picture he did "most" of the work credited to Edison. In the adulatory biography of the master that he published in 1894, and in various articles before and since then (despite their later estrangement), Dickson himself made no such assertions. The key ideas were always Edison's; he always passed on their execution by other hands, including Dickson's. After Dickson left West Orange, he achieved virtually nothing during the next forty years of his life.

However, he had worked with Edison in 1889 over one or two primitive projecting systems using a screen, though the first results were poor. He wanted passionately to go on with projecting, but the master thwarted him. Then the Lathams set up their independent laboratory, and—by a shadowy maneuver—first lured away one of Edison's technical assistants, a French photographer named Eugene Lauste. Next they made covert approaches to the more expert Dickson.

Now a new figure appears on the West Orange scene, a big bustling man named W. E. Gilmore, who in April, 1894, had been engaged by Edison to act as his financial manager and who was expected to set the chaotic Edison affairs in some order. Hearing rumors that Dickson entertained covert relations with the Lathams, Gilmore roundly accused him of dishonorable conduct and demanded that he be fired. Though Edison refused to believe such charges then, he ended by agreeing to his dismissal. And so Dickson went over to the Lathams, who, in the spring of 1895, were to launch their imperfect motion picture projector.[20] Later it was noticed by Edison that various notebooks relating to some of the early experiments they had made together on a projection camera and screen had also disappeared around the time of Dickson's departure.*[21]

* An article published by Dickson many years later indicated that he had retained possession of some drawings and manuscript notes of Edison's dating from the motion picture research work of 1888–1895. (*Journal of the Society of Motion Picture Engineers,* Dec., 1933, vol. xxi, no. 6.)

Dickson seems to have been unable to help the Latham concern find the right principle for their projection machine. After a while he left their service and worked as technical director for another group making peep-hole kinetoscopes like Edison's, but with slight mechanical variations. When suit for patent infringement was brought in 1897 against his company, the Mutascope & Biograph, Dickson placed his stock in that corporation in escrow and left for England, evidently unwilling

The Lathams' magic-lantern kinetoscope, called the "Pantoptikon," made a passing sensation in the press. It was nevertheless a crude affair, that probably helped to attach to the new art the colloquial name, "flickers." Their motion picture camera, however, obviously infringed upon Edison's patents, and he at once threatened suit. At the same time, he promised to build a better projecting machine himself.

The productions of technical workers in this field now appeared in torrents. Someone, at the Lathams' laboratory, hit upon the *reel,* as a superior device for handling the long film strips; in France, the gifted Lumière cut down the number of exposures per second from Edison's 46 to only 16, yet maintained an excellent effect of persistence of vision. Finally, in 1896, Thomas Armat, of Washington, D.C., an amateur of the camera, developed new gear for his projection lantern (the "vitascope") that permitted the positive film a longer period of rest and a shorter period of movement. Each individual picture was thus allowed more time to be illuminated on the screen and fill the eye, an important step toward good projection. The completion of Edison's promised "screen machine," meanwhile, had been delayed. When the distributing agents for the Kinetoscope Company, Raff and Gammon, saw what excellent results Armat had obtained with his projection device, they promptly bought an option on it. "We thought it would be a great deal better us for to control the machine than to have it fall into the hands of parties unfriendly to you and us," they wrote Edison in January, 1896. Since Armat's "screen machine" worked so well, and Edison's was not ready, why should they not join forces and combine Armat's projection device with the Edison motion picture camera? They proposed, moreover, that the Edison company should manufacture all the Armat projectors, or vitascopes, and sell them under the Edison label.

At first the veteran inventor made some difficulty about accepting this proposal; and Thomas Armat was loath to make an accommodation that would deny him credit for his important contribution to motion picture progress. But, as Raff argued in a letter to Armat, no matter how good a machine should be invented by another, the customers would prefer Edison's:

Kinetoscope and phonograph men and others have been watching and waiting for a year for the announcement of . . . the Edison machine which projects kineto-

---

to testify against his former chief. (His own published statements of earlier date could have been used against him.) His behavior, though highly ambivalent, seems not to have been knowingly faithless. In later years he and Edison were reconciled; when Dickson was in reduced circumstances he even obtained some small loans or gifts of money from Edison, who in reality was quite fond of him.

scopic views upon a screen or canvas. In order to secure the largest profit in the shortest time it is necessary that we attach Mr. Edison's great name to this new machine. While Mr. Edison has no desire to pose as the inventor of this machine, yet we think we can arrange with him for the use of his name and manufactory.[22]

It is a sorry episode in the great inventor's career; and it was with evident reluctance that he, who had contrived so much, now yielded and agreed to lend his name to the product of another. At this time there were persistent rumors that Edison was in financial difficulties, and they were not far from the truth. An agreement was reached that satisfied neither inventor. It was, in fact, terminated a year or so later when Edison devised his own projection gear, and Armat took back his patent rights. Several years later the whole affair was ventilated in court during the suit between Edison and the Mutascope Company. Edison then testified that there was little difference between his own earlier projection machine and Armat's, except for the latter's "egg beater," or intermittent, movement. Yet it was that difference in movement that was all-important.*

Meanwhile, in 1896, Edison and his associates anticipated big profits, as preparations were made to open theaters with wall screens. A first showing of the new vitascope projector was given before newspaper reporters in one of the large shops at West Orange. Annabelle, a great theatrical favorite of the day, with attendant dancing girls, appeared life-size, and in color too, on a screen 20 by 12 feet. It was a frosty morning; Edison, wrapped in two overcoats, walked about looking highly pleased, and said, chuckling, "This is good enough to warrant our establishing a bald-head row, and we will do it too." The press mistakenly publicized the new projection machine as "Edison's Latest Triumph."

The formal public presentation of the "enlarged Kinetoscope" took place on the night of April 23, at Koster & Bial's fashionable Music Hall on Herald Square, before a silk-hatted audience embracing many leading figures in the theatrical and business world. This event signalized the introduction of living pictures to the theaters of Broadway, and thereby to all the world. The presentation included ballet girls dancing with umbrellas, burlesque boxers, some vaudeville skits, and finally so realistic a scene of waves crashing upon a beach and stone pier that some of the viewers in the front row recoiled in fear. The audience was astonished

* Many years afterward, in 1921, Edison evidently tried to make amends to Thomas Armat by writing him a letter saying that in reviewing the early history of the motion picture (in the course of a statement to a newspaper) he had given credit to Armat as the inventor of the "first practical projection machine." While he had had a machine of his own, he said, he dropped it when Armat turned up with a better one. (N.Y. *Times,* June 9, 1921.)

and exhilarated by this newest "miracle of science," and sent up great cheers for Edison. Thomas Armat, who had agreed to stay in the background, was working in the projection booth to keep the machine in adjustment; the gray-haired Edison was in a box, but refused to come forward and respond to the applause of the crowd for the success of "his" vitascope.[23] From that time forward, the magic of the motion picture took hold of men everywhere. Charles Frohman, the dean of the American theater, who was present, expressed the fear that henceforth no one would want to look at the dead scenery of the stage, painted trees, frozen waves, and the like.

After the combined work of Edison and Armat, there was left for others only the refinement of motion picture technique. For many years, however, the "movies" remained limited in subject to short recordings of dances, prize fights, comic skits, and freak performers, which became an established feature of hundreds of music halls and "arcades" throughout America and Europe. Though Edison himself was an imaginative storyteller, he had no concept of how the motion picture could be made an art form. His attention remained confined to the improvement of its techniques and the profitable manufacture of cameras and projection machines. The "shooting" of films in the Black Maria was delegated to others.

With the coming of the motion picture, a medium almost unlimited in its scope, both as visual art and dramatic representation, had been created. A Chinese proverb holds that "one hundred tellings are not equal to one seeing." The motion picture brought a new power of seeing.

Its future development was first foreshadowed, though dimly, in 1903, when an Edison camera man at the Black Maria, after seeing some French experimenter's "story film," brought forth a work entitled *The Life of an American Fireman*. It was conceived in the form of a dramatic sequence and packed with sudden disasters and heroic rescues. This effort was quickly followed by *The Great Train Robbery*, distributed by Edison's studio in 1904, and now regarded as the classic prototype of the motion picture *play*. A flood of Westerns and thrillers, having the literary quality of the dime novel, quickly followed and generated a great new wave of popularity, even though the films ran to only 1,000 feet at most, so that the action had to take place at breakneck speed and came to a jarring halt after about fourteen minutes. Now, little motion picture theaters and nickelodeons mushroomed all over the country; by 1909 there were about 8,000 of them, a figure that was soon afterward doubled. From a relatively small amusement business, the movies within a few years were to be transformed into a major industry. Who could have known, from seeing the absurd little "shorts" of 1896 to 1903, that untold millions of human beings throughout the world would soon come to live vicariously in the darkened cinemas and share the joys and sorrows of the stars, Mary Pickford, Lillian Gish, and Charles Chaplin?

Despite vigorous competition, Edison's film company expanded rapidly and built a large glass studio in the Bronx at a cost of $100,000. The Thomas A. Edison, Inc., label became known throughout the world. But the conditions of this new mass entertainment industry were such that its administration was distasteful to its founder, and he delegated it to others. The Edison pictures, like those of most competitors, aimed at entertainment on a low level, artistic standards being set by the rampant commercialism of the film producers and distributors.

Though Edison was for some years the largest, or nearly the largest, manufacturer of motion picture machines and films, numerous strong competitors were now ranged against him—Biograph, Vitagraph, Essanay, Kalem, Lubin, B. F. Keith, and others. It must be remembered that by his blunder in failing to apply for European patents for his motion picture apparatus in 1891, Edison had lost the monopoly of the process of film making; his various competitors in the industry had purchased American patent rights to English and French machines that were obvious imitations or improved models of his invention. There were prolonged patent wars in court with the firms controlling the Latham, Armat, and various foreign patents. The selling and distribution of both films and cameras was for years a chaotic business; not only were manufacturing patent rights infringed upon, but films themselves were "duped," that is, duplicated, without payment of fees to their owners.

After ten years of litigation, Edison won a verdict upholding his original motion picture patents of 1891 in Federal court, at Chicago, in October, 1907. At this point some of the larger defendants, especially Jeremiah J. Kennedy, of the Biograph Company, made overtures to Edison's financial manager at the time, Frank L. Dyer, with a view to bringing about law and order in the industry. Kennedy's plan called for a pooling of patents and setting up a centralized control over both producing and distributing of films, by means of a national trust agreement under the Edison licenses. Dyer, Kennedy, and the representatives of eight other film-producing companies (two of them foreign concerns) soon afterward organized the Motion Picture Patents Corporation, whereby all the producers in the group recognized Edison's patents and guaranteed him payments of several hundred thousands annually in royalties for their licenses. Provision was also made for a national system of exchanges under which almost all screen theaters in the country were to pay the trust weekly fees for each projection machine, while the nationwide distribution of regular "packages" of film reels was also strictly controlled in the interests of the trust, by fixed allotment of the business among its members.

It was an affair of the most dubious legality in that era of "trust busting." Nevertheless, until the lawsuits of independents, such as William Fox and Carl Laemmle were settled, and before the suit of the United States Attorney General charging conspiracy fell upon their heads, the major film producers could look forward cheerfully to years of assured profit. The formal signing of the trust agreement in December, 1908, was made the occasion of a sumptuous banquet held in Edison's great library, in the laboratory at West Orange.

Edison received these former opponents most cordially, and in his highly informal manner. He ate quickly, then said, "You boys talk it over, while I take a nap." On a cot placed in a corner of the library he dozed off soundly, while discussion among the scheming film magnates raged on. Conceivably he did not care to know in detail what they were talking about. When it was over he was awakened. "All right, where do you want me to sign?" he asked. He made no pretense of reading the perfected document and, explaining that he had some difficult experiments under way, said, "Good-by boys. I have to get back to work."

The "patent trust" and its distributing branch, the General Film Corporation, endured for almost ten years, until the United States Supreme Court finally ordered its dissolution, in April, 1917, under a consent decree. But during that long interval Edison's film company, as one of the leading participants in the trust, derived profits of up to a million a year from royalty fees and allotments of its film productions.[24]

In the late nineties, after he had severed his connections with the General Electric Company, Edison had been on the verge of financial ruin, thanks to the failure of his ore-separating venture. In 1898, in response to a kindly letter from Henry Villard inquiring if he needed help, he had written cheerfully from his Ogdensburg mine that he hoped, at any rate, to be able to clear off the debts he had incurred. "My three companies, the Phonograph Works, the National Phonograph Company, and the Edison Manufacturing Company (making motion picture machines and films) are making a great amount of money, which gives me a large income." [25] At that time he was still pouring his revenues into the ore-milling works.

But soon after he gave up that stubborn campaign, he was in the black again! Edison cared nothing for mere money, as against the joy of a new battle to wrest from nature more of her secrets—but he could always earn his way. By his wits, with the help of his ingenious kinetoscope, and a few other tricks, he had saved himself again.

At this time of his life, when he was almost sixty, he told an interviewer:

The point in which I am different is that I have, beside the inventor's usual make-up, a bump of practicality . . . the sense of the business-money value of an invention. Oh, no, I didn't have it naturally. It was pounded into me by some pretty hard knocks.[26]

Once more, in the early years of the new century, he felt like the Count of Monte Cristo. With his recent winnings he could now afford to "plunge" into a new, a different, and a very hazardous experimental project.

chapter XX

A stern chase: the storage battery

1   When Edison had first gone up to the Jersey highlands in 1890 to crush magnetite rock, the world was given over to the Bicycle Craze. Everywhere, from St. Petersburg to San Francisco, people were tasting a new freedom of movement; young and middle-aged, bold men in knickerbockers and daring ladies in bloomers dashed about on their pneumatic-tired velocipedes, down the open roads, to the country, the fields and the flowers, and the open sky, at a speed of ten miles an hour; while those who plodded along in horse-drawn vehicles looked at them envyingly.

But a decade later, when Edison came down from his hills, beaten but undaunted, the Automobile Age had begun. The fascination of rapid movement was more intoxicating still at twenty, or even thirty miles an hour, thanks to the horseless carriage. In light coupés or phaetons powered by all sorts of steam and gasoline engines, or electric motors, some three thousand Americans were already "burning up the roads"—living man's dream of devouring distance in self-propelled chariots. The first crop of steam cars and gas buggies were noisy, hot, and bothersome; they stank and gave much trouble mechanically; yet every year they gained in following.

Edison, on occasion, tried the early gas buggies and steamers, and kindled to them at once. For him, the faster the better. Although these vehicles had been imported from Europe since the 1880s, one heard nowadays of "automobile inventors" who were attached to almost every carriage shop in the Middle West.

Edison allowed that the horse was doomed; the future would belong to the motored carriage. Like others, he foresaw a future in which millions would ride to their business or their pleasure in self-propelled carriages and trucks. He also ventured to predict that most of them would be driven by electric motors and storage batteries.

404

Why electric cars? A survey made in New York at the end of 1899 showed that of a hundred motorcabs in use in the downtown area, ninety were powered by storage batteries. Compared with cars using steam engines, which were heavy, or gas engines, then most unreliable, the electric runabout was clean, light, and quiet. Carriage builders were producing "electrics" in goodly number, by installing lead batteries and small motors in a Studebaker or Columbia vehicle. But who knew as much as Edison about electrical machines? It was the right time, he judged, to enter this field and contrive a product more practical than any other.

After his ore-separating venture was over, he had entertained all sorts of plans for new lines of activity. "I have enough ideas to break the Bank of England," he used to say. The phonograph shops and the motion picture studio were humming with business by 1900. Another project was the Edison Portland Cement Company, which would use his big rock crushers for grinding cement and limestone. He also contemplated a novel engineering scheme for the construction of low-cost concrete dwellings, a plan that needed several years of preparation. But, though he worked at or directed these and a good many other smaller affairs concurrently, there was always some one ruling idea that dominated his thoughts for a given period, and engaged his hopes above other ventures. Now it was the electric car—and its *sine qua non,* the storage battery. He saw the "commercial need" for the new product he had fixed upon. Like the incandescent lamp, it would conquer the masses for him once more. Hence the great plunge into the storage battery project. A cautionary thought should have come to him.

Three years earlier, at a festive gathering of members of the Association of Edison Illuminating Companies (the trade association of electric utilities using his system) at Manhattan Beach, New York, he met by chance one of that anxious breed of automobile inventors, a thin, long-legged young man with pale-blue eyes, named Henry Ford. Ford was then employed as chief engineer at the Detroit Edison Company's powerhouse, and had been introduced to Edison by the head of that company, Alexander Dow. "There's a young fellow," said Dow, "who has made a *gas* car." (The worthy Mr. Dow, in those days, kept pestering Henry Ford, who was a good electrician, to make an electric car instead.) The electrical men present had in fact been discussing with Edison the possible economies of charging storage batteries for the use of streetcars and private vehicles, during low-traffic hours at the lighting plants. But Ford was one of those who clung steadfastly to the gasoline engine. It was his first meeting with Edison.

For long years Henry Ford had fairly worshiped Thomas A. Edison as the greatest of inventors and regarded him as the example whom he, a mechanic of inventive disposition, desired above all to emulate. Now he was invited to sit next to this great man. He was shy, and usually dull in conversation. But when asked to describe the components of his first automobile model, the young engineer became animated, and Edison, cupping his ear, listened with keen interest.

The older man plied Ford with questions. Edison, who could be curt and discouraging to aspiring inventors, was impressed with what he heard. He brought his fist down upon the table and said, "Young man, that's the thing! You have it!—the self-contained unit carrying its own fuel with it. Keep at it." Such a simple gas engine powered with what Edison called "hydrocarbon," as cheap as kerosene, and soon to be available everywhere along the road, would be most practical if it were perfected. At that period the unlettered Ford did not even know, as he recalled, what the word hydrocarbon meant. But previously he had met with much discouragement in his part-time work on his "quadricycle." The fact that America's greatest inventive genius had given him prompt and positive encouragement was worth worlds to him.

Ford would go back to his machine shop in Detroit, filled with a new inspiration; he would soon leave the Detroit Edison Company plant, find some financial backers, and in 1899 launch his first small automobile manufacturing concern. Edison's own first impressions of Henry Ford seem to have been that the young man was hardly very bright—"a damn fool," aside from the subject of automobiles—though he afterward changed his mind about him. As for Ford, though they did not meet again for several years, he believed that this first encounter was a turning point for him; his gratitude toward Edison endured and became something like idolatry.

But while Edison had given Henry Ford excellent advice, he himself soon forgot all about it. In talking of automobiles he had remarked that electric runabouts had a radius limited to the vicinity of power stations; the lead-sulphur storage batteries then available were heavy, difficult to recharge correctly, and corroded quickly. But the years passed and those petrol engines were still as balky and stinking as ever. Edison, therefore, must develop an electrically driven car. As portable power plants for carriages the contemporary storage batteries also gave trouble. But Edison would build a better battery—who else? One that would weigh little, give great power, last forever! And small enough to be carried in a valise—a package of power that could be used in a thousand different ways to lighten man's burdens.

Edison "proceeded to each new undertaking, no matter how difficult, with the expectant joy of a naïve child," one of his expert chemists observed at about this time.[1] From 1900 on, he had eyes for nothing but the "miniature reservoir of electric force" that he must create.

The decade of the 1890s had been for him a period of frustration and financial adversity. But each time, after having been jilted by Nature at one approach, he would rebound and besiege her again on yet another flank, and try to wrest more secrets from her. For instinctively he thought of Nature as his first mistress, the most jealous and exacting of all. His courtship of her continued to be arduous and unending. How many wiles he employed, what sacrifices he made, what risks he ran, in order to win her favors and, in effect, to seduce, to conquer her. That he himself lived and thought in the metaphorical terms of this mythical drama is evident from the spontaneous phrases of one of his letters to a friend at this period, ". . . I don't think Nature would be so *unkind* as to withhold the secret of a good storage battery, if a real earnest hunt were made for it. I'm going to hunt."[2]

2   The whole idea of the "secondary" or storage battery had fascinated scientists for a century, ever since the time of Volta's primary cell. The voltaic cell, working by chemical action, changed chemical energy into electricity, but exhausted itself regularly and required renewal of its components. The possibility of a secondary cell, which, after having accumulated electrical force and having been discharged, could be recharged an almost indefinite number of times, by being "reversed," had also been present in men's minds for many years now. The first reversible, or "storage," battery had been invented in France, in 1859, by Planté, and marked a sensational advance in the electrical art; twenty years later the Faure battery of lead and sulphuric acid was perfected, a veritable "box of electricity," that stored energy and gave it back. On first inspecting the Faure storage battery, Lord Kelvin exclaimed that it embodied the realization of dreams he had scarcely hoped would be realized in his lifetime.

As soon as there were practicable generators producing current economically, electrical scientists saw in the storage battery a most useful auxiliary, a portable reservoir of power. But the lead and sulphuric acid storage battery of the eighties and nineties, despite the helpful energy it yielded, remained heavy and corrosive. Edison had experimented with lead batteries many years before and thrown them aside. The claims for them were exaggerated. He once remarked, "When a man gets on to accumulators [storage batteries] his inherent capacity for lying comes out."[3]

In looking at the problem of the horseless carriage in 1900 he had concluded that the gasoline engine was "unscientific" and wasteful compared with the electric motor. But in considering the batteries available as motive power, he was led to the conclusion that they also were inefficient. Their lead-acid components were scientifically incompatible and unnatural; they were too heavy; the corrosion of metal by acid constantly limited durability and efficiency. "I guess I'll have to make a battery," he said.

He defined the problem and the objectives for himself: to attempt an entirely new voltaic combination that would make his "box of electricity" light, durable, undeteriorating, quickly chargeable, and economical. Such an instrument, he felt in 1900, would "open up a new epoch in electricity." In his notebooks he wrote, "The object ... to produce a practicable battery which will permit the storage of a larger amount of energy per pound of material than possible with the type of battery using lead electrodes." [4]

His idea at the start was to use an alkaline solution as his electrolyte, so that his metals would not be corroded. The negative element would be some form of iron, or iron oxide. On the choice of the positive element he was less decided; he would hunt for it. In his first experiments he worked with copper oxide in various combinations, then examined hundreds of other materials.

As A. E. Kennelly, who assisted Edison in the early stages of this work, observed at the time, the lead batteries of the 1900 model gave four to six watt-hours per pound of battery, or required a weight of 124.5 to 186.5 pounds for each horsepower-hour delivered at the battery terminals. Edison's objective was to multiply such storage capacity and potential about threefold. [5]

"Grand science, chemistry. I like it best of all the sciences," Edison said. He had loved it since his boyhood. It was, in his view, the most vital science for the future, though "still a mere baby"; more was unknown than known of it. In truth, by modern standards, methods of chemical research were still primitive in 1900, and in the investigation of the chemical reactions within the cells of an electric storage battery one encountered a tremendous number of variables—as one does even now, more than half a century later.

In these later years Edison's research and development work required lengthy periods of intensive experimentation, particularly in the chemistry of both storage batteries and phonograph records. As in the early 1880s, during the electric light investigations at Menlo Park, he got together a staff of expert technicians to assist him; by 1901, he had a force of about ninety persons, among them several accomplished physicists and chemists, with Dr. J. W. Aylsworth as head of the chemical department. Without regard for cost, samples of every known material were assembled, and every possible tool or device that could be of aid. Having gathered personnel and equipment, he gave the signal for the hunt to begin. Hundreds of tests were made of various grades of copper and finely divided iron, and the results set down painstakingly in laboratory notebooks. This project was very different from ore separating, or from improving the motion picture camera; it was like the old days at Menlo Park, with its all-night sessions when time and labor meant nothing. The number of experiments mounted up into the hundreds, then to the thousands; at over ten thousand, Edison said, "they turned the register back to zero and started over again." A year, eighteen months went by, and they had not even a clue.[6]

Nature once more was being "unkind," and for a long season. As Kennelly pointed out, the chemical processes going on in those battery cells under the hundreds of different combinations tried were really of a tremendous complexity. If you tried, for instance, to make the negative electrodes (iron) lighter, you increased the rate of deterioration. If you added something else, such as copper oxide, it created new variables. Then there were the tricky problems posed by the electrolyte, an alkaline solution of 25 per cent caustic potash (KOH); this would not corrode metals, like sulphuric acid, but the cell capacity tended to drop off quickly. It seemed to need something to improve its conductivity. What that missing factor was would take a long time to find out. Then there was also a great drag hunt over a wide field for the right combination of metals to make up the positive electrode. They tried cadmium-copper, cobalt, magnesium, and, after many trials, nickel hydrate in various forms. In this area Edison and his associates began to get better results, and they felt a little more cheerful.

In the search for an improved positive electrode, Edison carried on his experiments by taking carbon strips and filling their pores with every conceivable substance that might serve as the positive element of an electrical couple. After charging, he tested for a galvanometer deflection on discharge, but usually got no results. Many photographs of the 1900s show him standing at a laboratory table about fifteen feet long and covered with some four hundred test tubes, each containing different electrolyte solutions, in which he tried his various electrical couples. That was what was so time-consuming. The tests of these different nickel-and-iron couples were made in terms of milliampere hours, and some of the early tests showed only around 300 milliamperes. Gradually, as the material was improved, the figure crept up to 500 milliamperes; and at last in 1903, while Edison was away on a trip to Canada seeking a good source of nickel, the figures went up to 1,000 milliamperes. Aylsworth, who was in charge of the tests, remembered afterward how vastly pleased Edison was to learn of this advance upon his return to West Orange.

Nickel hydrate, the most promising of materials tested for the positive electrode, was, however, a poor conductor; in order to get the desired electrical action, through an improved contact, Edison considered and tested a great variety of conducting substances that might be mixed and packed into the pockets of his battery cells with the nickel hydrates. But that was not easily found either. Sometimes the nickel gave trouble; sometimes it was the pure iron that was most mischievous. Very little was known then of any of these metals in their chemically pure state. In 1903 Edison built a small plant for the manufacture of batteries at Glen Ridge, a few miles outside of Orange, where he kept a team of trained men working for long months to refine various grades of iron and nickel, so as to learn what he could.

At sixty, stout and round-faced and deafer than ever, Edison had nonetheless changed very little in spirit and method since he had invented the carbon filament lamp thirty years earlier. There was the same abandonment to unending studies, no matter how difficult. His temper, to be sure, was a little shorter nowadays, sometimes violent when a clumsy mechanic upset things; but he continued to show an unfailing patience in the face of repeated disappointments. In fact, it was a rule with him never to show grief or bitterness, even when some long-prepared experiment proved to be a failure. The only difference in method was the larger organization he used; his staff of ninety, including trained scientists, technicians, draftsmen, and plain mechanics, was a pretty big laboratory around 1900.

Among his staff there was, for example, a stiff-necked Prussian chemist, who in later years told the biographer Emil Ludwig that he had a very poor opinion of Edison's knowledge of chemistry and physics. In his laboratory, moreover, there seemed to be little system and less discipline. Edison held no staff meetings, never seemed to give his research men any definite instructions as to what they were to do or how they were to do it, but left them on their own. He also told them rather little about what the object of their search was. The people employed in the laboratory either kept themselves busy, or pretended to be so, in order to avoid sudden dismissal. To a man with a German university background, Edison seemed to be wanting in dignity and allowed his assistants too great familiarity with him. The old hands joked with him, or told him off-color stories, or humored him by bringing him the kind of chewing tobacco, five-cent cigars, or pies he preferred. But even the German chemist admitted, in the end, that he had come to admire and enjoy Edison as a "character," remarkable in every way.[7]

A more sympathetic, yet realistic, portrait at this period of Edison's life is drawn by a Russian-born, Paris-trained chemist named Martin André Rosanoff, whom Edison engaged, on the spur of the moment, in 1903. Rosanoff reported for work early one morning, as ordered, and Edison introduced him to the other laboratory assistants whom he designated as "muckers," describing himself as their "chief mucker." He also alluded to himself sometimes as a sort of Don Quixote, who was tilting at scientific windmills in order to liberate the good storage battery. His chief draftsman, John Ott, long in his service, he nicknamed Santcho-Pantcho.

Respectfully, Rosanoff asked to be informed about the laboratory rules and regulations. Edison spat on the floor (he was chewing tobacco) and expostulated, "Hell, there ain't no rules around here! We're tryin' to accomplish somep'n." [8]

The new chemist was assigned to the problem of making an improved wax record, Dr. Aylsworth having recently retired. Where were the records of past experiments? he asked. Edison said there were none; he had forgotten all about them. "You'll have to start all over again." Rosanoff was astonished by this. On discussing the matter with the old hands, he was told by one that the "Old Man was simply not giving away any hard-earned trade secrets." But another drew the inference that the crafty Edison probably knew all about the old records and tests but was deliberately keeping Rosanoff in ignorance of them. "I think he is right. He and Aylsworth got into a mental rut for years.... He wants your fresh mind on the problem!" [9]

A sociologist wandered into the Edison Laboratory one day to inquire of the great inventor what methods of organization he used in conducting his research work. He represented one of the recently established philanthropic foundations, and asked permission to spend several days at Orange gathering the information needed.

"I'm the organization," Edison said roundly. The professor could discover that for himself by looking around for a few minutes, and could save himself much time thereby. In truth, the sociologist soon learned that though the inventor had numerous assistants and business managers, he kept control of everything in his own hands, and himself supervised all the different experiments and tests going on. Yet somehow the "system" worked.[10]

There were men at the laboratory with doctorates in physics and chemistry from the Universities of St. Petersburg, Berlin, Leipzig, Paris, and Oxford, men with considerably more formal scientific education than Edison possessed. He couldn't have cared less. He went about damning the whole breed of university-trained technicians and scientific theorists (whom he nevertheless employed), while they bore his jibes as good-humoredly as possible. His favorite preachments touched on the vice imputed to university scientists, of seeing only "that which they were taught to look for," and thus missing the great secrets of nature lying under foot. Many of his stories or reminiscences concerned successful inventions he had achieved in the face of warnings of more learned men, by pursuing his own helter-skelter methods of experiment. Did any one realize, he would boast, that the first promising clue leading to the carbon filament lamp had been discovered by him as a result of an experiment in "carbonizing a piece of limburger cheese?"

"Do you think I'd have amounted to anything if I went to school?" he would ask. No, he did not believe in the systematic disciplines of physicochemical analysis as taught in schools and universities. And, so far as he was concerned, he was right, after all, as the young Russian scientist realized. All the "hells" and "damns," all the crude pleasantries, the sloppy costume he sported, only reflected his fierce will to avoid the conventional and to be intellectually nonconformist at all costs. A conventional education, so vital to most ordinary mortals, might have tamed this "wild eagle," as Rosanoff reflected. "Had Edison been formally schooled, he might not have had the audacity to create such *impossible* things as the phonograph!" [11]

At the stage of technology reached toward 1900, inventions were being achieved by more and more lengthy processes of experimentation, carried on by whole teams under a scientific captain. This was true not only at the Bell or General Electric laboratories, but even at West Orange. Edison might continue to rail at "the-o-re-ti-cal scientists," but he worked with a good many university-trained men now as before. One of his chemists, for example, after working for over a year at improving the compound used for phonograph records, finally reached a solution. He reported to Edison that the task had been accomplished by systematic physico-chemical calculations—in short, by "theoretical chemistry." Though Edison could not have worked in that manner, he followed the chemist's explanation in this case with profound interest. He was content. "After all the important thing was the solution, not the way it had been reached."

A singular man indeed, and tireless. Often and again he would be heard exhorting or challenging his assistants to join him in an all-night session in order to lick their "damned problem." He did not believe in luck, or "wizardry"; things came to him nowadays, as in the past, mostly by hard work and taking the time to "try everything." He pretended that he never stopped to rest; but nearing sixty he would often be found taking more than cat naps, and in the most curious stances. For example he would curl up on his roll-top desk and wrap himself about several bulky volumes of Watts's *Encyclopedia of Chemistry,* whose contents, as Edison's men used to say, he would absorb during his sleep.

He still had his mischievous humor, with which he used to cheer or divert his disciples. When they worked late at night at Orange, there would be a shift in the power plant at midnight, and the lights would go off for a few moments before they came on again slowly to full candle power. During the interval of darkness Edison would close his eyes and fold his hands, as if he were laying himself down to sleep and saying a prayer. But when the lights turned up again, he would pretend to wake up suddenly and say cheerily, "Well boys, we've had a fine rest; now let's pitch into work again."

Walter Mallory, his loyal companion of the ore-milling adventure, used to come down about once a month from the new Edison cement works, and ask Edison whether he were making any progress in the storage battery experiments. When, after several months, he still heard that nothing had been accomplished, he thought he would offer his condolences. But, to his surprise, Edison flashed his fine, sudden smile and said emphatically, "Why, man, I've got a lot of results. I know several thousand things that won't work!" [12]

To his chief chemist Aylsworth, he once revealed something of his deep inward torment, however, when he said with much feeling, "In phonographic work we can use our ears and our eyes, aided by powerful microscopes; but in the battery our difficulties cannot be seen or heard, but must be observed by our mind's eye." [13]

He gave the utmost care to the design of an entirely new battery form, to the size and shape of the containing pockets of its plates; and, after making many of his rough sketches and having them drafted precisely and built up as models, he evolved a novel structure of nickel-plated steel, the positive pockets being packed with nickel hydrate and the negative being of iron oxide. The difficulty experienced in obtaining a good electrical contact in the positive element, it was remarked at the time, put lots of gray hairs in Edison's head. Finally, after many trials, he decided upon a certain form of graphite, that was mixed with the nickel hydrate filling the positive pockets, and found this satisfactory.

At last, in 1903, he was ready to test his batteries in actual usage. He installed them in carriages with a small electric motor and chain drive, and ran these vehicles over the rough country roads near Orange. Records were kept on test sheets and closely scanned. In his laboratory Edison also set up an electrical apparatus that jolted his battery cells up and down, day and night, in simulation of heavy road usage. He wanted his battery to be able to withstand abuse, to be foolproof.

Tom Robins, the conveyor belt inventor, dropped in at the library one day in 1904, just as Edison was getting ready to manufacture his new product. As the two men sat talking there was a loud crash just outside the window, as of some heavy object that had fallen; then another, and after a minute, another. Robins was alarmed, but Edison never turned a hair. After a while a workman came in and said, "Second floor O.K., Mr. Edison." Edison nodded and ordered, "Now try the third floor." To the wondering visitor he gravely explained that he was testing the endurance of his storage battery—by having packages of them thrown out of the windows of the upper floors of his laboratory. "For a scientist, Edison used some mighty peculiar methods," Robins reflected.[14]

At Silver Lake, in the summer of 1904, the manufacturing plant was made ready, special machine tools were set out, and 450 workmen were engaged. Production slowly got under way, and in the first season several thousand cells of the first battery model, the "E," were turned out. The demand far exceeded what Edison could supply; merchants and transport concerns had the first Edison storage batteries installed in light delivery wagons and, at the outset, found their electric carriages and trucks highly convenient.

With his usual self-confidence Edison had started a rousing publicity campaign for his "revolutionary" storage battery, even before he was ready to manufacture it. Once more, he did a thorough advance-agent job, whetting the public's curiosity to a high pitch, so that when "The Wizard's Newest Marvel" finally arrived—so the battery was heralded in the press —people really believed it was such. Thus, in authorized interviews during 1903, and early in 1904, he announced that there would soon be "a miniature dynamo in every home ... an automobile for every family." Men had said that no commercially successful battery could be made without lead. "The new battery means that another impossible thing has been accomplished." Lead batteries deteriorated rapidly even when they were not being used; they contained destructive acids; they leaked and were dead after a few months' use in an automobile. The new Edison battery had great power, weighed next to nothing, had neither acid nor lead, would not deteriorate when not in use, was almost endlessly rechargeable, and could withstand almost any mistreatment. He concluded:

Yes the new battery will settle the horse—not at once but by degrees. The price of automobiles will be reduced. . . . In fact I hope that the time has nearly arrived when every man may not only be able to light his own house, but charge his own machinery, heat his rooms, cook his food, etc., by electricity, without depending on anyone else for these services.

In short, he had "revolutionized the world of power" the newspapers reported; he had brought forth the "age of stored electricity," with independent power sources that were destined to change sea travel, land transport, warfare, and agriculture. The outlook for the ill-smelling gas buggies seemed dark indeed!

Edison posed before news photographers standing beside a spanking little red sulky, which had nickel-iron batteries and an electric motor. It was so easily run, he declared, that his twelve-year-old boy Charles could drive it. With that he leaped on board and sped away at twenty miles an hour.[15]

In more restrained terms, Dr. Kennelly, presenting a paper before the Institute of Electrical Engineers in New York, reported that storage capacity in the nickel-iron battery had been raised to 14 watt-hours per pound, or 233 per cent above the contemporary lead battery models; weight, at 53.3 pounds per electrical horsepower-hour, was reduced by about half.

However, bad reports soon began to come in from the first users; and then the reports grew steadily worse. After all the effulgent propaganda, it was a shock to learn that the battery containers leaked on being used in vehicles; that the cells were of uneven performance, and usually dropped about 30 per cent in power. Defective batteries returned to the factory were broken up and examined. The electrical contacts in the positive (nickel) element were found to be unreliable, although Edison had thought he had corrected this weakness by introducing graphite in the positive pockets.

The truth about this fiasco was spreading; it was even reported in European technical publications and newspapers. What was to be done? All the exuberant advance publicity, the habitual Edisonian optimism, had but made things worse. Moreover, Edison this time had raised $500,000 through friendly investors by giving a bond and mortgage on his other manufacturing properties; and these funds had all been exhausted by the work of research and the cost of setting up the factory.

This new crisis might have been something to laugh over, if it hadn't been so nearly disastrous for Edison. He recognized that his nickel-iron battery still needed prolonged study before it could be marketed. But if he shut down his plant and set to overhauling the whole job in his laboratory, he faced a very heavy financial loss. Still worse, and most humiliating to him, would be the public admission of failure, after his sweeping claims of success.

After a brief discussion with his financial associates, he made his decision swiftly. He would shut down production at once and take back all the batteries that had proved defective, at his own cost. Distribution of several thousand cells had already been initiated by agents in England and Germany as well as America. Some of the carriage companies, such as Studebakers', wanted him to continue delivering his batteries as long as they worked in some fashion. One of his associates questioned his wisdom in stopping manufacture, saying, "It is right to get an ideal thing, if possible, but the batteries, as we have had them, are so far in advance of lead batteries, that everybody is eager for the time when they will be on the market." [16]

To such objections Edison replied flatly, "I stopped manufacturing because the battery was not satisfactory to myself. . . ." And again, "The factory will not start up until I find out why the cells lose capacity." [17]

It was an act of courage. He had always shown pride in the quality and performance of the products he marketed. Now he not only reimbursed buyers for the defective batteries they turned in, but, since the Edison Storage Battery Company had used up all its capital, he financed it thereafter out of his own pocket. Fortunately the motion pictures—thanks to *The Great Train Robbery*—were now booming. At his command, virtually his whole laboratory force set to work to overcome the irregularities that had shown themselves in the first nickel-iron battery, the cause of which, at first, seemed extremely obscure. The new series of experiments brought forth more variables, and were to engage the labors of Edison and his staff for five years more.

3   The new succession of experiments and tests was "record-breaking," according to one of his principal laboratory assistants, Walter E. Holland, who adds, "I might almost say heart-breaking too, for of all the elusive, disappointing things one ever hunted for that was the worst." Edison vowed that he would find the right way to make a battery— "If it takes me seven years and $1,750,000"— alluding here to the cost of his recent education in iron ore milling.[18]

In chemical work he certainly followed his old methods of "cut and try," the way of the empirical inventor. Could he have used more precise theoretical calculations to shorten his labors? There was actually in the possession of chemical scientists very little exact knowledge of positive and negative electrode combinations, or of electrolytes, the liquid solution in which they were immersed. There was very little that could then be "predicted" theoretically. There was all too much "magic" in the storage battery—as there still is. If the field of possible electrode combinations has been narrowed a great deal by now, this is due largely to the work done *empirically,* and the information gathered, by Edison and his associates and by others who came after them.*

One chemist and his assistant were put to the task of improving the leaky containers, and working out a superior welding job. Another group worked over the refinement of the iron element; while still another, with Edison participating directly, was occupied in building up the conductivity of the nickel hydrate, the admixture of graphite particles having proved to be unsatisfactory. Many different materials were tested, by being given repeated charges. The results were awaited during long periods; some solutions would seem promising, then would "fade away." By the summer of 1905 Edison could report that he had again made a "vast number of experiments, now reading 10,296," and that he "had found out a great many things." [19]

In the winter of 1905, while working under great stress, he became very ill, and underwent what was then a fairly dangerous operation for mastoiditis. After this piece of radical surgery, he woke up the next day and growled for his newspapers. Henceforth he would be almost stone-deaf.[20] Though the annual winter vacation at Fort Myers would have benefited his health, and he longed to go, he returned instead to his laboratory to watch the testing of hundreds of new combinations for battery cells and examine the test sheets being compiled for them.

Though the road seemed to lead through all the turnings of an immense labyrinth, Edison insisted on preserving a mask of outward composure and good cheer.

---

* Edison's empirical methods in chemical research have often been assailed by theoretical scientists, the term "Edisonian" being applied to them as one of opprobrium. His associates, however, defended them at the time as the best possible strategy in this little-known field. For example, after Edison, at later stages of the battery investigation, determined upon the form of the positive element, it became necessary to ascertain what definite proportions and what quality of nickel hydrates and nickel flake would furnish the best results. He would try two materials in different proportions, running from nine parts hydrate to one part flake; eight parts hydrate to two of flake, and so on down. These were carefully tested through a long series of charges and discharges; from the tabulated results of hundreds of tests, he selected three that were most promising. After having determined the *proportions* most desirable, he then needed to find the best *quality* possible. Now there are several hundred variations in the quality of nickel flake, and about a thousand ways of making the nickel hydrate; hence an enormous amount of detail was involved before sufficient information could be obtained.

When an associate came to him one day and in forthright language declared that a long series of tests had proved negative and the whole venture was a waste of labor and money, Edison rounded on him and cried angrily, "Is that all you have to say for yourself?" and then walked out of the room. To another who also assured him that a certain series of experiments gave progressively worse results, and that they simply posed "a problem without solution," he replied with spirit, "I've been in the inventor business for thirty-three years, and my experience is that for every problem the Lord has made He has also made a solution. If you and I can't find the solution, then let's honestly admit that you and I are damn fools, but why blame it on the Lord and say He created something 'impossible.' " [21]

On his staff at this period there was a distinguished Swedish metallurgist, Dr. Roos, who had a room to himself in which he wrestled with the problem of making a seamless battery container that would never leak. The solution, however, proved to be highly elusive, and so the excellent Dr. Roos formed the habit of avoiding Edison at the hours when he made his rounds. But one day Edison encountered him by chance in the room of another technician and asked him what progress he had made. Improvising an expression of great joy, Roos shouted at him: "I've got it, I've got it at last! I just need another day or two to straighten out some small details."

Edison appeared delighted. "Didn't I tell you right along that you'd git it ... All you got to do is stick to it and work like hell. Sure you can have another day or two!" Roos then retreated. The other chemist thought he might congratulate his chief on the good news concerning one of their big trouble spots. But Edison laughed, and exclaimed, "Did *you* believe what Roos said?" And he explained, cheerfully enough, "He hasn't got a damn thing. But that's the way to talk!" [22]

Nevertheless, the ten years' hunt for a good storage battery was a prevailingly somber period, Edison's associates recalled, despite his ideas about keeping up morale. Age was coming upon him; his long, thick hair grew very white, while his well-defined eyebrows remained very dark, accentuating his pallor.

The year 1907 was one of profound business depression. Even the phonograph business was hard hit. The Edison company had clung to the cylinder record too long, while Berliner's disk record, produced by Victor Talking Machine, gained in popularity month by month.

In that year also, the newspapers reported that a Rolls-Royce car with a six-cylinder gasoline engine had passed a 10,000-mile endurance test. In Detroit, Henry Ford, working with a will as fierce as Edison's, perfected a cheap 15-horsepower 4-cylinder engine for his Model N car, precursor of the celebrated, raven-hued Model T. The Model N went twenty miles on a gallon of fuel and cost only $600. "Hundreds of persons," it was reported, rushed to buy Ford's small car. Had Edison missed the bus?

During the years of extreme tension created by the unrelenting campaign for the nickel storage battery the inventor's family found him rather more "difficult" than usual. The descriptive adjective is suggested by an article that appeared at a somewhat later period in *Collier's Weekly*, based on a lengthy interview with Mrs. Edison. The piece was entitled "She Married the Most Difficult Husband in America." To the interviewer Mrs. Edison declared that she had dedicated her life to the intimate service of her great man, "and thought it worth it a thousand times over." He had few friends and lived, she said, "a great deal by himself and in himself, shut out from the contacts open to most men." Neither hobbies nor amusements interested him. When his family made a little "conspiracy" to interest him in golf, it proved to be an utter failure. Mrs. Edison and the children, therefore, "always put his work first," and his home and family life were organized to that end.[23]

The identity of the reigning "mistress" of Edison's mind and heart has already been disclosed. Could any mortal woman really win him away from his old love, whose mysterious wiles and charms so obsessed him? Mina Edison tried more than any other, and in the end gained more empire over her iron-willed husband than any other. That he loved her greatly, and that he also esteemed her, was beyond doubt; he was capable of very genuine and strong affection. Mina Edison alone could sway his mind when he was obstinate; indeed, she alone, in later years, could speak to him and make herself understood without raising her voice. Though time pressed, he would linger with her for some moments each day, in her garden or in the big conservatory at Glenmont. And every day, no matter how distracted he might be, when he left the house and when he returned, he would go to her and kiss her. If she were not nearby as he went to work, he would send a servant to call her.

To the young children of his second marriage, he was by turns teasingly affectionate, or *distrait* and remote. "Sometimes it was as if he never saw us," his youngest son, Theodore, said long afterward. This distance from his children had an unfortunate effect on his two elder sons; his secret disappointment in them and their estrangement from him will be touched upon presently. But Mina Edison was determined that with *her* children, at least, it would be different, that there should be no such alienation from the father. She therefore encouraged her eldest son Charles (who was about fourteen in 1904) to be with his father as much as possible and, during school vacation, to work in his laboratory, even at the humblest tasks, such as the "bottle monkey's."

Charles was a vivacious and attractive boy and got on very well with his father. One summer night, after he had been washing bottles all day at the laboratory, the hour grew late, and he announced to his father and one of the chemical assistants, that he would like to take a nap, as his father so often did.

"Well, if you've got to sleep, go lie down under the table in the corner; nobody will step on you there," said his father casually.

It was long past midnight, 2 A.M. in fact, when the laboratory assistants witnessed a domestic scene unusual for those purlieus. Mrs. Edison had become greatly distressed at Charles's absence at such a late hour. She had evidently risen from bed, dressed, and ordered the carriage to take her to the laboratory.

When she came in and saw where Charles lay sleeping, she was shocked. The floor was no fit place for sleep; it was unclean. As her husband often chewed away at his old cigars or plugs of tobacco he used to expectorate freely all about him. Mina Edison now reminded him that she had offered to supply him with a spittoon, and repeated the offer. His assistant recalled: "He declined, saying that the floor itself was the surest spittoon, because 'you never missed it.' " Mrs. Edison roused the sleeping boy and carried him off to his proper bed.[24]

In the years 1905 to 1908, Edison applied for a long series of patents covering refinements in his storage battery, especially the process of making metallic film or flake out of the nickel by electroplating. He and his associates had gone far in the study of both iron and nickel in a state of very high purity. They had managed to obtain nickel film or flake as light as thistledown—about four one-hundred-thousandths of an inch in thickness. This was accomplished by electroplating thin layers of copper and nickel on a metal cylinder, then dissolving away the copper in a chemical bath, leaving nickel film so fine that tiny segments of it, about 630 layers, could be packed alternately with nickel hydrate in a cell pocket that was only four inches in height, the nickel flake replacing the graphite formerly used in the positive pockets as a conductor. The development of the nickel-flake process gave Edison an improved electrical contact and conductivity, and the lightness of weight he wanted, while extending capacity and wattage.

The whole structure of the storage cell was redesigned, so that instead of building its numerous "windows" out of flat pockets, Edison used seamless tubes of thin nickel-plated steel, perforated with minute holes through which the electrolyte could seep in freely. Alternate charges of the nickel flake and nickel hydrate were then fed by machine into the tubes, and packed or tamped down under pressure of four tons to the inch. Whereas formerly the active materials used to cause swelling of the pockets, the shift to the tubular structure made for a great improvement in battery performance.

In starting on this big experimental problem of the "completely reversible" battery in 1900, Edison had had a very good initial concept or "intuition" about using nickel as his positive element, and a noncorrosive electrolyte solution of potash. But refining upon that idea—"the last ten per cent"—had taken a fearful amount of time and study. Among chemical scientists there was then only a dim knowledge of how the tiny electrified atoms of oxygen called *ions* ("wanderers") detached themselves from the iron oxide, passed through the alkaline solution, and deposited themselves upon the nickel, thus producing the high oxide of nickel. By prolonged experimenting Edison and his co-workers brought about one improvement after another, and helped to clarify this problem.

His electrolyte, made of the caustic potash solution, at first had tended to drop off gradually in capacity, from cycle to cycle of charging and discharging. But some time in 1908, after they had tried hundreds of different additives, trials were made with a small amount of lithium hydroxide ($LiOH$); introducing lithium had the effect of raising capacity and then holding it steady over a long period. According to present-day chemical technicians, this step constituted "a real piece of magic" that could never have been calculated theoretically in Edison's time, and whose mechanism is still not clearly understood.[25] All that is known now (1959) is that lithium ions mysteriously improve conductivity and alter the oxidation tendency of the nickel flakes. Edison's empirical work, therefore, was successful in the end and helped to create for later men a wider base of information to calculate from. His ten-year campaign to master the imponderables of the storage battery has remained famous in the history of battery technology. It was really a beautiful work in applied science, carried out at the cost of a million dollars of his own money.

"At last the battery is finished," Edison wrote in the summer of 1909, with evident relief, to the head of the Adams Express Company, who had been waiting to use his improved product for delivery wagons. It seemed "an almost perfect instrument."[26] A year later factory production was resumed on a large scale; in the first twelve months thereafter a million dollars' worth of the new A battery cells was sold, beginning a profitable career for the Edison Storage Battery Company that has continued to this day.

Several important carriage works, among them Anderson's of Detroit, hastened to produce electric runabouts and light trucks, known as the Detroit Electric, that were powered by an array of Edison cells. In the Eastern states the similarly equipped Baker Runabout also had some vogue between 1911 and 1914. Department stores used such electric trucks for years. Light electric town cars were popular for a period because they were noiseless and fumeless; their radius was about sixty miles a day between chargings, and their Edison batteries—unlike lead batteries that needed long periods for recharging—could be fully recharged in seven hours.

As Edison wrote in a letter to Samuel Insull, now president of the Chicago Edison Company, the new storage battery promised "to add many electric Pigs to your big Electric Sow...." [27]

The nickel-iron-alkaline battery was generally acknowledged to possess special merits over the heavier lead-acid affairs, because it resisted decomposition, was well-nigh completely reversible, and showed a long life. Edison had hoped, above all, that it would be adaptable for automotive traction, and even for streetcars, a model of which was operated for a while on Main Street, West Orange, on a branch line of the Erie Railroad. However, to the inventor's deep disappointment, automobile traction proved to be just what it was *not* suited for. Nor was it useful as a starter battery. Its voltage capacity tended to be lower than that of contemporary lead batteries (about 1.19 against 1.50 volts per cell), so that one needed more Edison cells for a given task—which offset their advantage of lighter weight. The Edison battery was also affected by cold temperatures, its alkaline electrolyte becoming weak and slushy, though not freezing. When winter came it was common for many Edison-powered vehicles to become stuck in snowdrifts or on muddy roads. Meanwhile the far-ranging gas buggies of Ford and others, with their cheap, improved combustion engines, swept all before them.

On the other hand, the beautifully constructed Edison battery was found particularly useful where dependability and long life were important, as for stand-by purposes at power plants, and for railway signaling; or to provide current for miners' lamps, train lights, and other railway and marine appliances. Though the electric automobile had only a brief vogue, a remarkably wide field of usage was developed in industries such as mining and quarrying (for firing blasts), on merchant vessels, and ships of war. Edison batteries were also used at first to manipulate naval torpedoes; later a model was developed for driving submarines when submerged, with diminished risk of exposing the crew to injurious gases.

When demonstrating the use of his cells in a submarine, Edison told Navy officials, "Keep it clean, and give it water, and at the end of four years it will give its full capacity."

"Four years?" they queried in astonishment. "Yes," he replied, "four years, eight years; it will outwear the submarine itself." [28]

During World War I the Edison battery also entered into wide use in the growing field of radio telegraphy. In short it won acceptance as the most rugged of industrial storage batteries.

Was it worth the tremendous expenditure of effort? Edison used to profess that his inventive work was guided only by the criterion of commercial success or practicability. But the profits were rather on the moderate side, in the end, when measured against the "punishment" he had taken in that series of heart-breaking experiments lasting nearly ten years. As he himself sometimes suspected, whatever he might say overtly, he really "lacked the commercial temperament." The thought of calling a halt at some midway point and cutting his losses was never tolerated. Why did he go on and on? "I always invent to obtain money to go on inventing," he said simply. A few more of Nature's secrets had been wrested from her, though not as many as was hoped for. It was not for profit that he labored so hard—he scarcely needed more money nowadays —he was driven by his will, and by the "spirit of workmanship," to contrive and create. As one of his laboratory confreres remarked after they had finished their task, "Secrets have to be long-winded and roost high to get away when the Old Man goes hunting for them."

4   He "rested" at intervals, during the campaign for the storage battery, by turning his attention periodically to a number of highly different inventive tasks. One such new departure was the sizable Edison Portland Cement Company, established shortly after the closing of the Ogdensburg works. In such spare moments as he had, over several years, he studied this ancient industry, then drew up engineering plans for an entirely new type of cement kiln. A tract of limestone rock was purchased in eastern Pennsylvania, and the Giant Rolls were brought from Ogdensburg to crush the rock. His plant, when completed in 1907, was highly mechanized, being designed to grind out 1,000 barrels of cement every twenty-four hours, when the average cement kilns produced only 200 barrels a day.

Edison foresaw a large expansion of reinforced concrete construction in America when he first planned to enter this industry. True enough, cement output increased more than a hundred per cent within five years. By then Edison was ready with his novel type of long cement kiln, which was 150 feet in length and 9 feet wide—instead of 60 by 6, as was customary. His production gradually rose to a total of 1,100 barrels a day. Other cement manufacturers soon lengthened their own kilns and eventually made them even larger than his. He also devised machinery to produce a more finely ground cement than the current standard, by precise weighing and measuring of the mixture, and by use of a finer-mesh screen. Again he foresaw, correctly, that construction engineers in the future would require a portland cement of greater absorptive power.[29]

He had been having his difficulties with insurance companies, which refused to issue fire insurance for some of his wooden structures except at prohibitive rates. He therefore decided that after he had lowered his cement costs, he would make his own new factory buildings entirely of fireproof reinforced concrete. In the years after 1906 a cluster of twelve steel-and-concrete factories gradually reared themselves around his brick-walled laboratory at West Orange, six of them being of fairly large size.

To his fertile mind, the cement plant soon suggested other new devices. Having developed a very free-flowing compound of concrete, toward 1908, he conceived a highly original plan for making *poured cement* houses that could be cast in one piece and constructed in about six hours by means of specially designed machinery. The object was to produce cheap housing for workmen in the form of small detached dwellings, and on a mass production scale. Here, once more, we see Edison's mind preoccupied with the creation of objects of utility and convenience for general use.

"A decent house of six rooms, so far as the shell would go would cost only three hundred dollars," he calculated. When his cast-iron molds were taken away, after four days had been allowed for hardening, there would be revealed a complete house in one piece, with cellar, roof, floors, walls, stairways, doors, windows, bath, and conduits for electric and water service. Only the appliances for lighting, heating and plumbing would need to be installed. All told, counting the use of machinery, labor and overhead (or depreciation), a dwelling with a floor plan of 25 by 30 feet and three stories high might come to twelve hundred dollars. To those who objected that such housing would be monotonous in appearance, Edison answered, "We will give the workingman and his family ornamentation. They deserve it; and besides, it costs no more, after the pattern is made, to give decorative effects." [30]

His Patent No. 1,123,261 (December 22, 1908) specifies a complete system of cast-iron molds into which a cement house could be poured;

two sets of the molds were to be locked together from top to bottom, but would have an opening, or hopper, at the top of the roof. A chain of buckets was to carry the free-flowing mixture from a mixing tank down below to the open top and pour it into the molds. The empty interior space enclosed by the poured cement was to be the house.

This was a pioneering venture in prefabricated housing, conceived not long after the modern revival of reinforced concrete construction. In the opinion of historians of American architecture, Edison and the architect William Ransome (probably without knowledge of each other) were the first to experiment with poured cement houses. Again, construction experts said at the time that such a process as Edison's was "impossible," because it was thought that the heavier components of the concrete mixture, made of sand, cement, crushed stone, and water, would tend to separate out, or settle in the molds unevenly. The inventor, however, had devised a special additive for his mixture, a gluelike and jellylike colloid substance which rendered the flow uniform and even gave it a smooth finish.

A firm of New York architects was engaged to draw detailed plans, according to Edison's sketches, embracing mold plates of quite elaborate form, in which all stairways, pipes, and conduits would be set. Then, having obtained his cast-iron molds, and the mixing and pouring machinery he wanted, Edison put up several such houses in the vicinity of West Orange. They looked plain enough, though not more so than most low-priced dwellings of 1910. Some "bugs" showed themselves in his first poured cement houses, though he was certain they could easily be worked out. At all events, the scheme, which cost him about $100,000, failed to win acceptance at the time, and he dropped it. A third of a century later, when the country faced an acute housing shortage, construction men would have recourse to methods very similar to those used in Edison's "Poured Cement House."

At about this same period the spreading phonograph "craze" kept his factories at West Orange working on a two-shift basis, and, as we have seen, expanding in size. In May, 1906, the Edison house organ apologized to its several thousand dealers for having fallen behind their orders by more than two million records. Made then of hard wax, Edison's cylinder records, in his opinion and that of others, were superior in acoutistical value to Victor Talking Machine's disk records. Priced at only 35 cents, the Edison records' content was designed to appeal to the extensive cracker-barrel public in the country; they reproduced mainly "heart songs" and Negro melodies, as well as popular waltz tunes, such as "Pretty Peggy," and brass-band marches like the "Allagazam."

"Canned opera," however, was coming into fashion, thanks to the efforts of Edison's strongest competitors, Victor and Columbia. The Victor Red Seal record (having the lateral cut) came as high as $2.00, but one could hear the voice of Enrico Caruso for two minutes, and snatches of opera arias by Melba, Homer, Patti and Sembrich. The "star system" thus invaded the music record field, as it was soon to invade the motion picture. Victor and Columbia records moreover were soon lengthened to four minutes, thus avoiding the abbreviated versions of famous solo passages still used by Edison. The large rural following he had won remained faithful to the old-style Edison cylinders; but it was evidently high time to make a change.

"Father was intensely competitive," Edison's youngest son, Theodore, recalled afterward. "As soon as the other firms turned out an improved phonograph and record, he must surpass them at all costs."

Though the flat disks were more convenient for shipping and storing, Edison obstinately stuck to his cylinder form, because he thought it was technically superior to the disk; which was correct then. With the disk record the stylus, under sudden sound pressure, often jumped over the fine grooves. Edison finally reversed his policy with regard to serious music—formerly left for others to reproduce—announcing that his company would also distribute recordings of grand opera, to be made in a new studio in New York. Famous artists such as Antonio Scotti and Marie Rappold were hired at large fees, and their performances were reproduced on a much improved wax-compound record marketed in 1908 as the "Amberol," and designed by Edison and Aylsworth to run four minutes—but still a cylinder. A large, ornate concert model phonograph, called the "Amberola," was also developed by 1911 to compete with similar instruments of Victor and Columbia that were designed to appeal to the carriage trade. These machines cost from $200 to $800. A carefully fashioned diamond needle, by 1912, also gave the Edison phonographs the highest acoustical standard known in those days. The lion's share of the business, however, was in a cheap, $20 Edison phonograph which attained mass sales.

As early as 1904, Edison had become convinced that the chemical secret of his old wax compound had been stolen by a spy for one of the rival companies. One day he brought in a wax record made by the company he suspected and instructed one of his expert chemists to make a thorough analysis of it. When it was discovered that the other record was identical in composition with Edison's, the young chemist appeared highly agitated over the affair. But Edison remarked, with a twinkle in his eye, "What are you so excited about? Everybody steals in commerce and

industry. I've stolen a lot myself. But I *know how* to steal. They don't—
that's all that's the matter with them." [31]

There were lawful ways of evading another's patent, Edison suggested,
by devising an ingenious substitute; and there were crude, lawless ways
of doing this. At any rate he did not bother to sue his adversaries, but
worked to improve further the quality of his record and talking machine.

In 1912, he determined, at last, to change his policy, and make a disk
record of his own; when unexpected difficulties were encountered, he revived his "insomnia squads" and toiled with his assistants for about a
month, without going home more than once a week to change. At sixty-
five he still believed in the efficacy of the night attack. His recently installed time clock in the laboratory showed that during September, 1912,
he worked 112 hours a week. In the end he contrived a hard, smooth
plastic which largely eliminated the hissing sound of other types of records. Edison's mood was triumphant; he vowed that he would make the
phonograph "the greatest musical instrument."

One of the motives for his renewed zeal for the phonograph was revealed on the morning of February 17, 1912, when the white-haired
inventor, in a New York theater, demonstrated a whole "talking-picture-
play" on his new "kinetophone." From behind the screen his big Amberola clearly, and with considerable volume, reproduced the sound of actors'
voices in a passage from Shakespeare's *Julius Caesar*. Henceforth, he
predicted, grand opera would be available to everyone "for a dime." [32]

In this apparatus he used a sort of fish line with which the man in the
projection booth could regulate the pace of the phonograph if its sound
moved out of step with the facial expressions or gestures of the actors.
There was, however, a severe restriction imposed by the phonograph
record's capacity, still only of seven minutes' duration; the silent "feature
films" now ran for an hour or more. The response of the public and the
trade was an indifferent one; so that the inventor was given little encouragement to pursue the matter.

By 1914 the volume of Edison's phonograph business, stimulated by
the country-wide ragtime mania, surpassed 7 million dollars annually;
motion pictures, made immensely popular by the star system, brought
revenues that were even larger. A business dictating machine (the Edi-
phone), for which the inventor had perfected an ingenious device for
entering corrections automatically, demanded another new manufacturing unit. To this he added the "Telescribe," a machine combining the
dictating phonograph and telephone, which recorded both sides of telephone conversations.

In 1911, the sixty-six-year-old inventor reorganized his various industries and combined them in one corporation under the title of Thomas A. Edison, Incorporated. It was said of him at various times that he lacked "pecuniary ability." Henry Ford, who had limited himself to the development of a single invention, the cheap automobile, was reported to have said of Edison that he was "the world's greatest inventor and worst business man." Was this true? If Thomas A. Edison had confined his energies to only one of the many fields he cultivated, he might perhaps have become the richest man in the world. But he enjoyed life far more than did the obsessed money men; and even so, as a part-time avocation, managed to create a "family business" embracing about thirty different enterprises in the closing years of his life, whose gross annual sales amounted to between $20,000,000 and $27,000,000. By 1914 some 3,600 workers were employed in the big new factories of reinforced concrete that surrounded his laboratory on Valley Road, West Orange.

Probably no professional inventor was ever engaged in so many businesses, or in so much manufacturing, selling, and "meeting of payrolls." His son Charles, who was twenty-four in 1914, had studied at the Massachusetts Institute of Technology—"because father wanted me to be able to read a blueprint"—and was being trained to serve as an executive in various departments, so that he might eventually become head of the entire concern. In the twenties he was to be his father's lieutenant and chief managerial man. Nonetheless, while he lived, Thomas A. Edison ran the whole business and the laboratory as well, according to his own lights.

He was a thrifty, honest, and paternal capitalist of the old school. On one occasion he declared that his ideal in business was the old-time village general store, "with no overhead and steady profits." Instead of driving toward some large monopoly, he enjoyed the diversity of his affairs: while one line might be doing badly, another would prosper, and help carry him. He cared nothing for bookkeeping and accountants, though in later years he was forced to tolerate their ministrations in his establishment. His people were kept working for as many hours as possible, and he obliged them to punch a time clock; but he worked longer hours than anyone else. Nor did he Taylorize or "Fordize" his shops; for many old workers who were no longer strong or efficient, but whose services dated from old Menlo Park days, were retained by him out of sentiment. To be sure, when a depression came he would walk through his plants, firing people to right and left of him; but the old-timers, who knew his little ways, used to hide as he approached. And not all who were fired stayed fired. Stories are told of some of the old-timers who, after having been angrily dismissed by the Old Man, returned to work the next morning as usual. Edison saw them, but said nothing. There is no question but that he regarded his workers as individual human beings and maintained personal contacts with them possible only in the small businesses of other days. There were occasional strikes for wages and hours in some of his plants, especially during World War I; but the workers who knew Edison were undoubtedly fond of him.

All over his shops he had signs put up bearing a quotation attributed to Sir Joshua Reynolds: *"There is no expedient to which a man will not resort to avoid the real labor of thinking."* This would suggest that he believed workers had brains and should use them. On the other hand, it was sometimes said that he really had a low or cynical view of human nature, for he repeatedly declared that while he could "improve machines, he could not improve men." Though unaffected and democratic in spirit, he enjoyed personal service; some of his veteran employees, including his long-time private secretary, Meadowcroft, were quite subservient in humoring his whims, holding his coat for him, or fetching what he wanted when he wanted it.

At any rate, he was in later years the "laird" of the Valley Road, in the Oranges, where his industries were mainly centered. He owed money to no man. As he wrote Henry Ford in 1912, he kept control of his enterprises in his own hands nowadays. "Up to the present time I have only increased the [battery] plant with profits made in other things, and this has a limit. Of course I could go to Wall Street and get more, but my experience over there is as sad as Chopin's 'Funeral March.' I keep away." [33] After the "steamboat days" came for him, thanks to the phonograph and motion picture industries, he ranked as a member of the country's select group of multimillionaires. But while his industries earned, at times, as much as two million dollars of profit annually, he boasted that he "never paid any dividends." As fast as profits were realized from one invention that had come to market, he would plough the money earned into research and development work in new fields, for his curiosity about the facts of science remained undiminished to the end.

They were doing a land-office business at the Edison works in West Orange, in 1914, when hideous disaster struck without warning.

On the evening of December 9, 1914, Edison had gone home early for dinner, a clear sign of old age coming on. Just before 6 P.M. fire broke out in a small wooden structure located in the center of the factory quadrangle, where chemicals and inflammable motion picture film had been stored. From the outset the fire was intense. The factory fire squad quickly went into action, but water pressure in the village of West Orange was low. Since most of the adjacent factory buildings were of reinforced concrete, then considered fireproof, it was hoped that the fire would not spread far. But the burning chemical products shot forth great geysers of green flame which soon enveloped neighboring buildings and gained tremendous headway. Fire companies from eight nearby towns, including East Orange, Montclair, and Newark, soon rushed to the scene, but falling water pressure balked all their efforts. The several steel and concrete factories in the path of the wind were soon in full conflagration; one of them, Building 24, which was six stories high, contained large stores of chemicals used in manufacturing records, which sent up spectacular flames from big tanks on its upper floor. Numerous firemen and volunteers were overcome; one Edison worker was killed.

Charles Edison, then employed at the phonograph works, was on hand to help direct emergency measures. He was overwhelmed by the horror of the scene and his sense of despair, as he recalls. And where was his father? He had been called from the dinner table at Glenmont. At last Charles saw him come walking slowly into the yard, silent, impassive. All the Edison worldly goods, the whole "empire" he had gradually built up in the Orange Valley was going up in green and yellow flames; yet he stood and watched it all with folded arms.

He had made his new structures of reinforced concrete. Because of their "fireproof construction" he had had them insured at less than a third of their cost, which was more than a million dollars. The administration building at one side of the yard, containing stores of valuable instruments for making diamond needles and new disks, was partly on fire; and the firefighters and volunteers hastened to carry out such supplies as they could reach.

At last the Old Man spoke to his son Charles. "Where's Mother?" he asked. "Get her over here, and her friends too. They'll never see a fire like this again." [34]

Charles began to lament the incalculable loss. Fortunately the big library and main laboratory buildings were out of the path of the flames. But his father only said dryly: "Oh shucks, it's all right. We've just got rid of a lot of old rubbish."

However, he soon went into action, stayed up all night and directed the removal of bundles of phonograph records, and of such valuable merchandise and machinery as could be saved. A crowd of 10,000 persons had gathered to watch the fire. Late in the night connection was made with the East Orange reservoir, and the fire was gradually extinguished. At dawn the place was still smoldering. Most of the factory buildings had been gutted; but, as Edison pointed out, they were still standing.

Many persons came to him, or telegraphed, expressing their sorrow at this catastrophe whose cost, according to exaggerated reports, was reckoned at "four million dollars." * To one such sympathizer Edison replied, "I am sixty-seven; but I'm not too old to make a fresh start." And he added, "I've been through a lot of things like this. It prevents a man from being afflicted with *ennui*." [35] But to his son Charles he said: "I wonder what we will use for money?"

On the day after the fire he directed a crew of fifteen hundred men who cleared the ruins and removed the debris. He again worked with furious energy, as in his youth. Factory space was quickly rented in nearby towns; machinery was set up, and men were put to work. On the security of a large amount of receivables due him, he negotiated loans at banks. It happened that a few days before the fire, he and Mrs. Edison had been visiting the Henry Fords at Dearborn. At news of his misfortune, the "tycoon" of automobiles hastened to New Jersey and generously lent Edison $750,000 in order to speed the work of reconstruction. Edison furnished collateral in the form of stocks and bonds for most of this loan, and, over the years, repaid the principal in full, though Ford refused to accept interest.

Within three weeks the Edison factories had been restored to some semblance of order; and shortly after New Year's Day, 1915, most of the personnel were at work in two shifts producing the first of the new phonographs adapted to *disk* records; they were operating against a growing backlog of orders that would total almost ten million dollars that year. The speed of recovery was almost as spectacular as the disaster.

* T. A. Edison, Inc. showed claims for fire loss in 1914 at $919,788.00.

chapter **XXI**

Canonization

1   It was only after the turn of the century that the Electrical Age really arrived in America. Just prior to that time, through the nineties, the country as a whole, save for a few luminous city districts, still lived by the smoky glow of oil lamps or by gaslight. Then, thanks to long-distance transmission, power stations mushroomed everywhere. By the time of the World's Fair at St. Louis in 1904, with its glittering Palace of Electricity and its displays of heavy-current engineering, no further evidence was needed that the new form of energy would change men's lives at home and at work.

Electricity gave a new tempo and a new character to all industry. It was easily carried across mountain barriers and rivers from power sites, and was converted into a multitude of uses. The ponderous steam-powered mills of the nineteenth century had been darkened by their huge belts and shafts. Now the electric motor permitted the greatest flexibility in the design of the factory.[1] Motors large and small operated at almost any speed desired, and at varying distances. Introduced into assembly line plants during the first decade of this century, as at Henry Ford's River Rouge works, they had the effect of raising industrial efficiency, on the average, by 50 per cent.

When men looked about them nowadays, they saw that the buildings of their cities reached toward the clouds, because of the electric elevator; that 50,000 electric streetcars and railway cars carried them about with unheard-of ease and convenience; that incandescent lights blazed everywhere. And when they contemplated these changes they thought as with one accord of plain old Thomas A. Edison. That other brilliant inventors coming after him had—in his despite—developed high-voltage transmission and an a-c motor adaptable to manifold industrial use, mattered little. Leadership in inaugurating the Electrical Age was almost universally attributed to Mr. Edison. What James Watt had been to the Age of

432

Steam, Edison was to the new era of technology. Watt represented steam power—pounding piston rods, flywheels, roaring fires of coal. Edison symbolized electricity—thoughts and words soaring across great distances; energy freed from the engine and belt by smokeless motors; cities wreathed in light.[2]

Not only the man in the street, but scientists of high repute recognized in him "a central figure of this age of applied science." [3]

A new phase of glory as America's acknowledged folk hero opened for Edison as he approached his seventh decade, a glory that could, at times, be troubling and burdensome. The new giant among industries, the electric utilities, had unofficially adopted him as its patron saint some years earlier. In 1885, Samuel Insull, then his financial manager, had organized the Association of Edison Electric Illuminating Companies as a means of maintaining good business relations among local power companies using Edison equipment. Later on, a similar trade organization was established as the National Electric Light Association, which, however, served mainly as a lobbying agency for the utilities.

Though their lobbying activities before legislatures did not always do them honor, the men of the power companies, like the rest of the human tribe, instinctively enjoyed having a venerated "father-figure" for their industry. True, the inventor no longer stood at the head of their industry; nor did he draw, any longer, the smallest royalties from the many "Edison Companies" using his name; but with age he seemed to enjoy appearing before their annual conventions, where many of his "old boys," such as Sam Insull, John Lieb, and W. J. Hammer, now big men in their field, were reunited with him. He would address conventions, on occasion, "as a simple, democratic, old man," with only a few brief words of greeting, and receive the ovations of thousands of delegates. Moreover, his political views were usually very conservative, and thoroughly relished in these circles.

The year of the St. Louis Exposition, 1904, marked the twenty-fifth anniversary of Edison's invention of the carbon filament lamp and central station system. To commemorate these great events the Edison Association and various electric lighting companies collaborated in staging an impressive exhibit of "Edisonia" at the Fair. On view were the first incandescent lamps, the Jumbos of 1879, and other relics that made Edison appear even now a historical figure.

On February 11, a few weeks before the Exposition was to open, the American Institute of Electrical Engineers also paid homage to Edison, on the occasion of his birthday, at a banquet for five hundred guests that was held at the Waldorf-Astoria in New York. The after-dinner speakers vied with each other in paying homage to him. Replicas of the first practical incandescent light were given away as souvenirs to the guests; and for dessert, ices artfully shaped in the form of an Edison lamp were served up. The metropolitan newspapers made much of this birthday, publishing lengthy articles and interviews with Edison; from that day on the anniversary of his birth was regularly celebrated in the press each year, as that of a national hero, by similar articles.

There were, of course, many levels at which Edison appealed strongly to the popular mind as an object of hero worship. He was a man of science, yet had the "common touch." Almost every other American dreamed of making that "better mousetrap"; yet Edison's devices were no mere mousetraps, but creations that served both to transform society and to lighten men's burdens.

His legendary success story, like his expressive physiognomy—reproduced millions of times on his cylindrical records—was familiar to all men. In short, he was almost universally regarded as one of the real makers of America, one whose career millions would have liked to emulate, and so, well suited to serve as a folk hero. His very appearance, and his widely reported sayings, racy, humorous, and original in flavor, but strengthened the will of the multitude to idolize him.

In the two decades following the St. Louis Fair he was chosen repeatedly, in popular newspaper or magazine polls, as America's "greatest" or "most useful citizen." [4]

He had gained both weight and girth as he grew older, yet kept them under control. With his big head, white hair, inky-black eyebrows and large pale countenance, and clad soberly in dark clothes with a "rolled-collar" shirt and string necktie, he could pass easily for an old-fashioned American worthy out of the early years of the Republic. (The second Mrs. Edison, meanwhile, had managed to modify somewhat his earlier, negligent habits of dress.)

Newspaper men could never resist him. With the members of the fourth estate Edison had always been at his ease, ever since the days when he worked as a press telegrapher. His talk was usually that of a strong nonconformist who delighted in statements that would surprise, stun, or at least provoke his hearers. Over the years he had never lost his jesting humor, in which he sometimes descended to downright *blague*.

It is not surprising, therefore, that his regular birthday interviews became a standing feature on the calendar of the country's leading newspapers. He was invited to comment for the public on almost every topic of current interest, and without hesitation offered his opinions on the problems of the family, modern woman, clothing, diet, medicine, on the vice of cigarette smoking, on education, war, religion, progress, and the future of science. Since he was the greatest of inventors, was he not possessed of the "greatest mind" among Americans? Had he not, as one of his journalistic admirers wrote, outdone Joshua, by bidding the sun to stand still in Gideon, and the moon in the valley of Aijalon? [5] His press conferences provided both a forum and a medium of free advertisement. Over all the wires, and soon over the wireless as well, his words, wit, and wisdom on all subjects were profusely broadcast. At all events, he was, in old age, a prophet with honor in his own country—and usually an entertaining prophet.

Genius matures early and is through early, in the view of some psychologists. Though there have been famous exceptions, this maxim holds true for many scientists and inventors. In the second half of his career, Edison showed perceptibly failing powers. To the more advanced scientists, moreover, his methods of work and his attitude now showed a regressive spirit that was inhospitable to the new, more precise tools of modern science, which, it was already foreseen, would conquer new worlds of knowledge. Some of his learned contemporaries thought it regrettable that Edison was cursed, so to say, with adulation. Being a "great thinker" and being a great inventor may be quite different things, as it soon appeared.

At the anniversary interviews, one of the reporters usually opened up by saying, "Happy birthday, Mr. Edison. How are you feeling today?" The question having been relayed to him, on one occasion, by his old secretary, Meadowcroft, Edison wheeled about suddenly, executed a high kick clearing his big laboratory table and all its paraphernalia, then brought his leg down to the floor and shouted, "Fine as silk!"

Or on another occasion, he might begin by growling, "This is a hell of a thing—to congratulate a man on the fact that he is getting old."

To what did he attribute his success? was one of the most oft-repeated questions. Was he not a *genius?* One of his justly memorable phrases in reply to such queries was, "Genius is ninety-nine per cent perspiration and one per cent inspiration."

His gospel was "Work—bringing out the secrets of nature and applying them for the happiness of man." [6] He for his part would never retire "until the doctors brought in the oxygen cylinder." At seventy-five as at forty-five, he felt well because he worked "two shifts." Mrs. Edison, it seems, had cut down his hours to only sixteen a day.

The gospel of work, or of salvation through work, reflected perhaps a subconscious puritanical ethic, that may have been absorbed from his

Calvinist mother. Edison was always noticeably ashamed to be found not "busy, working." *

The subject of work brought him to that of rest. Most people accomplished too little because they spent too large a part of their lives in sleep. He insisted that he was constitutionally able to get more done with less sleep than others—usually four hours a day, or five hours in later life. He also ate little, but varied his diet. The American people, he warned, should "diminish their intake of food." But exercise, or calisthenics, or sport, he would have none of, saying, "I use my body just to carry my brain around."

One of the burning issues of the era was the gold standard versus free silver. On this theme Edison spoke like a Midwestern radical of the old school, who believed fervently in "Coin" Harvey's doctrines of money inflation. These were the views widely held among the agrarian Populists and Greenbackers of Michigan. Gold, he declared, was "a relic of Julius Caesar's days" and interest on money was "an invention of Satan." The gold standard was "a trick mechanism by which you can control money." Edison had lived through many depressions and money panics and now firmly believed that gold should be unseated from its financial throne and replaced by a greenback currency or "commodity dollar."

"We must get rid of this terrible money evil," he said. He was not opposed to bankers as such, but fiercely assailed "the money broker, the money profiteer, and the private banker" who gained power through the fictitious value given to gold. When the Federal Reserve Act of 1913 was passed, Edison expressed his approval of it, but said it did not go far enough.[7]

On the other hand, when he heard the capitalist system itself attacked by socialists—relatively numerous and influential in this country toward 1910—then he would answer that "as men are now, there will always be capitalists and workmen. Capitalism is only wealth created by generations of men. . . . Somebody must take charge of this wealth and hold it in trust for the rest of mankind." Socialism was no remedy for the evils of the day. He could testify that everyone was not equally endowed. As for the socialist contention that most men were "wage slaves," he argued that "human slavery will not be abolished until every task now accomplished by human hands is turned out by some machine."

Although he had long inveighed against the crass commercialism of

---

* Similarly, Henry Ford made of work and duty the cardinal tenets of his religion. A man could never leave his business, but must "think of it by day and dream of it by night," Ford wrote. "Work is the salvation of the race. The Lord is working and will clear the land of those who do not go ahead!"

moneylenders and stock jugglers who cheated the honest inventor and worker out of the fruits of their labor, he was for individualism to the bitter end. Paradoxically enough, while Edison, like his intellectual disciple Henry Ford, proclaimed his own individualism, his contributions to mass production and mass entertainment served to hasten the coming of the Machine Age, which would suppress individual differences between men and regiment their lives to a degree never before known. Nevertheless, an Edison, for his part, must continue, as Bertrand Russell phrased it, "to make instruments to make other instruments, to make still other instruments, ad infinitum."

In the late fall of 1910 a lively and prolonged controversy was provoked by Edison when he candidly professed himself to be a freethinker in religious matters. To the question put to him during an interview, "What does God mean to you?" he replied, "A personal God means absolutely nothing to me." In short the idea of God, he maintained, was "an abstraction." Moreover, he declared himself an enemy of all superstition, deplored the fact that most people were "incurably religious," and pointed out that "billions of prayers" had brought no mitigation of natural catastrophes such as great wars.[8]

Nonconformity was more widely respected in America, and religious freedom more honored, fifty years ago than now. Even so, the lay and religious press soon resounded with attacks on Edison, and his mail bag at the Orange postoffice became so swollen with angry letters from pious folk that, as it was said, he was glad to get away to Europe for a vacation in the summer of 1911.

Ministers in their pulpits and some of the people in the front church pews protested vigorously against the great inventor's materialistic and utilitarian bias. According to one of the country's leading churchmen, Cardinal Gibbons, Edison's judgments about the concept of God and the afterlife were worthy neither of a sage nor a scholar, but of "a mere mechanic." As in the case of Darwin, who had likewise been too long devoted to scientific investigations, the "sense of poetry and religion" alike had become "atrophied" in Edison.[9] Less eminent but no less earnest believers also wrote him impassioned letters beseeching him "not to take from them their God." Business associates pleaded with him not to destroy the prestige of his great name and injure the Edison Industries by blasphemous utterances. For a while the controversy threatened to be as heated as those provoked during the Victorian era in England by the skeptical writings of Darwin, Huxley, and Tyndall, which were the favorite reading of Edison's youth.

Some of Edison's remarks, reflecting a spirit of levity toward the superstitious, were given circulation and further irritated the orthodox. To a minister's query as to the value of lightning rods as protection for his church spire, Edison replied, "By all means, as Providence is apt to be absent-minded." [10] In the face of attack, he held his ground stanchly, saying:

The criticisms that have been hurled at me have not worried me. A man cannot control his beliefs. . . . I try to say exactly what I honestly believe to be the truth. . . . I have never seen the slightest scientific proof of the religious theories of heaven and hell, of future life for individuals, or of a personal God. . . . I work on certain lines that might be called, perhaps, mechanical. . . . Proof! Proof! That is what I have always been after. I do not know the soul, I know the mind. If there is really any soul I have found no evidence of it in my investigations. . . . But I have found repeatedly evidence of mind. . . . I do not believe in the God of the theologians; but that there is a Supreme Intelligence, I do not doubt.

He continued to speak in praise of Thomas Paine, whose freethinking books had first been given him to read, when he was a boy, by his father. In 1916 he wrote a eulogy of Paine, as an introduction to a new edition of his works, declaring that he embraced Paine's view that "the truth is governed by Nature's laws and cannot be denied"; and he honored him also for writing, "The world is my country, to do good is my religion."

Meanwhile, what of his wife, who had been brought up in the Methodist Church by one of its foremost laymen? Evidently nothing Edison said or did could sway her convictions in matters of religion. Some years earlier he had written her an affectionate letter, in the course of which he remarked on the fact that President McKinley in his Thanksgiving proclamation of 1898, had thanked God for our victory over Spain. "But the same God gave us yellow fever, and to be consistent McKinley ought to have thanked him for that also. Thus we see terrible contradictions everywhere about the mystery of life." [11]

Mina Edison would neither agree with her husband nor contend with him. She continued to give her services unstintedly to the Methodist Church, helping to collect considerable funds, as well as making donations of her own, for the building of a new, much-needed Methodist church in East Orange. She also continued to invite clerics to her home. One day, without warning her husband, she invited six Methodist bishops to Sunday dinner. Edison might have managed peaceably with one—but six was too much. The occasion was marked by a heated argument about the Bible, evidently provoked by Edison. In the end, to Mrs. Edison's great embarrassment, he threw up his hands, exclaimed, "I'm not going to listen to any more of this nonsense!" and stalked out of the room.

By dint of great tact, Mina Edison at length reached a friendly under-standing with her husband on the subject. Ministers were to continue to come to dinner (though not six at a time), but there was to be no discus-sion of religion; and all was peaceful again.[12]

After some years Edison qualified his profession of agnosticism by stressing his belief in the existence of a Supreme Intelligence. "It could almost be proved from chemistry," he said on one occasion. He also expressed the thought that it *might be possible* that the human personal-ity, in some changed form, survived after death.

At this period, toward 1920, intense interest had been aroused by the spiritualist teachings of the British scientist Sir Oliver Lodge. Though Edison cared little for the table-tipping theosophists, he now aired some views of his own about "life entities." He said he had long held the notion that the cells of the human body possessed "intelligence," and, taken to-gether, constituted "a community made up of its innumerable cells or inhabitants." A man, he concluded, was not merely an individual, but also "a vast collection of myriads of individuals." The intelligence of a man, then, consisted of the combined intelligence of all the cells, or "en-tities" within him, "as a city is made up of the combined intelligence of its inhabitants." After death those cells were separated and diffused, yet persisted in some new form, served over and over again, lived forever, and could no more be destroyed than matter. Thus, he demonstrated, he had burned his thumb; but the skin was perfectly reformed and replaced. "The life entities rebuilt that thumb with consummate care."

Such views were not much more palatable to orthodox religious souls than his earlier professions of agnosticism. There was something in the inventor that enjoyed the excitement and controversy he aroused; and, in a somewhat mischievous spirit, he returned to the same subject some years later when he issued statements hinting that he was actually en-gaged in serious investigations to determine whether the human personal-ity survived after death. "Edison Working on How to Communicate with the Next World!" was the sensational title of an article in a popular maga-zine in 1920. "I am at work," he said, "on the most sensitive apparatus I have undertaken to build, and await the results with the keenest inter-est." Such a scientific apparatus, he mused, "might be operated by per-sonalities in another existence or sphere who wish to get in touch with us; it will give them a better opportunity to express themselves than ouija boards or tilting tables." [13]

This mystifying declaration aroused new excitement, which soon died down, however, when Edison avowed that he had been misrepresented by some poor reporter. But Henry Ford was truly moved. He too had been going about posing before the newspapers as a "great thinker" and aping the admired Edison's methods. His great friend's irreligion had deeply troubled the much simpler soul of the automobile magnate. But when he heard about Edison's renewed search for those entities (which Ford called "enities") he remarked with an air of supreme relief, "The greatest thing that has occurred in the last fifty years is Mr. Edison's conclusion that there is a future life for all of us."

2   On the question of education, Edison's opinions aroused almost as fierce controversy as his idea on religion. What was abhorrent to him, he declared in a series of interviews and articles that appeared between 1911 and 1914, was all formal schooling and college training, such as he himself had managed so well to do without. School, for him, meant only "the dull study of hieroglyphs," teaching and learning by rote. Like Rousseau and his followers, he advocated that every child should be encouraged to develop his faculties by *making things,* and by observing nature for himself, instead of having his mind crammed with memorized facts.

I like the Montessori method. It teaches through play. It makes learning a pleasure. It follows the natural instincts of the human being.

The present system [he said] casts the brain into a mold. It does not encourage original thought or reasoning.... The seeing of things in the making is what counts.

There was a solid core of truth in these views, which were, to be sure, not at all original with Edison; in the United States John Dewey had for years advocated a thoroughgoing reform of our lower-school system that would eliminate the teaching of ready-made lessons by rote.

Edison, however, also assailed the whole idea of teaching the liberal arts in higher institutions.

What we need [he said] are men capable of doing work. I wouldn't give a penny for the ordinary college graduate, except those from institutes of technology. ... They aren't filled up with Latin, Philosophy and all that ninny stuff. America needs practical, skilled engineers, business-managers and industrial men. In three or four centuries, when the country is settled and commercialism is diminished there will be time for literary men.

Our colleges, he insisted, must teach "something useful."

During a discussion of this question with one of the university-trained scientists in his laboratory, strong objections were offered to his notion that the study of classical literature and the arts could contribute nothing useful to the mind. Had not Jesus asked long ago, "Who knows what is useful and what is not useful?" the other pointed out. No, Edison reiterated, Latin was a dead language used only in the Roman Catholic Church.[14]

One somewhat novel proposal he made at this time was to introduce the use of motion pictures in schoolrooms. To this end he prepared a few educational pictures which, as he promised, would teach "everything from mathematics to morality." Lessons would be rendered so vivid that children would "really want to go to school." When the pictures were ready, a group of educators were invited to see them at the Edison Laboratory, among them Professor Dewey of Columbia University. What they saw, however, made them shake their heads. Perhaps some day such methods could be perfected. But how, one educator asked, could algebra be taught through motion pictures?

Algebra, Edison answered, was of little practical use. "I can hire mathematicians at fifteen dollars a week, *but they can't hire me,*" he said.

He went on to say that he would give lessons in science graphically, and would even show steam going through an engine's cylinders.

"But how can you photograph steam, even through glass?" one visiting professor asked. To which the irrepressible inventor quickly answered, "Oh, I would color the steam with ammonia and a solution of hydrochloric acid!" [15]

Undaunted by the criticism that had greeted his proposals for the reform of the schools, he soon afterward took up another aspect of the problem of education and vocational training. The proponents of intelligence tests for I.Q. ratings had just recently begun to attract notice. Edison embraced their ideas with typical enthusiasm. He drew up his own questionnaire and announced to the newspapers that henceforth all applicants for jobs in his shops or laboratory would be subjected to intelligence tests by his system, the "ignoramometer." The key faculty, he declared, in a good worker or in a good executive was that of the memory. To be of real value a man must possess a reservoir of useful knowledge.

For many weeks the newspapers followed the results of Edison's intelligence tests, which he was giving to thousands of persons, and published questions included in them, such as:

> Where does the finest cotton grow?
> Who invented logarithms?
> Where is Korea?
> What voltage is used on street-cars?
> Who composed *Il Trovatore?*
> What is the weight of air in a room 20 x 30 x 10?
> How is leather tanned?

After a time he announced triumphantly that his "ignoramometer" had proved that out of 718 men of college education only about 10 per cent managed a "fair" or passing grade! People are so *ignorant!* he exclaimed. "Only two per cent of the people *think,* as I gather from my questionnaire. . . ." There were numerous college professors, he assured reporters, who did not know where Korea was, or what were the principal ingredients of white paint. He would never employ such persons, for when he called upon one of his men for information or for a decision, he wanted the answer right away, and could not wait for data to be looked up "tomorrow." [16] He added, shrewdly enough, that he did not really care whether a man knew the location of Timbuktu or not. "But if he ever knew any of these things and doesn't know them now, I do very much care about that in connection with giving him a job."

Again educational authorities girded at Edison for assigning too great importance to the mere remembering of facts and too little to the power to reason and find out facts. Meanwhile, some of the items blandly inserted in his questionnaires raised some suspicions that he was being less than serious in this inquiry. For example:

If you were desirous of obtaining an order from a manufacturer with a jealous wife, and you saw him with a chorus girl, what would you do?

Various journals and magazines joined in the fun and proposed that Edison himself be subjected to such tests. Willingly, he assented, saying, "I never give another man a dose of medicine that I wouldn't take myself." Thereupon he proceeded to rattle off answers to long lists of questions put to him that touched a variety of subjects, his answers averaging about 95 per cent in accuracy.

What was not known at that period was that his attempts to educate his own children, so far as they went, had been a complete failure. (I speak here of the elder children born of his first marriage; those of the second marriage were quite differently brought up by Mina Edison.)

On the strength of his own experience, he had discouraged his eldest son, Tom junior, from going to college. Tom had been sickly in youth; such schooling as he had did not reach beyond the high-school stage. As for William Leslie, the second son, whatever plans his father might have entertained for him, he showed no aptitude for learning and was not interested in going beyond a preparatory school. The spirit of the elder sons was undoubtedly affected by the early loss of their mother in 1884. It was not easy to be the son of such a father, so long away from them, so often obsessed by his work, and seemingly so self-centered or remote.

Edison's second marriage, moreover, had introduced a painful division in the family. When she became the mistress of Glenmont in 1886, Mina Edison was but a few years older than her stepdaughter Marion and her stepson Tom and naturally found it difficult to impose her authority over them. It is probable that the two boys and Marion had been allowed to run wild during the two years between their mother's death and the second marriage. Moreover, Mina Edison was somewhat reserved by nature. There can be no doubt that she tried earnestly to do her duty by her stepchildren. But, though concerned for their welfare, she herself sometimes realized that she was perhaps wanting in "the captivating ways" that might have won them to her; and on occasion told herself that she must try to be "less self-assertive." [17]

Her high-spirited stepdaughter Marion had shown from the outset an adolescent hostility, or jealousy, toward her young stepmother. Once she had been sent to Europe to study, she preferred to remain there, marrying a German officer; she did not return to this country until the end of World War I, when she was divorced from her husband. Always an attractive and independent personality, Marion Edison Oser was never a "problem" to her parents, and in her later years maintained very friendly relations with the younger Edison children.

Tom junior had much sweetness of personality and showed some little mechanical bent during the years when he stayed at West Orange and worked about his father's laboratory. Sad to say, he even tried to be an inventor on his own. Later, like his brother Will, he drifted away from the parental home. Though regularly supplied by his father with enough funds to live on, Tom was often in trouble, and for a period drank a good deal. In partnership with some raffish characters he tried a number of dubious commercial ventures. One of these, launched as the "Thomas A. Edison, Jr., Electric Company," organized in 1898, created much anxiety for the father, when he discovered that the Edison name was associated with promoting such alleged "new inventions" as a machine for "photographing thought." When this enterprise folded up, it was succeeded by an "Edison, Junior, Chemical Company," that sold an "Electric Vitalizer" purporting to "cure anything from catarrh to locomotor ataxia." The father investigated again, and found that his rather dreamy son was being exploited by dishonest associates, to whom he lent his name for small fees. In 1904 Edison *père* brought an injunction suit charging fraud against the new company, and forced it into liquidation. He then formally disowned Tom junior, principally as a measure of legal protection. But as he was, in reality, very fond of his eldest son, a reconciliation was effected, and the prodigal youth was, after a time, brought back again, to be employed in various departments of the Edison Industries.

William Leslie, a tall and athletic youth, created anxieties of a somewhat different nature for his parent. A clubman, he played the role of man about town for some years. Against his father's wishes, he enlisted for service in the Spanish-American War. (In 1918, when he was forty, he also served in a tank division of the AEF.) On returning from Cuba, he promoted various small business schemes, with the help of money furnished occasionally by his father; these proving unrewarding, he turned to being a gentleman farmer, living with his wife on a country estate in New Jersey. Will was sometimes described as headstrong and stormy; despite efforts on his father's and Mina Edison's part to restore good relations, he maintained an attitude of hostility toward his stepmother.*

The destiny of the later children, born of Mina Edison, proved to be far more fortunate than that of the elder ones. Their mother brought them up in her own evangelical faith, reared them with care and affection, and stoutly opposed her husband's counsels that their formal or religious education be limited. On the contrary, their daughter Madeleine was sent to Bryn Mawr College; in 1914 she married John Eyre Sloane, of South Orange, New Jersey. Charles and Theodore attended good private schools and completed their education at Massachusetts Institute of Technology.

It was sometimes thought that if Charles could have chosen his own way of life, he might have tried to be a writer. In his salad days, toward 1915, he composed some ballads and light verse which were published in *Bruno's Weekly,* one of the "arty" periodicals of the time that issued from Greenwich Village. At that period Charles Edison also showed some passing interest in owning and managing the Little Thimble Theatre, one of the "little theaters" of the Washington Square quarter. He did this in order to seek new talent for phonograph recording, he relates, and as a part-time job in his evenings, by commuting from West Orange to New York after work.

About a year after completing his schooling, in 1914, he had set to work in earnest learning the family business under his father, with whom he always got on very well. Several years later he was made executive vice-president; on his father's retirement (in 1927), he was to succeed

* T. A. Edison, Jr., suffering from chronic illness, committed suicide in 1936, five years after his father's death, at the age of sixty. W. L. Edison, in 1932, began a suit to have his father's will (by which he was left very little money) set aside, but a composition was affected. He died in 1941. Neither of the elder sons had children.

him as head of Thomas A. Edison, Inc.—which had been the fond wish of Mrs. Edison all along.

Theodore Miller Edison, born in 1898, made an excellent record as an engineering student at M.I.T., showing, in his own way, a passion for science. But to the elder Edison's dismay, his youngest child seemed determined to become a mathematical physicist—one of the breed that Edison had always belittled. This made for a little dissension in the family, for the tall, black-haired, blue-eyed young Theodore proved to be as strong-willed as his father, and in the face of paternal disapproval continued his studies of higher physics.

"Theodore is a good boy," the father said, "but his forte is mathematics. I am a little afraid . . . he may go flying off into the clouds with that fellow Einstein. And if he does . . . I'm afraid he won't work with me." [18]

"Is it not strange that a man who always did exactly what he pleased would not let his own children do anything that they wanted to do?" a member of the family is reported to have remarked at this time. [19]

3   The outbreak of war in Europe in August, 1914, caused much anxious soul-searching among Americans, though opinion here, at the outset, was largely disinterested and humane. The inventor and Mrs. Edison were visiting the Henry Fords at Dearborn, a few months after war had begun, when a Detroit newspaper reporter interviewed Edison and recorded his thoughts on the great conflict. Principally it made him "sick at heart" for mankind, as he said. "This war had to come. Those military gangs in Europe piled up armaments until something had to break." Reflecting American public sentiment, which was decidedly neutral in the first year of the war, he asserted that he would have no part in it. "Making things which kill men is against my fiber. I leave that death-dealing work to my friends the Maxim brothers," he added with a malicious thrust at his old rivals in electric lighting. [20]

Henry Ford was highly gratified by his great friend's professions of humane principles. Having already become one of the world's biggest industrialists, Ford now yearned to distinguish himself in other ways than as a manufacturer of horseless carriages—in short, as a public figure and "practical humanitarian." He too went about the country nowadays, expounding his views in the public press on a variety of current questions. At this period his heart was set on keeping America out of war; he also contemplated a remarkable scheme, praiseworthy if not practical, for bringing the warring powers of Europe to the peace table.

In the late spring of 1915, when a vigorous agitation for "preparedness" was being sponsored by patriotic and other groups, Henry Ford, who publicly opposed this movement, was suddenly filled with consternation on reading in the newspapers that his fellow humanitarian, Edison, had changed his mind and allied himself with groups who were urging that we expand armaments and train military personnel for the eventuality of war.

Still worse, the press reported in July, 1915, that Edison had accepted the invitation of the government to head a board of scientists and inventors advising the military on the latest and most effective instruments of war. If true, it was a complete reversal of Edison's earlier attitude.

At first Ford refused to believe these reports and stated publicly that he knew Thomas A. Edison "would never use his great brain to invent weapons for the destruction of human life and property." Edison made no comment on this statement. But in private he and Ford talked things over in friendly fashion on several occasions in 1915.

Henry Ford had conceived the idea of a "peace ship," in which he was to sail for Europe toward Christmas, 1915, at the head of a remarkable assortment of peace advocates, in an effort to persuade the world's most powerful nations, in the full course of battle, to lay down their arms. It was a wonderfully quixotic enterprise, but not to Edison's taste. Ford was a persistent man; he came to New Jersey and tried hard to enlist the great inventor in his peace expedition. Edison is reported to have put him off with "I can't get away, I've got too much to do." Ford was extremely eager to have some national figures join him. William Jennings Bryan had refused to come, and Miss Jane Addams was held back by illness. There is a (perhaps apocryphal) story to the effect that when Edison saw Ford off at the dock, Ford cried to him that he would give him "a million dollars" if he would come along, and had to repeat his offer in a shout so that Edison could hear. But the old inventor just shook his head.[21] Ford returned, a while later, from that much-ridiculed, but well-intentioned winter cruise to Stockholm, a bitterly disillusioned man and no longer a liberal. But these incidents did not affect the almost idolatrous friendship he had conceived for Edison.

As late as the month of May, 1915, at the time of the sinking of the *S.S. Lusitania* by a German U-boat, with heavy loss of life among its American passengers, Edison said that if Germany continued to commit such outrages, we might find some way of retaliating that was short of war, remarking, "How can we help by going into the war? We haven't any troops, we haven't any ammunition, we are an unorganized mob."

The expression "unorganized mob" aroused much controversy. But Edison also added at the time that although he did *not* want us to fight, "I believe we ought to be ready for it . . . We ought to have on hand equipment for an army." [22] A few days later, he referred to Germany's advanced technical organization (which he had observed during a journey to Europe in 1911) and suggested that the United States might well emulate Germany's intensive use of scientists in preparing for war. He promised that if his country needed his inventive skill to make her safe from attack, he would willingly respond to any call upon him.[23]

The military services were already making plans for expansion. Secretary of the Navy Josephus Daniels read Edison's latest pronouncements and was inspired by them, as he said later, to attempt to organize an advisory board of scientists and inventors whose minds would be at the service of the military.

Since nothing could be more natural than that Edison himself should head such a board, Daniels wrote him on July 7, 1915, inviting him to undertake "this very great service for the Navy and the country at large."

President Woodrow Wilson approved of the proposal; it was Wilson's declared purpose to enlist "the best minds" in government services. At the time there was also much learned discussion (perhaps inspired by Thorstein Veblen) of the predestined leadership of the engineer and technologist in our society. Daniels was no doubt thinking along these lines when he appealed to Edison as "the man above all others who can turn dreams into realities and who has at his command, in addition to his own wonderful mind, the finest facilities for such work." That Daniels also wanted to appropriate Edison as a symbol, is indicated clearly in his letter when he writes, "I feel that our chances of getting the public interested and back of this project will be enormously increased if we can have some man, whose inventive genius is recognized by the whole world, to assist us . . ." Edison replied promptly with the equivalent of a cheerful "Aye, aye, sir." [24] As he wrote to a person who protested at his lending himself to the work of human destruction, he now believed that he was but helping in the defense of his government and people against the danger of an attack.

On receiving his acceptance, Secretary Daniels hastened to West Orange and discussed the plan of the proposed board with Edison, who shrewdly advised that instead of just naming two or three famous inventors, such as himself and Wilbur Wright, as its members, Daniels should address himself to the country's leading scientific and engineering societies, and invite them to select outstanding persons in their different fields. This was done, and a rather impressive group was gathered together for the Navy Consulting Board, including L. H. Baekeland, Willis R. Whitney, Frank J. Sprague, Hudson Maxim, Peter Cooper Hewitt, Elmer Sperry, and Thomas A. Robins, Edison's friend of ore-milling days, who was to serve as secretary of the Board. "It might be well to have a mathematical fellow to do the figuring," Edison remarked; and so they invited a professor from M.I.T. to come in. Later Arthur Compton, Robert Millikan, and Lee De Forest were associated with the original group.

The news that the "Wizard" himself had been enlisted for the national defense was soon bruited in the newspapers, together with fanciful rumors that he had gone to a hiding place in the New Jersey mountains to invent some infernal machine that would end the German submarine menace.

In reality military preparations got under way very slowly, while the Wilson administration pursued its wavering line of neutrality prior to the Presidential election contest of 1916. Over three months passed before the civilian experts of the "Inventor's Board," from which so much was expected, were called to their first meeting at the Navy Department in Washington, mainly for the purpose of establishing liaison with the Navy's bureau chiefs and with its Assistant Secretary, Franklin D. Roosevelt. The admirals had no precedent for dealing with a body of civilian scientists and at first treated them with the utmost reserve. What could mere landlubbers know about maritime gunnery and ballistics? The group of scientists were taken aboard the *S.S. Mayflower* for a little cruise around the Navy proving grounds; then they were politely told to wait until the Navy bureau chiefs asked for their advice.

Because of the publicity given the so-called "Edison Board," thousands of would-be inventors swamped the Navy Department with models or drawings of new military devices. There was no organization to handle such things, and they were all buried away in the capacious files of the Navy Department. The civilian board, which had no authority in any case, met only at rare intervals thereafter, up to the autumn of 1916.

Meanwhile Edison's activities reflected the shortages and the opportunities of a world at war. Importations of industrial chemicals from Germany and England had been halted, and there was little domestic supply. Edison himself was desperately in need of carbolic acid, benzol, and aniline oils, required in the manufacture of his phonograph records; and he was short of potash for storage batteries. He decided to make some of these chemicals himself and very soon, by the beginning of 1915, had thrown together several impromptu chemical factories at Silver Lake, New Jersey.

In one he made carbolic acid, in another benzol, a basis for aniline oils and paraphenylenediamine, which he not only used in making records but also supplied profitably to the textile and fur industries as the basis for dyes. The chemicals he manufactured were also basic materials for explosives—and his shops would be ready to make those too in the hour of need. Thus Edison flourished more than ever during the war boom, the number of personnel on his payrolls mounting to about 6,000.

In the late autumn of 1915 he was off with Mrs. Edison and his son Charles, on a vacation trip to San Francisco, where the Panama-Pacific Exposition was then being held. Ford joined them, and they appeared in San Francisco together in time to witness the special celebration of "Edison Day" at the Exposition on October 21, anniversary of the incandescent lamp invention. This was just before Ford's Peace Ship fiasco; Ford was as much a folk hero then as Edison. Together the two heroes of American industry rode in an open car through the streets of the city, in a procession officially held in honor of Edison, while fifty thousand persons stood in the rain and fog to cheer them.[25]

Tension between the United States and Germany rose in the early months of 1916. To support the preparedness movement, Secretary Daniels asked Edison and other members of the Navy Consulting Board to march in a great Preparedness Parade that was to be held in New York on May 13, 1916. Edison promptly assented. But he had been the recipient of a number of letters from presumed pro-Germans or cranks who threatened his life. Mrs. Edison, therefore, earnestly begged him not to run the risk of appearing in such an exposed place as at the head of that parade, among the crowds of New York. Only after an escort of government secret-service men was assigned to accompany him did she give her consent.

"Even so, and in spite of these precautions," as a friend of the family who marched in that procession recalled,

Mrs. Edison herself decided to accompany the parade from its start at Washington Square until the end of its route at Central Park. She was tall and wore a large violet-colored hat, and I remember seeing her forcing her way through the crowds on the sidewalk as we marched uptown, frequently turning to look at her husband. As she forged along, I could see a wave of spectators being pushed off the sidewalk into the street. It was a feat which called for great physical strength, while showing her devotion to Mr. Edison.[26]

Now nearly seventy, with his thick white hair, deep-set eyes, stalwart and deep-chested as an old oak, America's Most Useful Citizen was still one of the most imposing-looking of men; volley after volley of cheers greeted his appearance as he marched up Fifth Avenue that day with vigorous strides.

During that summer's hard-fought election campaign, Edison gave his public blessings to President Wilson's candidacy, though he had always hitherto voted as an independent Republican. This was considered a great boon to the Democratic Party; but still the party treasury was bare of funds. Secretary Daniels, therefore, arranged to meet with Edison and his friend Henry Ford, at lunch in a private dining room of a large New York hotel. Vance McCormick, Democratic National Chairman, was also on hand, all eagerness to extract a large check from Mr. Ford.

All his life Edison professed mistrust of politicians, and he knew well what was afoot; instead of helping, he mischievously delayed the proceedings. When their repast was over, he challenged the long-legged Ford to a trial of strength, as Daniels has related:

Pointing to a large chandelier with many globes ... in the middle of the room, Edison said: "Henry, I'll bet you anything you want to bet that I can kick that globe off that chandelier." It hung high toward the ceiling. Ford said he would take that bet. Edison rose, pushed the table to one side ... took his stand in the center and with his eye fixed on the globe made the highest kick I have ever seen a man make, and smashed the globe into smithereens. The automobile manufacturer took aim, but missed the chandelier.... Edison had won and for the balance of the meal was crowing over Ford: "You are a younger man than I am, and yet I can outkick you." He seemed prouder of that high kick than if he had invented a means of ending U-Boat warfare.[27]

During all this comedy, McCormick was perspiring with impatience to put the bite on Henry Ford. But Ford, when asked for donations, refused to give any cash; giving money to politicians was "the bunk," he said. However, he undertook to help Mr. Wilson in his own way by inserting paid advertisements in newspapers, favoring the Democratic candidate.

4   The scientists of the Navy Consulting Board grew more active as the war came closer to America. At the first Board meeting, at Washington, in October, 1915, Edison had been formally appointed its president; but, as he could hear almost nothing at the proceedings, W. H. Saunders, a well-known engineer, was made acting chairman. Edison participated by having someone next to him tap messages in Morse code on his knee, or writes notes for him of what was being said.

There were some who feared that the imperious old inventor would get on badly with his scientific peers. But Robert Millikan (the discoverer of cosmic rays) related afterward that Edison turned out to be "a much greater man than I expected to find, simple, direct, intelligent, and un-spoiled," and showed more "essential modesty" than one had suspected in him in view of "the disease to which he was exposed early—publicity and adulation." [28]

Since the main problem, in the event of transatlantic warfare, was to stop the German submarines, Edison, with penetrating common sense, called for maps showing the locations of all ships sunk, and thus indicat-ing the habitual radius of the U-boats. Though the undersea attacks had been raging for more than a year the U.S. Navy had no such record.

Edison said he had a theory, based on his own observations during numerous ocean crossings, that most of the merchant-ship officers hardly ever used navigation techniques, but guided themselves by various light-houses and familiar coastal points. The German U-boat commanders, knowing this, lay in wait for them in those localities. After many months' delay, a large map giving the locus of the submarine sinkings was ob-tained from the British Admiralty, and it supported Edison's theory completely.[29] Before one could plan to fight the "sea wolves," he reasoned, it was necessary to study their habitat and *modus operandi*. It was, in truth, by such logical and informed measures that the U-Boat peril was gradually overcome.

An even more significant proposal of Edison's, made as early as 1915, was that the Navy establish a scientific research laboratory for the de-velopment of new weapons. He himself, a private individual, had spent millions for scientific research; several industrial firms, such as Bell Telephone and General Electric, now boasted of much greater laboratories than his own. Yet the nation's military services, upon which all her se-curity depended, had never evinced any interest or spent any money on technical research since the days of John Paul Jones. The Navy Depart-ment also lacked facilities for studying the proposed inventions of scien-tific workers, which were simply filed away. Improved weapons were adopted in haphazard fashion, usually after experimental work had been carried on by private ordnance manufacturers.

Having observed how loath the Navy bureaucrats were to try new ideas, Edison formed a very poor opinion of them and, so, proposed that the Naval Research Laboratory he had in mind should be run only by civilian experts, under the Secretary of the Navy. "No Naval officer," he wrote, "should have anything to do with it. Annapolis produces only students who immediately enter for life into a system that takes away every incentive by which superior men can advance." [30]

With characteristic tenacity, and in the face of stout bureaucratic resistance by the admirals and bureau chiefs, Edison pushed his proposal to establish a Naval Research Laboratory, so that eventually, after the war was over, such an institution was authorized by Congress—though not under the civilian rule he advocated.

The Naval Research Laboratory was to be Edison's most vital contribution to the national defense; between the two world wars, it would be the only American institution for organized scientific research in weapons, hence of enormous importance in preparing for the technological warfare of the 1940s.*

Late in December, 1916, Secretary Daniels called on Edison and told him in confidence that the military prospects of the Allied Powers were now very dark; according to the British Admiralty, the Allies were fated to lose the war unless they could stop the German submarines. He now asked Edison to give all his time to inventing antisubmarine devices. Edison promptly turned over his business affairs to other hands and plunged into this work.

Giving free rein to his fine mechanical imagination he contrived a whole series of defensive instruments against U-boats: a device for the quick turning of ships; one for locating guns and torpedoes; another for raising smoke clouds to cover merchant vessels; antitorpedo nets; hidden lights; turbine heads for projectiles; collision mats; and finally a sonic apparatus employing a submerged phonograph diaphragm, to detect submarines.

In April, 1917, while Edison was thus occupied, the United States entered the war against Germany. Later in the summer he went sailing out of New London on an old yacht that had been converted into a "chaser," in order to test some of his schemes. To Ford, who by now had "enlisted" his factories for war service, he sent a message in September:

I am at sea nearly all the time in a 200 foot submarine chaser. Am around inland waters near New London, and have a large number of experiments still to finish.

---

* Edison had a number of imbroglios with the Navy "brass"—as did also Baekeland, Hudson Maxim, and other scientists on the Navy Consulting Board. The country's most famous chemical engineer, Baekeland, declared at a meeting of the Navy Consulting Board that the typical Annapolis and West Point graduates who headed military bureaus "trembled with fear that something new might go wrong" and thus their own careers might be jeopardized. Edison was convinced at the time that no progress in military research would be possible if the admirals ran the proposed Naval Research Laboratory. As he wrote Daniels shortly after the war ended:

"I do not believe there is more than one creative mind produced at Annapolis in three years ... If Naval officers are to control it the result will be zero ... When you are no longer Secretary I want to tell you a lot of things about the Navy that you are unaware of."

(To Josephus Daniels, November 7, 1919, Edison Laboratory Archives.)

Hope Ford gets a chance to come on yacht and help out. The experiments very interesting.[31]

He had a phonograph diaphragm held in the water by an outrigger which he hoped would record the noise of torpedoes being discharged several miles away. Then there was a contraption which—upon the detection of a torpedo discharge—would immediately turn a merchant vessel completely around, at an angle of 90 degrees, to evade the missile. As he related after the war:

I constructed a conical bag, resembling a fool's cap, nine feet wide, eighteen feet long. The bag was made of canvas and the rim of iron ... attached to four-inch hawsers. I rigged up four of these bags on a tramp steamer, two bags on each side. When the steamer was traveling full speed, I threw over the bags and the vessel came to a halt within 237 feet.[32]

Edison himself declared afterward that during such trials his heart was in his mouth, for fear that the sudden stopping and veering of an old freighter going at full speed might break it in two—though fortunately this did not happen. To obtain the use of that old converted yacht had required a private "war," lasting four months, between Edison and the Navy bureau chiefs. That, he said to intimates, angered him a great deal. He was also made ill at about the same time, as a result of a laboratory accident while experimenting with some phenolic acid which exploded and temporarily injured his eyes.

Because he seemed so ill, and because she feared that he faced serious danger—when so many ships were being sunk by submarines off our coasts —the ever loyal Mina Edison determined to accompany him on board ship. She was subject to seasickness; the officers considered her presence a "nuisance"; and the discomforts of life on this small ship would have been daunting to many another woman. But, for her, as she wrote to her daughter, "father was the only consideration," and he "does seem to need me." [33]

In reality the most effective work on submarine detection seems to have been done by a team of electronics experts headed by Dr. Willis R. Whitney, of the General Electric Laboratory, and Lee De Forest. Their "listening device," used in conjunction with the first De Forest radio telephones, which were set up on submarine chasers, gave some promise of improved defense against U-boats. When successful tests were completed and reported to Edison, then at West Orange, he was beside himself with excitement, and expressed the highest admiration for Whitney. He immediately telephoned the Secretary of the Navy in Washington, over an amplifying long-distance telephone he now had.

"Friend Dan'ls," he shouted in a voice that the Germans might have heard, "we have got the submarine *cured!* You must come to New London at once." Daniels and his staff came on the next day, and measures were soon taken to put the detection apparatus to good use.

War service was a frustrating experience for an aging inventor. Official Navy reports, as of 1920, show that thirty-nine inventions were submitted by Edison. He himself has said:

I made about forty-five inventions during the war, all perfectly good ones, and they pigeon-holed every one of them. The Naval officer resents any interference by civilians. Those fellows are a close corporation.[34]

As the war drew to its end, he carried on an agitation in government circles in favor of having the proposed Naval Research Laboratory not only placed under exclusively civilian control, but also located somewhere near New York, or at any rate as far as possible from Washington or Annapolis. To this end he strongly urged his point of view in testimony before a Congressional committee in 1919. But the political strength of the Navy leaders was too much for him. To Edison's intense disappointment, the country's first institution for scientific research in the military field, by act of Congress in 1920, was located in Washington and placed under the complete control of Navy officers. The admirals had had their way; Edison was always thereafter very unforgiving toward them. Nevertheless, the Navy, at Secretary Daniels's wish, tendered the Distinguished Service Medal to Edison in 1920, a rare honor for a civilian. He refused it, saying that he "did not wish to hurt anyone's feelings," and was aware that other civilian scientists who had helped in the war effort would believe they were entitled to equal honors.[35]

When the armistice came he turned his attention again to his own business, then, of course, greatly expanded. After the wartime orgy of inflation, a depression was under way by 1919. The Thomas A. Edison Works, like other "war babies" needed to be trimmed down. It even boasted now of a new and large Personnel Department for hiring labor, that had been recently organized by Charles Edison.* Edison *père* ordered Charles to reduce this department, and all but five of its employees were eliminated. But that proved to be not enough for the Old Man, who exclaimed: "Hell, I'm doin' the hirin' and firin' around here,"

---

* In an unpublished memoir dated 1953, deposited with the Columbia University Oral History Project, Charles Edison has stated that he tried earnestly to introduce improved labor relations at Thomas A. Edison, Inc., despite the opposition of his father. When he had first come to work in 1914 and was assigned to the storage battery plant at Silver Lake, New Jersey, Charles found conditions and hours of work extremely depressing—though no worse than in the average chemical fac-

whereupon he dismissed all the rest of the Personnel Department. Closing up his wartime chemical factories, as no longer needed, he removed many hundreds of workers from his payrolls. But the old-timers, even the feeble ones, he insisted, must remain, especially in the laboratory.

Charles, who was doing much of the administrative work nowadays, counted fifteen such superannuated persons in the laboratory "who weren't doing a damn thing," but whom he dared not fire, because of his father's sentiment about them.

---

tory of those times. The men labored up to twelve hours a day and seemed to him sodden with fatigue. The street opposite the Edison Storage Battery Company's shops was lined with cheap gin mills, where the workers would drink themselves into forgetfulness.

A raging dispute with his father over labor policy and the mass dismissals of 1920 caused the only serious strain that ever arose between them, he states, though this soon passed, for Charles fairly worshiped the Old Man.

chapter XXII

Sunset, with electric illuminations

1   The story of the intimate friendship between Henry Ford and Edison forms an important chapter in both men's later careers and strongly illuminates the special qualities of each. They were similar in social origin and background, shared many interests in common, and yet were strikingly different in mind and temperament. Ford was extremely sentimental; he kept the most trivial memorabilia of his boyhood days, loved rustic dances, old tunes and country folkways, interest in which he tried to revive among the helots of the assembly lines. Though one of the chief architects of the Machine Age, he amused himself by re-creating, at immense cost, an old-fashioned country village in Dearborn, such as he had known in his youth. He was endowed with a fine engineering talent as well as a fierce will to power; nevertheless he was an exceedingly simple being at bottom, and dangerously ignorant. Edison's was a complex personality, but a unitary one; if limited on some sides, he gave himself a tremendous education on others; far from being sentimental, he was by disposition a realist and in his thinking showed great detachment and critical power.

Since their first encounter in 1896, which had been so inspiring to Ford, the automobile magnate had regarded Edison with a reverence that was almost superstitious. Their next meeting came about in 1909, when Ford was embroiled in his long court fight to break the monopoly of the Detroit manufacturing group which owned the automobile engine patents of G. B. Selden. At a critical point in this affair, Ford is said to have walked in on Edison at his laboratory unannounced and sought his advice. The Model T was under way, and the future of his mass production plans depended on his decision. Edison advised him to stay out of the automobile trade association and fight their licensing monopoly. It was counsel that reflected Edison's objectivity, for he himself was then partner to a motion picture monopoly, based, however, on his own patents. When Ford triumphed over his opponents in court in 1911, he felt again that the uncanny old inventor had guided him well.

Edison called him "Henry"; but Ford never addressed the inventor save as "Mr. Edison." [1] Edison was quick of thought and speech, while Ford was a dull talker. At first the older man had felt the other to be something of a simpleton, but after a while he said he was "afraid of him, for I find him most right where I thought him most wrong." [2]

When the automobile manufacturer achieved world fame early in 1914 by announcing the $5 minimum wage, Edison (who was still paying $2.50 a day) after some soul searching, ended by expressing approval of Ford's program, which he said was equivalent to "an industrial revolution" in itself. Edison never professed to be a lover of mankind en masse. Ford did, in those early days; but, on the other hand, he had no intention of losing money because of his high-wage policy. His plants were equipped with the world's most advanced assembly lines and used electric motors on a very large scale. The whole scheme of "rationalization," he said, was inspired by Edison's pioneering. [3]

Ford was a latter-day prophet proclaiming the new gospel of cheap goods and high wages for labor; his volume output and winnings became fabulous. Unlike the Protean Edison, he was single-minded, in fact, a one-invention man. It has been suggested that Edison, sometimes badly frustrated in his attempts to become a big industrial captain, envied the ruthless business efficiency and vast commercial triumph of Ford. [4] If this was true, the emotion of envy did not remain with him very long; he could never have been happy in Ford's manner.

On the other hand Ford could be most generous in friendship. Late in 1912, he decided to equip his austere black Model T, up to then started by a hand crank, with a storage battery, self-starter, and electric lamps. His first thought was to use Edison batteries, and he soon contracted for a big supply of them. "I will design a starter, new dynamo, motor, new rigging and proper battery," Edison wrote in a letter of agreement in 1914. If tests proved the mechanism satisfactory, the Edison Storage Battery Company was to manufacture 100,000 battery-generator sets especially patterned for the Ford car. To finance this undertaking, the Ford Motor Company advanced altogether $1,150,000 on account. [5]

The alkaline battery, however, showed a relatively low voltage and did not function well in circuit with an automobile self-starter, Ford being as keenly disappointed as Edison in these results. Then, still eager to help his friend (after the West Orange fire), Ford undertook to build a small electric car powered by the Edison batteries he had already ordered. After visiting Edison in Florida in the winter of 1915, he returned to Detroit to find his engineers experimenting with an electric car designed to be driven by ordinary lead-acid batteries. According to his assistants, "he raised the devil. . . . They weren't to build a car for lead batteries; they were to use Edison batteries, he insisted." [6] But the nickel-iron model proved to be inadaptable for such use and, by mutual agreement, was abandoned.

Their relations, however, continued to be increasingly intimate on other than business grounds. In February, 1914, Mr. and Mrs. Ford had first come to Edison's winter retreat in Fort Myers for a long visit. Under the tutelage of John Burroughs, Ford had become so enamored of nature studies that on returning from a recent journey to England he had brought with him a collection of 380 songbirds, with which he hoped to stock America's forests. On this occasion Burroughs was invited to join Edison and Ford in southern Florida. "We'll go down to the Everglades and revert back to Nature," the inventor promised. "We will get away from fictitious civilization. If I get an idea I will leave my companions temporarily and go to my laboratory there to work it out." [7]

Between them Edison and Ford had done more perhaps than any other living men to foster, if not a fictitious civilization, then a highly mechanized one and its mass culture. There was a touching incongruity, therefore, in their tour of the Everglades and the cypress forests, under the guidance of John Burroughs, in their bird watching and their examination of exotic flowers. In those days the tall, thin, bearded Burroughs, "a philosopher who worshiped God's truth in Nature," by his pen and his example led mankind to the outdoors. Ford enjoyed himself so heartily that not long afterward he bought a winter home of his own near Edison's Seminole Lodge and next winter came back to the enchanting southern wilderness.

The Edisons had a boat dock on the Caloosahatchee River. Ford had the whim to send for fiddlers from rural Michigan and proceeded to hold square dances and Virginia reels on the dock in moonlight, while he prevailed upon Mrs. Edison and her grown children to go rustic.

Several months later, when Edison and Ford were visiting the Panama-Pacific Exposition at San Francisco, Ford took Edison, together with Harvey Firestone, out to Luther Burbank's plantation at Santa Rosa. The inventor had long admired the work of the California botanist and now conceived the idea of "botanizing" for himself in the beautiful gardens surrounding his Florida home. As with other wise old men, it was to be the happy recreation of his late years.

His own methods of experimentation, he pointed out, were similar to those followed by Luther Burbank. "He plants an acre and when it is in bloom he inspects it and picks out a single plant, of which he saves the seed. He has a sharp eye and can pick out of thousands a plant that has promise of what he wants." From this Burbank could propagate an improved variety with fair certainty of success. That was the essence of Edison's empirical method in chemical research.[8]

During the California trip, Edison, who found touring the countryside in an open automobile a splendid distraction, made plans with Ford, Burroughs, and Firestone, Ford's supplier of rubber tires, to go on a long camping trip by automobile through New England and New York State during the following summer. According to Firestone, the idea—inspired by Burroughs' preachments—was "to lure us away from the busy world of material and business developments." Curious, that Firestone should have described their plan in such spiritual terms, for he was the kind of man who could not stop selling rubber tires even when he was dreaming.

Edison himself made all the logistical arrangements for their tour in the summer of 1916, providing for a "six" touring car to carry the party and a Model-T Ford truck to follow it, bearing tents and camping equipment, camp servants and drivers. At the last moment, owing to the pressure of business affairs, Ford found himself unable to join that first expedition, which was a fairly short and simple affair. Edison, accompanied by Firestone, left West Orange on August 28, 1916, and stopped overnight in the Catskill Mountains, after covering 82 miles of unpaved country road; then joining Burroughs the next day at his homestead in Roxbury, New York, he went on with him along the shore of Lake George, over the Adirondacks, to upper Vermont, and returned home after ten days on the road. Edison always sat up in front with the driver. No matter how rough the road or how deep the holes, he enjoyed bouncing along at speeds up to forty miles an hour and never showed fatigue. The "shaking-out" was good for him, he believed. Burroughs, who was seventy-nine and very bony, pointed out, however, that Edison could bear the jolts easily because he was "well-cushioned."

On his return, Burroughs wrote Firestone and Edison:

That was a fine trip you gave us. John Muir would have called it a glorious trip. You arranged the weather just right and you begot in all of us the true holiday spirit. We were out on a lark. . . . My health had been so precarious during the summer that I feared I could not stand more than two or three days of the journey. But the farther I went the farther I wanted to go.[9]

Though Burroughs had suffered from the bad roads and the vibrations of the automobile, thanks to Edison's stories, he said, he had enjoyed "vibrations and convulsions . . . in the diaphragm around the campfire at night." He gives a diverting picture of Edison in the Adirondack Mountains playing prospector, pottering about with a little hammer with which he broke up pieces of granite and feldspar. Ah, there was a possible source of potash, if only it could be extracted economically! Burroughs continues in his letter:

It was a great pleasure to see Edison relax and turn vagabond so easily, sleeping in his clothes and dropping off to sleep like a baby, getting up to replenish the fire at daylight or before, making his toilet at the wayside creek or pool . . . and practicing what he preaches about excessive eating, taking only a little toast and a cup of hot milk. The luxuries of the "Waldorf-Astoria on Wheels" that followed us everywhere had no attractions for Mr. Edison. One cold night, you remember, he hit on a new way of folding his blankets; he made them interlock so and so, got into them, "made one revolution" and the thing was done. Do you remember with what boyish delight he would throw up his arms when he came upon some particularly striking view? I laugh when I think of the big car two girls were driving on a slippery street in Saranac . . . and when they put on the brakes suddenly, how the car suddenly changed ends and stopped, leaving the amazed girls looking up the street instead of down. Mr. Edison remarked: "Organized matter sometimes behaves in a strange manner." [10]

The next "gypsy" tour took place two years later, in the summer of 1918, after the war had taken a favorable turn. This time Ford made it his business to join Edison, Burroughs and Firestone and was very much present at all their subsequent trips to what they called "Nature's laboratory." Why, one asks, did these millionaire industrialist-inventors turn "vagabond"? Edison's idea of these vacation trips was expressed in a fragment of manuscript he wrote in 1916:

Mr. Ford has asked me to write something to be placed in the cornerstone of his new bird fountain (at Fairlawn in Dearborn), which cornerstone is to be laid by that lover of nature, John Burroughs.

I am greatly pleased to do so, because, while mankind appears to have been gradually drifting into an artificial life of merciless commercialism, there are still a few who have not been caught in the meshes of this frenzy and who are still human, and enjoy the wonderful panorama of the mountains, the valleys and the plain, with their wonderful content of living things, and among these persons I am proud to know my two friends, John Burroughs and Henry Ford.[11]

Unfortunately the newspapers, with hue and cry, soon followed the wanderings of America's Most Useful Citizen and his friend the Flivver King from one camp site to another, on that second trip. To Edison's annoyance, reporters persistently intruded themselves between him and the panorama of mountains and valleys, to press their queries about his war machines, or to describe or photograph him as he cooked potatoes over the fire, or took a siesta under a tree.

That second camping tour had been organized by Ford and Firestone on a much more elaborate scale than the first, and its itinerary had been publicly announced in advance! The "irrepressible boy of seventy-one," eager for his annual "shaking-up," hurried off to pick up John Burroughs in the Catskills, then sped on to Pittsburgh to meet Ford and Firestone. The itinerary for this journey was longer and took them through the Great Smoky Mountains.

Clad in linen dusters and soft caps they roared along in their motor caravan, startling the sleepy mountain hamlets of West Virginia and North Carolina. Soon they were in deep forgotten valleys where the country people sometimes gathered in little knots to stare at them as they stopped. Some of these people had never seen an electric light, but they knew and recognized Edison as "Mr. Phonograph." On one occasion Henry Ford was evidently discomfited when no one seemed to have heard of him or even to have seen any motorcars before. "Good," said Edison, "we shall have a good time here."

When they pitched their camp in an open meadow, Firestone and the bustling Ford would sometimes engage in a scything match or a "cradling" contest. Edison, however, as Burroughs recalled, was content to settle down in his car and read or meditate, while Ford swung an ax to cut wood for their campfire. To be sure there were numerous attendants about them now, a luxurious kitchen truck and several supply trucks, providing a sybaritic fare the like of which poor John Burroughs had never known on his own camping trips.

Edison, at any rate, was without affectation; in the morning, one would see him washing in the brook, with his coat off, in his braces, but still wearing his black string tie. "He is a good camper-out and turns vagabond easily," Burroughs remarks, "for he slept soundly in all his clothes, as his unkempt appearance in the morning revealed. He can rough it week in and week out and be happy."

In contrast with the fiercely energetic but undereducated Ford, Edison seemed to Burroughs, as he so often appeared to others in the Indian summer of his life, genial, relaxed, philosophical, going off by himself to read a book at every halt; "a trenchant and original thinker" to whom the other members of the party "instinctively deferred," while delighting in his witty sayings.

Yet like many other philosophers he could be found acting in complete contradiction to his own professed doctrines. In his journal for the 1919 camping trip, Burroughs wrote:

O Consistency, thy name is not Edison! Ten A.M. Edison not up yet—the man of little sleep! He inveighs against cane sugar, yet puts two heaping teaspoonfuls into each cup of coffee, and he takes three or four cups a day. He eats more than I do, yet calls me a gourmand. He eats pie by the yard and bolts his food.[12]

When their lead car broke down in some remote corner of West Virginia, it was Ford who took off his coat and repaired the radiator, using only some old metal obtained from a village blacksmith. Out of this incident, the journalists who followed in their trail worked up an apocryphal story, the burden of which was that when their car broke down, some village mechanic suggested that the trouble might be with the motor. "I am Henry Ford," spoke up the tall man, "and I say the motor is running perfectly." Then the rustic suggested that the electric spark distribution might not be working. "I am Thomas A. Edison," spoke up the stout man in the front seat, "and I say the wiring is all right." Whereupon the village mechanic, pointing to John Burroughs and his long white beard, remarked, "And I suppose that must be Santa Claus."

Though Ford was usually brief-spoken, he was easy to draw out on mechanical problems, which he discussed as a master of his art, and with unfailing enthusiasm. On the other hand he seemed much less interested in the natural beauty or historical associations of the countryside than Burroughs or even Edison. To Ford, a pretty waterfall was a potential hydroelectric site. Around this time he conceived his plan to develop small rural power stations to serve village factories and craftsmen's shops; the idea, according to Burroughs, was implanted in his mind originally by Edison. When he dilated on such schemes, Ford talked as if he had "taken the people into partnership with him," and meant to put all sorts of inventions and conveniences within the reach of the common man.

Around the campfire at night they talked of war and politics, of industry and labor and public life. It was plain to see that, having conquered the world of industry, Ford thought of crowning his career by winning some high political office. In 1918 there was much talk of Ford's intention to run for the Senate, as a step toward higher office still. Touching on this matter one evening, Edison (who often teased his friends) indicated that he held a poor opinion of Ford's political talents. "When Ford goes to the Senate," he remarked wryly, "he will be mum; he won't say a damn word." As for himself, Edison wanted no political honors, but, like Faraday, who had refused ennoblement, declared that as he was born "plain Thomas A. Edison" so he would die.

For a man who aspired to be the country's Chief Magistrate, Ford was decidedly lacking in mental stability. Only the year before he had spoken forth as a pacifist who vowed that he would never produce the materials of war. At that time, in 1917, Edward N. Hurley, a high official in President Wilson's Administration, had written Edison urging that he use his good offices in having Ford's great factories enlist for war service. In reply Edison wrote:

You ask *where is Ford?* Let me explain. He is an inventor. Inventors must be poets so that they may have imagination. To be commercially successful they must have the practicability of an Irish contractor's foreman and a Jewish broker.

These wild children of nature necessarily are a puzzle to a Captain of Industry like yourself. Don't try to understand him. Get Ford in somehow with the happy shipping board help.

Your trouble will be to get him started. At present he is pushing Liberty Motors.[13]

By 1917, Ford, with one of his sudden changes of heart, had dropped his pacifism and devoted himself to serving the war industries as the most militant of patriots.

Edison used to describe himself as a "progressive" in politics, one who naturally favored change and experiment in this as in other fields. Yet in old age he seemed to prefer that social change proceed as slowly as possible. Toward 1919 the trade-union movement experienced a cycle of rapid growth; there were many violent strikes; and there were not a few socialist agitators about. Neither Edison nor Ford cared much for the organization of labor, save under paternalistic systems of their own. The struggles of capital and labor, according to Edison, were "inherent"; captains of industry arose through the process of evolution. As for the charges of reformers and socialists that they were "exploiters," or mere accumulators of wealth, Edison denied this, so far as he was concerned, and Ford echoed his views.

In fact, both these millionaires professed the deepest contempt for mere money-getting, though even "poor" Edison almost in spite of himself earned profits of more than a million dollars annually. In a speech of his that was read for him before a convention of motion picture producers in 1923, he said, "Remember, you must never think of profits, but only of public service." [14] In the margin of a book by an author who expressed wonder at the breathless tempo of American life, and at its continued speed, he wrote: *"Yes, with a metronome of money!"*

Money was nothing, gold was nothing to him, but commodities, industrial products, inventions represented true wealth. And far from being an exploiter, Edison, as inventor and engineer, regarded himself as a creator of new wealth for all men, new tools, new industries. By 1922 the value of America's electric light and power industry alone was reckoned at more than fifteen billion dollars, and then people stopped counting.

Ford, the engineer of the poor man's chariot, ardently embraced Edison's ideas about the inventor's mission to create wealth. He said in 1930:

Our prosperity leads the world, due to the fact that we have an Edison. His inventions created millions of new jobs. . . . Edison has done more toward abolishing poverty than all the reformers and statesmen.

At all events, Ford fully shared Edison's abiding mistrust of the "money-changers," and his earnest wish that they should some day be dethroned. Like Edison, he thought of himself as a wealth creator and of the bankers and the "money men" as parasites who fed upon the productive elements in society. Both men, it must be remembered, had grown up in rural Michigan and may have absorbed the emotional prejudices of that former center of Greenback agitation against "Eastern bankers" and "goldbugs." The monetary radicals of the farm belt, in earlier times, also tended to associate the Jews, who were proverbially active in commerce and as middlemen, with their enemies the "goldbugs." A mild form of anti-Semitism has sometimes been traced in this country to the followers of "Coin" Harvey and William J. Bryan. But with Ford, anti-Semitism soon became rabid.

As time went on, Ford, according to his most recent biographers, tended to attribute his extremely frustrating experiences in the Peace Ship enterprise of 1915–1916 to the machinations of "the International Jew." After having played the liberal and humanitarian, he suddenly turned bigot. In May, 1920, *The Dearborn Independent*, a daily newspaper that was the house organ of the Ford Motor Company, suddenly launched a sensational anti-Semitic campaign and achieved nationwide distribution. This new "crusade" earned the owner of the newspaper much public ignominy, and led to his being exposed in public court, during suits for damages and libel, as one of the most abysmally ignorant of men. After two years Ford abruptly halted his anti-Jewish propaganda, paid damages to persons libeled, and retracted all the charges against the Jews that had been published in his newspapers.

Some reproaches have been directed at Edison, Ford's most esteemed and influential friend at the time, for not having counseled the automobile magnate against such unwisdom. It has been suggested also that Edison condoned, if he did not encourage, this misbegotten agitation.[15] It has been remarked that in December, 1920, when E. G. Liebold, managing editor of *The Dearborn Independent*, sent Edison some of its anti-Semitic articles and invited his comments, Edison wrote briefly, as if indicating approval, "Liebold—they don't like publicity." [16] But to deduce from this that Edison himself was ever a fanatic of race prejudice or "racial supremacy," or that he "inspired" Ford in this undertaking, seems wholly unreasonable.

In his eighth decade, as an aging man, to be sure, he showed many crotchets; like other gifted men he permitted himself on occasion thoughtless utterances in public, and seemed to enjoy provoking controversy. Several years earlier, in October, 1914, Edison, after having outraged many Christians by his frank professions of agnosticism, seems to have offended the Jews also by some unconsidered reflections that were quoted in the newspapers. He remarked then that an underlying cause of the outbreak of world war was the rising commercial power of Germany, which threatened England, and added that the Jews were largely responsible for Germany's commercial prosperity. The German military, he suggested, were influenced by the men of business in their country.

These opinions evoked strong protests, one of them by Jacob H. Schiff, the eminent New York banker and philanthropist. In replying to Schiff, Edison denied that he had meant to accuse the German Jews of any wrongdoing, or of having "started" the war. What he had meant was that the Jews deserved greater credit than they had been given as yet for Germany's industrial ascendancy. Then he went on to insist that "the military group that ruled Germany had the brains to take the advice of the great Jewish bankers and businessmen and give the captains of industry a free hand." [17] Such a view was, of course, the direct antithesis of official anti-Semitic doctrine of the Hitler type which always insisted that the Jews, far from aiding Germany, "betrayed" or "stabbed her in the back." In any case such unguarded ruminations on Edison's part, though they may have been inept and in poor taste, are in no sense to be equated with the monstrous, genocidal dogmas propounded many years later in Nazi Germany—which Edison would have abhorred with all his heart.

He was—outbursts of temperament aside—predominantly rational and humane; although bias against Catholics as well as Jews was openly professed in his time in business and social life by many Americans, who did not regard themselves as barbarians for that reason, Edison employed Jews in his laboratory as assistants.

It happened that about six months before he gave the interview touching on Germany's industrial ascendancy, in March, 1914, the case of Leo Frank, a citizen of Georgia who was hastily convicted of rape, was being widely debated in the press. There had been strong evidence given attesting to his innocence, and to the intimidation of the court by Ku Klux Klan mobs—which later broke into the jail and lynched him. But while Frank had been alive and waited for his case to be appealed, Edison, who was then in Florida, was asked for his opinion of this affair. In reply, he stated publicly his belief that "Frank should have a new trial." [18] Though, at times, he entertained some very queer notions about medicine, diet, education, women, and even science—in which he showed a decided "spirit of contradiction"—Edison was nothing like the superstitious Ford; he had far more detachment, as well as intelligence, and would never have set off on a crusade to save all America from one of her minority groups.

And while Henry Ford trembled at the alleged menace of the "mercenary" Jews, and pretended to flee "the artificial life of commercialism," in a motor caravan out in the open country, a spirit of crass commercialism, day by day, pervaded his well-advertised camping tours. Movie news cameras and publicity agents, who carried on promotion stunts of all sorts, gathered in the train of the Edison-Ford-Firestone motor caravan, which, by 1919, boasted fifty cars. Politicians arrived—even President Warren Harding, on one occasion—to bask in the limelight of its campfire. Trucks followed behind, bearing large placards reading: *"Buy Firestone Tires."* When the caravan arrived in some community the festive occasion was exploited by dealers as an opportunity to sell more Ford cars. Henry Ford and Firestone seemed to enjoy it all. But in 1921, John Burroughs died; after that, Edison, who was greatly bothered by the crowds, the dust, and the ballyhoo, decided to come no more.

2   In the autumn of 1922, Edison visited the General Electric Company's huge plant and laboratory at Schenectady after an absence of twenty-five years. The factories that had grown up out of the old Edison Machine Works now extended "for miles" and employed 18,000 workers who, at the order of President Gerard Swope, assembled to welcome and cheer Edison. A little ceremony was made of putting up a bronze plaque in his honor at the door of the great laboratory he had initiated. Here "hundreds" of scientists and technicians were at work in fields of technology that Edison had only dimly imagined. Dr. Steinmetz showed him around the laboratory and demonstrated his "lightning bolts," which were capable of discharging 120,000 volts at a bar of tungsten, causing it to "vanish" and turn into gas. Edison got along well with Steinmetz, he said, because "he never mentioned mathematics when he talked to me." W. R. Whitney, director of the laboratory, showed him the new process, perfected by General Electric's W. D. Coolidge, for converting tungsten into lamp filaments. The aged inventor sighed, and recalled what great troubles he had encountered in his own attempts to experiment with tungsten. Dr. Irving Langmuir also displayed new vacuum tubes for long-distance transmission; also a 100,000-candle-power lamp, and photoelectric mechanisms that reproduced sound on tape

for motion pictures and phonographs. Edison declared that he had greatly enjoyed seeing "the old place, and the old boys," but admitted that it had changed, had grown up.[19]

Some time after his visit to the General Electric Laboratory, Edison, old free lance that he was, shook his head and declared that the "corporation laboratory" would not do. The inventor was now a "hired person" for the corporation, assigning to it all his patents; such men now worked in large groups and held many conferences. The "weight of organization" was too great, Edison observed shrewdly; and the results, he predicted, would not be as rich as in the case of individual inventors working in small organizations. In truth, inventive and development work at the great corporate laboratories of the United States between 1900 and 1940 has been judged unremarkable by scientists themselves. However, the richest results seem to have been reached where some leading personality, both inspired and tenacious, directed a small group toward the objective.

In his late seventies Edison gave most of his attention to the improved phonographs, business machines, and secondary and primary batteries. The fortunes of the Edison Industries were at full tide during the postwar boom, toward 1923, then ebbing as radio came into general use and competing manufacturers placed on the market newly invented electronic phonographs, or even phonographs and radio receivers in combination. The creator of the mechanical phonograph, however, could not abide the new recording instruments first developed by a team of Bell Laboratory physicists. Nor would he concede that such instruments were acoustically superior to his 1913 phonograph. Instead, he announced, as late as 1925, that T. A. Edison, Inc., would market an improved mechanical phonograph adapted for long-playing records running twenty minutes.

As for radio, he detested it, predicting that the people would want to hear music of their own choice and not that which was imposed on them by the limited repertory of the broadcasting stations. The "radio craze" would soon pass, he said. "The present radio . . . is certainly a lemon. . . . It will in time cure the dealer of any desire to handle any kind of radio," he stated in a memorandum addressed to his staff in 1925.

The old phonograph business, however, was falling off rapidly; Edison's nationwide organization of 13,000 phonograph dealers, built up during forty years, threatened to disappear. His sons, Charles—who by then was playing a leading role in the family business—and Theodore both begged him to allow them to try new ideas and make both radio receivers and electronic phonographs. According to Theodore Edison, his father opposed the new ventures until it was too late to enter upon them.

His youngest son, Theodore, considered the most talented of his children, he regarded with mixed feelings of puzzlement and respect. There had been, at times, some lively disputes between them on the score of Theodore's addiction to mathematics and the method of theory and calculation. On one occasion, when no agreement seemed possible, the young man had packed his trunks and prepared to leave West Orange to look for a job somewhere else; but his father capitulated and agreed to have him work in his own way.[20] To this end, an old shed was first turned over to Theodore for use as a laboratory. After a while his work appeared impressive enough to warrant his transfer to a fully equipped laboratory, and eventually—two years after his father's retirement—to the post of Technical Director of the Edison Laboratory.

What the younger Edisons were especially eager to undertake was the development of a new electronic phonograph under the Edison label. They were alarmed at the lead their company's competitors had taken in this field and, therefore, in the autumn of 1925 began to conduct experiments on an electronic phonograph model in an old building in the "compound," this work being done in secret and without their father's approval. A while later, after the Old Man had got wind of their activities, he decided to start an electronic phonograph project of his own, in competition with his sons!

The aged Edison had as his helper, early in 1926, a gifted young Swedish engineer named Bertil Hauffman, who, after having won high honors in his country and a traveling fellowship, had come to the United States to work at the Edison Laboratory. Hauffman, by his fine experimental and theoretical work, helped the Edison Company's radio engineers to perfect a phonograph reproducing electronically recorded sound, and having a greatly expanded volume and frequency range. But the young engineer pursued modern scientific and mathematical methods in his experiments, such as he had been taught. Edison, however, insisted on handling their problems in his old "empirical" way. Though he could now hear almost nothing, he tried to test their phonograph models with the aid of an earphone; but found the effects of sound, to his ear, "distorted" and "terrible." He therefore ordered that Hauffman be fired at once.

Such punishment would have meant disgrace in his own country for the visiting engineer, and the younger Edisons were aware that he was actually doing brilliant work which the Old Man could not follow. Therefore, Theodore Edison arranged for the young Swede to continue his experimental work secretly during the rest of his stay in America, in a part of the Edison Storage Battery Plant which the old inventor was not likely to visit.[21]

"My father's past experience had simply got in his way," the youngest son said of this incident. After considerable delay, an electronic phonograph of clear and powerful tone, and almost completely free of surface noise, was brought out at the end of 1928, as the last product of the Edison Phonograph Works.[22]

Meanwhile, Charles Edison finally brought his father around to the idea of manufacturing radio receivers. The Old Man had given his consent very reluctantly, warning his son that he would lose "millions" and "make a damn fool of himself." Owing to long delays in obtaining licenses from the patent owners, principally the General Electric, Westinghouse and Bell Telephone group, and their common "patent pool," the Radio Corporation of America, Charles Edison was unable to enter the radio field before the end of 1928, on the eve of a great depression.

In the autumn of 1926, when he was well along in his eightieth year, Edison at last announced that he had decided to retire and would limit himself thereafter to laboratory experiments. His son Charles now succeeded him as the head of Thomas A. Edison, Inc.; three years later, Theodore was made a member of the company's Executive Board as well as its chief technical officer.*

Thomas A. Edison's heart was never in the radio receiver or electronic phonograph business. In 1930 the Edison Company, after producing phonographs successfully for more than forty years, announced that it would shut down all manufacture of both phonographs and records, con-

---

* Charles Edison thus assumed chief managerial control of the family business, from 1927 on. During the economic crisis of 1933, Charles, who then showed some leanings to liberal politics and public life, served in President Roosevelt's National Recovery Administration and later was appointed Assistant Secretary of the Navy, succeeding Claude Swanson as Secretary of the Navy for a short interval. In 1940 Charles Edison was elected, under the Democratic Party's banner, as Governor of New Jersey, serving for one term of three years. After he returned to private business, T. A. Edison, Inc., under his management made a strong recovery, during the prosperous World War II and postwar eras.

In May, 1931, Theodore, of whom it was said that he preferred science to business, resigned from all the offices and directorships he held in various Edison companies, in order to devote all his time to research and experiment in his own laboratory. He also formed an engineering consulting firm of his own some years later. Meanwhile, he remained, by his father's will, owner of a large interest in the Edison company's stock. After his mother's passing, in 1947, Theodore Edison made an outright gift of $1,260,000 in stock and cash (as then valued), which constituted half of his total inheritance, to 2,700 employees of the Thomas A. Edison Industries. By virtue of a trust fund established by him at the time, each employee was provided with a share of Edison stock for every year of service in excess of two, up to a maximum of sixteen shares. This social welfare fund created by Theodore Edison helped somewhat to right the balance of things and give partial recognition to the role of the workers in the Edison Industries, as distinguished from that of the great individualist who founded them.

fining itself to radio receivers and business machines. But after another year of profound business depression, the venture in radio manufacturing, which had lost about two million dollars, was also abandoned.

For several years before that, the interest of the Old Man himself had been directed to a new field of operations, one that was entirely different from those he had formerly cultivated in his ivy-clad laboratory at West Orange. It was to be a last "campaign" of experimentation to which he would surrender himself with all his unfailing enthusiasm. "I have had sixty years of mechanics and physics," he remarked on the occasion of his eighty-first birthday, and he was now turning to another science, as much for his own pleasure and amusement as for any profit that might come of it.[23]

3   In those last years the white-haired Edison might have been found more often than not in the gardens of his Florida home, where he remained for a much longer season than formerly. In addition to the thousands of tropical plants set out in the ground around Seminole Lodge, there was a special botanical laboratory filled with potted plants that he and a staff of botanists were developing from crossbred strains. He too was botanizing; but as always there was a practical object in view.

Rubber had recently become one of the most important commodities for the modern industrial world, thanks to the automobile. It was produced in distant tropical regions and in wartime became scarce and extremely dear; during the recent war its price had risen from about twenty cents to more than two dollars a pound. While visiting Burbank's plantation in California with Ford and Firestone in 1915, Edison had remarked that if the United States entered the war, rubber would be the first product to be cut off. Ford had asked him to do something about creating a domestic supply, or a substitute, to which Edison replied, "I will—some day." [24]

When rubber became costly again toward 1924–1925, owing to the British Far Eastern rubber-restriction scheme, Ford and Firestone renewed their pleas that Edison undertake a serious investigation of domestic sources of rubber, which they offered to finance. The Edison Botanic Research Company was thereupon organized in 1927, with the Ford Motor Company and the Firestone Tire & Rubber Company each advancing $93,500 to cover the costs of research, while Edison agreed to contribute his labor. Thus he was engaged once more in one of his famous "hunts" for some plant, either existing or to be developed by crossbreeding, which contained sufficient rubber to be processed on a large scale.

Once more the Old Man had a happy look in his eyes as he began his new studies at Fort Myers in the winter of 1927. He commenced as usual by collecting an extensive library on the subject. For four centuries, ever since the Spanish conquerors had brought back balls of wild rubber from the South American jungle, some of the world's greatest scientists had been fascinated by this extraordinary plant. Faraday, for one, in 1826 had made extensive studies of the chemical formula of India rubber, which was a step toward the ultimate invention of a synthetic rubber compound. Was it mere coincidence that Edison took up the same study pursued by his idol, Faraday, a century before him?

In addition to his own reading, a member of his staff named Baruch Jonas, who could read in a dozen languages, was assigned to search for information available in Spanish, Portuguese, German, and other tongues. A chief chemist and a staff of six botanists were engaged for the duration of the project; and, according to his usual custom, emissaries were sent forth to discover and collect specimens of latex-bearing plants native to this and other countries, not omitting the Everglades close to his Florida home and botanical laboratory.

Of the period that followed, Mrs. Edison said afterward, "Everything turned to rubber in the family. We talked rubber, thought rubber, dreamed rubber. Mr. Edison refused to let us do anything else." [25] He was enjoying himself, he was doing "something different." Hitherto he had dealt with things physical, chemical, and mineral, but not much with things vegetable, save for his "flier" in bamboo. Unlike other inventors of his age, he had a mind of infinite variety which turned from electrical work to sonic and optical problems, to mining engineering, to chemistry, and now to the biochemical mysteries of plant life. Now he was busily "ransacking the world," as his associates reported, gathering and dissecting every class of weed, vine shrub and bush that grew, not only in warmer climates but also in the North Temperate Zone. It was a large assignment, and one calling primarily for Edison's empirical method of work—which may have been why he enjoyed himself so greatly.

Prior to a severe illness suffered in 1929, his mental powers showed no perceptible decline. He "got the whole subject of rubber into his head" so he could see every phase of his problem. The main source of supply was the *Hevea brasiliensis,* a tree native to Brazil, but successfully transplanted on a large scale to Malaya, Ceylon, and Africa. Coolie labor in the moist equatorial lands had made rubber cheap and abundant, though the rubber tree required five years to mature and give forth its milky fluid.

It was well known to scientists that numerous other plants, even those common to the North Temperate Zone, contained caoutchouc latex in varying quantity: the giant milkweed, the Southern honeysuckle vine, and shrubs such as the Mexican guayule, already being cultivated in Southern California. But the rubber extracted from the guayule was unsatisfactory, being of a highly resinous quality. What Edison hoped to find was a "sowable and mowable" crop native to the United States, and capable of being cultivated, harvested, and processed within a year or eighteen months, so as to provide a source of supply in the event of war or other emergency. There was never any prospect of surpassing the economy of colonial labor in Far Eastern rubber, save when the market was artificially high.

In less than a year Edison reported to Henry Ford that he had collected 3,227 wild plants and shrubs from points ranging between New Jersey and Key West, and from various arboretums, and that 7 per cent were found to have rubber in various quality and amounts. "Everything looks favorable to a solution . . ." he concluded with his unvarying cheerfulness.

At the end of a second year's search, when some 14,000 plants had been examined and tested, of which about 600 contained some rubber, Edison, after flirting with honeysuckle and milkweed, fixed on the domestic goldenrod as the most promising plant of all. Remembering that goldenrod used to grow in empty lots near the railroad depots where he worked as a telegrapher in his youth, he had the notion of sending a questionnaire to thousands of railway station agents throughout North America, inviting them to mail him specimens. A mountainous mail, filled with dried weeds, soon engulfed him; but it was a way of covering the ground.

Goldenrod yielded about 5 per cent rubber. Edison selected the varieties that seemed most promising, divided the roots, planted them separately, divided them again, and crossbred. It was time-consuming; but a giant goldenrod about fourteen feet tall yielding about 12 per cent latex was ultimately developed. Ford was so greatly encouraged by Edison's reports that he purchased an extensive acreage in southern Georgia for the raising of goldenrod.

Edison worked on at his Florida botanical laboratory even in the heat of spring and early summer. Under his supervision a botanist would train a small hose on the anthers of selected plants in order to wash off the pollen; then taking pollen from another species he would place it on the dry pistils of the first flower in order to effect crossbreeding. This work had to be controlled under a microscope, with infinite care. Planted in January, a variety of goldenrod could be harvested from May to November. The crop was processed by means of a small still. The plant having been ground up and turned into a liquid solution, the tiny rubber particles separated themselves from the solid matter, and rose to the surface of the tank. Edison hoped eventually to be able to produce 100 to 150 pounds of rubber from an acre of goldenrod, and to process it economically by special machinery. But thus far such rubber could not be made for less than two dollars a pound; moreover, it was inferior to the rubber imported from the tropics.

In the late summer of 1929, while at West Orange again, Edison fell ill; his whole digestive apparatus seemed affected and there was some indication of kidney malfunction and of diabetes. But he rose from his bed saying, "Give me five years and the United States will have a rubber crop." But who, now, could give Edison five years?

Oddly enough, not only Edison, but Henry Ford, the Dupont Company, and Standard Oil of New Jersey, after 1925, received information about the new German chemical process for converting coal or petroleum derivatives into synthetic rubber of the butadiene and sodium type perfected by I. G. Farben around that time. But large-scale operations were not to be carried out until a decade later, prior to World War II. Edison also might have turned his eyes in this direction—which was to be the most profitable for systematic research—but for the fact that the synthetic process was known to require an enormous investment in a special chemical plant, which even our biggest rubber tire and chemical corporations refused to risk at that period. In 1940, Federal government subsidies alone would make synthetic rubber production feasible. In that year government scientists also thoroughly explored the possibilities of using vegetable materials available to us, such as Edison's variety of goldenrod and guayule, but reached the conclusion that processing such plants would be more difficult and costly than making synthetic, and would yield a product inferior to natural India rubber or the new synthetics.

Edison's goldenrod rubber project, therefore, was foredoomed; it was one more of the forlorn hopes of his later years. If Henry Ford suspected this—and his busy factotum Liebold seems to have been aware of the truth—the motor king would not have moved a hand to stop Edison's dying effort. Where Edison was concerned, the ruthless Ford was all sentiment.

In 1929, Harvey Firestone, having received a shipment of the costly goldenrod latex from Fort Myers, obligingly turned out four rubber tires for the inventor's Model-A Ford touring car.

Edison did not recover quickly from his illness of that year. He was suffering from uremia, yet worked on in his botanical laboratory, turning a smiling countenance to the world and assuring all and sundry that the solution of the rubber supply problem was at hand. "We are just beginning. . . ." he said.[26]

4   The ninth decade of the inventor's life was a time for the erection of monuments in his honor, a time for the bestowal of medals, ribbons, decorations, and other honorific ornaments, that came to him from all quarters of the world. He was touched with bronze; he was as a walking monument himself, in fact, an immortal. People used to come and address him with formal eulogies, generally dilating upon his benefactions to the human race. Solemn little delegations arrived frequently in West Orange, bearing their offerings of plaques or medals, and waited patiently in the library until, as often as not, word came to them that the great man was too busy to accept their gifts in person. He used to say that he had a whole bushel basket of such baubles in the mansion at Llewellyn Park. His library at the Edison Laboratory also gradually became cluttered up with sculptures, paintings, old photographs, and other memorabilia of his career. One of the most spectacular of these artifacts was a life-sized bronze figure of a nude youth by Lorado Taft, waving aloft a phonograph record. The veterans of Edison's motion picture enterprises had it that this bronze statue really represented the custard pie thrower of slapstick days in the movies.

If, perchance, the octogenarian Edison was in good humor, he would send word that he would come out to meet a delegation arriving to pay him tribute, as on the occasion of the award of the John Scott Medal in December, 1929. The award committee and the Edison Industries executives would be drawn up in the library, a ceremonious group, formally attired for the occasion. Then they would hear steps coming along the hall—scuff, scuff, scuff—as one witness relates. "It was someone with shoes too large for him."

The door opened. "Any rags, bottles, or old medals today?" sang out the famous man, in his high voice, as he beamed upon the visitors, his keen old eyes full of mischief. An ancient, white-haired but elegant figure rose from his desk at the far side of the library and trotted over slowly to greet his chief. This was Meadowcroft, the private secretary and faithful servitor of half a century, nearly blind, but with sharp hearing.

"Good morning, Dr. Johnson," he said gravely.

"Good morning, Boswell," Edison replied. It was an allusion to W. H. Meadowcroft's having written *The Boy's Life of Thomas A. Edison,* and to his having been co-author, with F. L. Dyer and T. C. Martin, of an "official" two-volume biography as well. Boswell-Meadowcroft was very loyal and very deferent, and, it was said, he truly loved the role of attendant to a great lion. The lion was also evidently fond of his "keeper," though on days when he suffered from physical pain, or moral disappointment over some experiment, he would sound as if he were going to make a quick meal of the old factotum. Yet Meadowcroft could be so soothing that he often acted as a sort of buffer between the inventor and his associates, or even some members of the family.

There were always little intrigues among the old-timers in the laboratory, who sought to humor the Chief, or "get around him" by their competitive gestures of solicitude. On this slippery terrain the tactful Boswell-Meadowcroft usually maneuvered his way very well. John V. Miller, Mrs. Edison's brother, who nowadays helped keep watch over the exchequer at the Edison Works and Laboratory, found it useful to operate through Meadowcroft; as even Charles Edison did, at times. It used to amaze people also to see how the old secretary, during an interview, would place his lips close to Edison's ear and roar at him, then, with perfect composure turn and ask the interlocutor, in a low voice, for his next query.[27]

Edison was more difficult than ever to his associates in these later years, as he heard almost nothing and his strength declined. When his laboratory assistants came to his desk to report on experiments, he would ply them with question after question, until they became confused. Or it would appear that someone had forgotten something, which was the unpardonable sin; a phonograph, for instance, had not been wound up. Then Edison's arguments would become fulminations, his eyes flaming, his great brows twitching menacingly, while he pounded the table and shouted. The "opponent" would be driven to raise his voice to the highest pitch, "so that what Edison considered a mild debate sounded like the hottest kind of row." A visitor to the laboratory recalls Edison pacing up and down his library angrily denouncing certain "boneheaded moves" on the part of his executives, while the latter, unheard, vainly shouted their explanations into his deaf ears.[28]

A signal event in Edison's old age was Henry Ford's decision to build an immense museum of the history of industry and invention, as a monument not only to himself but also to his friend Edison, in Ford's native town of Dearborn, Michigan. Detroit's man of destiny, now in his middle sixties, neglected his motor business and amused himself in an imperial manner.

The untutored man who had once declared in a public courtroom that

"History is the bunk" and that he could hire all the historians he wanted, now seemed wholly dominated by his sense of past times, and especially of his own past. He laid out Greenfield Village in Dearborn as a reconstruction of the rural setting of his early life, stuffing it with nineteenth-century farm cottages, churches, and taverns. Here was the jigsaw-styled frame house of a dentist who had been good to him, the original chapel his wife had attended in girlhood, and the Little Red Schoolhouse of the lady who had first sung "Mary Had a Little Lamb" (probably not authentic). Next to the nineteenth-century Americana was the vast museum of industry and invention holding everything in the line of machinery from ancient looms to the giant locomotives of 1925. But over all this reconstruction of the past there loomed the superhuman figure of Edison, whose glory Ford was bent on preserving *in toto*. As he told his intimates, he feared that our young people would grow up without knowing what had gone before them. But history to him was mainly industrial history, in fact a hodgepodge so unpredictable in its composition as to make his museum in its way impressive enough.

Hence Ford began to gather up such relics as he could find of the legendary Menlo Park Laboratory, now in ruins. He not only scoured the New Jersey countryside for the very planks that had fallen off Edison's old sheds, but also gathered together a notable collection of original models of Edison's inventions, installing them in a section of Greenfield Village that was at first known as the "Edison Institute." These models were so well restored, or, in some cases, reproduced (with the advice of Francis Jehl, of the original Menlo Park team), that everything worked as before, the paper-carbon lamps as well as the 1882 dynamos. Wherever he found the debris of Edison's life and labor he went digging—even in Fort Myers, where he removed the small electrical laboratory of 1887, now fallen into disuse.

"Dear me," Mrs. Edison said plaintively one day, when she spied him from her window, scouring about her grounds at Seminole Lodge, "I do wish Mr. Ford would keep out of our backyard." [29]

About three millions were thus lavished by Ford upon his collection of Edisonia alone, the total cost of Greenfield Village being more than ten million dollars. When it was all done, the restored Menlo Park laboratory stood in its native red New Jersey clay (transported thither by train) and even had a heap of old metal junk lying outside. For had not Edison once said that what the good inventor needed was "imagination and a scrap heap"?

From the governments of Japan, Russia, Chile, and many of the nations lying between, honors had been heaped on the old "Titan"; yet no official honors had been awarded him by his own government—save the Distinguished Service Medal, which he had refused. At length, at the suggestion of one of the latterday tycoons of the electrical industry, Secretary of the Treasury Andrew Mellon took the initiative in appealing to Congress to make amends for such neglect; and, by a resolution of May 21, 1928, the Congressional Medal of Honor was awarded to Edison. Mr. Mellon himself came to the Edison Laboratory to make the presentation, in a dignified little ceremony held on the anniversary of the lamp invention, in October.

Some twenty years earlier an association had been formed that was called "Edison Pioneers," made up of the master's old associates at the Menlo Park and Orange laboratories, as well as old friends and fellow inventors. Its main function was to hold annual luncheon meetings in honor of Edison on the occasion of his birthday. But for October, 1929, the Edison Pioneers also planned an imposing celebration of the fiftieth anniversary of Edison's incandescent lamp invention. On finding that they lacked funds with which to support a great public festival, representatives of the association approached General Electric and persuaded its officers to take over the whole program.

The General Electric Company, which had long ago absorbed and vastly expanded Edison's lamp business, still invoked his name in its advertising matter as founding father of the industry. Since this company, as well as its chief competitor, Westinghouse, had recently been under Congressional attack for alleged monopolistic practices, it was considered good public relations policy to sponsor a grand program in celebration of Edison's miracle, to be known as "Light's Golden Jubilee," ostensibly in honor of America's most famous inventor, but also with a view to stimulating the sale of light bulbs. The site selected for the staging of this pageant was originally supposed to be the headquarters of General Electric at Schenectady, New York. A rising young publicity man, Edward L. Bernays, was drafted for the duration of the campaign, and began to draw up impressive plans for public banquets in various cities, to be concluded with a "colossal" affair at Schenectady, on October 21, at which the venerable Edison would reenact the 1879 invention and lighting of his first lamp.

The trouble was that Edison himself had not been consulted, in the first place, about these elaborate plans, but was simply informed that he was to appear at certain public gatherings and play the part assigned to him. Moreover, he saw no reason why he should be called upon to furnish so much publicity and free advertising for General Electric.

The heads of that great corporation, Chairman Owen D. Young and

President Gerard Swope, were of the generation that came after Edison's and had had no part in the old controversies of 1892, when his lamp business and machine works were taken over. But in the recent matter of the Edison Industries' negotiations for radio patents General Electric, as one of the principal patent owners, had scarcely been sentimental or helpful to the old inventor.

At this point one of Edison's intimates communicated with Henry Ford and informed him of General Electric's scheme to take over the Edison Jubilee of 1929 and thus "commercialize" Edison's fame. Ford declared that it was a "shameful action" and vowed that he would do something about it.

The Ford Museum and Greenfield Village were, by then, nearing completion. Ford therefore decided to combine the formal dedication of his own institution with the celebration in honor of the electric light. He ordered his builders to rush the job on the great museum building, a vast replica of Independence Hall in Philadelphia, for it was to be the stage setting of the opening-day banquet. Not long after one of the Edison Industries' executives had telephoned him in the early winter of 1929, he suddenly appeared in the library of his venerable friend at West Orange, and waited for him, walking up and down restlessly, and muttering to himself: "I'll show 'em. I'll kidnap the whole party." Edison agreed to be "kidnapped." [30]

It was a spectacle indeed to see one giant of industry snatching a "Golden Jubilee of Light" and its hero from the other. Messrs. Young and Swope, greatly taken aback, appealed to Edison and to Ford to accept their original program as scheduled for the Schenectady site. But Edison was stubborn, and Ford was adamant in his determination to have all the trouble and as much of the publicity as possible for himself. In the end a compromise agreement was reached to make the affair appear as a combined operation. However, the site of the main festival was to be Dearborn, Michigan.

When Edison stepped from the train at Dearborn, two days before the Jubilee was to open, he looked, it was noticed, "like a benevolent old wreck," for he had been so gravely ill, in August, 1929, with pneumonia, that his life was feared for. (In great anxiety, Henry Ford had rushed from Detroit to Edison's bedside at West Orange to see him.) Now as Edison beheld Greenfield Village and the transplanted "Menlo Park," he smiled his broadest smile. Here were all the old bulbs, telegraph instruments and stock tickers, the "Long-waisted Mary Ann" dynamos, the old generating plant of Pearl Street, and even an old mortar and pestle he had used and thrown away. There was not only the old "tabernacle," but also Mrs. Jordan's plain boardinghouse across the road from it; even the old railroad station at Smith's Creek, Michigan, where he had worked, and beside it a reproduction of a little Grand Trunk Railway train, including the baggage car with Tom Edison's little laboratory-on-wheels!

After Ford had shown him this truly monumental restoration and asked for his opinion, Edison said, "Well, you've got this just about ninety-nine and one half per cent perfect."

"What is the matter with the other one half per cent?" Ford asked.

"Well, we never kept it as clean as this!" Edison drawled.[31]

As one journalist reported, "No emperor could have been more fastidious in the transplanting of an exotic civilization than Mr. Ford has been in assembling nineteenth-century America near the factories that are the world-wide symbol of twentieth-century America."[32] What had started out as an "advertising stunt" by General Electric had turned into an elaborate historical pageant of a most singular character. Henry Ford claimed that he got no advertising benefit out of it. His assistant, Liebold, put it somewhat differently. "Mr. Ford naturally received favorable publicity," Liebold wrote, "but it had not been done for that purpose."[33]

President and Mrs. Hoover, and Secretary Mellon, with attendant secret-service men, arrived on the morning of the twenty-first of October, at the head of a delegation including the nation's most eminent political and financial personages, among them Owen D. Young, Thomas W. Lamont, J. P. Morgan (the younger), Charles M. Schwab, and Otto H. Kahn; there were also scientists and inventors such as Orville Wright and Madame Curie. As President Hoover approached Detroit in his train, his party was met by Mr. and Mrs. Ford and Mr. and Mrs. Edison at a transfer point, where all changed to a little train of the Abraham Lincoln era, drawn by a wood-burning locomotive. In this they traveled over a spur line for half a mile to the restored "Smith's Creek Junction," where, seventy years before, Edison was supposed to have been given the bounce out of his baggage-car laboratory. Here a trainboy came on with a basket of merchandise; Edison took it from him and made an effort to walk about for a few moments, crying, "Candy, apples, sandwiches, newspapers!"—offering them to President Hoover. It was absurd, and yet also sheer symbolic drama; the American Dream reenacted before the world's newspapers and movie cameras. But the principal actor was now feeble and his voice weak.

At nightfall, after all the sights had been displayed to the distinguished guests, Edison appeared on the second floor of the "Menlo Park" laboratory to demonstrate how he made a carbonized thread and vacuum globe in 1879 and at a given moment turned it on.

His chief associates of that great hour, Kruesi, Johnson, Batchelor, and Upton, were all dead. But fortunately Francis Jehl, one of the few survivors of Menlo Park days, had been found, somewhere in Europe, and brought back to help in the restoration and to assist his old master again with the pumping. As many of the guests as could be crowded into the laboratory were on hand to watch Edison re-create what seemed a scientific fable. But millions more throughout the world were now sitting at their radios, listening to the announcer reporting the event:

The lamp is now ready, as it was a half century ago! Will it light? Will it burn? Edison touches the wire.

Ladies and gentlemen—*it lights!* Light's Golden Jubilee has come to a triumphant climax!

As the model of the old carbon filament lamp was turned up, all over "Monlo Paik," all over Dearborn, and Detroit, and in other great cities across the country special lamps blazed up suddenly with an immense yellow refulgence along the main avenues, as the voice on the radio continued:

And Edison said: *"Let there be light!"*

The excitement had been wearing for the Old Man. But he had more to come. The festival was to be topped off with a banquet for five hundred guests in that inflated model of Independence Hall; Owen Young was to be toastmaster, and President Hoover was to give the principal address in dedication of what was afterward called Ford's "Old Curiosity Shop." But at the door of the banquet hall Edison faltered and all but collapsed. Led to a settee in the corridor, he sat down and wept, overcome with emotion and fatigue.

"I won't go in," he said to Mrs. Edison. Only she could have overcome his resistance. They brought him some warm milk; he revived a little, entered the hall and took his place at the seat of honor. Messages from many nations were read, tributes were offered, while he heard nothing and ate nothing. President Hoover spoke.

I have though it fitting for the President of the United States to take part in paying honor to one of our great Americans. . . . Mr. Edison . . . has repelled the darkness . . . has brought to our country great distinction throughout the world. He has brought benefaction to all of us. . . .

Edison spoke briefly, but with feeling, in reply. He was happy, he said, that tribute was being paid to scientific work.

This experience makes me realize as never before that Americans are sentimental and this crowning event of Light's Golden Jubilee fills me with gratitude. As to

Henry Ford, words are inadequate to express my feelings. I can only say to you, that in the fullest and richest meaning of the term—he is my friend. Good night.

Then he slumped into his chair and turned as white as death. Mrs. Edison and President Hoover's physician at once helped him to a room in the rear of the speaker's table, and laid him on a sofa. Drugs were administered, and he came to; then he was taken to the Ford residence and put to bed for several days. "I am tired of all the glory, I want to get back to work," he said.

EDISON COLLAPSES AT JUBILEE BEFORE NOTABLES AT DEARBORN, the newspaper headlines reported.

During his recent illness in the preceding summer he had permitted doctors, for all his skepticism about them, to examine him thoroughly. They had found that his long-enduring constitution had been completely undermined; he suffered from a combination of uremic poisoning, Bright's disease, diabetes, and gastric ulcer. It was a mystery to the medical men how he had lasted so long. His days in the laboratory were now definitely over, though none was supposed to mention that to him or anyone else. "Keep him going. Don't change anything, have the same people, don't let him stop," the family doctor advised.

In these last days many of the things being said about him were, of course, exaggerated compliments, conceived in the spirit of compassion. Yet much of the growing legend attached to him reflected the innate ideas that many typical Americans of his day had of what was best and strongest in themselves and in the form of civilization they had made, materialistic, and yet also meliorative and charitable. It was after all true, as was so often said on those honorific occasions arranged for Edison in his later days, that no other of our workers in science had appealed so much to the imagination of everyman; no other inventor's name had become so universally known and popular as his. "He revolutionized industry," Dr. Le Compte du Noüy observed at the end, and "as one of the greatest of technical geniuses" became "a symbol of our Machine Age, one of the founders of our mechanical civilization, for better or for worse." [34]

During the span of his life, another kind of fame was accorded to the purely scientific discoverers, unjustly limited or retarded in the case of the Americans Henry and Gibbs, but imposing enough in that of a Planck, a Rutherford, or an Einstein. Yet their great contributions to the knowledge of nature's laws were fated to be remote from the understanding of common men, who were so captivated on the other hand by the practical experimenter and contriver Edison. There were other notable American inventors in his time; but they were more usually lonely, troubled, or diffident souls, and his personality outshone theirs. By 1929, two whole generations of Americans had grown accustomed to watching the progress of the self-educated man who worked in a laboratory that had its windows virtually open to all the world. He had known how to *dramatize* his inventions, had known how to engender interest, faith, and hope in the success of his projects. That the career of this plain-spoken, rough-hewn man made so deep an impression on multitudes of Americans speaks well for them, after all. To modern workers in science, in the corporation laboratories and universities, he typified the independent, lone-handed inventor of the nineteenth-century's "heroic age of invention." He had begun his work in the age of gas and kerosene lamps and was leaving it with the cities throughout the world garlanded with his lights, and music and voices sounding everywhere.

5   He absented himself for longer and longer intervals from his laboratory in the two last years that remained, during which it seemed only his will kept him alive, for he ate next to nothing. Often he stayed abed or sat in an easy chair at home, but still kept in close touch with his technical assistants, who daily brought him news of the progress of the goldenrod rubber experiment. Mrs. Edison watched patiently over his rest and diet, and maintained an almost daily routine of motor drives with him along the country roads of New Jersey.

In those last two years of ill health and weakness Edison still tried to work—though usually at his home, in a large room upstairs that served as his den, with Mina Edison always there also, occupied with her own domestic, social, and religious activities, at a large flat-topped desk placed at the other end of the room. Mrs. Edison had two telephones at her desk and a battery of push buttons with which she could "buzz" the servants. She might be working over anything from a program for a future Chautauqua session to the affairs of the School Board at West Orange, or those of a local temperance committee seeking to curb Sunday amusements in the town, including beer drinking. She also busied herself in helping to beautify the streets and gardens of the place, and in aiding the Negroes living in slums on the outskirts of the Edisons' winter home at Fort Myers, Florida. Facing her big desk, there hung on the wall a large framed photograph of Edison in his twenties, a youth with the expression of a dreamer. Beside it was a portrait of her eldest son, Charles, whose features—though softer and less rugged—strongly resembled his father's.

Edison's interest in new "campaigns" never flagged. Having recently met Colonel Charles A. Lindbergh, he insisted on being taken to Newark Airport to learn something about the problems of airplane landing and take-off. "The aviators tell me that they must find a means to see through a fog," he said. "I have an idea about it. I am waiting for a real fog—a water fog—and I will see if I can't penetrate it." Perhaps a rocket would do the trick?

At about this period he ventured the prophecy that "There will one day spring from the brain of Science a machine or force so fearful in its potentialities, so absolutely terrifying, that even man, the fighter who will dare torture and death in order to inflict torture and death, will be appalled, and so will abandon war forever...." He was greatly interested in the impact of Albert Einstein's formulations, but admitted that he couldn't understand any of it. "I am the zero of mathematics," he conceded ruefully. In one of his last interviews, he went on to say, "I am much interested in atomic energy, but so far as I can see we have not yet reached a point where this exhaustless force can be harnessed and utilized." [35]

On August 1, 1931, he had a sudden sinking spell and lay at the point of death. He could absorb virtually no food at all—but, to the astonishment of his physicians, he rallied and tried to rise from bed again. As the intense summer heat oppressed him, one of the new air-conditioning machines was installed in his bedroom, and he remained indoors, resting and reading. But as soon as he understood the real state of affairs, that he would never be well enough to go back to work, he seemed to lose the desire to live.

Now, for the first time, he ceased to fight; as he lay in bed reading or writing in one of his notebooks, the words or drawings wavered, the notebooks slipped from his hand. Meadowcroft, he was told, was also dying; John Ott, the last of the ancient co-workers, was sinking fast. From Detroit, Henry Ford made a hurried trip in September to look on his idol for a last time. Edison rallied enough to speak to Ford with considerable spirit and ask him many questions.

In October, when the sharp-thrusting hills of the Orange Valley were truly daubed in orange, he sank again. In the great library below the hill, everyone spoke in hushed tones; the big chair at the huge old roll-top desk was empty; the great carved clock ticked on.

Gloom hung over the mansion in Llewellyn Park. Edison slept, waked a little, and drowsed again, he who had resented every moment so lost. As a patient he was both difficult and courageous. There were several eminent medical men attending him, and he persisted in discussing their method of treatment, inquiring of them what medicines they were trying, how his body was reacting—and why. When they prescribed medicine, he insisted on measuring out his own doses. After blood tests were taken he would show the keenest interest, insisting that Dr. Hubert S. Howe, the chief physician, bring him slides and microscope so that he too might examine them. When the chart was set up before him with its stabilized line, he could read it and follow his progress; when the line turned downward, after the first week of October, he could see very clearly how the "campaign" was going. His interest in applied science, one might truly say, never flagged and ended only with the termination of this "last experiment."

One who ministered to him asked if "he had thought of a life hereafter."

"It does not matter," he replied in a low voice. "No one knows." [36]

In the early days of October it became generally known that only a short time was left to him. His mind became befogged at last, his great eyes dimmed, and from time to time he sank into a coma. At the brief intervals when he was conscious he was placed in a chair by one of the tall windows of his bedroom, overlooking a great sweep of lawn and handsome beeches. Could he still see? Mina Edison was the last person he appeared to recognize. She bent over the pale invalid and placed her mouth to his ear so that she could communicate with him. "Are you suffering?" she asked. "No, just waiting," he replied. Once he looked toward the window and the last audible words he uttered were: "It is very beautiful over there."

Newspaper reporters maintained a deathwatch in the vicinity of Glenmont day and night. The room upstairs was kept dark at night, with a nurse sitting beside the patient; if the lights went on—then all the world must be told. In the last hours many of Edison's laboratory associates waited in the hall downstairs, while Charles Edison would go up the great stairway, then down, to make his report.

Fifty-two years earlier almost to the day, the men associated with Edison were sitting up with him at night to watch how many hours his carbon filament lamp would live. Now, his son Charles, after his periodic visits to the sickroom, used the same phrase spoken long ago by the watchers at the Menlo Park Laboratory in October, 1879: "The light still burns."

On October 17 his pulse dropped steeply; in the early morning hours of Sunday the 18th, at 3:24 A.M., the lights of his room went on, and the doctors and nurse came out to announce the end. The electromagnetic telegraph, the telephone, the radio, with all of which his life had been bound up, flashed the news to all the corners of the world. Funeral services were set up for the morning of Wednesday, the twenty-first.

That day some thought was given in high places to the idea of commemorating his passing in some unusual way, in a manner worthy of such a hero's death; the President of the United States, it was proposed, should order all electric current to be turned off for a minute or two, in streets, factories, public places, and homes throughout the nation. But no sooner was the thought uttered than it was realized that such action was unthinkable. Owing to the very nature of Edison's contribution to the technical organization of modern society, his now so vital system of electric power distribution—the blood circulation of the community, as it were—would have been arrested; there was risk of incalculable disaster in halting, even for an instant, the great webs of transmission lines, and the whole monster mechanism of power that had grown in a half century out of his discoveries. The idea of a momentary "blackout" was therefore abandoned as entirely impracticable. Instead the President suggested that lights be dimmed for a few minutes, voluntarily, where possible, as in private residences, at 10 P.M. of the day of Thomas A. Edison's funeral. Many paid this last tribute, silently, and then the lights were turned on again.

References to notes in text

chapter I (pages 1–12)

1. W. Ogden Wheeler, *The Ogden Family in America, Elizabethtown Branch* (Newark, N.J., 1907), p. 343.
2. Affidavit and Petition of John Edison before the Royal Commission, Digby, Nova Scotia, Dec. 8, 1776; photostat from Public Records Office, London, in Edison Laboratory National Monument Archives. (The Edison Laboratory Archives will be referred to hereinafter under the abbreviation ELA.)
3. Minutes of the Council of Safety of New Jersey, July 2, 1777. John Edison is reported as having been transferred from the New Ark to the Morristown Jail, and as having been brought before the Council for military trial on Jan. 12, 1778.
4. A. W. Savary, "Connections of the Family of Edison the Inventor with Digby, Nova Scotia," *New England Historical and Genealogical Record*, vol. XLVII, 1894, p. 199.
5. William A. Simonds, *Edison, His Life, His Work, His Genius* (Indianapolis, 1934), p. 32.
6. "Mr. Edison's Notes for W. H. Meadowcroft" (MS.), Book II; ELA. (These are autobiographical notes dictated by Edison in 1908 for use in the authorized biography by F. L. Dyer, T. C. Martin and W. H. Meadowcroft; they are more complete than published quotations from them, and they are unedited.)
7. Simonds, *op. cit.*, p. 35.
8. Alexis de Tocqueville, *Democracy in America* (New York, 1899), vol. II, p. 46.
9. *Ibid.*, p. 52.
10. James G. Crowther, *Famous American Men of Science* (New York, 1938), p. 309.
11. Dirk L. Struik, *The Origins of American Science* (2nd ed., New York, 1957), p. 37.
12. Joseph Rossman, *The Psychology of the Inventor* (Washington, 1929), p. 28.

chapter II (pages 13–40)

1. F. L. Dyer, T. C. Martin and W. H. Meadowcroft, *Edison, His Life and Inventions*, (New York, 1929), vol. I, pp. 18–19.
2. New York *Herald Tribune*, Edison obituary, Oct. 19, 1931.
3. C. F. Ballantine, "The True Story of Edison's Childhood," *Michigan Historical Collection*, vol. IV, 1926, pp. 168–92.
4. J. B. McClure, *Thomas A. Edison and His Inventions* (Chicago, 1879), p. 35.
5. Simonds, *Edison, His Life, Work, Genius*, p. 38.
6. Marion Edison Wheeler, "Edison's Boyhood," Cleveland *Plain Dealer*, May 15, 1955.
7. Edison's Notes for Meadowcroft, Book II, p. 9.
8. Dyer, Martin and Meadowcroft, *op. cit.*, vol. I, p. 18.
9. William K. L. and Laura Dickson, *The Life and Inventions of Thomas A. Edison* (London, 1894), pp. 6–7.
10. Ballantine, *op. cit.*, p. 180.
11. New York *Herald Tribune*, Oct. 19, 1931.
12. Thomas A. Edison, *The Diary and Sundry Observations of Thomas A. Edison*, ed. Dagobert D. Runes (New York, 1948), pp. 112–13.
13. G. B. Engles to T. A. Edison, Aug. 13, 1885; ELA.
14. McClure, *op. cit.*, p. 35.
15. J. F. Talbot, in the Port Huron, Mich., *Commercial* (undated); ELA.
16. New York *Herald Tribune*, Oct. 19, 1931.
17. Edison's Notes for Meadowcroft, Book II, p. 8.
18. T. A. Edison, Introduction to the *Collected Works of Thomas Paine* (New York, 1925).
19. Dickson, *op. cit.*, p. 36; citing the statements of Samuel Edison.
20. Edison's Notes for Meadowcroft, Book II, pp. 7–8.
21. Mrs. Marion Edison Oser to author, Oct. 15, 1956.
22. R. L. Thompson, *Wiring a Continent* (Princeton, N.J., 1952), p. 25.

23. Edison's Notes for Meadowcroft, Book II, p. 9.

24. *Ibid.*, Book I, p. 6.

25. Dickson, *op. cit.*, p. 5.

26. McClure, *op. cit.*, p. 37.

27. Edison's Notes for Meadowcroft, Book II, pp. 4–5; Charles Edison to author.

28. New York *World*, Feb. 11, 1921 (interview with Edison).

29. Edison's Notes for Meadowcroft, Book II, pp. 14–15.

30. Dyer, Martin and Meadowcroft, *op. cit.*, vol. I, p. 37.

31. Henry Ford and Samuel Crowther, *Edison as I Knew Him* (New York, 1930), p. 24.

32. Edison, *Diary*, p. 47.

33. *Ibid.*, p. 48.

34. Membership card of Edison in the Henry Ford Museum Library, Dearborn.

35. Edison, *Diary*, p. 45.

36. Andrew Ure, *Arts, Manufactures, and Mines* (London, 1856), vol. II, pp. 489–90.

37. G. P. Lathrop, "Talks with Edison," *Harper's Magazine*, February, 1890, p. 426.

38. *Ibid.*, p. 424.

39. James Clancy, of Port Huron (undated letter); ELA.

40. *The Weekly Herald*, Feb. 3, 1862; copy in Henry Ford Museum Library.

chapter III (pages 41–58)

1. McClure, *Edison and His Inventions*, p. 48.

2. Edison's Notes for Meadowcroft, Book I, p. 12.

3. *Ibid.*, p. 23.

4. Walter P. Phillips, *Sketches Old and New* (Boston, 1897), p. 65.

5. *Ibid.*, p. 66.

6. Edison's Notes for Meadowcroft, Book I, p. 20.

7. *Ibid.*, p. 21.

8. Dyer, Martin and Meadowcroft, *Edison, His Life and Inventions*, vol. I, p. 92.

9. Dickson, *Life and Inventions of Edison*, p. 34.

10. Edison's Notes for Meadowcroft, Book I, p. 18.

11. *Ibid.*, p. 22.

12. Jot Spencer to T. A. Edison, Apr. 6, 1878; ELA.

13. Simonds, *Edison, His Life, Work, Genius*, p. 59.

14. Lathrop, "Talks with Edison," *Harper's Magazine*, February, 1890, p. 423.

15. *Ibid.*

16. Simonds, *op. cit.*, p. 60. A copy of George Tucker's *Thomas Jefferson*, autographed by Edison, "Memphis, March 11, 1866," fixes the time of the inventor's stay in that city; the volume is in the possession of Mrs. Carrie Edison Morse, Detroit, Mich.

17. Dyer, Martin and Meadowcroft, *op. cit.*, vol. I, p. 83.

18. *Ibid.*, pp. 80, 92.

19. Dickson, *op. cit*, p. 33.

20. Lathrop, *op. cit.*, p. 425.

21. McClure, *op. cit.*, p. 51.

22. Edison's Notes for Meadowcroft, Book I, p. 20.

23. C. Temple, "Edison in Louisville," Louisville *Courier*, Feb. 4, 1947.

24. Edison letter, headed "Memphis" (undated, but probably written in 1866); in Henry Ford Museum Library.

25. Simonds, *op. cit.*, p. 63.

26. James Symington, of Port Huron, to Edison, May 18, 1877; ELA.

27. Edison's Notes for Meadowcroft, Book I, pp. 20–21.

chapter IV (pages 59–71)

1. Dickson, *Life and Inventions of Edison*, p. 40.

2. *Ibid.*, p. 41.

3. Phillips, *Sketches Old and New*, p. 178.

4. Memorandum of G. P. Lowrey, in Henry Villard's unpublished notes on Edison; Villard Papers, Houghton Library, Harvard University.

5. Dickson, *op. cit.*, pp. 49–50.

6. *Journal of the Telegraph* (Boston), June 14, 1868; also Dec. 12, 1868.

7. Phillips, *op. cit.*, pp. 179–80.

8. *The Telegrapher* (New York), Jan. 30, 1869.

9. Jeremy Bentham's maxims, cited in Sir James A. Salter, *Modern Mechanization and Society* (London, 1933), p. 22.

10. Lathrop, "Talks with Edison," *Harper's Magazine*, February, 1890, p. 426.

11. Francis Jehl, *Reminiscences of Menlo Park* (Dearborn, 1937), vol. I, p. 8.

12. *Atlantic & Pacific Telegraph Company v. G. B. Prescott, T. A. Edison, Western Union, et al.;* an appeal before the Secretary of the Interior, Mar. 20, 1875; with reply by Western Union Telegraph Co.

13. *The Telegrapher* (New York), Apr. 17, 1869.

chapter V (pages 72–83)

1. Edison to J. Hanaford, June 10, 1869; ELA.
2. Edison's Notes for Meadowcroft, Book I, pp. 32–33.
3. *Ibid.*, Book II, pp. 1–2.
4. *Ibid.*, Book II, p. 4.
5. *Ibid.*, Book I, p. 34.
6. Undated autograph letter of Edison, probably of May, 1870; Henry Ford Museum Library, Dearborn.
7. Phillips, *Sketches Old and New*, p. 185.
8. McClure, *Edison and His Inventions*, p. 48.

chapter VI (pages 84–104)

1. Dickson, *Life and Inventions of Edison*, p. 67.
2. Autograph letter of Edison in winter of 1871; Henry Ford Museum Library.
3. A. E. Harlow, *Old Wires and New Waves* (New York, 1936), pp. 333–34.
4. Dickson, *op. cit.*, p. 68.
5. Mary C. Nerney, *Edison, Modern Olympian* (New York, 1934), p. 53.
6. Dickson, *op. cit.*, p. 47.
7. Edison's Notes for Meadowcroft, Book I, p. 40.
8. Dickson, *op. cit.*, pp. 88–89.
9. Edison's Laboratory Notebooks, 1871, No. 1678; ELA.
10. *Ibid.*, Feb. 2, 1872, No. 1776; ELA.
11. Jehl, *Reminiscences*, vol. I, p. 51.
12. Dyer, Martin and Meadowcroft, *Edison, His Life and Inventions*, vol. I, p. 146.
13. McClure, *Edison and His Inventions*, pp. 17–18.
14. Edison Laboratory Notebooks, Sept. 15, 1872; ELA.
15. *Atlantic & Pacific Telegraph Co. v. Prescott et al.*, Mar. 20, 1875.
16. *Edison and Western Union*, brochure published by the Western Union Telegraph Co. (New York, 1931).
17. Dickson, *op. cit.*, p. 52.
18. *Ibid.*, p. 88.
19. McClure, *op. cit.*, p. 68.
20. New York *Sun*, Oct. 19, 1931.
21. McClure, *op. cit.*, p. 68.
22. Edison, *Diary*, p. 54.
23. Marion Edison Oser, "Early Recollections" (voice-written), March, 1956.
24. Dickson, *op. cit.*, p. 71.

25. Phillips, *Sketches Old and New*, p. 186.
26. Marion Edison Oser to author (interview), Oct., 15, 1956.
27. T. A. Edison to Samuel Edison, Jan. 29, 1877; Henry Ford Museum Library.
28. Edison Laboratory Notebooks, Feb. 14, 1872; ELA.
29. Edison's Notes for Meadowcroft, Book I, p. 43.
30. J. D. Reid, *The Telegraph in America* (New York, 1877), pp. 588–89.
31. Edison's Notes for Meadowcroft, Book II, p. 13.

chapter VII (pages 105–130)

1. *Atlantic & Pacific Telegraph Co. v. G. B. Prescott, T. A. Edison, Western Union, et al.*, New York Superior Court, 1875; testimony of J. T. Murray, p. 51.
2. Jehl, *Reminiscences*, vol. I, p. 10; also U.S. Patent No. 180,857, "Autographic Printing," executed Mar. 7, 1876.
3. Thompson, *Wiring a Continent*, pp. 446–47.
4. A. E. Harlow, *Old Wires and New Waves*, pp. 368–69.
5. *Atlantic & Pacific v. Prescott;* W. Orton's testimony, pp. 107ff.
6. Edison's Notes for Meadowcroft, Book II, p. 12.
7. *Atlantic & Pacific v. Prescott;* Exhibit No. 11, for the defense.
8. *Ibid.*, Exhibit No. 15, for the defense.
9. Dickson, *Life and Inventions of Edison*, p. 87.
10. Matthew Josephson, *The Robber Barons* (New York, 1934), pp. 83, 205.
11. Edison to G. P. Prescott, May 19, 1874; ELA.
12. Edison's Notes for Meadowcroft, Book. II, p. 12.
13. Phillips, *Sketches Old and New*, p. 183.
14. *Ibid.*, p. 185.
15. *Atlantic & Pacific v. Prescott*, p. 34.
16. Harlow, *op. cit.*, p. 363.
17. *Atlantic & Pacific v. Prescott;* cross-examination of J. G. Reiff, pp. 45ff.
18. Jay Gould to T. A. Edison, Jan. 20, 1877; ELA.
19. *Atlantic & Pacific v. Prescott;* cross-examination of J. G. Reiff, pp. 47–48; brief of E. N. Dickerson for plaintiff, pp. 72–73.
20. Jehl, *op. cit.*, vol. I, p. 68.

21. G. H. Sandison, "The Real Edison," reminiscences of a former telegrapher of the A. & P. system, *Columbian Magazine*, vol. III, 1911.

22. Edison's Notes for Meadowcroft, Book II, pp. 13–14.

23. T. A. Edison to Samuel Edison, Nov. 8, 1877; Henry Ford Museum Library.

24. *Atlantic & Pacific v. Prescott;* Exhibit No. 12, for the defense, Dec. 10, 1874.

25. Edison's Notes for Meadowcroft, Book II, pp. 13–14.

26. *Ibid.*, Book II, p. 14.

27. Petition of Western Union to the Secretary of the Interior, Mar. 20, 1875.

28. *Telegraphic Journal* (London), Sept. 1, 1874.

29. Theodore M. Edison to author.

30. U.S. Patent No. 480,567, "Duplex Telegraphs," executed Aug. 19, 1874.

31. T. A. Edison to Jay Gould; draft of a letter of 1877; ELA.

32. Edison's Notes for Meadowcroft, Book II, pp. 18–19.

33. Edison Laboratory Notebooks, Dec. 11, 1875.

34. *Ibid.*, Dec. 12, 1875 (entry on "Etheric Force").

35. New York *Herald*, Dec. 2, 1875.

36. E. J. Houston, "Etheric Force of Edison," *Scientific American*, vol. XXXIV (1876), Supplement No. 5.

37. Sir J. Ambrose Fleming, *Fifty Years of Electricity* (London, 1924).

chapter VIII (pages 131–155)

1. G. M. Shaw, "Sketch of Edison," *Popular Science Monthly*, August, 1878, pp. 489–90.

2. L. Stieringer, *The Life and Inventions of Thomas Edison* (Burgoyne, 1890), p. 37.

3. Edison to Frank Royce, "Menlo Park, June, 1876," Henry Ford Museum Library.

4. Jehl, *Reminiscences*, vol. I, p. 226.

5. Nerney, *Edison, Modern Olympian*, p. 64.

6. Norbert Wiener, *The Human Use of Human Beings, Cybernetics and Society* (2d ed., revised, New York, 1954), p. 115.

7. R. C. McLaurin, "Edison's Services to Science," *Science*, June 14, 1915, p. 813.

8. Werner Sombart, *Die Entstehung der Moderne Kapitalismus* (Tübingen, 1924), vol. III, pp. 74 *ff*.

9. Crowther, *Famous American Men of Science*, p. 363.

10. *American Bell Telephone Company v. P. A. Dowd and American Speaking Telegraph Co.;* U.S. Circuit Court, District of Massachusetts, 1878; Testimony of T. A. Edison; vol. III, pp. 105 *ff*.

11. Marion Edison Oser to author, October 15, 1956.

12. Dyer, Martin, and Meadowcroft, *Edison, His Life*, vol. I, p. 179.

13. Edison to Samuel Edison (undated); Henry Ford Museum Library.

14. Edison to P. A. Dowd, May 14, 1877; Henry Ford Museum Library.

15. Shaw, *op. cit.*, p. 490.

16. Laboratory Notebooks, 1877; "Telephone."

17. Shaw, *op. cit.*, p. 490.

18. Edison to T. B. David (dated only "1878"); letter in possession of S. P. Grace, of Bell Laboratories, New York.

19. J. E. Kingsbury, "Edison's Carbon Transmitter" (Ms.); Library of Western Electric Co., New York.

20. Dyer, Martin and Meadowcroft, vol. II (Appendix), p. 875.

21. Simonds, *Edison, His Life*, p. 118.

22. A. W. Robertson, *Story of the Telephone in England* (London, 1947), pp. 11–12.

23. Lathrop, "Talks with Edison," *Harper's Magazine*, February, 1890, p. 437.

24. Jehl, *op. cit.*, vol. I, p. 175 (citing a letter of Samuel Insull).

25. Edison's Notes for Meadowcroft, Book II, p. 44.

26. George Gouraud to Edison, January 15, 1879; ELA.

27. Edison's Notes for Meadowcroft, Book II, pp. 44 *ff*.

28. Jehl, *op. cit.*, vol. I, p. 278.

29. Samuel Insull's Autobiography (unpublished), p. 20; quoted by permission of Samuel Insull, Jr.

30. George Bernard Shaw, *The Irrational Knot* (New York, 1905), Preface, ix–xi.

31. E. H. Johnson to Edison, December 2, 1879; ELA.

32. Edison to E. H. Johnson, December 7, 1879; ELA.

33. Edison's Notes for Meadowcroft, Book II, pp. 25–26.

34. Lathrop, *op. cit.*, p. 437.

chapter IX (pages 156–174)

1. Lathrop, "Talks with Edison," *Harper's Magazine*, February, 1890, p. 427.

2. *Ibid.*
3. Jehl, *Reminiscences*, vol. I, p. 181.
4. Laboratory Notebooks, "Telegraphic Repeater," 1877; ELA.
5. E. H. Johnson to Edison, July 17, 1877; ELA.
6. Lathrop, *op. cit.*, p. 428.
7. Laboratory Notebooks, "Telephone," July 18, 1877; ELA.
8. E. H. Johnson, article in *Electrical World*, February 22, 1890.
9. General Ben Butler to Edison, October 23, 1877; ELA.
10. Laboratory Notebooks, "Phonograph," 1877; File A 54-13; also, "The Invention of the Phonograph" (MS), by Norman R. Speiden, May 5, 1949 (a study of the order of events in the invention of the phonograph, based on Edison's laboratory notes and sketches); copy in ELA.
11. Edison's Notes for Meadowcroft, Book II, p. 19.
12. Norman R. Speiden, *op. cit.*
13. *Leslie's Weekly*, April 2, 1878.
14. R. C. McLaurin, "Edison's Service to Science," *Science*, June 4, 1915.
15. *Leslie's Weekly*, April 2, 1878.
16. *Scientific American*, Supplement, December, 1878.
17. *Nature* (London), March 20, 1879.
18. McClure, *Edison and His Inventions*, p. 720.
19. New York *Tribune*, August 31, 1879.
20. Dyer, Martin and Meadowcroft, *Edison, His Life and Inventions*, vol. I, p. 211.
21. *Scientific American*, June 12, 1878.
22. W. H. Bishop, "A Night with Edison," *Scribner's Magazine*, August, 1878.
23. Edison's Notes for Meadowcroft, Book II, p. 20.
24. "An Afternoon with Edison," New York *Daily Graphic*, April 2, 1878.
25. Bishop, *op. cit.*
26. Dickson, *Life and Inventions of Edison*, p. 101.
27. New York *Daily Graphic*, April 2, 1878.
28. Thomas A. Edison, "The Phonograph and Its Future," *North American Review*, June, 1878.
29. Roland Gelatt, *The Fabulous Phonograph* (Philadelphia, 1954), pp. 30–31.
30. Philadelphia *Record*, February 12, 1880.

1. Edison's Notes for Meadowcroft, Book I, pp. 1–3.
2. *Ibid.*, p. 5.
3. Edison's Memorandum, "Beginning of the Incandescent Lamp" (MS.), 1926; ELA.
4. *Telegraphic Journal* (London), June 1, 1878.
5. Letter of Benjamin Silliman, Jr., in St. Louis *Globe-Democrat*, July 27, 1878.
6. New York *Sun*, September 10, 1878; and *Electrical Engineering*, February 1, 1893.
7. Edison's Memorandum, "Beginning of the Incandescent Lamp," 1926; ELA.
8. New York *Sun*, October 25, 1878.
9. Edison to Theodore Puskas (cablegram), September 22, 1878; ELA.
10. New York *Tribune*, September 28, 1878.
11. Laboratory Notebooks, No. 184, "Electricity vs. Gas," 1878; ELA.
12. Harold C. Passer, *The Electrical Manufacturers: 1875–1900* (Cambridge, 1953), p. 80.
13. Edison's Memorandum, "Beginning of the Incandescent Lamp," 1926; ELA.
14. J. W. Howell and Henry Schroeder, *History of the Incandescent Lamp* (Schenectady, 1927), p. 48.
15. Grosvenor P. Lowrey to H. M. Twombly, October 1, 1878; ELA.
16. New York *Sun*, September 16, 1878.
17. New York *Tribune*, September 28, 1878.
18. New York *Sun*, October 20, 1878.
19. *Engineering* (London), October 25, 1878.
20. F. R. Upton speech before Edison Pioneers, February 11, 1918; copy in ELA.
21. Edison to S. L. Griffin, November 14, 1878; ELA.
22. G. P. Lowrey to Edison, October 1, 1878; ELA.
23. Edison to G. P. Lowrey, October 2, 1878; ELA.
24. G. P. Lowrey to Edison, October 2, 1878; ELA.
25. Edison to Theodore Puskas, November 18, 1878; ELA.
26. H. Stafford Hatfield, *The Inventor and His World* (New York, 1933), p. 67.
27. New York *Tribune*, November 15, 1878; interview with Lowrey.
28. New York *Commercial-Advertiser*, November 25, 1878.

29. S. L. Griffin to G. P. Lowrey, November 1, 1878; ELA.

30. Jehl, *Reminiscences,* vol. I, pp. 256–57.

31. F. R. Upton, Memorandum, 1909 (MS.), pp. 2–3; ELA.

32. Howell and Schroeder, *op. cit.,* pp. 53–54.

33. *Engineering* (London), February 1, 1879.

34. *Ibid.,* October 25, 1878; article by Silvanus Thompson.

35. T. C. Martin, *Forty Years of Edison Service* (New York, 1922), pp. 3–4.

36. Passer, *op. cit.,* p. 83.

37. Nerney, *Edison, Modern Olympian,* p. 63; quoting an interview with John Ott.

38. Edison to T. Puskas, November 18, 1878; ELA.

39. A. W. Churchill, "Edison's Early Work," *Scientific American Supp.,* April 1, 1905.

40. Edison Memorandum, "Beginning of the Incandescent Light," p. 6; ELA.

41. Laboratory Notebooks, "Electric Light," February 8, 1879; ELA.

42. Edison's Memorandum of 1926, p. 6; ELA.

43. New York *Herald,* April 27, 1879.

44. G. P. Lowrey to Edison, April 14, 1879; ELA.

45. Edison to T. Puskas, June 3, 1879; ELA.

46. G. P. Lowrey to Edison, January 2, 1879; ELA.

47. Jehl, *Reminiscences,* vol. I, pp. 246–47.

48. F. Jehl, Memorandum to W. H. Meadowcroft, 1908, describing visit of the E.E.L.C. directors; ELA.

49. New York *Times,* Edison obituary, October 19, 1931.

50. New York *Herald,* April 27, 1879.

51. New York *Tribune,* January 2, 1880; New York *Herald,* January 3, 1880.

52. Edison, *Diary,* p. 43.

53. F. Jehl, Memorandum to W. H. Meadowcroft, 1908; ELA.

chapter XI (pages 205–227)

1. A. W. Churchill, *Scientific American Supp.,* April 1, 1905.

2. *Journal of Gaslighting* (London), February 18, 1879.

3. New York *Tribune,* January 2, 1880.

4. F. R. Upton, speech before Edison Pioneers, February 11, 1918; copy in ELA.

5. Lathrop, "Talks with Edison," *Harper's Magazine,* February, 1890, p. 492.

6. *Engineering News,* December 25, 1880.

7. *Scientific American,* October 18, 1879.

8. Edison to T. Puskas, January 28, 1879; ELA.

9. Upton, *op. cit.*

10. F. Jehl, Memorandum to W. H. Meadowcroft, 1908; also Dyer, Martin and Meadowcroft, *Edison, His Life,* vol. I, p. 292.

11. Jehl, *Reminiscences,* vol. I, pp. 295, 310.

12. George Weston (letter), in *Scientific American,* November 1, 1879.

13. Laboratory Notebooks, December 23, 1879; Edison's Memorandum, 1926, p. 7; ELA.

14. Laboratory Notebooks, "Electric Light," No. 0731, October 21, 1879; ELA.

15. New York *Herald,* January 3, 1880.

16. E. M. Fox, quoted in "Edison's Light" (special supplement), New York *Herald,* December 21, 1879.

17. Howell and Schroeder, *History of the Incandescent Lamp,* p. 55.

18. *Ibid.,* p. 56.

19. J. G. Crowther, *Famous American Men of Science,* pp. 387–88.

20. Jehl, *op. cit.,* vol. I, p. 304.

21. Laboratory Notebooks, "Electric Light," No. 79,073, 1879; ELA.

22. Upton, *op. cit.*

23. M. A. Rosanoff, "Edison in His Laboratory," *Harper's Magazine,* September, 1932, p. 409.

24. New York *Herald,* December 31, 1879.

25. A. O. Tate, *Edison's Open Door* (New York, 1938), pp. 108–9.

26. Jehl, *Reminiscences,* vol. II, p. 530.

27. C. L. Clarke, in "Edisonia," *Early History of the Edison Electric Light* (New York, 1904), p. 86.

28. New York *Herald,* December 21, 1879.

29. Francis A. Jones, *Life Story of Thomas A. Edison* (New York, 1907), p. 106.

30. Laboratory Notebooks, "Electric Light," No. 0731, p. 107; ELA.

31. "Edison's Light" (supplement), New York *Herald,* December 21, 1879.

32. New York *Tribune,* January 2, 1880.

33. S. L. Griffin to W. A. Bailey, December 3, 1879; ELA.

34. G. P. Lowrey to Edison, November 13, 1879; ELA.
35. New York *Herald*, January 2, 1880.
36. New York *Tribune*, January 2, 1880.
37. T. C. Martin, *Forty Years of Edison Service*, pp. 21–22.
38. New York *Tribune*, January 18, 1880.

chapter XII (pages 228–250)
1. Laboratory Notebooks, 1878, No. 4184, "Electricity versus Gas"; ELA.
2. D. O. Woodbury, *Beloved Scientist: Elihu Thomson* (New York, 1944), pp. 108–9.
3. Nerney, *Edison, Modern Olympian*, pp. 86–87.
4. Dyer, Martin and Meadowcroft, *Edison, His Life and Inventions*, vol. II, Appendix, p. 914.
5. New York *Times*, October 19, 1931.
6. Edison's Memorandum of 1926, p. 10; ELA.
7. Dickson, *Life and Inventions of Edison*, pp. 198 *ff*.
8. New York *Sun*, May 2, 1889.
9. G. P. Lowrey to Mrs. G. P. Lowrey, April 20, 1880; copy in ELA.
10. *Scientific American*, May 22, 1880.
11. *Ibid.*, August 21, 1880.
12. Jehl, *Reminiscences*, vol. II, p. 504.
13. G. P. Lowrey to Mrs. Lowrey, June 5, 1880; ELA.
14. Dyer, Martin and Meadowcroft, *op. cit.*, vol. I, p. 459.
15. Frank Julian Sprague, "The Electric Railway," *Century Magazine*, July, 1905.
16. H. C. Passer, *The Electrical Manufacturers*, pp. 218 *ff*.
17. T. C. Martin, "Edison's Pioneer Work in the Electric Railway," *Scientific American*, November 18, 1911.
18. *Electric World*, August 4, 1884.
19. New York *Mirror*, December 21, 1878.
20. "Edisonia," p. 33.
21. New York *Herald*, November 18, 1880; New York *Star*, December 22, 1880.
22. Jehl, *op. cit.*, vol. I, p. 72.
23. Sarah Bernhardt, *Memories of My Life* (New York, 1907), pp. 393–96.
24. Marion Edison Oser to author. October 15, 1956.
25. New York *Herald*, December 22, 1880.
26. *Ibid.*, January 2, 1880.

27. J. W. Hammond, *Men and Volts* (Philadelphia, 1941), pp. 44–45.
28. "Edisonia," pp. 85, 166.
29. *Edison Electric Light Co. v. U.S. Electric Lighting Co.* (the Edison Filament Case) 1886–1891; testimony of T. A. Edison; U.S. Circuit Court of the Southern District of N.Y., vol. IV, pp. 2570 *ff*.
30. Dyer, Martin and Meadowcroft, *op. cit.*, vol. II, p. 719.
31. "Edisonia," pp. 149–51; A. A. Bright, *The Electric Lamp Industry* (New York, 1949), pp 76–77.
32. Samuel Insull's Autobiography (unpublished) 1934, p. 22; quoted with permission of Samuel Insull, Jr.
33. Nerney, *op. cit.*, p. 109; statement of "two former colleagues."
34. C. L. Clarke, "Menlo Park in 1880," in Jehl, *Reminiscences of Menlo Park*, vol. II, p. 862.

chapter XIII (pages 251–267)
1. New York *Tribune*, December 23, 1880.
2. *Ibid.*, February 14, 1881.
3. Samuel Insull's Autobiography (unpublished), 1934, p. 260.
4. Dyer, Martin and Meadowcroft, *Edison, His Life*, vol. I, p. 370.
5. Edison to T. Puskas, November 13, 1878; ELA.
6. Jehl, *Reminiscences*, vol. II, pp. 749–50.
7. Samuel Insull's Autobiography, p. 22.
8. Samuel Insull, Address, 1902; copy in ELA.
9. Samuel Insull's Autobiography, p. 25.
10. New York *Tribune*, August 4, 1881.
11. Edison's Notes for Meadowcroft, 1908, Book II, p. 14.
12. Martin, *Forty Years of Edison Service*, p. 36.
13. Memorandum on "Conduits," 1887; ELA.
14. Walter Edison Kruesi, "A Memoir of John Kruesi" (MS.); Henry Ford Museum Library.
15. Dyer, Martin and Meadowcroft, *op. cit.*, vol. I, p. 393.
16. Walter Edison Kruesi, *op. cit.*
17. Edison's Notes for Meadowcroft, 1908, Book II, p. 40; ELA.
18. Charles Batchelor to Edison (cablegram), October 25, 1881; ELA.
19. E. H. Johnson to Edison, January 19, 1882; ELA.

20. London *Standard*, January 20, 1882.

21. *Edison Electric Light Company Bulletin No. 3*, August 28, 1882; ELA.

22. Edison to Theodore Waterhouse, July 24, 1883; ELA.

23. Edison's Notes for Meadowcroft, 1908, Book II, p. 38; ELA.

24. H. L. Satterlee, *J. Pierpont Morgan* (New York, 1939), p. 156.

25. *Edison Electric Light Company Bulletin No. 1*, July 5, 1882; ELA.

26. Martin, *Forty Years of Edison Service*, pp. 56–57.

27. "Edisonia," p. 47.

28. Boston *Herald*, January 28, 1883.

29. *Edison Electric Light Co. v. U.S. Electric Lighting Co.*, 1890; testimony of Thomas A. Edison.

30. *Edison Electric Light Company Bulletin No. 14*, October 14, 1882; ELA.

31. S. B. Eaton to Edison, September 18, 1882; ELA.

32. Martin, *Forty Years of Edison Service*, pp. 65–66.

33. Crowther, *Famous American Men of Science*, p. 305.

34. New York *Sun*, September 4, 1882.

chapter XIV (pages 268–290)

1. McClure, *Edison and His Inventions*, pp. 1–2.

2. *Electrical World*, August 3, 1883.

3. Nerney, *Edison, Modern Olympian*, p. 105.

4. Samuel Insull's Autobiography (unpublished), pp. 39–40.

5. Dyer, Martin and Meadowcroft, *Edison, His Life and Inventions*, vol. I, p. 429.

6. Louisville *Courier*, August 3, 1883.

7. Martin, *Forty Years of Edison Service*, p. 63.

8. New York *World*, October 10, 1883; *Edison Electric Light Company Bulletin No. 21*, October 31, 1883.

9. Jehl, *Reminiscences*, vol. II, p. 504.

10. W. C. White, "Electrons and the Edison Effect," *General Electric Review*, October, 1943.

11. Edison to Clayton Sharp, *Journal of the American Institute of Electrical Engineers*, January, 1922, vol. XLI, pp. 68–78.

12. Laboratory Notebooks, 1883 ("Edison Effect"); ELA.

13. W. C. White, *op. cit.*

14. Edison to Clayton Sharp, *loc. cit.*

15. E. J. Houston, "Notes on Phenomena in Incandescent Lamps," *Transactions of American Institute of Electrical Engineers*, vol. I, No. 1, 1884.

16. Admiral H. G. Bowen, *The Edison Effect* (Edison Foundation, W. Orange, N.J., 1950), quoting letter of W. D. Coolidge, p. 52.

17. *Electrical World*, August 3, 1883.

18. MS. of letter by Edison, December 1884, on "Edison Effect"; ELA.

19. Dyer, Martin and Meadowcroft, *op. cit.*, vol. II, pp. 577–78.

20. *Ibid.*, vol. II, p. 576.

21. *Ibid.*, vol. II, p. 578.

22. A. O. Tate, *Edison's Open Door*, p. 126.

23. H. G. Bowen, *op. cit.*, pp. 7, 57.

24. Nerney, *Edison, Modern Olympian*, p. 235; citing "Edison's Golden Book."

25. Jehl, *Reminiscences*, vol. II, p. 506.

26. David T. Marshall, *Recollections of Edison* (New York, 1931), p. 56.

27. Marion Edison Oser, "Early Recollections" (voice-written), March, 1956, p. 2; ELA.

28. Dickson, *Life and Inventions of Edison*, p. 88.

29. Marion Edison Oser to author, October 15, 1956.

30. Marshall, *op. cit.*, p. 58.

31. A. O. Tate, *op. cit.*, p. 84.

32. Jehl, *Reminiscences*, vol. II, p. 511.

33. Edison, *Diary*, p. 7; MS. in ELA.

34. Edison to Samuel Insull, January 5, 1884; ELA.

35. Marion Edison Oser, "Early Recollections," p. 6.

chapter XV (pages 291–312)

1. H. C. Passer, *The Electrical Manufacturers*, p. 85.

2. Satterlee, *Pierpont Morgan*, pp. 207 *ff.*

3. *Ibid.*, p. 212.

4. *Electrical World*, November 1, 1884.

5. Satterlee, *op. cit.*, p. 217.

6. Edison's Notes for Meadowcroft, 1908, vol. II, p. 35; ELA.

7. Edison to Insull ("St. Augustine, Fla."), undated; probably January, 1884; ELA.

8. New York *Tribune*, October 26, 1884.

9. *Edison Electric Light Company Bulletin No. 20*, October 31, 1883, p. 45.

10. Edison to S. B. Eaton, November 13, 1882; ELA.

11. Edison, *Diary*, July 13, 1885; (MS.) in ELA.

12. Dyer, Martin and Meadowcroft, *Edison, His Life,* vol. I, p. 436.

13. Edison to Eugene Crowell, November 13, 1882; ELA.

14. Calvin B. Goddard (Secretary of E.E.L.C.) to Edison, May 28, 1883; ELA.

15. J. G. Chapman of St. Louis to S. Insull, October 25, 1884; ELA.

16. G. Barker to Edison, October 26, 1884; ELA.

17. S. Insull to A. O. Tate, October 27, 1884; ELA.

18. *Ibid.*

19. *Electrical World,* May 30, 1885.

20. J. W. Hammond, *Men and Volts,* pp. 60–61.

21. Marion Edison Oser to author, October 15, 1956.

22. Nerney, *Edison, Modern Olympian,* pp. 136–37.

23. Edison, *Diary,* p. 16.

24. Edison to S. Insull (telegram), June 5, 1885; ELA.

25. Akron (O.) *Times,* March 4, 1886.

26. Florence Fritz, *Bamboo and Sailing Ships* (Ft. Myers, Fla., 1949), pp. 7–8.

27. Edison, *Diary,* July 18, 1885, p. 22.

28. *Ibid.,* p. 8.

29. Marshall, *Recollections of Edison,* p. 66.

30. Edison, *Diary,* p. 54.

31. Edison to Lewis Miller, September 30, 1884; autograph letter in possession of Charles Edison.

32. Edison to Messrs. Hulsenkamp and Cranford of Ft. Myers, October 18, 1885; ELA.

33. Simonds, *Edison, His Life, Work, Genius,* p. 244.

34. W. A. Crofutt, in New York *Mail and Express,* October 8, 1887.

35. Emil Ludwig, "Edison," *Neue Rundschau* (Berlin), January, 1932, pp. 63 *ff.*

36. A. O. Tate, *Edison's Open Door,* p. 140.

chapter XVI (pages 313–338)

1. T. C. Martin, "A Day with Edison," *Electrical World,* August 25, 1888.

2. *Electrical Engineering,* August 12, 1890.

3. New York *World,* November 6, 1921.

4. Laboratory Notebooks, 1886; ELA.

5. New York *Mail and Express,* October 8, 1887.

6. A. O. Tate, Memorandum to Edison, May 25, 1887 ("Phonograph"); ELA.

7. Edison to G. Gouraud, August 1, 1887 (cablegram); ELA.

8. New York *Globe-Democrat,* January 19, 1889.

9. R. Gelatt, *The Fabulous Phonograph,* p. 41.

10. New York *Post,* October 18, 1887; E. T. Gilliland to Edison, May 19, 1888; ELA.

11. A. O. Tate, *Edison's Open Door,* p. 157.

12. Laboratory Notebooks, 1888–1896 (passim); notes of Edison undated.

13. Emil Ludwig, "Edison," *Neue Rundschau* (Berlin), p. 67.

14. A. O. Tate, *Edison's Open Door,* p. 164.

15. Dickson, *Life and Inventions of Edison,* pp. 132–33.

16. Transcript of record by Sir Arthur Sullivan, dated October 31, 1888; ELA.

17. U. H. Painter to E. H. Johnson, February 12, 1888; ELA.

18. Nerney, *Edison, Modern Olympian,* p. 138.

19. Copy of Edison's affidavit, September 21, 1888; ELA.

20. Jesse Lippincott's affidavit. September 21, 1888; ELA. New York *Herald,* January 18, 1889.

21. Jesse Lippincott's affidavit, September 21, 1888.

22. New York *Tribune,* May 18, 1889.

23. Edison to E. T. Gilliland (cablegram), September 11, 1888; ELA.

24. New York *World,* May 13, 1889.

25. *Proceedings, N.A. Phonograph Dealers' Association,* 1891, pp. 21–22; ELA.

26. Edison to Jesse Lippincott, September 21, 1888; ELA.

27. A. O. Tate, Memorandum to Edison, September 30, 1893; ELA.

28. New York *Times,* April 21, 1889.

29. New York *Tribune,* August 19, 1889.

30. Marion Edison Oser to author, October 15, 1956.

31. Edison's Notes for Meadowcroft, 1908, Book II, p. 39.

32. Dyer, Martin and Meadowcroft, *Edison, His Life and Inventions,* vol. II, p. 742.

33. A. O. Tate, *Edison's Open Door,* p. 237.

34. R. Vallery-Radot, *Vie de Pasteur* (Paris, 1900), pp. 152–53.

35. Dickson, *op. cit.,* pp. 233–34.

chapter **XVII** (pages **339–366**)

1. Dyer, Martin and Meadowcroft, *Edison, His Life and Inventions*, vol. I, p. 357.

2. *Ibid.*, vol. I, pp. 381–82.

3. S B. Eaton to Edison, August 22, 1889; ELA.

4. Edison to Henry Villard, February 8, 1890; Villard Papers, Houghton Library, Harvard University. The Villard Papers are quoted by permission of Houghton Library.

5. Dyer, Martin and Meadowcroft, *op. cit.*, vol. II, p. 665.

6. D. O. Woodbury, *Beloved Scientist: Elihu Thomson*, p. 157.

7. *Ibid.*, p. 156.

8. *A Warning*, pamphlet of E.E.L.C., 1887; ELA.

9. T. A. Edison, "Dangers of Electric Lighting," *North American Review*, November, 1889.

10. Edison, Memorandum to E. H. Johnson, 1886, on Siemens and Halske's Report on "Z.B.D." (a-c) system; ELA.

11. Dickson, *Life and Inventions of Edison*, p. 330.

12. Francis G. Leupp, *George Westinghouse* (New York, 1918), pp. 133, 145 ff.

13. T. A. Edison, "Dangers of Electric Lighting," *North American Review*, November, 1889.

14. New York *Times*, August 7, 1890.

15. Edison to Edward Dean Adams, February 2, 1889; Villard Papers, Houghton Library, Harvard University.

16. F. J. Sprague to E. H. Johnson, September 13, 1886; Sprague Papers, Engineering Societies Library, New York.

17. H. C. Passer, *The Electrical Manufacturers*, p. 174.

18. Edison to H. Villard, February 2, 1889; Villard Papers, Houghton Library, Harvard University.

19. "The Edison General Electric," notes by Henry Villard, New York *Post*, December 31, 1888; Villard Papers, Houghton Library, Harvard University.

20. S. Insull, Memorandum on E.G.E., August 18, 1888; ELA.

21. S. Insull's Autobiography, 1934 (unpublished), pp. 50–51.

22. *Ibid.*

23. New York *Times*, December 16, 1891.

24. New York *Post*, December 31, 1888.

25. C. H. Coster to S. B. Eaton, April 19, 1889; ELA.

26. H. Villard to Edison, February 13, 1891; Villard Papers, Houghton Library, Harvard University.

27. Edison to H. Villard, February 8, 1890; Villard Papers, Houghton Library, Harvard University.

28. T. A. Edison, "My Forty Years of Litigation," *Literary Digest*, September 13, 1913.

29. Edison to A. O. Tate, July 2, 1888; ELA.

30. *Consolidated Electric v. McKeesport Electric Co.*, the "Westinghouse-Edison Case"; U.S. Circuit Court, Pittsburgh, Pa., October 5, 1889; opinion of Justice Bradley.

31. *Edison Electric Light v. U.S. Electric Co.*, Circuit Court of the Southern District of N.Y., IV, 2571–73; Hearings, June 19, 1890.

32. Edison to H. Villard, February 24, 1890 and April 1, 1889; Villard Papers, Houghton Library, Harvard University.

33. J. W. Hammond, *Men and Volts*, p. 191.

34. Edison to S. Insull, March 4, 1891; ELA.

35. Edison to H. Villard, February 8, 1890; Villard Papers, Houghton Library, Harvard University.

36. A. O. Tate, *Edison's Open Door*, p. 278.

37. New York *Times*, December 16, 1891.

38. A. A. Bright, *The Electric Lamp Industry*, p. 93.

39. H. S. Fraser, "Thomas Edison," unpublished MS.; ELA.

40. D. O. Woodbury, *op. cit.*, p. 204.

41. New York *Times*, December 16, 1891, and February 21, 1892.

42. New York *World*, February 20, 1892.

43. A. O. Tate, *op. cit.*, p. 261.

44. New York *World*, February 20, 1892.

45. A. O. Tate, *op. cit.*, p. 278.

chapter **XVIII** (pages **367–379**)

1. Tom A Robins, Sr., "Friends in a Lifetime," 1944 (MS.); in possession of Hewitt-Robins Company, Darien, Conn.

2. Theodore Waters, "Edison's Revolution in Iron Mining," *McClure's Magazine*, November, 1897, pp. 77–89.

3. *Ibid.*, p. 81.

4. Dyer, Martin and Meadowcroft, *Edison, His Life,* vol. II, p. 475.

5. Edison to H. M. Livor, January 21, 1890; ELA.

6. M. A. Rosanoff, in *Harper's Magazine,* September, 1932.

7. T. A. Robins, Sr., *op. cit.,* p. 8.

8. H. Ford and S. G. Crowther, *Edison As I Knew Him,* p. 28.

9. Nerney, *Edison, Modern Olympian* (quoting interview with Dan Smith), p. 149.

10. Edison to A. O. Tate, April 19, 1893; ELA.

11. Dyer, Martin and Meadowcroft, *op. cit.,* vol. II, p. 495.

12. "Anonymous" to Edison, March 16, 1898; ELA.

13. Letter of "A Visitor to Edison, N.J." in 1897 and 1899, December 13, 1899; ELA.

14. Nerney, *op. cit.,* p. 155.

15. New York *Journal,* January 14, 1898.

16. Edison to Mrs. Mina Edison, December 1, 1898; in possession of Mrs. Madeleine Edison Sloane.

17. R. C. Beck, in Newark *Star-Ledger,* September 4, 1955; interviews at Ogdensburg.

18. Simonds, *Edison, His Life, His Work, His Genius,* p. 290.

chapter XIX (pages 380–403)

1. New York *Sun,* February 27, 1903.

2. Edison's notes: "Progress," 1890; ELA.

3. M. Pupin to Edison, March 28, 1896; ELA.

4. G. S. Bryan, *Edison, the Man and His Work,* p. 258.

5. W. K. L. Dickson, *History of the Kinetoscope and Kinetograph,* brochure, (1895); in ELA.

6. Dickson, "Edison's Invention of the Kineto-Phonograph," *Century Magazine,* June, 1894.

7. Edison to E. B. Seeley, October 17, 1888; ELA.

8. Dickson, in *Century Magazine,* June, 1894.

9. Edison's Caveat No. 4, November 2, 1889; Kinetograph; ELA.

10. G. S. Bryan, *op. cit.,* p. 188.

11. Terry Ramsaye, *A Million and One Nights; A History of the Motion Picture* (New York, 1926, 2 vol.), vol. I, p. 78.

12. Dickson, in *Century Magazine,* June, 1894.

13. New York *Sun,* June 3, 1891; *Harper's Weekly,* June 13, 1891.

14. T. Ramsaye, *op. cit.,* vol. I, p. 76.

15. Note by Edison in a copy of E. Muybridge's "Zoepraxography," probably late in 1893; ELA.

16. Simonds, *Edison, His Life, Work, Genius,* p. 265.

17. T. Ramsaye, *op. cit.,* vol. I, p. 109.

18. T. A. Edison, "The Motion Picture," *Century Magazine,* June, 1894.

19. T. Ramsaye, *op. cit.,* vol. II, p. 119.

20. New York *Sun,* April 22, 1895.

21. *Edison v. Mutascope and Biograph Cos.,* U.S. District Court, Southern District of N.Y.; Edison's testimony; January 29, 1900.

22. N. C. Raff to W. E. Gilmore, January 13, 1896; and N. C. Raff to G. Armat, March 5, 1896. In *Edison v. Mutascope and Biograph Cos.*

23. New York *Journal,* April 4, 1896; New York *Times,* April 24, 1896.

24. *T. A. Edison Industries Annual Report,* February 29, 1912; in files of T. A. Edison, Inc., West Orange, N.J.

25. Edison to H. Villard, December 10, 1898; Villard Papers, Houghton Library, Harvard University.

26. French Strother, "Edison," *World's Work,* June, 1906.

chapter XX (pages 404–431)

1. M. A. Rosanoff, "Edison in His Laboratory," *Harper's Magazine,* September, 1932.

2. Dyer, Martin and Meadowcroft, *Edison, His Life and Inventions,* vol. II, p. 554.

3. W. E. Ayrton, in the *Times* (London) *Engineering Supplement,* July 22, 1908.

4. Laboratory Notebooks, "Storage Battery," 1901; ELA.

5. A. E. Kennelly, "The New Edison Storage Battery," *Journal of the American Institute of Electrical Engineers,* May 21, 1901.

6. New York *World* (interview), February 15, 1903.

7. Emil Ludwig, "Edison," *Neue Rundschau,* January, 1932.

8. M. A. Rosanoff, *op. cit.,* p. 410.

9. *Ibid.,* p. 411.

10. Maurice Holland, "Edison's Organization Method," pamphlet, National

Research Council (New York, April, 1927).

11. M. A. Rosanoff, *op. cit.*, p. 414.

12. Ray Stannard Baker, "Edison's Latest Marvel," *Windsor Magazine*, November, 1902.

13. Dyer, Martin and Meadowcroft, *op. cit.*, vol. II, p. 563.

14. Tom A. Robins, Sr., "Friends in A Lifetime," p. 7.

15. New York *World*, February 15, 1903.

16. H. E. Parshall to Edison, May 16, 1905; ELA.

17. Edison to L. C. Weir, February 27, 1905; ELA.

18. Edison to L. C. Weir, February 15, 1909; ELA.

19. Edison to L. C. Weir, July 7, 1905; ELA.

20. New York *Tribune*, January 25, 1905.

21. M. A. Rosanoff, *op. cit.*, p. 416.

22. *Ibid.*, p. 408.

23. Martha Coman and H. Weir, "She Married the Most Difficult Husband in the World," *Colliers Weekly*, July 18, 1925.

24. M. A. Rosanoff, *op. cit.*, p. 409.

25. Letter of Dr. F. T. Bonner of Arthur D. Little Institute to author.

26. Edison to L. C. Weir, June 6, 1909; ELA.

27. Edison to S. Insull, October 22, 1910; ELA.

28. *Technical World*, February, 1915.

29. George Iles, *Inventors at Work* (New York, 1906), p. 433.

30. Dyer, Martin and Meadowcroft, *op. cit.*, vol. II, p. 523.

31. M. A. Rosanoff, *op. cit.*, p. 405.

32. New York *World*, February 18, 1913.

33. Edison to Henry Ford, October 29, 1912; Ford Museum Library.

34. Charles Edison to author.

35. New York *Times*, December 10, 1914.

chapter XXI (pages 432–455)

1. Lewis Mumford, *Technics and Civilization* (New York, 1938), p. 224.

2. W. Kaempfert, "Edison" (obituary article), New York *Times*, October 19, 1931.

3. *The Independent*, May 1, 1913.

4. New York *Times*, December 14, 1922.

5. T. C. Martin, *op. cit.*, p. 41.

6. New York *World*, February 13, 1922.

7. New York *Times*, February 12, 1914; December 6, 1921.

8. *Current Literature*, December, 1910.

9. Cardinal Gibbons, "Edison on Immortality," *Columbian Magazine*, January, 1911.

10. M. C. Nerney, *Edison, Modern Olympian*, p. 252.

11. Edison to Mina Edison, December 1, 1908; letter in possession of Madeleine Edison Sloane.

12. H. R. Fraser, "Edison," 1954 (unpublished), p. 374; copy in ELA.

13. B. C. Forbes, "Edison Working to Communicate with the Next World," *American Magazine*, October, 1920.

14. M. A. Rosanoff, "Edison in His Laboratory," *Harper's Magazine*, September, 1932, p. 411.

15. *Harper's Weekly*, November 4, 1911.

16. New York *Times*, May 9, 13, 14, 16, 1921.

17. Letter of Mina Edison to Madeleine Edison Sloane, September 28, 1917.

18. New York *Times*, June 12, 1923.

19. M. C. Nerney, *op. cit.*, p. 234.

20. New York *Times*, October 25, 1914.

21. Allan Nevins and F. E. Hill, *Ford: Expansion and Challenge* (New York, 1957), vol. II, p. 41.

22. New York *Times*, May 22, 1915.

23. New York *Times*, May 30, 1915.

24. Edison to Secretary of the Navy Josephus Daniels, July 15, 1915.

25. San Francisco *Chronicle*, October 22, 1915.

26. T. A. Robins, Sr., "Friends in a Lifetime," pp. 18–19.

27. Josephus Daniels, *The Wilson Era* (Chapel Hill, N.C., 1944), pp. 464–65.

28. R. A. Millikan, "Edison's Contribution," *Science*, January 15, 1932.

29. T. A. Robins, Sr., to author, October 10, 1957.

30. Edison to T. A. Robins, Sr., February 4, 1919; ELA.

31. Note to W. Meadowcroft (for Henry Ford), August 17, 1917; Henry Ford Museum Library.

32. New York *Times*, December 13, 1921.

33. Mina Edison to Madeleine Edison Sloane, August 28, 1917.

34. New York *World*, February 13, 1923.

35. New York *Times*, February 8, 1920.

chapter **XXII** (pages 456–486)

1. Oral Record of Memoir of Charles Voorhees, p. 137; Ford Motor Company Archives. Also E. G. Liebold, Oral Record of Memoir, p. 1506; Ford Motor Co.

2. M. C. Nerney, *Edison, Modern Olympian,* p. 241.

3. New York *Times,* January 9 and January 10, 1914.

4. Keith Sward, *The Legend of Henry Ford* (New York, 1948), pp. 112–13.

5. Correspondence of T. A. Edison, Inc., and Ford Motor Co., July 31, 1914; Ford Motor Co. Archives.

6. E. G. Liebold, Oral Record of Memoir, p. 1558; Ford Motor Co. Archives.

7. New York *Times,* February 13, 1914.

8. W. A. Simonds, *Edison, His Life, Work, Genius,* p. 302.

9. John Burroughs to Harvey Firestone, December 11, 1916; Ford Motor Co. Archives.

10. *Ibid.*

11. Edison's note for the bird fountain at "Fairlane," 1916; copy in Henry Ford Museum Library.

12. *Our Vacation Days of 1918* (Akron, 1926).

13. Edison to Edward N. Hurley, January 12, 1918; Henry Ford Museum Library.

14. New York *Times,* February 16, 1923.

15. Nevins and Hill, *Ford: Expansion and Challenge,* p. 316 *ff.*

16. Edison to E. G. Liebold, December 22, 1920; Ford Motor Co. Archives.

17. Edison to Jacob H. Schiff, November 13, 1914; Ford Motor Co. Archives.

18. New York *Times,* April 1, 1914.

19. New York *Times,* October 17 and 18, 1922.

20. New York *Times,* June 22, 1922.

21. Theodore M. Edison to author.

22. R. Gelatt, *The Fabulous Phonograph,* p. 249.

23. New York *Times,* February 12, 1928 (interview with W. H. Meadowcroft).

24. Edison to E. G. Liebold, December 16, 1927; Ford Motor Co. Archives.

25. M. C. Nerney, *Edison, Modern Olympian,* pp. 237–38.

26. Bailly Millard, in *Technical World Magazine,* October, 1914.

27. Nerney, *op. cit.,* p. 242.

28. Bailly Millard, *loc. cit.*

29. M. C. Nerney, *op. cit.,* p. 242.

30. Charles Edison to author; also Edward L. Bernays to author.

31. E. G. Liebold, *op. cit.,* pp. 693 *ff.*

32. New York *World,* October 21, 1929.

33. E. G. Liebold, *op. cit.,* pp. 693 *ff.*

34. New York *Times,* October 19, 1931.

35. Shaw Desmond, "Edison's Views," *Strand Magazine* (London), August, 1922.

36. L. W. McChesney, "A Light is Extinguished," pamphlet, privately printed (West Orange, 1932).

Index

About the Author

A New Yorker—born in Brooklyn, February 15, 1899—Matthew Josephson was called by the late John Erskine "our ablest American biographer." After receiving an A.B. from Columbia University in 1920, he served briefly as a reporter for the *Newark Ledger*. He then went to Europe, like so many of his literary contemporaries, and there wrote a measure of surrealist poetry and helped edit the literary periodicals *Broom* and *Transition*.

This period of expatriate estheticism ended, he became (1924) a customers' man in Wall Street. With the success of his first biography, *Zola and His Times* (1928), he began the series of books that has continued to give him a wide reading public. Though *The Robber Barons*, his study of the great American financiers of the nineteenth century, is perhaps the best known of his seven books on American themes, Matthew Josephson has also been acclaimed for his biographies of Rousseau, Stendhal, and Victor Hugo. Granted a Guggenheim traveling fellowship in creative literature in 1933, Mr. Josephson was elected a member of the National Institute of Arts and Letters in 1948. Married, and the father of two sons, he lives in New York City.